# QUANTUM TRANSPORT IN MESOSCOPIC SYSTEMS: COMPLEXITY AND STATISTICAL FLUCTUATIONS

Mesoscopic Physics and Nanotechnology

SERIES EDITORS

Harold G. Craighead
Purusottam Jena
Charles M. Marcus
Allan H. MacDonald
Shaul Mukamel
Mark A. Reed
John C. Spence

1. Y. Imry: *Introduction to Mesoscopic Physics*
2. Y. Imry: *Introduction to Mesoscopic Physics, Second Edition*
3. K. Nakamura and T. Harayama: *Quantum Chaos and Quantum Dots*
4. P. A. Mello and N. Kumar: *Quantum Transport in Mesoscopic Systems: Complexity and Statistical Fluctuations*
5. D. Gatteschi, R. Sessoli and J. Villain: *Molecular Nanomagnets*
6. C. M. Mate: *Tribology on a Small Scale – A Bottom Up Approach to Friction, Lubrication, and Wear*

# QUANTUM TRANSPORT IN MESOSCOPIC SYSTEMS: COMPLEXITY AND STATISTICAL FLUCTUATIONS

A MAXIMUM-ENTROPY VIEWPOINT

**Pier A. Mello**
*Instituto de Física*
*Universidad Nacional Autónoma de México, México*

and

**Narendra Kumar**
*Raman Research Institute, Bangalore, India*

# OXFORD
**UNIVERSITY PRESS**

Great Clarendon Street, Oxford OX2 6DP

Oxford University Press is a department of the University of Oxford.
It furthers the University's objective of excellence in research, scholarship,
and education by publishing worldwide in

Oxford New York

Auckland Bangkok Buenos Aires Cape Town Chennai
Dar es Salaam Delhi Hong Kong Istanbul Karachi Kolkata
Kuala Lumpur Madrid Melbourne Mexico City Mumbai Nairobi
São Paulo Shanghai Taipei Tokyo Toronto

Oxford is a registered trade mark of Oxford University Press
in the UK and in certain other countries

Published in the United States
by Oxford University Press Inc., New York

© Oxford University Press, 2004

The moral rights of the author have been asserted
Database right Oxford University Press (maker)

First published 2004

First published in paperback 2010

All rights reserved. No part of this publication may be reproduced,
stored in a retrieval system, or transmitted, in any form or by any means,
without the prior permission in writing of Oxford University Press,
or as expressly permitted by law, or under terms agreed with the appropriate
reprographics rights organization. Enquiries concerning reproduction
outside the scope of the above should be sent to the Rights Department,
Oxford University Press, at the address above

You must not circulate this book in any other binding or cover
and you must impose this same condition on any acquirer

A catalogue record for this title is available from the British Library

Library of Congress Cataloging in Publication Data
(Data available)

ISBN 978-0-19-852582-0 (hbk.); 978-0-19-852583-7 (pbk.)

10 9 8 7 6 5 4 3 2 1

Typeset using the author's LaTeX files by Julie Harris

Printed in Great Britain
on acid-free paper by
CPI Antony Rowe, Chippenham, Wiltshire

*From PIER A. MELLO to his wife LAURA*


*From NARENDRA KUMAR to his wife ANN*

# PREFACE TO THE PAPERBACK EDITION

This paperback edition is essentially a re-issue of the hardback first edition (1994) of our book. We have corrected the typographical errors that had escaped our attention in the first edition. We have deleted a comment in Chapter 7 which, after some further thought, did not seem to be well justified.

A paragraph was added in Chapter 7 to emphasize the role of a generalized central-limit theorem in the description of quasi-one-dimensional transport, which was not done so explicitly in the hardback version.

A number of short additions were made in various places, in order to point out some relevant developments that had occurred prior to the hardback edition of 2004 and were not mentioned there, as well as some others that occurred after that publication date. Also, we have cited some additional references that were either missed out earlier, or had appeared after the first edition came out.

# PREFACE

The aim of this book is to present a statistical theory of certain complex wave-interference phenomena of considerable interest in mesoscopic physics, namely the statistical fluctuations of transmission and reflection of waves. The complexity here may derive from the chaotic nature of the underlying classical dynamics, as in the case of chaotic microwave cavities and quantum dots, or from the quenched randomness of scattering potentials, as in the case of disordered conductors. A theme that recurs throughout this book is that of universality of the statistical behavior, in the sense that it involves a relatively small number of relevant physical parameters, while the rest of the system details serves as mere scaffolding. Admittedly, there is an extensive literature on the subject of statistical wave scattering in physical systems such as atomic nuclei, disordered conductors and chaotic cavities; a powerful, non-perturbative approach is that of random-matrix theory. The treatment developed in this book is also frankly statistical; we follow a maximum-entropy approach, where the Shannon information entropy is maximized, subject to the symmetries and constraints that are physically relevant. We believe, however, that the present route to random-matrix theory is refreshingly novel. Indeed, this distinctive feature provides a partial justification for writing this book. Another, rather compelling, reason is that it collects and organizes in one place the material and notions which are scattered through an extensive literature on the subject. Much of the material presented here is derived from the published work of the authors in collaboration with several co-workers, and also from the work of others.

The book contains two rather long chapters, namely Chapters 2 and 3, on the quantum theory of scattering which is, of course, covered in several standard texts. Our emphasis, however, is somewhat different, as it is set specifically in the context of scattering in quasi-one-dimensional, multichannel systems. These are related directly to the scattering problems encountered in mesoscopic systems, which, as we mentioned earlier, are of great interest in this book. Furthermore, their inclusion makes the book sufficiently self-contained. We have also included a chapter on linear response theory, namely Chapter 4, which again is adapted to the context of mesoscopic systems, where it is the sample-specific conductance rather than the material-specific conductivity which is of relevance. These chapters, together with Chapter 5 on the maximum-entropy approach and Chapter 8 on localization theory, have been written in a pedagogical style, so that they can be used as part of a graduate course. Scattering and electronic transport through classically chaotic cavities and quasi-one-dimensional disordered systems are the subject matter of Chapters 6 and 7, respectively. There are many exercises, most of them worked out in detail, distributed throughout the book. This should help graduate students, their teachers and the research scholars who are interested generally in the subject of quantum transport through disordered and chaotic systems.

In the course of writing this book, we have interacted with many colleagues working in the field, whose help is gratefully acknowledged. We would specifically like to thank Harold Baranger, Rubén Barrera, Markus Büttiker, Yoseph Imry, Arun Jayannavar, Jean-Louis Pichard, Prabhakar Pradhan, Alberto Robledo and Boris Shapiro. Also, very useful and enjoyable were the discussions with Víctor Gopar, Moisés Martínez, Anantha Ramakrishna and Eugenia Várguez, some of our latest Ph.D. students, as well as with several other students who attended a number of graduate courses given by one of us (PAM) on the present material. In addition, we are grateful to the Instituto de Física, UNAM, the Conacyt, Mexico, the Raman Research Institute, India, and the Third World Academy of Sciences, whose financial support was essential for enabling the contact between the two authors.

This book has been long in the making. The initial encouragement for writing it came in about 1995 from Professor Herman Feshbach, since deceased. His teachings on the optical model of the atomic nucleus and the statistical theory of nuclear reactions has had, through the years, a profound impact on one of the authors (PAM).

*Mexico*     P. A. M.
*India*     N. K.
*June* 2003

# CONTENTS

**1 Introduction**   1
  1.1 Atomic nuclei and microwave cavities   2
  1.2 Wave localization and fluctuations   4
  1.3 Mesoscopic conductors: time- and length-scales   5
      1.3.1 Ballistic mesoscopic cavities   7
      1.3.2 Diffusive mesoscopic conductors   7
      1.3.3 Statistical approach to mesoscopic fluctuations   8
  1.4 Organization of the book   13

**2 Introduction to the quantum mechanical time-independent scattering theory I: one-dimensional scattering**   15
  2.1 Potential scattering in infinite one-dimensional space   16
      2.1.1 The Lippmann–Schwinger equation; the free Green function; the reflection and the transmission amplitudes   16
      2.1.2 The $T$ matrix   23
      2.1.3 The full Green function   25
      2.1.4 The $S$ matrix   30
      2.1.5 The transfer or $M$ matrix   44
      2.1.6 Combining the $S$ matrices for two scatterers in series   48
      2.1.7 Transformation of the scattering and the transfer matrices under a translation   51
      2.1.8 An exactly soluble example   53
      2.1.9 Scattering by a step potential   57
      2.1.10 Combination of reflection and transmission amplitudes for a one-dimensional disordered conductor: invariant imbedding equations   68
  2.2 Potential scattering in semi-infinite one-dimensional space: resonance theory   70
      2.2.1 A soluble model for the study of resonances   71
      2.2.2 Behavior of the phase shift   73
      2.2.3 Behavior of the wave function   78
      2.2.4 Analytical study of the internal amplitude of the wave function near resonance   84
      2.2.5 The analytic structure of $S(k)$ in the complex-momentum plane   87
      2.2.6 Analytic structure of $S(E)$ in the complex-energy plane   90

|  |  |  |  |
|---|---|---|---|
| | 2.2.7 | The $R$-matrix theory of scattering | 93 |
| | 2.2.8 | The 'motion' of the $S$ matrix as a function of energy | 109 |

## 3 Introduction to the quantum mechanical time-independent scattering theory II: scattering inside waveguides and cavities  120
- 3.1 Quasi-one-dimensional scattering theory  120
  - 3.1.1 The reflection and transmission amplitudes; the Lippmann–Schwinger coupled equations  120
  - 3.1.2 The $S$ matrix  138
  - 3.1.3 The transfer matrix  141
  - 3.1.4 Combining the $S$ matrices for two scatterers in series  145
  - 3.1.5 Transformation of the scattering and transfer matrices under a translation  146
  - 3.1.6 Exactly soluble example for the two-channel problem  147
  - 3.1.7 Extension of the $S$ and $M$ matrices to include open and closed channels  155
- 3.2 Scattering by a cavity with an arbitrary number of waveguides  168
  - 3.2.1 Statement of the problem  168
  - 3.2.2 The $S$ matrix; the reflection and transmission amplitudes  171
- 3.3 The $R$-matrix theory of two-dimensional scattering  181

## 4 Linear response theory of quantum electronic transport  187
- 4.1 The system in equilibrium  188
- 4.2 Application of an external electromagnetic field  191
- 4.3 The external field in the scalar potential gauge  193
  - 4.3.1 The charge density and the potential profile  194
  - 4.3.2 The current density  208
- 4.4 The external field in the vector potential gauge  209
- 4.5 Evaluation of the conductance  221

## 5 The maximum-entropy approach: an information-theoretic viewpoint  226
- 5.1 Probability and information entropy: the role of the relevant physical parameters as constraints  227
  - 5.1.1 Properties of the entropy  229
  - 5.1.2 Continuous random variables  233
- 5.2 The role of symmetries in motivating a natural probability measure  234
- 5.3 Applications to equilibrium statistical mechanics  235
  - 5.3.1 The classical microcanonical ensemble  236
  - 5.3.2 The classical canonical ensemble  236

|       | 5.3.3 The quantum mechanical canonical ensemble | 239 |
|---|---|---|
| 5.4 | The maximum-entropy criterion in the context of statistical inference | 241 |

# 6 Electronic transport through open chaotic cavities — 244

- 6.1 Statistical ensembles of $S$ matrices: the invariant measure — 245
- 6.2 The one-channel case — 249
- 6.3 The multichannel case — 251
- 6.4 Absence of prompt (direct) processes — 253
  - 6.4.1 Averages of products of $S$: weak localization and conductance fluctuations — 253
  - 6.4.2 The distribution of the conductance in the two-equal-lead case — 258
- 6.5 Presence of prompt (direct) processes — 262
  - 6.5.1 The case $\beta = 2$ — 262
  - 6.5.2 The case $\beta = 1$ — 264
- 6.6 Numerical calculations and comparison with theory — 264
  - 6.6.1 Absence of prompt (direct) processes — 265
  - 6.6.2 Presence of prompt (direct) processes — 268
- 6.7 Dephasing effects: comparison with experimental data — 270
  - 6.7.1 The limit of large $N_\phi$ — 272
  - 6.7.2 Arbitrary $N_\phi$ — 274
  - 6.7.3 Physical experiments — 276

# 7 Electronic transport through quasi-one-dimensional disordered systems — 279

- 7.1 Ensemble of transfer matrices; the invariant measure; the combination law and the Smoluchowski equation — 280
  - 7.1.1 The invariant measure — 284
  - 7.1.2 The ensemble of transfer matrices — 286
- 7.2 The Fokker–Planck equation for a disordered one-dimensional conductor — 288
  - 7.2.1 The maximum-entropy ansatz for the building block — 289
  - 7.2.2 Constructing the probability density for a system of finite length — 291
- 7.3 The Fokker–Planck equation for a quasi-one-dimensional multichannel disordered conductor — 298
  - 7.3.1 The maximum-entropy ansatz for the building block — 298
  - 7.3.2 Constructing the probability density for a system of finite length — 301
  - 7.3.3 The diffusion equation for the orthogonal universality class, $\beta = 1$ — 302
  - 7.3.4 The diffusion equation for the unitary universality class, $\beta = 2$ — 310

|  |  |  |
|---|---|---|
| 7.4 | A unified form of the diffusion equation for the various universality classes describing quasi-one-dimensional disordered conductors: calculation of expectation values | 318 |
|  | 7.4.1 The moments of the conductance | 319 |
| 7.5 | The correlations in the electronic transmission and reflection from disordered quasi-one-dimensional conductors | 325 |

**8 An introduction to localization theory**    330
    8.1 Strong localization    331
    8.2 Mobility edge    335
    8.3 Coherent back-scattering (CBS)    337
    8.4 Scaling theory    340
    8.5 Weak localization: quantum correction to the conductivity    342
       8.5.1 The Hamiltonian and the Green function    343
       8.5.2 Ensemble-averaged Green's function in the self-consistent Born approximation    349
    8.6 Electrical conductivity of a disordered metal and quantum corrections: weak localization    350
       8.6.1 Classical (Drude) conductivity    354
       8.6.2 Weak localization (WL) and quantum correction to the classical (Drude) conductivity: the maximally-crossed diagrams    357
       8.6.3 Scale dependence of the conductivity    360

**A The theorem of Kane–Serota–Lee**    362

**B The conductivity tensor in RPA**    367

**C The conductance in terms of the transmission coefficient of the sample**    376

**D Evaluation of the invariant measure**    380
    D.1 The orthogonal case, $\beta = 1$    382
    D.2 The unitary case, $\beta = 2$    385

**References**    389

**Index**    397

# 1
## INTRODUCTION

As a physical phenomenon, as an experimental probe and as a tool for theoretical analysis, the *complex* scattering of waves has captured the interest of physicists for a long time. The physical system here can be essentially a *one-body* problem of non-interacting electrons moving in a disordered medium with randomly distributed static scatterers, such as a metal containing impurities; it may be a chaotic cavity, open or closed, for which the classical motion, i.e., the particle dynamics or the geometrical (ray) optics, is chaotic. It may also be a *many-body* nucleus, or a complicated molecule, that acts as a target for an incoming nucleon, an electron or a photon. Thus, for instance, the problem of coherent multiple scattering of waves has long been of great importance in optics [147]. Interest in this problem has been revived, both for electromagnetic waves [161] and for electrons [7], in relation to the phenomenon of localization, weak or strong, that gives rise to a great many fascinating effects. In all these cases, the complexity of behavior is fundamentally due to the time-persistent wave interference of alternatives, namely the partial-wave amplitudes along alternative paths, well known, for instance, in wave optics, as causing the Young double-slit fringes, speckles and the diffracted caustics. The condition necessary for this is phase coherence in time. In the context of quantum mechanics, this demands absence of inelastic processes, or of entanglement with the environmental degrees of freedom, requiring low temperatures in general. These conditions are assumed to hold throughout this book: all scatterers are assumed to be static, allowing potential or boundary scattering only.

Excellent examples of quantum mechanical scattering by complicated many-body systems are found in nuclear physics, where the typical linear dimension is of the order of a few Fermis (1 fm = $10^{-15}$ m). Studies of such systems go as far back as the 1930s, when compound nucleus resonances were first discovered. In these cases the analysis is often statistical, inasmuch as the details of the many-body problem are intractably complicated, even unknown, and physically of little interest. In fact, new *universal* features may well emerge from a statistical treatment: these features would otherwise remain hidden or get lost in the particular details.

Interestingly, some 'simple' one-particle dynamical systems are known to show features in common with these many-body nuclear problems. A simple case in point here is the quantum mechanical scattering of just one particle by three circular disks in a plane, where the underlying classical dynamics is known to be fully chaotic. Such dynamical systems have been studied by the quantum-chaos community, in which the main question is how the nature of the underlying classical dynamics influences the quantum properties [32,162]. In contrast to the nuclear case, these systems can be solved numerically; however, when the results are analyzed statistically, they are closely related to those of the nuclear systems. In particular, two types of scattering processes have been studied experimentally in simple artificially created systems: electron transport through microstructures called 'ballistic quantum dots', whose linear dimensions are of the order of 1 $\mu$m, and scattering of microwaves through metallic cavities, with typical dimensions of 0.1 m.

In addition to the above artificially-created chaotic systems, complex interference phenomena in naturally occurring scattering systems have also been the subject of interest over the last two decades. Indeed, electron transport in disordered metals has been intensively investigated [7,24], as has the transmission of electromagnetic waves through disordered media [161]. Because these are also examples of scattering in complex environments, where the character of the disorder is not exactly known, a statistical approach which treats an ensemble of disordered potentials is therefore natural.

The statistical properties of *fluctuations* due to the underlying wave interference in these systems, whose dimensions span about fourteen orders of magnitude, show a universal character which turns out to depend only on some very general physical principles and constitute the central theme of this book. It is important to clarify here the operational meaning of the statistical fluctuations. In principle, they imply an ensemble of macroscopically identical, but microscopically (and hence necessarily uncontrollable) distinct, samples. Experiments are, of course, usually performed on a given sample. One can, however, measure a sample-specific quantity as a function of some tunable parameters, e.g., the chemical potential, the magnetic field, the incident wave frequency for the case of a microwave chaotic cavity, or even the shape of a quantum dot for the electronic case. It turns out that these parametric variations are locally smooth but, globally, have a statistical character, and thus effectively generate the ensemble. There is an implicit appeal to *ergodicity* here, a notion meaning, in this case, equality of the ensemble average and the average with respect to a suitably chosen parameter that has been varied.

## 1.1 Atomic nuclei and microwave cavities

Although our main interest throughout this book is electronic transport through disordered conductors, ballistic chaotic cavities or ballistic quantum dots, we wish to emphasize the generality of the ideas involved, by including in this section their application in the field of nuclear physics, where some of them were first introduced, and also in the field of microwave cavities.

In the 1950s, Feshbach, Porter and Weisskopf [72,73] invented a most successful model, the so-called optical model of the nucleus, that works very well over a wide range of energies. The optical model has also been applied successfully in the description of a number of chemical reactions, thus bringing us from the nuclear to the molecular-size scale.

In the nuclear context, the optical model describes the scattering of a nucleon by an atomic nucleus—a complicated many-body problem—in terms of the following two distinct time-scales.

(i) A *prompt* response arising from *direct processes*, in which the incident nucleon feels a mean field produced by the other nucleons. This response is described mathematically in terms of the *average* of the actual scattering amplitudes over an energy interval bracketing a given energy $E$; these averaged amplitudes, also known as *optical* amplitudes, show a much slower energy variation than the original ones.

(ii) A *delayed*, or *equilibrated*, response, corresponding to the formation and decay of the compound nucleus. It is described by the difference between the exact and the optical scattering amplitudes; it varies appreciably with the energy $E$ and is studied with *statistical* concepts using techniques known as *random-matrix theory* (RMT).

We now recall from statistical mechanics that one has to introduce the notion of *ergodicity*, so as to replace time averages, which are hard to get at, by *ensemble averages*, which are easier to calculate. Similarly, in the present context one also finds it advantageous to study energy averages in terms of ensemble averages through an ergodic property [3, 75, 135].

More than thirty years ago, Ericson and Mayer-Kuckuk [71] noted the connection of the above problems with the theory of waveguides and cavities. This is aptly summarized in their statement:

Nuclear reaction theory is equivalent to the theory of waveguides... We will concentrate on processes in which the incident wave goes through a highly complicated motion in the nucleus... We will picture the nucleus as a closed cavity, with reflecting but highly irregular walls.

The equivalence claimed above has been borne out by recent experiments with microwave cavities, with the important proviso, however, that the explicit *randomness* of the 'irregular walls' anticipated from the nuclear case is *not* necessary in order to see these features. The analogy between the nuclear reaction theory and the theory of waveguides holds for simple smooth cavities, so long as the corresponding classical dynamics is *chaotic* [29–31, 82]. This has been the focus of several recent experiments involving microwave scattering from metallic cavities [65, 66]. In this regard, it is also important to note the earlier experiments of Genack [80] and collaborators on the scattering of microwaves by an explicitly disordered dielectric medium. It is also appropriate to remark here that, quite independently of its connection to the complex scattering in nuclear physics, the use of statistical concepts to analyze electromagnetic scattering by waveguides started much earlier in the context of the propagation of radio waves [161].

## 1.2 Wave localization and fluctuations

It may be apt to identify three phases of development in the physics of quantum transport exhibiting complex wave-interference phenomena. Phase I began with the *oft quoted but seldom read* paper of Anderson [9] that established the strong (i.e., exponential) localization of all the one-electron eigenfunctions of a random electronic system for the disorder-to-bandwidth ratio exceeding a critical value, and its generalization by Mott [143] to the case of sub-critical disorders where a mobility edge separates the extended eigenstates from the localized ones. Disorder-induced localization (the Anderson transition), as well as the very existence of a sharp mobility edge in such a random one-body system, was indeed a surprise of theoretical physics. In phase II, the emphasis shifted from the qualitative nature of the eigenstates to a quantitative measure of transport (conductance) which is directly accessible experimentally. In an influential paper, Thouless [168] derived a simple expression for the conductance $G$ of a mesoscopic sample in terms of the sensitivity of the energy levels to the boundary condition variation, e.g., from the periodic to the anti-periodic, giving $G = (e^2/\pi\hbar)\delta/\Delta$, the ratio of the typical level shift $\delta$ to the typical level spacing $\Delta$ close to the Fermi energy. This led to the one-parameter scaling theory of localization by Abrahams *et al.* [1, 115], that described the Anderson transition effectively as a second-order (continuous) phase transition, with the sample conductance as the only relevant scaling parameter, as indeed observed in experiments (physical as well as numerical) in many a disordered system. The scaling theory revealed the crucial role of spatial dimensionality $d$. Thus, a remarkable result following from the scaling theory was the absence of a truly metallic state for a two-dimensional system (and, *a fortiori*, for $d = 1$, a fact known before) for arbitrarily small disorder. Another important development in phase II was the recognition of *weak localization* effects [25] in the metallic state just above the mobility edge, as also found in clean two-dimensional systems, leading to a calculable quantum correction to the conductivity, with magnetic field and temperature dependencies as actually observed. Phase III began with the realization that certain phase-coherent quantities such as the conductance of a disordered conductor, in striking contrast to others such as the density of states, are sample specific and non-self-averaging even in the thermodynamic limit of infinite sample size [21]. This is because of their phase-sensitive dependence on the microscopic details of disorder. (Displacement of an impurity scatterer by an amount of the order of the Fermi wavelength $\sim 1\,\text{Å}$ can alter the relative phases of wave amplitudes along the alternative paths by $\sim \pi$.) The statistics of fluctuations of these sample-specific quantities, however, turns out to be quite universal: e.g., one encounters the *universal conductance fluctuations* (UCF) in the good-metallic weak-localization regime [116].

This book belongs to phase III. It is devoted to the study of the statistics of fluctuations—arising from complex wave interference occurring in the scattering process—of certain physical quantities of interest, like the conductance of a disordered multichannel conductor and the transmission through and reflection from a chaotic cavity. The statistics of the associated scattering phase shifts and time delays are, however, not addressed here.

It must be noted that all our discussion is restricted to within the non-interacting particles (waves) picture. This is, of course, true for electromagnetic waves, but it is often a good approximation for strongly-disordered conductors, such as doped compensated semiconductors. Electron–electron interaction in clean systems is all-important and continues to be a dominant theme to date. Indeed, interplay of interaction and disorder, particularly near the mobility edge when screening is relatively ineffective, remains an unresolved problem in the field of quantum transport in random systems.

## 1.3 Mesoscopic conductors: time- and length-scales

In this section we define a phase-coherent transport system, often called a *mesoscopic* system. The term refers to microstructures in which the phase of the single-electron wave function—in an independent-particle approximation—remains coherent across the system of interest. This means that the phase-coherence length associated with processes that can change the state of the environment (i.e., the other electrons or the phonon field) to an orthogonal state exceeds the system size. We may note in passing that this transport-contextual definition of mesoscopic systems, however, differs from that originally introduced by van Kampen in the context of statistical mechanics, where the finite size effects dominate the thermal behavior.

A disordered metallic sample of linear size $L$, held at a sufficiently low temperature, is a canonical example with which to introduce the various time-scales and length-scales that define and characterize the different regimes of mesoscopic quantum transport and the statistics of the associated fluctuations. The single most important elementary length-scale in the system is the transport mean free path $l_e$ for elastic scattering from the random impurity potential. Together with the wavelength $\lambda$ ($= \lambda_F \equiv 2\pi/k_F$) for a disordered degenerate metallic system of Fermi wave-vector magnitude $k_F$, it defines the all-important dimensionless disorder parameter $(1/k_F l_e)$ that measures the degree of randomness. Thus, for $1/k_F l_e$ larger than a critical value (of order 1), we have the Anderson localization. Similarly, we may have the inelastic mean free path $l_{\text{in}}$ due to scattering by phonons, other electrons, etc. This is the time-scale on which the electron wave loses its phase coherence. (In general, however, $\tau_{\text{in}}(l_{\text{in}})$ must be replaced by a dephasing, or phase-breaking, time $\tau_\phi(l_\phi)$ inasmuch as all inelastic scatterings are not equally effective.) Next, we note that the elastic mean free path $l_e$ together with the Fermi speed $v_F$ ($= \hbar k_F/m$) defines the diffusion constant $D_e = (1/3)v_F l_e$ (for three dimensions), which in turn, together with the inelastic scattering time $\tau_{\text{in}}$ ($\equiv l_{\text{in}}/v_F$), defines a length $L_T = (D_e \tau_{\text{in}})^{1/2} \equiv \sqrt{(1/3)l_{\text{in}} l_e}$, the distance through which the particle typically diffuses before losing its phase coherence. This length-scale was introduced by Thouless and hence the name *Thouless length* ($L_T$). Note that for a typical mesoscopic conductor at low temperature we have $l_e < L_T < l_{\text{in}}$.

In terms of these length-scales one can define three experimentally distinct regimes, or limits, for coherent transport (when $1/k_F l_e \ll 1$).

(a) *Ballistic or waveguide limit* (no impurity scattering: $\lambda_F < L \ll l_e, l_{\text{in}}$)

The wave propagates through the sample without any elastic or phase-breaking scattering in the bulk. This is, for example, true of chaotic cavities for microwaves and for electronic microstructures such as quantum dots. All scattering is at the boundary, and one expects large statistical fluctuations over an ensemble generated by the cavity shape variations, or parametrically, e.g., by varying the wavelength.

(b) *Diffusive weak-localization limit* ($\lambda_F \ll l_e \ll L < L_T$)

Here the wave traverses the sample coherently. Scattering (reflection, transmission, conductance) is sample specific, and one observes the full statistical fluctuations which are not suppressed by self-averaging.

These two regimes (a) and (b) correspond to the mesoscopic systems of interest here, with the quantum (wave) coherent transport exhibiting complex interference effects. Phase-coherent transport through such mesoscopic systems is sample specific. Thus, for example, resistances are non-additive in series, and can even be sequence dependent. (This resistive encoding of the information about the sequence is a provocative thought!) Furthermore, the phase coherence introduces spatial non-locality in that the conductance probed between two points on the conductor (wire, say) may depend on the conditions elsewhere, as in a waveguide. Moreover, the measured transport is not only sample specific, but also probe specific. Thus, one speaks of the sample conductance rather than the material conductivity.

(c) *Macroscopic samples* ($\lambda_F \ll l_e \ll L_T \ll L$)

Here the sample effectively breaks up into mesoscopic subsamples of size $L_T$ of the diffusive type (b). Thus, the wave propagates coherently within each subsample but there is no phase coherence from subsample to subsample. As a result, the transport properties (e.g., conductances) compose incoherently among the subsamples, leading to averaging of the otherwise large intra-subsample fluctuations known from (a) and (b).

It is important to note that static disorder, no matter how strong, cannot cause loss of coherence. One needs inelastic scattering, or entanglement with the environment (with or without energy exchange), for decoherence.

Other interesting physical scales are the ergodic time scale $\tau_{\text{erg}} = L^2/2D_e$, which is the time taken to diffuse over the whole sample of size $L$ in the absence of any phase breaking, and the Heisenberg time $\tau_H = \hbar/\Delta$, which is the longest time-scale in the problem (the analogue of the classical Poincaré or recurrence time). For time-scales greater than $\tau_H$ we are up against the quantum limit where diffusion under elastic scattering alone becomes an ill-defined process—we have to introduce some damping so as to ensure the diffusive behavior.

In this book we will be concerned with the statistics of mesoscopic fluctuations in the above limits (a) for ballistic cavities, and (b) for diffusive conductors, which will now be described in greater detail.

### 1.3.1  Ballistic mesoscopic cavities

With the rapid advancement in submicron microfabrication technology, it is now possible to have zero- or low-dimensional quantum confined micro/nano-structures in semiconductors, referred to as quantum dots (or *artificial atoms*, or also *designer atoms*), where the size, shape and the carrier number can be controlled through gate voltages, as in a MOSFET (metal oxide semiconductor field effect transistor). An open mesoscopic cavity is such a microstructure with waveguide leads attached to it. In the so-called *ballistic cavities*, or *quantum dots*, the electron motion is practically ballistic, except for specular reflections from the walls; thus the elastic mean free path exceeds the system size. In the most favorable material system, GaAs heterostructures, this condition can be realized for cavities of size at most $1\,\mu$m and at temperatures less than 1 K.

Experimentally, in a typical two-probe measurement, an electric current is driven through the leads connected to the cavity (quantum dot) and the potential drop across the cavity with the same leads is measured, from which the conductance $G$ is calculated. In an independent-electron picture, one thus aims to understand the quantum mechanical single-electron scattering by the cavity, while the leads play the role of waveguides. It is the multiple scattering of the waves reflected by the various portions of the cavity that gives rise to interference effects. As indicated earlier, three important experimental probes of the interference effects are an external magnetic field $B$, the Fermi energy $\epsilon_F$ and the shape of the cavity. When these are varied the relative phase of the various partial waves changes, and so does the interference pattern. The changing interference pattern in turn causes the conductance to fluctuate; this sensitivity of $G$ to small changes in parameters through quantum interference gives rise to the conductance fluctuations referred to earlier.

The connection between scattering by chaotic cavities and mesoscopic systems was first made theoretically [95]. Subsequently, cavities in the shape of a stadium, for which the single-electron classical dynamics would be chaotic, were reported. More recently, several other types of structures have been investigated, including experimental *ensembles* of shapes [26, 27, 56, 58, 89].

### 1.3.2  Diffusive mesoscopic conductors

A $d$-dimensional sample of size $L$ containing a finite average concentration $n_{\text{imp}}$ ($\equiv N_{\text{imp}}/L^d$) of impurities scattering incoherently (i.e., without any coherent multiple scattering, as in a classical conductor) would give not only an impurity-averaged mean resistance, but also random fluctuations about the mean. In such a case, the ordinary Poissonian statistics of concentration fluctuations translates into a root-mean-square conductance fluctuation $\delta G_d^{\text{cl}} \propto L^{(d-4)/2}$, which for $d < 4$ vanishes in the thermodynamic limit $L \to \infty$, $N_{\text{imp}} \to \infty$ and

$N_{\text{imp}}/L^d = n_{\text{imp}} =$ constant. For this incoherent (classical) transport, therefore, the scatterers contribute ohmically, making the conductance self-averaging. A quantum-coherent transport in a mesoscopic system is, however, different in that it is dominated by wave interference. This makes the individual impurity contributions to the resistance non-additive, leading to a conductance which is non-self-averaging. Indeed, it is now known that the conductance fluctuation is not only relatively larger than the classical value, but it is also universal. Thus, as noted in Section 1.2, we have the universal (in magnitude) conductance fluctuations (UCF) [116], namely that $\delta G_d^Q = C_d e^2/h \gg \delta G_d^{\text{cl}}$, with $C_d$ a dimensionality $d$-dependent numerical constant. These fluctuations have been observed in numerical simulations on random lattices that compare well with theoretical calculations. Here the statistics is done on the ensemble generated parametrically by varying the chemical potential, the magnetic field, or through different realizations of the randomness.

### 1.3.3 Statistical approach to mesoscopic fluctuations

The complexity of coherent wave interference in explicitly disordered (i.e., through random potentials) or in dynamically chaotic systems calls for a statistical treatment. It turns out, as will be seen later, that the very complexity of the system washes out the particularity of the microscopic details (of the underlying Hamiltonian) and leads to a certain generality of macroscopic behavior. The latter is best expressed through a statistical law for the fluctuations as belonging to a certain universality class [8, 67]. Indeed, this universality represents a simplicity which reflects the *symmetries* of the system in question, depends on a rather limited number of *relevant parameters* (generally of a macroscopic nature and with a clear physical significance), and is, of course, insensitive to the tyranny of other numerous details. Such an approach, replacing the complex dynamics by simpler statistics, was pioneered by Wigner in the 1950s in the context of nuclear physics, for the study of the 'sea' of resonances observed in the scattering of neutrons as a function of energy by a heavy atomic nucleus, where, as noticed earlier, the underlying many-body Hamiltonian matrix is much too large and intractably complex, even unknown. This statistical approach, now celebrated as *random-matrix theory* (RMT) [40, 124, 128, 152], has since been applied successfully to a host of problems outside nuclear physics, e.g., to the mesoscopic fluctuations in chaotic cavities and disordered conductors, which are of main interest in this book.

In his original statistical analysis of isolated nuclear resonances, Wigner approached the problem from the standpoint of scattering theory. It was in those early papers that the notion of a statistical $R$ function was introduced. Since the construction of the $R$ function was conceived in terms of its poles and residues—associated with the bound states inside a box—the attention of a number of people became concentrated on the statistical properties of the discrete spectrum of closed systems, which were later made use of in the study of statistical nuclear reactions in various regimes.

The universality classes of classical RMT are determined by the symmetries underlying the system [40, 124, 128, 152] as follows. (However, we shall not touch upon the new universality classes discovered more recently.)

Consider first a closed system invariant under the operation of time inversion, performed by an *anti-unitary* operator that we designate by $\theta$, so that $H = \theta H \theta^{-1}$. We call this property *time-reversal invariance* (TRI). We have two possibilities, namely $\theta^2 = +1$ and $\theta^2 = -1$, corresponding to integer and half-integer spin, respectively (see, for instance, [142, 176]). We recall that, for the spinless case and in the coordinate representation, $\theta$ is just the operator for complex conjugation of the wave function operated upon. Take a reference TRI Hamiltonian $H_0$, that provides a basis, i.e.,

$$(\epsilon_\alpha - H_0)\phi_\alpha = 0. \tag{1.1}$$

We contemplate the case $\theta^2 = +1$ first. Given the states $\phi_\alpha$, we can always find states $\psi_\alpha$ so that $\theta\psi_\alpha = \psi_\alpha$. One can then easily show that the matrix elements of a TRI Hamiltonian $H$ with respect to this last basis are real; the resulting matrix is thus *real symmetric*, a property that is preserved by a *real orthogonal* transformation. This is the *orthogonal* case.

Next we contemplate the case $\theta^2 = -1$. Now the states $\phi_\alpha$ and $\chi_\alpha = \theta\phi_\alpha$ are degenerate and orthogonal. This degeneracy, Kramer's degeneracy, is present even in the absence of any other symmetry. The matrix of a TRI Hamiltonian $H$ in this basis has the general structure

$$H = \begin{bmatrix} h & -a^* \\ a & h^* \end{bmatrix}, \tag{1.2}$$

where $h$ is Hermitian and $a$ is anti-symmetric. Such a general structure is preserved by a subset of the unitary transformations $U$, called *symplectic*, that gives the name to the present class. This case will not be discussed further in the present book. One can show that, if, additionally, the system is invariant under rotations, then this case reduces to the orthogonal one [124, 152].

Finally, when TRI is broken, we have the *unitary* case; here, the Hermitian Hamiltonian matrices can be transformed among themselves (an automorphism) by a general unitary transformation.

The nuclear Hamiltonian is TRI, commutes with the total angular momentum $J$ and thus belongs to the orthogonal case.

In the problem of electronic transport in an independent-electron picture, assume that the electron is subject to a non-rotationally-invariant real potential. If the spin–orbit interaction is neglected, the spin-up and spin-down electronic subsystems are decoupled; one can then ignore the electronic spin, except for a twofold degeneracy. A spinless-particle theory is then sufficient in order to describe quantum transport, and two cases can occur: either the system is TRI (the orthogonal case), or this symmetry is removed by an applied magnetic field (the unitary case). In the presence of strong spin–orbit scattering, spin components

TABLE 1.1. The universality classes of random-matrix theory discussed in the text.

| $[H,\theta]$ | Spin | $[H,J]$ | Ensemble | $\beta$ |
|---|---|---|---|---|
| 0 | integer | | orthogonal | 1 |
| 0 | 1/2 integer | 0 | orthogonal | 1 |
| 0 | 1/2 integer | $\neq 0$ | symplectic | 4 |
| $\neq 0$ | | | unitary | 2 |

are coupled and a spin-1/2 theory is then necessary. Also, two cases can occur: in the presence of time-reversal symmetry one has the symplectic case, and the removal of this last symmetry by a magnetic field again gives the unitary case.

The three universality classes are summarized in Table 1.1. In the table, $\theta$ and $J$ indicate the operators of time reversal and total angular momentum, respectively. The symbol $\beta$ was introduced by Dyson to designate the various universality classes.

The universality class in question also has important consequences for the structure of the scattering matrix $S$ of an open system [124, 128, 152]. It is precisely this object that we shall be dealing with most of the time in this book. From flux conservation, $S$ is a *unitary* matrix:

$$SS^\dagger = I. \tag{1.3}$$

In the absence of any other symmetry, i.e., in the *unitary* case, this is the only general property of $S$. In the *orthogonal* case, $S$ is, in addition, a *symmetric* matrix:

$$S = S^\top. \tag{1.4}$$

In the *symplectic* case, $S$ is called, in the language of quaternions, a *self-dual* matrix.

It may be noted here in passing that the set of three Dyson universality classes (the 'three-fold way') has now been enlarged to ten [5, 53, 182]. Applications of this extension in the context of quantum transport through quasi-1-dimensional disordered conductors can be found, for instance, in [54, 92].

Returning to the closed-system case, i.e., in the absence of connecting leads, it turns out that the nearest-neighbor level-spacing distribution $p(s) \sim s^\beta$ as $s \to 0$, where the exponent $\beta$ labels the universality class, with $\beta = 1$ for the orthogonal case, $\beta = 2$ for the unitary case and $\beta = 4$ for the symplectic case. Here $s \equiv (E_{i+1} - E_i)/\Delta$ is the ratio of nearest-neighbor level spacings to the mean level spacing $\Delta$. All of the three universality classes capture the essential physics of level repulsion, or avoided level crossing, for a complex quantum system with complete 'mixed-up-ness'. This is true when the underlying classical system is chaotic. In sharp contrast to this, for an integrable system, where the classical many-degrees-of-freedom problem is essentially decomposable into many independent one-degree-of-freedom problems, nothing forbids the quantum level crossing, and indeed the level-spacing spectrum is Poissonian, i.e., $\beta = 0$.

The universality above refers only to the statistics of *fluctuations* of the level spacings about the mean spacing, while the latter (the mean level spacing, or the level density associated with it), of course, has a smooth secular parametric variation which is *not* universal.

It is the universality classes introduced above—as dictated by the symmetries in question—that make the whole topic so appealing. Indeed, whenever a universal behavior has been discovered in physics, its understanding has represented a great challenge. Let us just cite, as a beautiful example, the universality of the critical-point phenomena in statistical mechanics whose comprehension in the past has been of far-reaching consequences. The universality observed in the purely *statistical* regularity of complex mesoscopic systems of interest here, and that observed in the *statistical mechanics* of the critical-point phenomena, may both be reducible ultimately to some central limit, of which the Gaussian limit is just the simplest (normal) one. The limiting irrelevance of details here arises from the very complexity of the system.

Our interest in this book centers on the statistical fluctuations of mesoscopic *transport* involving scattering through open systems, e.g., a cavity with waveguide leads attached to it, rather than the spectral statistics of a closed system. This corresponds to the statistical analysis of the original scattering problem addressed by Wigner, and undertaken subsequently by several groups, though not always from the same point of view [40]. Sometimes a phenomenological approach was taken; in some papers the description was done in terms of the statistics of the poles and residues of the scattering matrix $S$, or those of the $R$ or $K$ matrix; in others, statistics on the $S$ matrix itself was done. A somewhat related statistical treatment of scattering through complex mesoscopic systems is the information-theoretic one, which is based on the general idea of maximum entropy [97, 102, 104, 117, 158, 159], adapted to the problem of statistical wave scattering, as in this book [131–133, 137, 164]. The central idea here is to seek out the most probable probability distribution for the quantity of interest, which is consistent with the physically relevant constraints of symmetry and of values of the relevant parameters. This is implemented through a maximization of the Shannon entropy $S[p] \equiv -\sum_i p_i \ln p_i$, subject to the constraints $\langle f^{(\alpha)} \rangle = \sum_i f_i^{(\alpha)} p_i = $ given, say, to be imposed by means of Lagrange multipliers. Here $p_i$ is the probability associated with the $i$th possible event and $f_i^{(\alpha)}$ is a given function of $i$. Such a maximization of uncertainty can be shown to uniquely pick the most probable distribution, i.e., the broadest or the one as random as allowed to be and, therefore, the most unbiased distribution (having the minimum information), from among the several alternatives otherwise degenerate with respect to the constraints imposed. It is appropriate to mention at this point that one of the first applications of the idea of maximizing Shannon's entropy in the context of RMT is due to Balian [10].

In the context of statistics, the maximum-entropy method (MEM) has been widely used for organizing or refining statistical data, hypothesis testing and for the solution of ill-posed problems where the available information is partial—the number of unknowns exceeds the number of equations, say. Its probabilistic underpinning is the Bayesian hypothesis that allows one to update the probabilistic description as more and more data become available. The Shannon entropy generalizes the concept of entropy to statistics—outside of its well-known original domain of validity, namely that of statistical mechanics—and was introduced as such by Jaynes as a general constructive criterion. It carries the philosophical dictum of William of Occam (1300–1349): '*Essentia non sunt multiplicanda praeter necessitatem*', commonly known as *Occam's razor* (literally meaning that entities do not have to be multiplied beyond necessity). It certainly has the logic of Laplace's *principle of insufficient reason* behind it—of assigning equal a priori probabilities unless there is reason to think otherwise. Jaynes' viewpoint is, however, *subjective* in spirit—we are asked to maximize the Shannon entropy subject to whatever constraint is known to us. We prefer to replace the latter by the statement 'subject to the physically relevant constraints' and call the method the *maximum-entropy approach* (MEA). Of course, operationally, the MEM and MEA are implemented in the same manner. In a number of cases one has discovered a generalized central-limit theorem (CLT) responsible for the success of this approach [125, 128, 136]. In all of these cases, a microscopic calculation, if one is possible at all, might end up being a *scaffolding* because of the final insensitivity of the results to most of the details. Sometimes the physically relevant constraints determine the statistical distribution uniquely; when this is not the case, the MEA helps make the choice unique. The MEA has a proven track record of success in mesoscopic physics, as can be seen in the references given above—it is not a *caveat emptor*.

In addition to the above statistical approaches, which are macroscopic, inductive and essentially non-perturbative, there are other dynamical approaches which are microscopic, deductive and systematically perturbative in nature. Thus, we have the impurity-diagrammatic technique that enables us to calculate the mean, the variance and higher moments of quantities of interest to various orders in the disorder parameter $1/k_F l_e$ [6]. There is also the supersymmetric field-theoretic approach that reduces the *random* microscopic Hamiltonian to an effectively *non-random* field-theoretic nonlinear $\sigma$-model which is then treated perturbatively [70]. Yet another microscopic approach, specially applicable to one-dimensional and quasi-one-dimensional systems, is based on the method of invariant imbedding [154] that directly addresses the evolution of the emergent quantities, e.g., reflection/transmission amplitude coefficients, with the sample length [85, 109]. The latter also allows a treatment of coherent amplification/attenuation, and admits approximate generalization to $d > 1$ [110].

Most of our approach in this book to treating mesoscopic fluctuations is frankly statistical—in the spirit of RMT and MEA. This approach has been demonstrably successful in treating scattering in chaotic cavities, and quantum diffusion in disordered quasi-one-dimensional conductors. Novel applications are steadily multiplying. Its adoption in this book demands no apology. None is offered.

## 1.4 Organization of the book

There are eight chapters and four appendices in this book, with the present Chapter 1 serving as a general introduction to the subject of coherent wave transport through disordered or chaotic mesoscopic systems and to a statistical approach to these problems based on the maximum-entropy approach. Chapters 2 and 3 are devoted to basic potential scattering theory, specialized to the case of a quasi-one-dimensional system with $N$ channels, or a chaotic cavity with $N$-channel leads connected to it. The $N$-channel scattering matrix $S$ and the related transfer matrix $M$ are introduced, properly parametrized, and their composition laws and symmetry properties discussed. An invariant imbedding approach giving the evolution of the reflection and the transmission amplitudes with the one-dimensional sample length is also included. Special attention has been paid to the evanescent (closed) channels as distinct from the propagating (open) channels. For completeness, we also introduce the Wigner $R$-matrix scattering formalism. Chapter 4 concerns the linear response theory of Kubo and its specialization to calculating the conductance of a narrow constriction connected to reservoirs through widening horns. The Landauer two-probe conductance formula [52, 111], that treats the entire sample as a single scatterer and relates its conductance to the total transmission coefficient, is derived and discussed. Chapter 5 motivates and discusses the basic idea of information-theoretic entropy, the Shannon entropy and the maximum-entropy approach which is based on it. The concepts of prior and of the physically relevant constraints are introduced through a number of simple examples. Chapter 6 is devoted to scattering from classically chaotic cavities. A maximum-entropy approach is employed to derive the probability distribution for the $S$ matrix, with and without the presence of direct processes. The latter are included as a constraint through the average $S$ matrix, in addition to its analytic properties. Theoretical calculations are compared with the results of computer simulations and with experimental data. Chapter 7 treats the problem of finding the probability distribution of quantities related to quantum transport through a strictly one-dimensional (i.e., 1-channel) and an $N$-channel quasi-one-dimensional disordered system. We use the MEA, wherein the distribution of the transfer matrix for an elementary building block is determined by maximizing the associated Shannon entropy subject to the physically relevant constraints. The composition law for the transfer matrices then leads to the evolution equation of the full probability distribution in the sample length—essentially a diffusion equation [132] (see also the review [21]). Various physical quantities of interest are then calculated from the $N$-channel diffusion equation; in particular, the *universal* conductance fluctuations in the good metallic limit (as $N \to \infty$). Chapter 8 treats the weak localization effect due to coherent back-scattering and the resulting quantum correction to the conductivity. To this end, the impurity diagram technique is introduced in a simplified manner. This is included for the sake of completeness.

Chapters 2 and 3 on scattering theory, Chapter 4 on linear response theory and Chapter 8 on weak localization, as well as Chapter 5 on the maximum-entropy approach, are frankly pedagogical, reasonably self-contained texts, and may be read or taught as such as part of a graduate course. Chapters 6 and 7, by and large, contain the research material due originally to the authors and their collaborators. Some of these are, admittedly, algebraically involved—but most of the details are there, or given in the appendices, at hand just in case the reader cares for them. Judicious skipping is always an option. A number of solved exercises have been included for training of the interested reader.

# 2

# INTRODUCTION TO THE QUANTUM MECHANICAL TIME-INDEPENDENT SCATTERING THEORY I: ONE-DIMENSIONAL SCATTERING

In this and the next chapter we present the notions of quantum mechanical scattering theory that will be needed throughout most of the book. On the one hand, this material is relevant to the study of scattering of classical scalar waves. On the other, it is very relevant to the quantum mechanical calculation of the conductance of mesoscopic systems. In fact, the study of the electronic transport through such systems carried out in Chapter 4 within the approach of linear response theory leads us to an understanding of the physical conditions under which the conductance calculation can be reduced to a quantum mechanical single-electron scattering problem; hence the relevance of the concepts that we shall discuss in these two chapters.

The motivation for this summary is to make the book reasonably self-contained. Although most of the material we cover has been discussed in a number of textbooks on quantum mechanics (an excellent presentation is given, for instance, in [141, 147, 155]) and in the nuclear physics literature [112, 123, 153], we thought it would be convenient for the reader to find a coherent presentation, with an emphasis on our future needs. The style in these two chapters is clearly pedagogical, so that the expert reader can skip them and go on to the next chapters.

We only present the time-independent scattering formalism, which is of interest in the stationary state. In the present chapter we illustrate the main concepts of scattering theory in one dimension. We first deal in Section 2.1 with an infinite domain, from $-\infty$ to $+\infty$, and introduce the Lippmann–Schwinger equation, the concepts of reflection and transmission amplitudes in the scattering process, the $T$ matrix, the scattering or $S$ matrix and the transfer or $M$ matrix. Next, in Section 2.2 we analyze in detail a soluble example in a semi-infinite domain consisting of an impenetrable wall and a delta potential at some distance from it; this problem serves the purpose of illustrating the concept of resonances and the analytic structure of the resulting $S$ matrix. We take advantage of this example to introduce the scattering formalism known as 'Wigner's $R$-matrix theory'.

Throughout the book we confine ourselves to the case of spinless particles, i.e., we ignore spin-dependent interactions.

## 2.1 Potential scattering in infinite one-dimensional space

### 2.1.1 *The Lippmann–Schwinger equation; the free Green function; the reflection and the transmission amplitudes*

The time-independent Schrödinger equation for a *spinless* particle subject to the one-dimensional potential $V(x)$ is

$$-\frac{\hbar^2}{2m}\frac{\partial^2 \psi(x)}{\partial x^2} + V(x)\psi(x) = E\psi(x). \tag{2.1}$$

The local potential $V(x)$ is real (and hence Hermitian) and is taken to be of finite support, or at least of short range. The scattering problem to be considered here is illustrated schematically in Fig. 2.1, where, for simplicity, $V(x)$ is taken to be positive (thus it will admit no bound states).

The most general potential allowed by quantum mechanics must certainly be Hermitian, but not necessarily local or real. For instance, it might be a function of the position and momentum, i.e., $V(x,p) = V(x,-i\hbar\partial_x)$. This more general possibility is not contemplated in the main text, but is discussed in a number of exercises to be found within this section.

In the region outside the range of the potential the wave function $\psi(x)$ is a linear combination of plane waves. Two physical situations are depicted in the figure. In (a) a plane wave is incident from the left, giving rise to a reflected and a transmitted wave, with amplitudes denoted by $r$ and $t$, respectively. In (b) a plane wave is incident from the right, $r'$ and $t'$ denoting the corresponding reflection and transmission amplitudes.

In what follows we find these amplitudes in terms of the potential. We choose the normalization of the plane waves to be

$$\phi_s(E;x) = \frac{e^{iskx}}{\sqrt{2\pi\hbar^2 k/m}}, \quad k > 0, \tag{2.2}$$

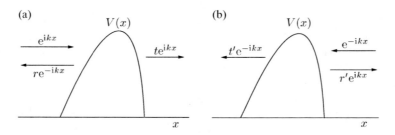

FIG. 2.1. Scattering by a one-dimensional potential. In (a) a plane wave is incident from the left and produces a reflected wave to the left and a transmitted one to the right. In (b) we have a similar situation, with the incident wave impinging from the right. The normalization factor $(2\pi\hbar^2 k/m)^{-1/2}$ of eqn (2.2) is common to all terms and is not indicated.

with $E = \hbar^2 k^2/2m$ being the energy; $k$ is defined to be *positive*, so the direction of propagation is specified by $s = \pm$. We then have

$$\langle \phi_s(E) | \phi_{s'}(E') \rangle = \delta_{ss'} \delta(E - E'). \tag{2.3}$$

Notice that the plane waves (2.2) give rise to the $k$-independent particle current

$$J = \frac{s}{2\pi\hbar}. \tag{2.4}$$

Denoting by $H_0$ the kinetic energy operator, our plane waves are solutions of the unperturbed equation

$$(E - H_0) \phi_s(E; x) = 0, \tag{2.5}$$

while the Schrödinger equation (2.1) can be written as

$$(E - H_0) \psi(x) = V \psi(x), \tag{2.6a}$$

or

$$\left( \frac{\partial^2}{\partial x^2} + k^2 \right) \psi(x) = U \psi(x), \tag{2.6b}$$

where

$$U = \frac{2m}{\hbar^2} V. \tag{2.7}$$

We shall need below the *unperturbed Green function* $G_0^{(\pm)}(x, x')$, an object of fundamental importance in the present context, defined through the equation

$$(E - H_0) G_0^{(\pm)}(x, x') = \delta(x - x'). \tag{2.8a}$$

More explicitly, $G_0^{(\pm)}(x, x')$ satisfies the differential equation

$$\left( \frac{\partial^2}{\partial x^2} + k^2 \right) G_0^{(\pm)}(x, x') = \frac{2m}{\hbar^2} \delta(x - x'). \tag{2.8b}$$

The indices $(\pm)$ indicate *outgoing wave* $(+)$ or *incoming wave* $(-)$ boundary conditions, implying, respectively, an outgoing or incoming asymptotic current. It will be useful to consider the function $g_0^{(\pm)}(x, x')$, which satisfies the differential equation

$$\left( \frac{\partial^2}{\partial x^2} + k^2 \right) g_0^{(\pm)}(x, x') = \delta(x - x') \tag{2.8c}$$

and is defined as

$$g_0^{(\pm)}(x, x') = \frac{\hbar^2}{2m} G_0^{(\pm)}(x, x'). \tag{2.9}$$

Equation (2.8c) will be solved below.

Considering the right-hand side of eqn (2.6b) as the inhomogeneous part of the otherwise homogeneous eqn (2.5), we can use the unperturbed Green function $g_0^{(\pm)}(x,x')$ to write, for the *perturbed* wave function $\psi_s^{(\pm)}(E;x)$, the integral equation

$$\psi_s^{(\pm)}(E;x) = \phi_s(E;x) + \int_{-\infty}^{+\infty} g_0^{(\pm)}(x,x')U(x')\psi_s^{(\pm)}(E;x')\,\mathrm{d}x', \qquad (2.10)$$

which can be verified by introducing it into the Schrödinger equation (2.6) and using the defining equation for the Green function, eqn (2.8c). Equation (2.10) is known as the *Lippmann–Schwinger equation*. The wave function $\psi_s^{(+)}(E;x)$ consists of the incident plane wave $\phi_s(E;x)$, plus *outgoing* waves at infinity. Not of direct experimental interest, but of great theoretical importance, are those solutions $\psi_s^{(-)}(E;x)$ that contain, in addition to the incident plane wave $\phi_s(E;x)$, *incoming* waves at infinity.

We note, additionally, that iteration of the Lippmann–Schwinger equation (2.10) generates the familiar Born series of potential scattering, whose first few terms are

$$\psi_s^{(\pm)}(E;x) = \phi_s(E;x) + \int_{-\infty}^{+\infty} g_0^{(\pm)}(x,x')U(x')\phi_s(E;x')\,\mathrm{d}x' + \cdots. \qquad (2.11)$$

One can show (see, for instance, [155, p. 298]) that the perturbed states $\psi_s^{(\pm)}(E;x)$ satisfy the same orthonormality relation as the unperturbed ones, namely eqn (2.3), i.e.,

$$\left\langle \psi_s^{(\pm)}(E) | \psi_{s'}^{(\pm)}(E') \right\rangle = \delta_{ss'}\delta(E - E'). \qquad (2.12)$$

Note that the set $\{|\psi_s^{(+)}(E)\rangle\}$ and the set $\{|\psi_s^{(-)}(E)\rangle\}$ *separately* form complete sets of orthonormal functions.

We now solve eqn (2.8). Considering the unperturbed Green function as a function of $x$, we see that for $x \neq x'$ the solutions of eqn (2.8) are plane waves, i.e.,

$$g_0^{(\pm)}(x < x') = f_1(x')\mathrm{e}^{\mp ikx}, \quad g_0^{(\pm)}(x > x') = f_2(x')\mathrm{e}^{\pm ikx}, \qquad (2.13)$$

where $f_1(x')$ and $f_2(x')$ are two functions of $x'$ to be determined. Continuity at $x = x'$ implies that

$$f_1(x')\mathrm{e}^{\mp ikx'} = f_2(x')\mathrm{e}^{\pm ikx'} \equiv A_{\pm}(x'). \qquad (2.14)$$

Thus

$$g_0^{(\pm)}(x < x') = A_{\pm}(x')\mathrm{e}^{\pm ik(x'-x)}, \quad g_0^{(\pm)}(x > x') = A_{\pm}(x')\mathrm{e}^{\pm ik(x-x')}. \qquad (2.15)$$

The discontinuity of the slope at $x = x'$ is obtained from eqn (2.8) as

$$\left[\frac{\partial g_0^{(\pm)}(x-x')}{\partial x}\right]_{x=x'-\epsilon}^{x=x'+\epsilon} = 1, \qquad (2.16)$$

so that

$$A_\pm(x') = \pm\frac{1}{2ik}. \qquad (2.17)$$

The unperturbed Green function is thus (as before, $k$ is defined to be a *positive number*)

$$g_0^{(\pm)}(x,x') = \pm\frac{1}{2ik}e^{\pm ik|x-x'|}, \qquad (2.18)$$

and this clearly has the symmetry properties

$$\left[g_0^{(+)}(x,x')\right]^* = g_0^{(-)}(x',x), \qquad (2.19a)$$

$$g_0^{(\pm)}(x,x') = g_0^{(\pm)}(x',x). \qquad (2.19b)$$

Introducing the Green function (2.18) into the Lippmann–Schwinger equation (2.10), we find

$$\psi_s^{(\pm)}(E;x) = \phi_s(E;x) \pm \frac{1}{2ik}\int_{-\infty}^{\infty} e^{\pm ik|x-x'|}U(x')\psi_s^{(\pm)}(E;x')\,dx'. \qquad (2.20)$$

From eqn (2.20) and the fact that $U(x)$ is real, we see that the incoming and outgoing wave solutions are related by

$$\psi_s^{(-)}(E;x) = \left\{\psi_{-s}^{(+)}(E;x)\right\}^* = \theta\psi_{-s}^{(+)}(E;x), \qquad (2.21)$$

where we have introduced the *anti-unitary operator of time reversal* $\theta$, which, in the present case of spinless particles and in the coordinate representation, is just the operator of complex conjugation [142,176]. In this case, the operation of time reversal physically corresponds to interchanging the incoming and the outgoing waves together with complex conjugation of the amplitudes. The potential $U(x)$ being real implies that the Hamiltonian is invariant under the operation of time reversal (*time-reversal invariance* (TRI)), i.e.,

$$[H,\theta] = 0. \qquad (2.22)$$

Equation (2.22) implies that, if $\psi_{-s}^{(+)}(E;x)$ is a solution of the Schrödinger equation for the energy $E$, then $\psi_s^{(-)}(E;x) = \theta\psi_{-s}^{(+)}(E;x)$ of eqn (2.21) is a solution of the *same* equation, i.e., for the *same potential* and the *same energy*. Equation (2.21) and the last argument can be clearly extended to a general Hermitian, but

TRI, potential $V(x, x')$. Let us recall that in this case TRI implies that $V(x, x')$ is *real*, so that Hermiticity requires

$$V(x, x') \equiv \langle x|V|x'\rangle = V(x', x). \tag{2.23}$$

In this notation, the local potential we have been using throughout this chapter is written as

$$V(x, x') = V(x)\delta(x - x'). \tag{2.24}$$

Notice that $V(x, x')$ has dimensions of energy/length.

**Exercise 2.1** The most general potential appearing in the Schrödinger equation that is allowed by quantum mechanics must be Hermitian, but not necessarily local or real. For instance, $V$ might be velocity dependent. Hermiticity requires $V(x, x') = V^*(x', x)$. Write the Lippmann–Schwinger equation of eqns (2.10), (2.20) and the time-reversal relations (2.21) for a general Hermitian potential.

Following the same steps as above, we find

$$\psi_s^{(\pm)}(E; x) = \phi_s(E; x) + \int_{-\infty}^{\infty}\int_{-\infty}^{\infty} g_0^{(\pm)}(x, x')U(x', x'')\psi_s^{(\pm)}(E; x'')\,dx'\,dx'' \tag{2.25}$$

and, introducing eqn (2.18) into (2.25), we have

$$\psi_s^{(\pm)}(E; x) = \phi_s(E; x) \pm \frac{1}{2ik}\int_{-\infty}^{\infty}\int_{-\infty}^{\infty} e^{\pm ik|x-x'|}U(x', x'')\psi_s^{(\pm)}(E; x'')\,dx'\,dx''. \tag{2.26}$$

From this last equation we see that

$$\psi_s^{(-)}[E; U; x] = \left\{\psi_{-s}^{(+)}[E; U^*; x]\right\}^* = \theta\psi_{-s}^{(+)}[E; U^*; x], \tag{2.27}$$

where we have included in the notation the functional dependence of the wave function on the potential. Since eqn (2.22) is not valid in this case, application of the $\theta$ operation to the wave function does not give a solution of the *same* Schrödinger equation, but of the equation in which $U$ is replaced by $U^*$.

*The asymptotic form of the wave function*

We can now find the asymptotic form of the wave function. Take, for example, $\psi_+^{(+)}(E; x)$. For $x \to -\infty$ we have

$$\psi_+^{(+)}(E;x) \sim \frac{e^{ikx}}{\sqrt{2\pi\hbar^2 k/m}}$$
$$- 2\pi i \frac{e^{-ikx}}{\sqrt{2\pi\hbar^2 k/m}} \int_{-\infty}^{\infty} \frac{e^{ikx'}}{\sqrt{2\pi\hbar^2 k/m}} V(x')\psi_+^{(+)}(E;x')\,dx'$$
$$= \phi_+(E;x) - 2\pi i \left\langle \phi_-(E) \,|V|\, \psi_+^{(+)}(E) \right\rangle \phi_-(E;x). \tag{2.28}$$

For $x \to +\infty$ we have

$$\psi_+^{(+)}(E;x) \sim \frac{e^{ikx}}{\sqrt{2\pi\hbar^2 k/m}}$$
$$- 2\pi i \frac{e^{ikx}}{\sqrt{2\pi\hbar^2 k/m}} \int_{-\infty}^{\infty} \frac{e^{-ikx'}}{\sqrt{2\pi\hbar^2 k/m}} V(x')\psi_+^{(+)}(E;x')\,dx'$$
$$= \phi_+(E;x) - 2\pi i \left\langle \phi_+(E) \,|V|\, \psi_+^{(+)}(E) \right\rangle \phi_+(E;x)$$
$$= \left[ 1 - 2\pi i \left\langle \phi_+(E) \,|V|\, \psi_+^{(+)}(E) \right\rangle \right] \phi_+(E;x). \tag{2.29}$$

---

**Exercise 2.2** Write the asymptotic eqns (2.28) and (2.29) for a general Hermitian potential.

We find

$$\psi_+^{(+)}(E;x) \sim \frac{e^{ikx}}{\sqrt{2\pi\hbar^2 k/m}}$$
$$- 2\pi i \frac{e^{-ikx}}{\sqrt{2\pi\hbar^2 k/m}} \iint_{-\infty}^{\infty} \frac{e^{ikx'}}{\sqrt{2\pi\hbar^2 k/m}} V(x', x'')\psi_+^{(+)}(E;x'')\,dx'\,dx''$$
$$= \phi_+(E;x) - 2\pi i \left\langle \phi_-(E) \,|V|\, \psi_+^{(+)}(E) \right\rangle \phi_-(E;x) \tag{2.30}$$

and

$$\psi_+^{(+)}(E;x) \sim \frac{e^{ikx}}{\sqrt{2\pi\hbar^2 k/m}}$$
$$- 2\pi i \frac{e^{ikx}}{\sqrt{2\pi\hbar^2 k/m}} \iint_{-\infty}^{\infty} \frac{e^{-ikx'}}{\sqrt{2\pi\hbar^2 k/m}} V(x', x'')\psi_+^{(+)}(E;x'')\,dx'\,dx''$$
$$= \phi_+(E;x) - 2\pi i \left\langle \phi_+(E) \,|V|\, \psi_+^{(+)}(E) \right\rangle \phi_+(E;x)$$
$$= \left[ 1 - 2\pi i \left\langle \phi_+(E) \,|V|\, \psi_+^{(+)}(E) \right\rangle \right] \phi_+(E;x). \tag{2.31}$$

---

From eqns (2.28) and (2.29) we read off the reflection and the transmission amplitudes as

$$r = -2\pi i \left\langle \phi_-(E) \,|V|\, \psi_+^{(+)}(E) \right\rangle, \tag{2.32}$$

$$t = 1 - 2\pi i \left\langle \phi_+(E) \,|V|\, \psi_+^{(+)}(E) \right\rangle, \tag{2.33}$$

respectively, in terms of the unperturbed wave function $\phi_s(E)$, the potential $V$ and the perturbed wave function $\psi_+^{(+)}(E)$. From the above exercise we see that $r$ and $t$ retain this last form for a general Hermitian potential. Similarly, in the physical situation of a plane wave incident from the right, the reflection and the transmission amplitudes can be expressed as

$$r' = -2\pi i \left\langle \phi_+(E) \,|V|\, \psi_-^{(+)}(E) \right\rangle, \tag{2.34}$$

$$t' = 1 - 2\pi i \left\langle \phi_-(E) \,|V|\, \psi_-^{(+)}(E) \right\rangle. \tag{2.35}$$

To summarize, we have found the asymptotic behavior of the two linearly independent solutions $\psi_+^{(+)}(E;x)$ and $\psi_-^{(+)}(E;x)$ to be

$$\psi_+^{(+)}(E;x) \sim \begin{cases} \phi_+(E;x) + r\phi_-(E;x), & x \to -\infty, \\ t\phi_+(E;x), & x \to +\infty, \end{cases} \tag{2.36a}$$

$$\psi_-^{(+)}(E;x) \sim \begin{cases} t'\phi_-(E;x), & x \to -\infty, \\ \phi_-(E;x) + r'\phi_+(E;x), & x \to +\infty, \end{cases} \tag{2.36b}$$

the coefficients $r$, $t$, $r'$ and $t'$ being given in eqns (2.32)–(2.35). Since the Schrödinger equation is *linear*, the linear combination of the above two solutions

$$\psi(E;x) = a^{(1)}\psi_+^{(+)}(E;x) + a^{(2)}\psi_-^{(+)}(E;x) \tag{2.37}$$

is also a solution of the Schrödinger equation for the same energy. Its asymptotic behavior is

$$\psi(E;x) \sim \begin{cases} a^{(1)}\phi_+(E;x) + \left(ra^{(1)} + t'a^{(2)}\right)\phi_-(E;x) \\ \quad = a^{(1)}\phi_+(E;x) + b^{(1)}\phi_-(E;x), & x \to -\infty, \\ a^{(2)}\phi_-(E;x) + \left(a^{(1)}t + a^{(2)}r'\right)\phi_+(E;x) \\ \quad = a^{(2)}\phi_-(E;x) + b^{(2)}\phi_+(E;x), & x \to +\infty. \end{cases} \tag{2.38}$$

This wave function represents the *most general solution of the Schrödinger equation*, which is illustrated, in the asymptotic region, in Fig. 2.2. Here $a^{(1)}$ and $a^{(2)}$ denote the amplitudes of the incoming waves and we have introduced the notation $b^{(1)}$ and $b^{(2)}$ to indicate the amplitudes of the outgoing waves. Equation (2.38) shows that the two sets of amplitudes are *linearly related*, according to the equation

$$\begin{bmatrix} b^{(1)} \\ b^{(2)} \end{bmatrix} = \begin{bmatrix} r & t' \\ t & r' \end{bmatrix} \begin{bmatrix} a^{(1)} \\ a^{(2)} \end{bmatrix}. \tag{2.39}$$

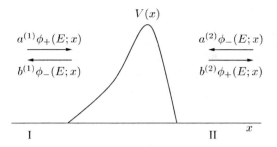

FIG. 2.2. Scattering by a one-dimensional potential. The most general solution of the scattering problem by the potential $V(x)$ is indicated in the asymptotic region. The $\phi_s(E;x)$ are the plane waves of eqn (2.2).

We are thus free to specify the two incoming amplitudes $a^{(1)}$ and $a^{(2)}$; the two outgoing amplitudes $b^{(1)}$ and $b^{(2)}$ are then dictated by the Schrödinger equation and are given by eqn (2.39) (see also [141, p. 96]).

### 2.1.2 The T matrix

In what follows we express the reflection and the transmission amplitudes $r$ and $t$ as matrix elements involving the *unperturbed* wave functions only. This will be at the expense of having to deal with a more complicated operator than $V$, which is called the transition operator $T$.

To this end, it is useful to write some of the above expressions in operator form. We first prove the following theorem.

**Theorem** *The Green function of eqn (2.18) can be written in the alternative form*

$$G_0^{(\pm)}(x,x') = \lim_{\eta \to 0^+} \left\langle x \left| \frac{1}{E \pm i\eta - H_0} \right| x' \right\rangle, \qquad (2.40)$$

*the $\pm$ ensuring outgoing or incoming boundary conditions.*

**Proof** To prove this statement we insert, on the right-hand side of (2.40), a complete set of eigenstates of $H_0$ to obtain

$$\begin{aligned}
G_0^{(\pm)}(x,x') &= \lim_{\eta \to 0^+} \sum_s \int_0^\infty \frac{\langle x|\phi_s(E')\rangle \, dE' \, \langle \phi_s(E')|x'\rangle}{E \pm i\eta - E'} \\
&= -\frac{m}{\pi \hbar^2} \lim_{\eta' \to 0^+} \int_{-\infty}^{+\infty} \frac{e^{ik'(x-x')}}{k'^2 - (k^2 \pm i\eta')} \, dk',
\end{aligned} \qquad (2.41)$$

with $\eta' = 2m\eta/\hbar^2$. Let us consider, specifically, $G_0^{(+)}$. We have

$$G_0^{(+)}(x,x') = -\frac{m}{\pi \hbar^2} \lim_{\varsigma \to 0^+} \int_{-\infty}^{+\infty} \frac{e^{ik'(x-x')}}{[k' - (k + i\varsigma)][k' + (k + i\varsigma)]} \, dk', \qquad (2.42)$$

where $\varsigma \to 0$ as $\eta' \to 0$, and $k > 0$ as usual. This last integral can be performed by extending the variable $k'$ to the complex plane and using Cauchy's contour integration. The poles of the integrand are shown in Fig. 2.3.

We observe that for $x - x' > 0$ we ought to close the contour of integration in the upper half-plane, and in the lower half-plane for $x - x' < 0$. In the first case we pick up the pole at $k + i\varsigma$ and in the second case the pole at $-(k + i\varsigma)$. We thus find that

$$G_0^{(+)}(x > x') = \frac{2m}{\hbar^2} \lim_{\varsigma \to 0^+} \frac{e^{i(k+i\varsigma)(x-x')}}{2i(k+i\varsigma)}, \qquad (2.43a)$$

$$G_0^{(+)}(x < x') = \frac{2m}{\hbar^2} \lim_{\varsigma \to 0^+} \frac{e^{-i(k+i\varsigma)(x-x')}}{2i(k+i\varsigma)}. \qquad (2.43b)$$

We see that these two cases can be condensed in the expression (2.18), thus proving our statement (2.40) for $G_0^{(+)}$. A similar procedure proves it for $G_0^{(-)}$. □

As a result, we have achieved our goal of writing the unperturbed Green function in operator form, i.e.,

$$G_0^{(\pm)} = \lim_{\eta \to 0^+} \frac{1}{E \pm i\eta - H_0}, \qquad (2.44)$$

for which we shall sometimes use the abbreviated notation

$$G_0^{(\pm)} = \frac{1}{E^{(\pm)} - H_0}. \qquad (2.45)$$

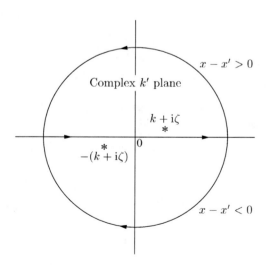

FIG. 2.3. Analytic properties of the integrand in eqn (2.42).

The Lippmann–Schwinger equation (2.10) can thus be expressed in the representation-independent form

$$\left|\psi_s^{(\pm)}(E)\right\rangle = |\phi_s(E)\rangle + \frac{1}{E^{(\pm)} - H_0} V \left|\psi_s^{(\pm)}(E)\right\rangle, \quad (2.46a)$$

the wave function $\psi_s^{(\pm)}(E;x)$ that we have employed so far in the coordinate representation being

$$\psi_s^{(\pm)}(E;x) = \left\langle x \mid \psi_s^{(\pm)}(E) \right\rangle. \quad (2.46b)$$

From eqn (2.46) it is clear that the *transition operator* $T(E)$ satisfying the integral equation

$$T(E) = V + V \frac{1}{E^{(+)} - H_0} T \quad (2.47)$$

has the property

$$T |\phi_s(E)\rangle = V \left|\psi_s^{(+)}(E)\right\rangle. \quad (2.48)$$

One can thus write $r$, $t$, $r'$ and $t'$ of eqns (2.32)–(2.35) as matrix elements of the $T$ operator between unperturbed states as

$$r = -2\pi\mathrm{i} \left\langle \phi_-(E) |T| \phi_+(E) \right\rangle \equiv -2\pi\mathrm{i} T_{-+}(E), \quad (2.49a)$$
$$t = 1 - 2\pi\mathrm{i} \left\langle \phi_+(E) |T| \phi_+(E) \right\rangle \equiv 1 - 2\pi\mathrm{i} T_{++}(E), \quad (2.49b)$$
$$r' = -2\pi\mathrm{i} \left\langle \phi_+(E) |T| \phi_-(E) \right\rangle \equiv -2\pi\mathrm{i} T_{+-}(E), \quad (2.49c)$$
$$t' = 1 - 2\pi\mathrm{i} \left\langle \phi_-(E) |T| \phi_-(E) \right\rangle \equiv 1 - 2\pi\mathrm{i} T_{--}(E), \quad (2.49d)$$

where we have defined

$$T_{s's}(E) \equiv \left\langle \phi_{s'}(E) |T| \phi_s(E) \right\rangle. \quad (2.50)$$

Equations (2.46)–(2.50) retain their form for a general Hermitian potential, not necessarily TRI.

### 2.1.3 The full Green function

The full Green function $G^{(\pm)}(x,x')$ for the scattering problem discussed above is defined by the equation

$$(E - H) G^{(\pm)}(x,x') = \delta(x - x'), \quad (2.51a)$$

where now $H$ is the total Hamiltonian. As usual, the indices $(\pm)$ indicate outgoing wave $(+)$ or incoming wave $(-)$ boundary conditions. More explicitly, for a real, local potential, $G^{(\pm)}(x,x')$ satisfies the differential equation

$$\left[\frac{\partial^2}{\partial x^2} + k^2 - U(x)\right] G^{(\pm)}(x,x') = \frac{2m}{\hbar^2} \delta(x - x'), \quad (2.51b)$$

or

$$\left[\frac{\partial^2}{\partial x^2} + k^2 - U(x)\right] g^{(\pm)}(x,x') = \delta(x - x'), \quad (2.51c)$$

where we have related $g$ and $G$ just as in eqn (2.9).

In operator form we have that the equivalent of eqn (2.44) is

$$G^{(\pm)} = \lim_{\eta \to 0^+} \frac{1}{E \pm i\eta - H}. \tag{2.52}$$

Here, $H = H_0 + V$, where $V$ could be a general Hermitian potential. This last form of $G$ can be written as (where $E^{(\pm)} = E \pm i\eta$ and the $\lim_{\eta \to 0^+}$ is always understood)

$$\begin{aligned} G^{(\pm)} &= \frac{1}{(E^{(\pm)} - H_0)\left[1 - \left(1/(E^{(\pm)} - H_0)\right)V\right]} \\ &= \frac{1}{1 - G_0^{(\pm)} V} G_0^{(\pm)}, \end{aligned} \tag{2.53}$$

for which we assume that the expansion

$$G^{(\pm)} = G_0^{(\pm)} + G_0^{(\pm)} V G_0^{(\pm)} + \cdots \tag{2.54}$$

in powers of the potential exists.

The unperturbed Green function is given in eqn (2.18); it fulfills the symmetry property

$$G_0^{(\pm)}(x, x') = G_0^{(\pm)}(x', x). \tag{2.55}$$

To find the symmetry property for the full Green function for a real, local potential $V(x)$, we make use of the expansion (2.54) to write

$$G^{(\pm)}(x, x') = G_0^{(\pm)}(x, x') + \int G_0^{(\pm)}(x, x_1) V(x_1) G_0^{(\pm)}(x_1, x') \, dx_1 + \cdots, \tag{2.56}$$

from which we find

$$\begin{aligned} G^{(\pm)}(x', x) &= G_0^{(\pm)}(x', x) + \int G_0^{(\pm)}(x', x_1) V(x_1) G_0^{(\pm)}(x_1, x) \, dx_1 + \cdots \\ &= G_0^{(\pm)}(x, x') + \int G_0^{(\pm)}(x, x_1) V(x_1) G_0^{(\pm)}(x_1, x') \, dx_1 + \cdots, \end{aligned} \tag{2.57}$$

and so

$$G^{(\pm)}(x', x) = G^{(\pm)}(x, x'). \tag{2.58}$$

---

**Exercise 2.3** Show that eqn (2.58) is valid for a more general Hermitian, but TRI, potential (the potential thus satisfies eqn (2.23)).

From eqn (2.54) we can write the series expansion for $G^{(+)}$ in the coordinate representation as

$$G^{(\pm)}(x,x') = G_0^{(\pm)}(x,x') + \int G_0^{(\pm)}(x,x_1) V(x_1,x_2) G_0^{(\pm)}(x_2,x') \, \mathrm{d}x_1 \, \mathrm{d}x_2 + \cdots, \quad (2.59)$$

from which we find

$$\begin{aligned} G^{(\pm)}(x',x) &= G_0^{(\pm)}(x',x) + \int G_0^{(\pm)}(x',x_1) V(x_1,x_2) G_0^{(\pm)}(x_2,x) \, \mathrm{d}x_1 \, \mathrm{d}x_2 + \cdots \\ &= G_0^{(\pm)}(x,x') + \int G_0^{(\pm)}(x,x_2) V(x_2,x_1) G_0^{(\pm)}(x_1,x') \, \mathrm{d}x_1 \, \mathrm{d}x_2 + \cdots \\ &= G_0^{(\pm)}(x,x') + \int G_0^{(\pm)}(x,x_1) V(x_1,x_2) G_0^{(\pm)}(x_2,x') \, \mathrm{d}x_1 \, \mathrm{d}x_2 + \cdots \\ &= G^{(\pm)}(x,x'). \end{aligned} \quad (2.60)$$

We have used the symmetry properties (2.55) for $G_0$ and (2.23) for $V$. In the last but one equality, the labels of the integration variables $x_1$ and $x_2$ were interchanged.

**Exercise 2.4** Generalize the symmetry property of the full Green function to a general Hermitian, but not necessarily TRI, potential.

The potential satisfies the Hermiticity requirement

$$V(x_2,x_1) = V^*(x_1,x_2). \quad (2.61)$$

From eqn (2.54) we can write the series expansion for $G^{(+)}$ in the coordinate representation as

$$G^{(+)}(x,x') = G_0^{(+)}(x,x') + \int G_0^{(+)}(x,x_1) V(x_1,x_2) G_0^{(+)}(x_2,x') \, \mathrm{d}x_1 \, \mathrm{d}x_2 + \cdots, \quad (2.62)$$

from which we find

$$\begin{aligned} G^{(+)}(x',x) &= G_0^{(+)}(x',x) + \int G_0^{(+)}(x',x_1) V(x_1,x_2) G_0^{(+)}(x_2,x) \, \mathrm{d}x_1 \, \mathrm{d}x_2 + \cdots \\ &= G_0^{(+)}(x,x') + \int G_0^{(+)}(x,x_2) V^*(x_2,x_1) G_0^{(+)}(x_1,x') \, \mathrm{d}x_1 \, \mathrm{d}x_2 + \cdots \\ &= G_0^{(+)}(x,x') + \int G_0^{(+)}(x,x_1) V^*(x_1,x_2) G_0^{(+)}(x_2,x') \, \mathrm{d}x_1 \, \mathrm{d}x_2 + \cdots. \end{aligned} \quad (2.63)$$

We have used the symmetry properties (2.55) for $G_0$ and (2.61) for $V$. In the last equality, the labels of the integration variables $x_1$ and $x_2$ were interchanged. We see that the last expression in eqn (2.63) coincides with $G^{(+)}(x,x')$ of eqn (2.62), with $V(x_1,x_2)$ replaced by $V^*(x_1,x_2)$. Including the functional dependence of $G$ on the potential, we have thus found that

$$G^{(+)}\left[x',x;V\right] = G^{(+)}\left[x,x';V^*\right]. \quad (2.64)$$

Making use of the symmetry relation

$$G_0^{(+)}(x,x') = \left[G_0^{(-)}(x,x')\right]^* \qquad (2.65)$$

for $G_0$ (see eqns (2.19)), we can write the last expression in eqn (2.63) as

$$G^{(+)}(x',x) = \left[G_0^{(-)}(x,x') + \int G_0^{(-)}(x,x_1)V(x_1,x_2)G_0^{(-)}(x_2,x')\,dx_1\,dx_2 + \cdots\right]^*.$$

The expression in the square brackets is recognized as the series expansion of $G^{(-)}(x,x')$, for the *same* potential. Thus we have the symmetry property

$$G^{(+)}(x',x) = \left[G^{(-)}(x,x')\right]^*. \qquad (2.66)$$

The potential is the same on both sides and is thus not indicated.

**Exercise 2.5** Write the Lippmann–Schwinger equation (2.46) in terms of the full Green function.

Expanding $|\psi\rangle$ in a Born series we have

$$\begin{aligned}G_0V\,|\psi\rangle &= G_0V\left[1 + G_0V + G_0VG_0V + \cdots\right]|\phi\rangle \\ &= G_0\left[V + VG_0V + VG_0VG_0V + \cdots\right]|\phi\rangle \\ &= \left[G_0 + G_0VG_0 + G_0VG_0VG_0 + \cdots\right]V\,|\phi\rangle.\end{aligned}$$

The expression in the last square bracket is the series expansion (2.54) of the full Green function. We thus find the identity

$$G_0V\,|\psi\rangle = GV\,|\phi\rangle, \qquad (2.67)$$

which allows us to write the Lippmann–Schwinger equation (2.46) as

$$\left|\psi_s^{(\pm)}(E)\right\rangle = |\phi_s(E)\rangle + \frac{1}{E^{(\pm)} - H}V\,|\phi_s(E)\rangle. \qquad (2.68)$$

We now solve eqn (2.51c), concentrating on the situation of a local, and hence real, potential $U(x)$, and for outgoing wave boundary conditions. For $x \neq x'$, the right-hand side of eqn (2.51c) vanishes, i.e.,

$$\left[\frac{\partial^2}{\partial x^2} + k^2 - U(x)\right]g^{(+)}(x,x') = 0, \qquad (2.69)$$

so that $g^{(+)}(x,x')$, considered as a function of the variable $x$, must be proportional to the complete wave function $\psi_s^{(+)}(E;x)$ discussed in the previous sections, with $x'$-dependent coefficients, to be determined. In fact, the functions

$$g^{(+)}(x < x') = f_1(x')\psi_-^{(+)}(E;x) \tag{2.70}$$
$$\sim f_1(x')t'(k)e^{-ikx} \quad \text{as } x \to -\infty$$

and

$$g^{(+)}(x > x') = f_2(x')\psi_+^{(+)}(E;x) \tag{2.71}$$
$$\sim f_2(x')t(k)e^{ikx} \quad \text{as } x \to +\infty$$

satisfy the differential equation (2.69) and the correct outgoing wave boundary conditions as a function of the variable $x$. Continuity at $x = x'$ implies that

$$f_1(x')\psi_-^{(+)}(E;x') = f_2(x')\psi_+^{(+)}(E;x'), \tag{2.72}$$

or

$$\frac{f_1(x')}{\psi_+^{(+)}(E;x')} = \frac{f_2(x')}{\psi_-^{(+)}(E;x')} \equiv A(x'). \tag{2.73}$$

Thus

$$g^{(+)}(x < x') = A(x')\psi_-^{(+)}(E;x)\psi_+^{(+)}(E;x'), \tag{2.74a}$$
$$g^{(+)}(x > x') = A(x')\psi_-^{(+)}(E;x')\psi_+^{(+)}(E;x). \tag{2.74b}$$

According to eqn (2.51c), the slope of $g^{(+)}(x, x')$ must have, at $x = x'$, the discontinuity

$$\lim_{\epsilon \to 0^+} \left[\frac{\partial g^{(+)}(x,x')}{\partial x}\right]_{x=x'-\epsilon}^{x=x'+\epsilon} = 1. \tag{2.75}$$

Using eqns (2.74), we can thus determine $A(x')$ to be

$$\frac{1}{A(x')} = \psi_-^{(+)}(E;x')\left[\frac{\partial \psi_+^{(+)}(E;x)}{\partial x}\right]_{x=x'} - \left[\frac{\partial \psi_-^{(+)}(E;x)}{\partial x}\right]_{x=x'}\psi_+^{(+)}(E;x')$$
$$= W\left[\psi_-^{(+)}(E;x'), \psi_+^{(+)}(E;x')\right], \tag{2.76}$$

the last equation defining the Wronskian of the two functions $\psi_-^{(+)}(E;x')$ and $\psi_+^{(+)}(E;x')$. Since these functions are two solutions of the same differential equation (2.6b), their Wronskian is independent of position, and thus $A(x')$ must be a constant. We calculate the Wronskian as $x' \to +\infty$. In this limit, $\psi_+^{(+)}(E;x')$ behaves as shown in eqn (2.29), i.e.,

$$\psi_+^{(+)}(E;x') \sim t(k)\phi_+(E;x'), \tag{2.77}$$

and similarly

$$\psi_-^{(+)}(E;x') \sim \phi_-(E;x') + r'(k)\phi_+(E;x'), \tag{2.78}$$

so that, for the Wronskian (2.76), we obtain

$$\begin{aligned} W\left[\psi_-^{(+)}(E;x'),\psi_+^{(+)}(E;x')\right] \\ = W\left[\phi_-(E;x')+r'(k)\phi_+(E;x'),t(k)\phi_+(E;x')\right] \\ = t(k)W\left[\phi_-(E;x'),\phi_+(E;x')\right] \\ = 2\mathrm{i}kt(k)\frac{m}{2\pi\hbar^2 k}. \end{aligned} \qquad (2.79)$$

From eqn (2.76) we thus obtain the constant $A$ as

$$A = \frac{1}{2\mathrm{i}kt(k)}\frac{2\pi\hbar^2 k}{m}, \qquad (2.80)$$

so that, finally, the Green function $g^{(+)}(x,x')$ of eqns (2.74) is given by

$$g^{(+)}(x,x') = -\frac{\mathrm{i}\pi\hbar^2}{mt(k)}\psi_-^{(+)}(E;x_<)\psi_+^{(+)}(E;x_>), \qquad (2.81\mathrm{a})$$

and $G^{(+)}(x,x')$ by

$$G^{(+)}(x,x') = -\frac{2\pi\mathrm{i}}{t(k)}\psi_-^{(+)}(E;x_<)\psi_+^{(+)}(E;x_>). \qquad (2.81\mathrm{b})$$

Here, $x_<$ and $x_>$ denote, respectively, the lesser and the greater of $x$ and $x'$. Notice that the symmetry relation (2.58) is satisfied.

### 2.1.4 The S matrix

At this point we introduce a concept of great importance in scattering theory that we shall deal with throughout most of the book, namely the *scattering matrix*, or *S matrix*.

We indicated earlier, immediately below eqn (2.12), that the states $|\psi_s^{(+)}(E)\rangle$, corresponding to outgoing boundary conditions, form a complete orthonormal set, and so do the states $|\psi_s^{(-)}(E)\rangle$, corresponding to incoming boundary conditions. We now demonstrate that the scalar product of two states, when one is taken from each of these two sets, is of particular interest, as it is related to the scattering amplitudes $r$, $r'$, $t$ and $t'$, which are shown in Fig. 2.1 and given explicitly in eqns (2.32)–(2.35); they relate the incoming and the outgoing amplitudes, according to eqn (2.39).

From the Lippmann–Schwinger equations (2.46) and (2.68), we write the states $|\psi_s^{(+)}(E)\rangle$ and $|\psi_{s'}^{(-)}[E')\rangle$ as (see, in particular, [155, Section 4.2])

$$\left|\psi_s^{(+)}(E)\right\rangle = |\phi_s(E)\rangle + \frac{1}{E-H_0+\mathrm{i}\eta}V\left|\psi_s^{(+)}(E)\right\rangle, \qquad (2.82)$$

$$\left|\psi_{s'}^{(-)}(E')\right\rangle = |\phi_{s'}(E')\rangle + \frac{1}{E'-H-\mathrm{i}\eta}V\left|\phi_{s'}(E')\right\rangle. \qquad (2.83)$$

Using the adjoint of (2.83), we proceed to calculate the scalar product of interest as

$$\left\langle \psi_{s'}^{(-)}(E') \mid \psi_s^{(+)}(E) \right\rangle$$
$$= \left\langle \phi_{s'}(E') \mid \psi_s^{(+)}(E) \right\rangle + \left\langle \phi_{s'}(E') \middle| V \frac{1}{E' - H + i\eta} \middle| \psi_s^{(+)}(E) \right\rangle$$
$$= \left\langle \phi_{s'}(E') \mid \psi_s^{(+)}(E) \right\rangle + \frac{\left\langle \phi_{s'}(E') \middle| V \middle| \psi_s^{(+)}(E) \right\rangle}{E' - E + i\eta}. \tag{2.84}$$

Using eqn (2.82), we have

$$\left\langle \psi_{s'}^{(-)}(E') \mid \psi_s^{(+)}(E) \right\rangle$$
$$= \left\langle \phi_{s'}(E') \mid \phi_s(E) \right\rangle + \left\langle \phi_{s'}(E') \middle| \frac{1}{E - H_0 + i\eta} V \middle| \psi_s^{(+)}(E) \right\rangle$$
$$+ \frac{\left\langle \phi_{s'}(E') \middle| V \middle| \psi_s^{(+)}(E) \right\rangle}{E' - E + i\eta}$$
$$= \left\langle \phi_{s'}(E') \mid \phi_s(E) \right\rangle$$
$$+ \left\langle \phi_{s'}(E') \middle| V \middle| \psi_s^{(+)}(E) \right\rangle \left[ \frac{1}{E - E' + i\eta} + \frac{1}{E' - E + i\eta} \right]. \tag{2.85}$$

Now

$$\frac{1}{E - E' + i\eta} + \frac{1}{E' - E + i\eta}$$
$$= -2\pi i \frac{\eta/\pi}{(E' - E)^2 + \eta^2} \rightarrow -2\pi i \delta(E - E') \quad \text{as } \eta \rightarrow 0^+, \tag{2.86}$$

so that

$$\left\langle \psi_{s'}^{(-)}(E') \middle| \psi_s^{(+)}(E) \right\rangle$$
$$= \delta_{s's} \delta(E - E') - 2\pi i \left\langle \phi_{s'}(E') \middle| V \middle| \psi_s^{(+)}(E) \right\rangle \delta(E' - E)$$
$$= \left[ \delta_{s's} - 2\pi i \left\langle \phi_{s'}(E') \middle| V \middle| \psi_s^{(+)}(E) \right\rangle \right] \delta(E' - E). \tag{2.87}$$

Using the property (2.48) of the $T$ operator and the definition (2.50), we find

$$\left\langle \psi_{s'}^{(-)}(E') \middle| \psi_s^{(+)}(E) \right\rangle = \left[ \delta_{s's} - 2\pi i \left\langle \phi_{s'}(E') \middle| T \middle| \phi_s(E) \right\rangle \right] \delta(E' - E)$$
$$= \left[ \delta_{s's} - 2\pi i T_{s's}(E) \right] \delta(E' - E), \tag{2.88}$$

or

$$\left\langle \psi_{s'}^{(-)}(E') \middle| \psi_s^{(+)}(E) \right\rangle \equiv S_{s's}(E)\delta(E'-E), \tag{2.89}$$

where we have defined the matrix elements $S_{s's}(E)$ of the *on-shell scattering matrix*, or *S matrix*, as

$$S_{s's}(E) = \delta_{s's} - 2\pi i T_{s's}(E). \tag{2.90}$$

The right-hand side of eqn (2.89) is the overlap matrix of the two complete orthonormal sets $|\psi_s^{(+)}(E)\rangle$ and $|\psi_{s'}^{(-)}(E')\rangle$, and it is therefore an infinite-dimensional *unitary* matrix. This matrix is block diagonal as on-shell $S$ matrices; each block corresponds to one $E$ value and is itself unitary. Thus the states $|\psi_s^{(+)}(E)\rangle$ can be expressed as linear combinations of the states $|\psi_{s'}^{(-)}(E)\rangle$ with the *same energy*, the expansion coefficients being the on-shell $S$-matrix elements given by eqn (2.90), i.e.,

$$\left|\psi_s^{(+)}(E)\right\rangle = \sum_{s'} \left|\psi_{s'}^{(-)}(E)\right\rangle S_{s's}(E), \tag{2.91}$$

with

$$SS^\dagger = I. \tag{2.92}$$

Notice that the above derivation of the $S$ matrix is valid for a general Hermitian potential, not necessarily TRI.

We now see from eqn (2.90) that the reflection and the transmission amplitudes found in eqns (2.49) can be expressed in terms of the $S$-matrix elements in the particularly simple form

$$r = S_{-+}, \tag{2.93a}$$
$$t = S_{++}, \tag{2.93b}$$
$$r' = S_{+-}, \tag{2.93c}$$
$$t' = S_{--}. \tag{2.93d}$$

As a result, the matrix that appears in eqn (2.39) and relates the incoming to the outgoing amplitudes can be constructed from the $S$-matrix elements as

$$S = \begin{bmatrix} S_{-+} & S_{--} \\ S_{++} & S_{+-} \end{bmatrix} = \begin{bmatrix} r & t' \\ t & r' \end{bmatrix}, \tag{2.94}$$

so that we can write eqn (2.39) as

$$\begin{bmatrix} b^{(1)} \\ b^{(2)} \end{bmatrix} = S \begin{bmatrix} a^{(1)} \\ a^{(2)} \end{bmatrix}. \tag{2.95}$$

In the absence of a potential, this matrix goes over into the Pauli matrix $\sigma_x$. A different choice of the order of the matrix elements would give the unit matrix $I_2$ in the absence of a potential. The present choice has the advantage that, in the presence of time-reversal invariance, $S$ is a symmetric matrix, as we shall see in eqn (2.120) below.

Unitarity of the $S$ matrix ensures *flux conservation*. In fact, the incoming flux for the situation illustrated in Fig. 2.2 is given by $|a^{(1)}|^2 + |a^{(2)}|^2$ and the outgoing flux by $|b^{(1)}|^2 + |b^{(2)}|^2$, i.e., by the norm of the vectors on the right- and left-hand sides of eqn (2.39), respectively, and this norm is conserved by a unitary matrix $S$.

We present in the following exercise an alternative procedure to show the relation between the scattering amplitudes and the scalar product of a $\psi^{(+)}$ and a $\psi^{(-)}$ state, that employs the Schrödinger equation instead of the Lippmann–Schwinger equation, which was the starting point of the above proof (eqns (2.82) and (2.83)). The procedure presented below will be found useful in later sections (see Sections 2.1.9 and 3.2).

**Exercise 2.6** Show the relation between the scattering amplitudes and the scalar product of a $\psi^{(-)}$ and a $\psi^{(+)}$ state using the Schrödinger equation directly.

For simplicity, we restrict the discussion to a local, and hence real, potential $U(x)$. The Schrödinger equation for $\psi_s^{(+)}(E;x)$ and for $[\psi_{s'}^{(-)}(E';x)]^*$, in the notation of eqn (2.6b), is

$$\left(\partial_x^2 + k^2\right) \psi_s^{(+)}(E;x) = U(x)\psi_s^{(+)}(E;x),$$
$$\left(\partial_x^2 + k'^2\right) \left[\psi_{s'}^{(-)}(E';x)\right]^* = U(x) \left[\psi_{s'}^{(-)}(E';x)\right]^*. \qquad (2.96)$$

We multiply the first of these equations by $[\psi_{s'}^{(-)}(E';x)]^*$, the second one by $\psi_s^{(+)}(E;x)$ and subtract the resulting two equations to obtain

$$\left(k'^2 - k^2\right) \left[\psi_{s'}^{(-)}(E';x)\right]^* \psi_s^{(+)}(E;x)$$
$$= \partial_x \left\{ \left[\psi_{s'}^{(-)}(E';x)\right]^* \partial_x \psi_s^{(+)}(E;x) - \psi_s^{(+)}(E;x)\partial_x \left[\psi_{s'}^{(-)}(E';x)\right]^* \right\}. \qquad (2.97)$$

The potential has dropped out from the last equation, and there will be no further reference to it in what follows, except that it must be of finite range, which is needed after eqn (2.100). Its existence will also be implicit in the asymptotic form of the wave functions, as in eqn (2.103), where we assume, in general, that $r \neq 0$ and $t \neq 1$.

We multiply the two sides of the above equation by the convergence factor $e^{-\epsilon|x|}$ and integrate from $-\infty$ to $+\infty$; due to the convergence factor the integrals are well defined. At the end of the calculation we take the limit when $\epsilon \to 0$. We obtain

$$\left(k'^2 - k^2\right) \lim_{\epsilon \to 0} \int_{-\infty}^{+\infty} e^{-\epsilon|x|} \left[\psi_{s'}^{(-)}(E';x)\right]^* \psi_s^{(+)}(E;x)\, dx$$
$$= \lim_{\epsilon \to 0} \int_{-\infty}^{+\infty} e^{-\epsilon|x|} \partial_x \left\{\cdots\right\} dx \equiv R, \qquad (2.98)$$

where $\{\cdots\}$ indicates, for brevity, the curly bracket on the right-hand side of eqn (2.97). Also, the right-hand side of eqn (2.98) is denoted by $R$.

Integration of $R$ by parts gives

$$R = \lim_{\epsilon \to 0} \left\{ \left[ e^{-\epsilon |x|} \{\cdots\} \right]_{-\infty}^{+\infty} + \epsilon \int_{-\infty}^{+\infty} \operatorname{sgn}(x) e^{-\epsilon |x|} \{\cdots\} \, \mathrm{d}x \right\}, \quad (2.99)$$

where $\operatorname{sgn}(x)$ is the sign function, which is negative (positive) for negative (positive) $x$. The quantity $\{\cdots\}$ is everywhere finite, so that $\left[ e^{-\epsilon |x|} \{\cdots\} \right]_{-\infty}^{+\infty} = 0$ and

$$R = \lim_{\epsilon \to 0} \left\{ \epsilon \int_{-\infty}^{+\infty} \operatorname{sgn}(x) e^{-\epsilon |x|} \{\cdots\} \, \mathrm{d}x \right\}. \quad (2.100)$$

The potential $U(x)$ is nonzero from $x = -L_1$ to $x = L_2$, say. In the above integration from $-\infty$ to $+\infty$ we can omit the finite interval $(-L_1, L_2)$, as the integral over $(-L_1, L_2)$ is certainly a finite quantity, and so

$$\lim_{\epsilon \to 0} \left\{ \epsilon \int_{-L_1}^{L_2} \operatorname{sgn}(x) e^{-\epsilon |x|} \{\cdots\} \, \mathrm{d}x \right\} = 0. \quad (2.101)$$

Thus

$$R = \lim_{\epsilon \to 0} \left\{ \epsilon \left[ -\int_{-\infty}^{-L_1} e^{\epsilon x} \{\cdots\} \, \mathrm{d}x + \int_{L_2}^{+\infty} e^{-\epsilon x} \{\cdots\} \, \mathrm{d}x \right] \right\}. \quad (2.102)$$

Now the integration is performed over the potential-free region only, where we know the structure of the wave function.

From now on we shall treat the particular case $s = s' = +$. The wave functions we need have the form

$$\psi_+^{(+)}(E; x) = \sqrt{\frac{m}{2\pi \hbar^2 k}} \times \begin{cases} e^{ikx} + r(k) e^{-ikx}, & x < -L_1, \\ t(k) e^{ikx}, & x > L_2, \end{cases} \quad (2.103)$$

$$\left[ \psi_+^{(-)}(E'; x) \right]^* = \psi_-^{(+)}(E'; x)$$
$$= \sqrt{\frac{m}{2\pi \hbar^2 k'}} \times \begin{cases} t'(k') e^{-ik'x}, & x < -L_1, \\ e^{-ik'x} + r'(k') e^{ik'x}, & x > L_2, \end{cases} \quad (2.104)$$

with

$$k, k' > 0. \quad (2.105)$$

Substituting these expressions for the wave functions into eqn (2.102), we find

$$R = \frac{m}{2\pi \hbar^2 \sqrt{kk'}}$$
$$\times \lim_{\epsilon \to 0} \epsilon \left\{ -i t'(k') \int_{-\infty}^{0} \left[ (k+k') e^{i(k-k')x} - r(k)(k-k') e^{-i(k+k')x} \right] e^{\epsilon x} \, \mathrm{d}x \right.$$
$$\left. + i t(k) \int_{0}^{\infty} \left[ (k+k') e^{i(k-k')x} + r'(k')(k-k') e^{i(k+k')x} \right] e^{-\epsilon x} \, \mathrm{d}x \right\}, \quad (2.106)$$

where, for convenience, we have added the integrals from $-L_1$ to 0 and from 0 to $L_2$, which vanish as $\epsilon \to 0$, as can be seen from the same argument used in connection with eqn (2.101).

Using the original equation (2.98), we find

$$\lim_{\epsilon \to 0^+} \int_{-\infty}^{+\infty} e^{-\epsilon|x|} \left[\psi_+^{(-)}(E';x)\right]^* \psi_+^{(+)}(E;x) \, dx$$

$$= \frac{m}{2\pi\hbar^2 \sqrt{kk'}} \lim_{\epsilon \to 0} \epsilon \left\{ t'(k') \left[\frac{1}{k-k'} \frac{1}{k-k'-i\epsilon} + r(k) \frac{1}{k+k'} \frac{1}{k+k'+i\epsilon}\right] \right.$$

$$\left. + t(k) \left[\frac{1}{k-k'} \frac{1}{k-k'+i\epsilon} + r'(k') \frac{1}{k+k'} \frac{1}{k+k'+i\epsilon}\right]\right\}$$

$$= \frac{m}{2\pi\hbar^2} \lim_{\epsilon \to 0^+} F_\epsilon(k,k'), \qquad (2.107)$$

where we have defined the function

$$F_\epsilon(k,k') = \frac{1}{\sqrt{kk'}} \frac{\epsilon}{k-k'} \left[t'(k') \frac{1}{k-k'-i\epsilon} + t(k) \frac{1}{k-k'+i\epsilon}\right]. \qquad (2.108)$$

We have used the fact that the last term in the second line and the last one in the third line of eqn (2.107), being finite due to the inequality (2.105), do not contribute in the limit. From the time-reversal invariance result found in eqn (2.120) we can write $t'(k') = t(k')$, so that

$$F_\epsilon(k,k') = \frac{1}{\sqrt{kk'}} \frac{\epsilon}{k-k'} \frac{(k-k'+i\epsilon)t(k') + (k-k'-i\epsilon)t(k)}{(k-k')^2+\epsilon^2}$$

$$= \frac{1}{\sqrt{kk'}} \left\{\pi\left[t(k')+t(k)\right] \frac{\epsilon/\pi}{(k-k')^2+\epsilon^2} - i\frac{t(k')-t(k)}{k'-k} \frac{\epsilon^2}{(k-k')^2+\epsilon^2}\right\}$$

$$= F_\epsilon^{(1)}(k,k') + F_\epsilon^{(2)}(k,k'). \qquad (2.109)$$

We now analyze separately the two terms of the previous equation. For the first term we find, in the limit $\epsilon \to 0^+$,

$$F_\epsilon^{(1)}(k,k') = \frac{\pi}{\sqrt{kk'}} \left[t(k')+t(k)\right] \frac{\epsilon/\pi}{(k-k')^2+\epsilon^2}$$

$$\xrightarrow[\epsilon \to 0^+]{} \frac{\pi}{\sqrt{kk'}} \left[t(k')+t(k)\right] \delta(k'-k)$$

$$= \frac{2\pi}{k} t(k) \delta(k'-k). \qquad (2.110)$$

The second term in eqn (2.109) is

$$F_\epsilon^{(2)}(k,k') = -\frac{i}{\sqrt{kk'}} \frac{t(k')-t(k)}{k'-k} \frac{\epsilon^2}{(k-k')^2+\epsilon^2}. \qquad (2.111)$$

Consider $F_\epsilon^{(2)}(k,k')$ as a function of $k'$, for small but finite $\epsilon$. The second factor on the right-hand side of eqn (2.111) is a continuous function of $k'$ for all $k' \neq k$, and as

$k' \to k$ it tends to the finite limit $\partial t(k)/\partial k$. The last factor is continuous and finite for all $k'$; it takes the value 1 at $k' = k$, with a half-width at half-maximum $\epsilon$ around $k' = k$; this region becomes a point, i.e., a set of zero measure, as $\epsilon \to 0$. Thus, if we construct the integral

$$I = \int_a^b F_\epsilon^{(2)}(k,k')\phi(k')\,\mathrm{d}k', \tag{2.112}$$

where $k$ is contained inside the interval $(a,b)$ and $\phi(k')$ is an arbitrary but continuous function of $k'$, then we find that $I$ exists and, moreover, vanishes in the limit $\epsilon \to 0$. It is in this sense that we say that

$$\lim_{\epsilon \to 0} F_\epsilon^{(2)}(k,k') = 0. \tag{2.113}$$

We are thus left with $F_\epsilon^{(1)}(k,k')$ only. Since in the limit $\epsilon \to 0^+$ the integral on the left-hand side of eqn (2.107) tends to the scalar product $(\psi_+^{(-)}(E';x), \psi_+^{(+)}(E;x))$, we finally find

$$\left(\psi_+^{(-)}(E';x), \psi_+^{(+)}(E;x)\right) = t(k)\,\delta\left(E - E'\right), \tag{2.114}$$

which is identical to the result of eqns (2.88) and (2.49) for $s = s' = +$. A similar procedure can be carried out for the other values of $s$ and $s'$.

We now present a somewhat different argument to arrive at the conclusion (2.113), since it will be applied in Exercise 3.6 of the next chapter, that deals with the $S$-matrix elements for a ballistic cavity. As we mentioned above, the integral $I$ of eqn (2.112) exists; it can be written as the principal value integral

$$\mathcal{P}\int_a^b F_\epsilon^{(2)}(k,k')\phi(k')\,\mathrm{d}k' = \mathcal{P}\int_a^b \left[F_\epsilon^{(2)}(k,k')\right]_1 \phi(k')\,\mathrm{d}k' \\ + \mathcal{P}\int_a^b \left[F_\epsilon^{(2)}(k,k')\right]_2 \phi(k')\,\mathrm{d}k', \tag{2.115}$$

where the two terms arise from having split the integrand into the two terms

$$\left[F_\epsilon^{(2)}(k,k')\right]_1 = -\frac{\mathrm{i}}{\sqrt{kk'}}\frac{t(k')}{k'-k}\frac{\epsilon^2}{(k-k')^2+\epsilon^2}, \tag{2.116a}$$

$$\left[F_\epsilon^{(2)}(k,k')\right]_2 = \frac{\mathrm{i}}{\sqrt{kk'}}\frac{t(k)}{k'-k}\frac{\epsilon^2}{(k-k')^2+\epsilon^2}. \tag{2.116b}$$

Notice that the principal value, which could have been omitted from the left-hand side of eqn (2.115), is essential for each one of the two terms on its right-hand side (where the integrand diverges at $k' = k$, as the above eqn (2.116) makes clear) to exist. For instance, the first integral on the right-hand side of eqn (2.115) can be written as

$$\mathcal{P}\int_a^b \left[F_\epsilon^{(2)}(k,k')\right]_1 \phi(k')\,\mathrm{d}k' = -\mathcal{P}\int_a^b \frac{\mathrm{i}}{\sqrt{kk'}}\frac{t(k')}{k'-k}\frac{\epsilon^2}{(k-k')^2+\epsilon^2}\,\mathrm{d}k' \\ \equiv \mathcal{P}\int_a^b \frac{g_\epsilon(k,k')}{k'-k}\,\mathrm{d}k'$$

POTENTIAL SCATTERING IN INFINITE ONE-DIMENSIONAL SPACE    37

$$= \lim_{\eta \to 0^+} \left[ \int_a^{k-\eta} \frac{g_\epsilon(k,k')}{k'-k} \, dk' + \int_{k+\eta}^b \frac{g_\epsilon(k,k')}{k'-k} \, dk' \right]$$
$$\xrightarrow[\epsilon \to 0^+]{} 0. \qquad (2.117)$$

The principal value integral in the above eqn (2.117) exists for any finite $\epsilon$. In this case the function $g_\epsilon(k, k')$ of the above equation is non-negligible and almost constant in a region $O(\epsilon)$. Thus that principal value integral vanishes as we take the limit $\epsilon \to 0$ at the end, as indicated in the last line. A similar argument applies for the second integral in eqn (2.115) and we thus arrive at the conclusion (2.113).

We now investigate another important property of the $S$ matrix, which is fulfilled when the Hamiltonian is invariant under the operation $\theta$ of time reversal (TRI), eqn (2.22). For the present argument, the potential could be, in general, non-local, as in eqn (2.23). Inside the scalar product of eqn (2.89) defining the $S$-matrix element, we apply the anti-unitary operator $\theta$ and obtain (see, for instance, [142, 176])

$$\begin{aligned}
\left(\psi_{s'}^{(-)}(E';x), \psi_s^{(+)}(E;x)\right) &= \left(\theta\psi_{s'}^{(-)}(E';x), \theta\psi_s^{(+)}(E;x)\right)^* \\
&= \left(\psi_{-s'}^{(+)}(E';x), \psi_{-s}^{(-)}(E;x)\right)^* \\
&= \left(\psi_{-s}^{(-)}(E;x), \psi_{-s'}^{(+)}(E';x)\right), \qquad (2.118)
\end{aligned}$$

where we have used eqn (2.21), valid under TRI. We thus find

$$S_{s's}(E) = S_{-s,-s'}(E). \qquad (2.119)$$

As a result, the $S$ matrix of eqn (2.94) is *symmetric*, i.e.,

$$S = \begin{bmatrix} r & t \\ t & r' \end{bmatrix}. \qquad (2.120)$$

**Exercise 2.7** Prove the symmetry relation (2.119) using the definition (2.90) and the property of the $T$-matrix elements under TRI.

We expand eqn (2.47) as

$$T = V + VG_0V + VG_0VG_0V + \cdots. \qquad (2.121)$$

$G_0$ satisfies the relation (2.19b). Under TRI, $V$ satisfies eqn (2.23). We thus find

$$T(x, x') = T(x', x). \qquad (2.122)$$

As a consequence, $T_{++}$ used in eqn (2.49) can be shown to equal $T_{--}$. In fact,

$$\tilde{T}_{++} = \iint e^{-ikx} T(x,x') e^{ikx'} \, dx \, dx'$$
$$= \iint e^{-ikx'} T(x',x) e^{ikx} \, dx \, dx'$$
$$= \iint e^{ikx} T(x,x') e^{-ikx'} \, dx \, dx' = \tilde{T}_{--}, \qquad (2.123)$$

where, for convenience, we have defined

$$\tilde{T} \equiv \frac{2\pi\hbar^2 k}{m} T. \qquad (2.124)$$

Thus
$$T_{++} = T_{--}; \qquad (2.125)$$
however, $T_{+-}$ and $T_{-+}$ may, in general, be different. We thus find eqn (2.119) for the $S$-matrix elements.

---

An alternative way to arrive at the symmetry of the $S$ matrix from the property of time-reversal invariance is as follows. In Fig. 2.2 we denote by I and II the free regions on the left and the right of the potential, respectively. Aside from a normalization factor, the wave function on either side can be written as

$$\psi^{(1)}(x) \propto a^{(1)} e^{ikx} + b^{(1)} e^{-ikx}, \qquad (2.126a)$$
$$\psi^{(2)}(x) \propto b^{(2)} e^{ikx} + a^{(2)} e^{-ikx}, \qquad (2.126b)$$

respectively. Because of time-reversal invariance, the complex conjugate of the wave function, i.e.,

$$\left\{\psi^{(1)}(x)\right\}^* \propto (b^{(1)})^* e^{ikx} + (a^{(1)})^* e^{-ikx}, \qquad (2.127a)$$
$$\left\{\psi^{(2)}(x)\right\}^* \propto (a^{(2)})^* e^{ikx} + (b^{(2)})^* e^{-ikx}, \qquad (2.127b)$$

is a solution of the same Schrödinger equation with the same energy. Therefore, the coefficients must be related precisely by the same $S$ matrix as in eqn (2.95), i.e.,

$$\begin{bmatrix} (a^{(1)})^* \\ (a^{(2)})^* \end{bmatrix} = S \begin{bmatrix} (b^{(1)})^* \\ (b^{(2)})^* \end{bmatrix}, \qquad (2.128)$$

from which we find

$$\begin{bmatrix} b^{(1)} \\ b^{(2)} \end{bmatrix} = (S^*)^{-1} \begin{bmatrix} a^{(1)} \\ a^{(2)} \end{bmatrix}. \qquad (2.129)$$

Comparing with eqn (2.95), we find

$$S = (S^*)^{-1} \qquad (2.130)$$

or

POTENTIAL SCATTERING IN INFINITE ONE-DIMENSIONAL SPACE 39

$$SS^* = I. \tag{2.131}$$

This equation, together with the unitarity relation (2.92), gives the desired symmetry property

$$S = S^\top. \tag{2.132}$$

The transmission amplitudes for a one-dimensional Hermitian potential not fulfilling TRI are computed, using a first-order Born approximation, in the following exercise.

**Exercise 2.8** Analyze, using a first-order Born approximation, the transmission amplitudes $t$ and $t'$ of eqns (2.49b) and (2.49d), respectively, for a Hermitian potential, i.e., one satisfying eqn (2.61), but not necessarily fulfilling TRI, i.e., not satisfying eqn (2.23). Apply the results to the particular case of the velocity-dependent potential

$$\hat{V} = v(x)\hat{p} + \hat{p}v(x), \tag{2.133}$$

where $v(x)$ is a real function of $x$ and $\hat{p} = -i\hbar\partial/\partial x$. This potential is to be taken as a formal, illustrative example; although it is not obvious how to realize it in practice, it is in principle admissible by quantum mechanics [157].

The first term in the Born series, eqn (2.121), gives (see the definition (2.124))

$$\tilde{T}_{++} = \iint e^{-ikx} V(x,x') e^{ikx'} \, dx \, dx' + \cdots, \tag{2.134a}$$

$$\tilde{T}_{--} = \iint e^{ikx} V(x,x') e^{-ikx'} \, dx \, dx' + \cdots$$

$$= \iint e^{-ikx} V^*(x,x') e^{ikx'} \, dx \, dx' + \cdots. \tag{2.134b}$$

We thus see that for a non-real potential, $\tilde{T}_{++}$ and $\tilde{T}_{--}$, and hence $t$ and $t'$, do not in general coincide. The last equations must be contrasted with eqns (2.123) and (2.125).

The velocity-dependent potential of eqn (2.133) gives

$$V(x,x') = \langle x|\hat{V}|x'\rangle = -i\hbar \left[v(x) + v(x')\right] \delta'(x-x'), \tag{2.135a}$$

$$V^*(x,x') = i\hbar \left[v(x) + v(x')\right] \delta'(x-x'), \tag{2.135b}$$

where the prime in the delta function denotes its derivative with respect to the argument. We have, to first order,

$$\tilde{T}_{--} = -\tilde{T}_{++}. \tag{2.136}$$

One finds, explicitly,

$$\tilde{T}_{++} = 2\hbar k \int_{-\infty}^{\infty} v(x) \, dx + \cdots, \tag{2.137}$$

so that

$$T_{++} = \frac{1}{2\pi} \int_{-\infty}^{\infty} u(x) \, dx + \cdots. \tag{2.138}$$

We have defined

$$u(x) = \frac{2mv(x)}{\hbar}, \tag{2.139}$$

which has the same dimensions as $k$. Finally, the transmission amplitudes $t$ and $t'$ are found from eqns (2.49b) and (2.49d) to be

$$t = 1 - \mathrm{i}\int_{-\infty}^{\infty} u(x)\,\mathrm{d}x + \cdots, \tag{2.140a}$$

$$t' = 1 + \mathrm{i}\int_{-\infty}^{\infty} u(x)\,\mathrm{d}x + \cdots. \tag{2.140b}$$

Thus the resulting $S$ matrix is not symmetric.

The scattering wave function for the velocity-dependent potential (2.133) is studied in the following exercise.

**Exercise 2.9** Study the scattering solution of the Schrödinger equation for the velocity-dependent potential defined in eqn (2.133). As an example, apply the general result to the particular case in which $u(x)$ is a barrier given by

$$u(x) = \begin{cases} 0, & x < 0, \\ u_0, & 0 < x < a, \\ 0, & x > a. \end{cases} \tag{2.141}$$

For the required potential of eqn (2.133) we can write the time-independent Schrödinger equation as

$$\left(\frac{\partial^2}{\partial x^2} + k^2\right)\psi(x) = -\mathrm{i}\left[u(x)\frac{\partial}{\partial x} + \frac{\partial}{\partial x}u(x)\right]\psi(x), \tag{2.142}$$

where $u(x)$ is given in eqn (2.139). It may be noted in passing that the corresponding classical Lagrangian equation of motion is self-adjoint [157]. If we propose for the wave function $\psi(x)$ the structure

$$\psi(x) = \phi(x)\mathrm{e}^{-\mathrm{i}\int^{x} u(x')\,\mathrm{d}x'}, \tag{2.143}$$

then we find that $\phi(x)$ must satisfy the equation

$$\left(\frac{\partial^2}{\partial x^2} + k^2\right)\phi(x) = -u^2(x)\phi(x), \tag{2.144}$$

which is a Schrödinger equation with the *attractive* 'TRI potential' $-u^2(x)$. We could interpret this result by saying that the gauge transformation (2.143) reduces the problem for $\psi(x)$ to one for $\phi(x)$ obeying TRI. Still, we shall find below that for the original wave function $\psi(x)$ the $S$ matrix is not symmetric.

We apply the above result to the particular case (2.141). We have two situations.

## POTENTIAL SCATTERING IN INFINITE ONE-DIMENSIONAL SPACE

(i) *Incidence from the left*

Then $\phi(x)$ has the structure

$$\phi(x) = \begin{cases} e^{ikx} + r_\phi e^{-ikx}, & x < 0, \\ Ae^{i\bar{k}x} + Be^{-i\bar{k}x}, & 0 < x < a, \\ t_\phi e^{ikx}, & x > a, \end{cases} \quad (2.145)$$

where

$$\bar{k}^2 = k^2 + u_0^2. \quad (2.146)$$

Here, $t_\phi$ can be taken from eqn (2.263) below, which will be deduced for a positive barrier, by replacing $\bar{k}$ in that equation by the value given in the above eqn (2.146), which is appropriate for a negative barrier, with the result

$$t_\phi = \frac{e^{-ika}}{\cos \bar{k}a - i\left[(k^2 + \bar{k}^2)/2k\bar{k}\right] \sin \bar{k}a}. \quad (2.147)$$

If we choose the lower limit of integration in eqn (2.143) as 0, i.e.,

$$\psi(x) = \phi(x) e^{-i\int_0^x u(x')\,dx'}, \quad (2.148)$$

then we have

$$\int_0^x u(x')\,dx' = \begin{cases} 0, & x < 0, \\ u_0 x, & 0 < x < a, \\ u_0 a, & x > a, \end{cases} \quad (2.149)$$

and the wave function $\psi(x)$ takes the form

$$\psi(x) = \begin{cases} e^{ikx} + r_\phi e^{-ikx}, & x < 0, \\ t_\phi e^{-iu_0 a} e^{ikx}, & x > a. \end{cases} \quad (2.150)$$

(ii) *Incidence from the right*

Then $\phi(x)$ has the structure

$$\phi(x) = \begin{cases} t_\phi e^{-ikx}, & x < 0, \\ Ce^{i\bar{k}x} + De^{-i\bar{k}x}, & 0 < x < a, \\ e^{-ikx} + r'_\phi e^{ikx}, & x > a. \end{cases} \quad (2.151)$$

We have used the fact that $t_\phi = t'_\phi$, since the 'potential' appearing in the equation for $\phi(x)$, eqn (2.144), is TRI. If we choose the lower limit of integration in eqn (2.143) as $a$, i.e.,

$$\psi(x) = \phi(x) e^{-i\int_a^x u(x')\,dx'}, \quad (2.152)$$

then we have

$$\int_a^x u(x')\,dx' = \begin{cases} -u_0 a, & x < 0, \\ u_0(x-a), & 0 < x < a, \\ 0, & x > a, \end{cases} \quad (2.153)$$

and the wave function $\psi(x)$ takes the form

$$\psi(x) = \begin{cases} t_\phi \mathrm{e}^{iu_0 a} \mathrm{e}^{-ikx}, & x < 0, \\ \mathrm{e}^{-ikx} + r'_\phi \mathrm{e}^{ikx}, & x > a. \end{cases} \quad (2.154)$$

From eqns (2.150) and (2.154) we thus find that the transmission amplitudes for the original problem defined by the wave function $\psi(x)$ are

$$t = t_\phi \mathrm{e}^{-iu_0 a}, \quad (2.155a)$$

$$t' = t_\phi \mathrm{e}^{iu_0 a}. \quad (2.155b)$$

It is easy to see that, using a first Born approximation, these last results reduce to those of the previous exercise.

One other example of a non-TRI Hamiltonian is

$$H = \frac{1}{2m} \left[ p + 2mv(x) \right]^2, \quad (2.156)$$

which could be interpreted as having added to a free particle the longitudinal vector potential

$$A(x) = \frac{2mc}{e} v(x) \equiv \frac{\partial W(x)}{\partial x}. \quad (2.157)$$

This Hamiltonian differs from the one treated in Exercise 2.9 by the addition of the term $2mv^2(x)$. We can again write the wave function in the form (2.143) and find, for $\phi(x)$, the *free* equation

$$\left( \frac{\partial^2}{\partial x^2} + k^2 \right) \phi(x) = 0, \quad (2.158)$$

for which $t_\phi = t'_\phi = 1$ and $r_\phi = r'_\phi = 0$. For the particular $u(x)$ given in eqn (2.141), the $t$ and $t'$ of eqn (2.155) reduce to

$$t = \mathrm{e}^{-iu_0 a}, \quad (2.159a)$$

$$t' = \mathrm{e}^{iu_0 a}. \quad (2.159b)$$

*The polar representation of the S matrix*

Since the matrix elements of $S$ are not independent of one another, because of the requirement of unitarity and, if applicable, symmetry, it is convenient to express $S$ in terms of independent parameters. For instance, in the case of TRI, a unitary and symmetric $S$ matrix can be written in the so-called 'polar representation' [14, 87, 132, 134] as

$$S = \begin{bmatrix} \mathrm{e}^{i\phi} & 0 \\ 0 & \mathrm{e}^{i\psi} \end{bmatrix} \begin{bmatrix} -\sqrt{1-\tau} & \sqrt{\tau} \\ \sqrt{\tau} & \sqrt{1-\tau} \end{bmatrix} \begin{bmatrix} \mathrm{e}^{i\phi} & 0 \\ 0 & \mathrm{e}^{i\psi} \end{bmatrix}, \quad (2.160)$$

where $\tau$, $\phi$ and $\psi$ are three independent parameters. The parameter $\tau$ coincides with the transmission coefficient $T$, and thus $0 \leqslant \tau \leqslant 1$, while $\phi$ and $\psi$ are arbitrary phases.

*Relationship between the S matrix and the full Green function*

We show below how the elements of the $S$ matrix can be expressed in terms of the full Green function discussed in Section 2.1.3. Specifically, we make use of eqn (2.81), obtained for a local, and hence real, potential. Using the asymptotic form of the wave functions as $x \to -\infty$ and $x' \to +\infty$, eqn (2.81a) gives

$$\lim_{\substack{x \to -\infty \\ x' \to +\infty}} \left[ e^{ikx} g^{(+)}(x,x') e^{-ikx'} \right] = \frac{t'(k)t(k)}{2ikt(k)}. \qquad (2.161)$$

Then, using the fact that $t'(k) = t(k)$ for a system with TRI, we have

$$t(k) = 2ik \lim_{\substack{x \to -\infty \\ x' \to +\infty}} \left[ e^{ik(x-x')} g^{(+)}(x,x') \right], \qquad (2.162a)$$

or

$$t(k) = i\hbar v \lim_{\substack{x \to -\infty \\ x' \to +\infty}} \left[ e^{ik(x-x')} G^{(+)}(x,x') \right], \qquad (2.162b)$$

where $v = \hbar k/m$ is the particle velocity. This is the desired relation between the transmission amplitude and the full Green function.

Similarly, using the asymptotic form of the wave function for $x < x' \to -\infty$ we find, from eqn (2.81),

$$g^{(+)}(x < x' \to -\infty) \sim \frac{t'(k) e^{-ikx} \left[ e^{ikx'} + r(k) e^{-ikx'} \right]}{2ikt(k)}.$$

By again using $t'(k) = t(k)$, under TRI, we find

$$r(k) = \lim_{x<x' \to -\infty} \left\{ e^{ik(x+x')} \left[ 2ik g^{(+)}(x,x') - e^{ik(x'-x)} \right] \right\}, \qquad (2.163a)$$

or

$$r(k) = \lim_{x<x' \to -\infty} \left\{ e^{ik(x+x')} \left[ i\hbar v G^{(+)}(x,x') - e^{ik(x'-x)} \right] \right\}, \qquad (2.163b)$$

which is the desired relation between the reflection amplitude and the full Green function.

**Exercise 2.10** Obtain the reflection amplitude $r'(k)$ in terms of the full Green function.

Using the asymptotic form of the wave function for $x' > x \to +\infty$ we find, from eqn (2.81),

$$g^{(+)}(x' > x \to +\infty) \sim \frac{\left[ e^{-ikx} + r'(k) e^{ikx} \right] t(k) e^{ikx'}}{2ikt(k)},$$

so that

$$r'(k) = \lim_{x'>x \to +\infty} \left\{ e^{-ik(x+x')} \left[ 2ik g^{(+)}(x,x') - e^{ik(x'-x)} \right] \right\}, \qquad (2.164a)$$

or

$$r'(k) = \lim_{x' > x \to +\infty} \left\{ e^{-ik(x+x')} \left[ i\hbar v G^{(+)}(x, x') - e^{ik(x'-x)} \right] \right\}. \tag{2.164b}$$

### 2.1.5 *The transfer or M matrix*

Instead of the two linear solutions $\psi_+^{(+)}(E;x)$ and $\psi_-^{(+)}(E;x)$, whose asymptotic behavior is given in eqn (2.36), we could, alternatively, use the following two linear combinations of these solutions:

$$\psi_{L,+}(E;x) = \psi_+^{(+)}(E;x) - \frac{r}{t'}\psi_-^{(+)}(E;x) \tag{2.165a}$$

$$\sim \begin{cases} \phi_+(E;x), & x \to -\infty, \\ \left(t - \frac{rr'}{t'}\right)\phi_+(E;x) - \frac{r}{t'}\phi_-(E;x), & x \to +\infty, \end{cases} \tag{2.165b}$$

$$\psi_{L,-}(E;x) = \frac{1}{t'}\psi_-^{(+)}(E;x) \tag{2.165c}$$

$$\sim \begin{cases} \phi_-(E;x), & x \to -\infty, \\ \frac{r'}{t'}\phi_+(E;x) + \frac{1}{t'}\phi_-(E;x), & x \to +\infty. \end{cases} \tag{2.165d}$$

On the left of the potential, $\psi_{L,+}(E;x)$ contains only a unit-amplitude plane wave traveling to the right, while $\psi_{L,-}(E;x)$ contains only a unit-amplitude plane wave traveling to the left. These are two linearly independent solutions for the energy $E$, from which we can construct the linear combination

$$\psi(E;x) = a^{(1)}\psi_{L,+}(E;x) + b^{(1)}\psi_{L,-}(E;x). \tag{2.166}$$

For arbitrary values of $a^{(1)}$ and $a^{(2)}$, this represents the most general solution of the Schrödinger equation for the energy $E$ (see Fig. 2.2). Its asymptotic behavior is

$$\psi(E;x) \sim \begin{cases} a^{(1)}\phi_+(E;x) + b^{(1)}\phi_-(E;x), & x \to -\infty, \\[1ex] \left[\left(t - \frac{rr'}{t'}\right)a^{(1)} + \frac{r'}{t'}b^{(1)}\right]\phi_+(E;x) \\ + \left[-\frac{r}{t'}a^{(1)} + \frac{1}{t'}b^{(1)}\right]\phi_-(E;x) \\ = b^{(2)}\phi_+(E;x) + a^{(2)}\phi_-(E;x), & x \to +\infty. \end{cases} \tag{2.167}$$

We have thus shown that, as a consequence of the *linearity* of the Schrödinger equation, the amplitudes on the right of the potential, i.e., $b^{(2)}$ and $a^{(2)}$, are linearly related to the ones on the left, i.e., $a^{(1)}$ and $b^{(1)}$, according to the equation

$$\begin{bmatrix} b^{(2)} \\ a^{(2)} \end{bmatrix} = \begin{bmatrix} \alpha & \beta \\ \gamma & \delta \end{bmatrix} \begin{bmatrix} a^{(1)} \\ b^{(1)} \end{bmatrix}, \tag{2.168}$$

where

$$\alpha = t - \frac{rr'}{t'} = \frac{1}{t^*}, \qquad \beta = \frac{r'}{t'}, \qquad (2.169a)$$

$$\gamma = -\frac{r}{t'}, \qquad \delta = \frac{1}{t'}. \qquad (2.169b)$$

We can thus fix arbitrarily the amplitudes in region I of Fig. 2.2; the two amplitudes in region II are then uniquely determined by the Schrödinger equation, according to the linear relation (2.168).

The matrix that appears in eqn (2.168) is called the *transfer matrix* [141, Chapter 6] and it will be denoted by $M$ as follows:

$$\begin{bmatrix} b^{(2)} \\ a^{(2)} \end{bmatrix} = M \begin{bmatrix} a^{(1)} \\ b^{(1)} \end{bmatrix}, \qquad (2.170a)$$

$$M = \begin{bmatrix} \alpha & \beta \\ \gamma & \delta \end{bmatrix}. \qquad (2.170b)$$

The relations (2.169) can be inverted to give the $S$-matrix elements in terms of the elements of the $M$ matrix as

$$r = -\frac{\gamma}{\delta}, \qquad t' = \frac{1}{\delta}, \qquad (2.171a)$$

$$t = \frac{1}{\alpha^*}, \qquad r' = \frac{\beta}{\delta}. \qquad (2.171b)$$

The transfer matrix $M$ must satisfy certain conditions in order to fulfill the properties of flux conservation and, when applicable, time-reversal invariance.

First, *current conservation* requires the equality of the currents in regions I and II of Fig. 2.2, i.e.,

$$J_{\mathrm{I}} \propto \begin{bmatrix} a^{(1)} & b^{(1)} \end{bmatrix}^* \begin{bmatrix} 1 & 0 \\ 0 & -1 \end{bmatrix} \begin{bmatrix} a^{(1)} \\ b^{(1)} \end{bmatrix} \qquad (2.172a)$$

and

$$J_{\mathrm{II}} \propto \begin{bmatrix} b^{(2)} & a^{(2)} \end{bmatrix}^* \begin{bmatrix} 1 & 0 \\ 0 & -1 \end{bmatrix} \begin{bmatrix} b^{(2)} \\ a^{(2)} \end{bmatrix}$$

$$= \begin{bmatrix} a^{(1)} & b^{(1)} \end{bmatrix}^* M^\dagger \begin{bmatrix} 1 & 0 \\ 0 & -1 \end{bmatrix} M \begin{bmatrix} a^{(1)} \\ b^{(1)} \end{bmatrix} \qquad (2.172b)$$

must be equal. Thus

$$M^\dagger \sigma_z M = \sigma_z, \qquad (2.173)$$

$\sigma_z$ being the Pauli matrix

$$\sigma_z = \begin{bmatrix} 1 & 0 \\ 0 & -1 \end{bmatrix}. \qquad (2.174)$$

The current-conservation condition, eqn (2.173), implies, for the matrix elements of $M$, that

$$|\alpha|^2 - |\gamma|^2 = 1, \tag{2.175a}$$

$$|\delta|^2 - |\beta|^2 = 1, \tag{2.175b}$$

$$\alpha^*\beta = \gamma^*\delta. \tag{2.175c}$$

Matrices satisfying eqn (2.173) are called *pseudounitary* (in contrast with the definition of unitary matrices, for which the Pauli matrix $\sigma_z$ is absent). We thus see that the $M$ matrices satisfying the condition of flux conservation form a non-compact group, which is called the *pseudounitary group* $U(1,1)$.

Next, if *TRI* applies, the same transfer matrix must relate the coefficients of the time-reversed state (2.127), i.e.,

$$\begin{bmatrix} (a^{(2)})^* \\ (b^{(2)})^* \end{bmatrix} = M \begin{bmatrix} (b^{(1)})^* \\ (a^{(1)})^* \end{bmatrix}. \tag{2.176}$$

Taking the complex conjugate of this expression, we can write

$$\begin{bmatrix} b^{(2)} \\ a^{(2)} \end{bmatrix} = \sigma_x M^* \sigma_x \begin{bmatrix} a^{(1)} \\ b^{(1)} \end{bmatrix}, \tag{2.177}$$

where $\sigma_x$ is the Pauli matrix

$$\sigma_x = \begin{bmatrix} 0 & 1 \\ 1 & 0 \end{bmatrix}. \tag{2.178}$$

Comparing eqn (2.177) with the general definition (2.170), we thus find the relation

$$M^* = \sigma_x M \sigma_x, \tag{2.179}$$

as the condition to be satisfied by the transfer matrix $M$ in order to fulfill the property of time-reversal invariance. The condition (2.179) for TRI implies, for the matrix elements of $M$, that

$$\delta = \alpha^*, \tag{2.180a}$$

$$\gamma = \beta^*, \tag{2.180b}$$

so that $M$ of (2.170) must have the form

$$M = \begin{bmatrix} \alpha & \beta \\ \beta^* & \alpha^* \end{bmatrix}. \tag{2.181}$$

The current-conservation conditions (2.175) then imply that

$$|\alpha|^2 - |\beta|^2 = 1, \tag{2.182}$$

so that $\det M = 1$.

The $M$ matrices satisfying the current-conservation condition (2.173) and the TRI restriction (2.179) are *unimodular pseudounitary matrices* that form the subgroup $SU(1,1)$ of $U(1,1)$. We note in passing that the group $SU(1,1)$ can be shown to be isomorphic to the *real symplectic group* $Sp(2,\mathcal{R})$ and homomorphic to the Lorentz group $SO(2,1)$ [18, 178].

When TRI applies, the relations (2.169) reduce to

$$\alpha = \frac{1}{t^*}, \quad \beta = -\frac{r^*}{t^*}, \tag{2.183}$$

and the relations (2.171) to

$$r = -\frac{\beta^*}{\alpha^*}, \quad t = t' = \frac{1}{\alpha^*}, \quad r' = \frac{\beta}{\alpha^*}. \tag{2.184}$$

(Notice that, as a result, $S = S^\top$.)

For the TRI case, any transfer matrix can be given the polar representation [14, 87, 132, 134]

$$M = \begin{bmatrix} e^{i\mu} & 0 \\ 0 & e^{-i\mu} \end{bmatrix} \begin{bmatrix} \sqrt{1+\lambda} & \sqrt{\lambda} \\ \sqrt{\lambda} & \sqrt{1+\lambda} \end{bmatrix} \begin{bmatrix} e^{i\nu} & 0 \\ 0 & e^{-i\nu} \end{bmatrix}, \tag{2.185}$$

where $\lambda$, $\mu$ and $\nu$ are now three independent parameters. The parameters $\mu$ and $\nu$ are arbitrary phases, while $\lambda$ is real and non-negative and is related to $\tau$ of eqn (2.160) by

$$\tau = \frac{1}{1+\lambda}. \tag{2.186}$$

From eqns (2.184) and (2.185) we find the reflection and the transmission amplitudes in terms of the polar parameters $\lambda$, $\mu$ and $\nu$ to be

$$r = -\left(\frac{\lambda}{1+\lambda}\right)^{1/2} e^{2i\nu}, \tag{2.187a}$$

$$r' = \left(\frac{\lambda}{1+\lambda}\right)^{1/2} e^{2i\mu}, \tag{2.187b}$$

$$t = \frac{1}{(1+\lambda)^{1/2}} e^{i(\mu+\nu)}. \tag{2.187c}$$

The most useful property of the transfer matrices $M$ is *serial multiplicativity*. Consider two scatterers giving rise to two non-overlapping potentials $V_1(x)$ and $V_2(x)$. This is illustrated in Fig. 2.4, where the amplitudes of the various plane waves are also shown. We denote by $M_1$ and $M_2$, respectively, the transfer matrices for the two potentials. The transfer matrix $M_1$ takes the coefficients $a^{(1)}$ and $b^{(1)}$ to $b^{(2)}$ and $a^{(2)}$, and $M_2$ from $b^{(2)}$ and $a^{(2)}$ to $b^{(3)}$ and $a^{(3)}$; the product

$$M = M_2 M_1 \tag{2.188}$$

takes $a^{(1)}$ and $b^{(1)}$ directly to $b^{(3)}$ and $a^{(3)}$, and is thus the transfer matrix of the combined system.

Combining the $S$ matrices of the two scatterers is not as simple as combining their transfer matrices; this is undertaken in what follows.

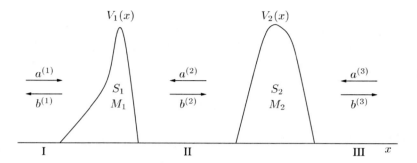

FIG. 2.4. Two potentials $V_1(x)$ and $V_2(x)$, with a potential-free region in between; their associated $S$ and $M$ matrices are called $S_1$, $S_2$ and $M_1$, $M_2$, respectively. In the various regions, the amplitudes of the planes waves (2.2) appearing in the most general solution of the Schrödinger equation are indicated.

### 2.1.6 Combining the S matrices for two scatterers in series

The scattering matrices $S_1$, $S_2$ and $S$ of the two scatterers and of the combined system, respectively, relate the various amplitudes shown in Fig. 2.4 as follows:

$$\begin{bmatrix} b^{(1)} \\ b^{(2)} \end{bmatrix} = S_1 \begin{bmatrix} a^{(1)} \\ a^{(2)} \end{bmatrix}, \quad \begin{bmatrix} a^{(2)} \\ b^{(3)} \end{bmatrix} = S_2 \begin{bmatrix} b^{(2)} \\ a^{(3)} \end{bmatrix}, \quad \begin{bmatrix} b^{(1)} \\ b^{(3)} \end{bmatrix} = S \begin{bmatrix} a^{(1)} \\ a^{(3)} \end{bmatrix}. \quad (2.189)$$

These matrices have the structure

$$S_1 = \begin{bmatrix} r_1 & t'_1 \\ t_1 & r'_1 \end{bmatrix}, \quad S_2 = \begin{bmatrix} r_2 & t'_2 \\ t_2 & r'_2 \end{bmatrix}, \quad S = \begin{bmatrix} r & t' \\ t & r' \end{bmatrix}. \quad (2.190)$$

It is useful to keep the various $S$ matrices in their general form, although for TRI they are symmetric, as explained above.

We can expand the first two eqns (2.189) to obtain

$$\begin{aligned} b^{(1)} &= r_1 a^{(1)} + t'_1 a^{(2)}, & a^{(2)} &= r_2 b^{(2)} + t'_2 a^{(3)}, \\ b^{(2)} &= t_1 a^{(1)} + r'_1 a^{(2)}, & b^{(3)} &= t_2 b^{(2)} + r'_2 a^{(3)}, \end{aligned} \quad (2.191)$$

or

$$\begin{aligned} \begin{bmatrix} b^{(1)} \\ b^{(3)} \end{bmatrix} &= \begin{bmatrix} r_1 & 0 \\ 0 & r'_2 \end{bmatrix} \begin{bmatrix} a^{(1)} \\ a^{(3)} \end{bmatrix} + \begin{bmatrix} t'_1 & 0 \\ 0 & t_2 \end{bmatrix} \begin{bmatrix} a^{(2)} \\ b^{(2)} \end{bmatrix}, \\ \sigma_x \begin{bmatrix} a^{(2)} \\ b^{(2)} \end{bmatrix} &= \begin{bmatrix} t_1 & 0 \\ 0 & t'_2 \end{bmatrix} \begin{bmatrix} a^{(1)} \\ a^{(3)} \end{bmatrix} + \begin{bmatrix} r'_1 & 0 \\ 0 & r_2 \end{bmatrix} \begin{bmatrix} a^{(2)} \\ b^{(2)} \end{bmatrix}, \end{aligned} \quad (2.192)$$

where $\sigma_x$ is the Pauli matrix (2.178). From these last two equations we eliminate $[a^{(2)}, b^{(2)}]$ to obtain a relation between $[b^{(1)}, b^{(3)}]$ and $[a^{(1)}, a^{(3)}]$, from which, using also the last of eqns (2.189), we identify the total $S$ matrix to be

POTENTIAL SCATTERING IN INFINITE ONE-DIMENSIONAL SPACE    49

$$S = \begin{bmatrix} r_1 & 0 \\ 0 & r'_2 \end{bmatrix} + \begin{bmatrix} t'_1 & 0 \\ 0 & t_2 \end{bmatrix} \frac{1}{\sigma_x - \begin{bmatrix} r'_1 & 0 \\ 0 & r_2 \end{bmatrix}} \begin{bmatrix} t_1 & 0 \\ 0 & t'_2 \end{bmatrix}. \qquad (2.193)$$

Taking the inverse indicated in the last term, we finally find

$$S = \begin{bmatrix} r_1 + t'_1 r_2 t_1/(1 - r'_1 r_2) & t'_1 t'_2/(1 - r'_1 r_2) \\ t_2 t_1/(1 - r'_1 r_2) & r'_2 + t_2 r'_1 t'_2/(1 - r'_1 r_2) \end{bmatrix}. \qquad (2.194)$$

The reflection and the transmission amplitudes of eqn (2.194) can be interpreted physically from the following construction. Figure 2.5 indicates, schematically, the multiple reflections and transmissions suffered by an incident wave with unit amplitude, impinging from the left. It is clear that summation of the series gives $r$ and $t$ of eqn (2.194). Similarly, impinging from the right we obtain the series expansion for $r'$ and $t'$.

A more compact way to arrive at the above result is as follows. We combine the first two eqns (2.189) as

$$\begin{bmatrix} b^{(1)} \\ b^{(2)} \\ a^{(2)} \\ b^{(3)} \end{bmatrix} = S^0_{12} \begin{bmatrix} a^{(1)} \\ a^{(2)} \\ b^{(2)} \\ a^{(3)} \end{bmatrix}, \qquad (2.195)$$

where

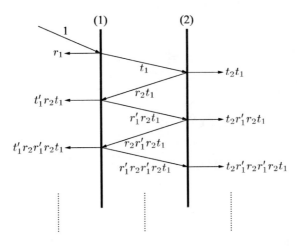

FIG. 2.5. A schematic representation of the multiple scattering series produced by two scatterers. The sum of the reflected (transmitted) terms gives the reflection (transmission) amplitude of eqn (2.194).

$$S_{12}^0 = \begin{bmatrix} S_1 & 0 \\ 0 & S_2 \end{bmatrix} = \begin{bmatrix} r_1 & t_1' & & 0 \\ t_1 & r_1' & & \\ & & r_2 & t_2' \\ 0 & & t_2 & r_2' \end{bmatrix}. \tag{2.196}$$

We reorder the rows and columns of this last matrix and define

$$S_{12} \equiv \begin{bmatrix} r_1 & 0 & t_1' & 0 \\ 0 & r_2' & 0 & t_2 \\ t_1 & 0 & r_1' & 0 \\ 0 & t_2' & 0 & r_2 \end{bmatrix} \equiv \begin{bmatrix} S_{12}^{PP} & S_{12}^{PQ} \\ S_{12}^{QP} & S_{12}^{QQ} \end{bmatrix}. \tag{2.197}$$

Here, $P$ projects unto the 'external' region (regions I and III in Fig. 2.4) and $Q$ unto the 'internal' region (region II in Fig. 2.4). Thus, $S_{12}^{PP}$ is the submatrix that connects the external region to itself, $S_{12}^{PQ}$ connects the internal region to the external region, $S_{12}^{QP}$ connects the external region to the internal region and $S_{12}^{QQ}$ connects the internal region to itself. Using the matrix $S_{12}$ of eqn (2.197), we can relate the amplitudes of the wave functions by

$$\begin{bmatrix} b^P \\ \sigma_x c^Q \end{bmatrix} = \begin{bmatrix} S_{12}^{PP} & S_{12}^{PQ} \\ S_{12}^{QP} & S_{12}^{QQ} \end{bmatrix} \begin{bmatrix} a^P \\ c^Q \end{bmatrix}, \tag{2.198}$$

where

$$a^P = \begin{bmatrix} a^{(1)} \\ a^{(3)} \end{bmatrix}, \quad b^P = \begin{bmatrix} b^{(1)} \\ b^{(3)} \end{bmatrix}, \quad c^Q = \begin{bmatrix} a^{(2)} \\ b^{(2)} \end{bmatrix}. \tag{2.199}$$

From eqn (2.198) we obtain the pair of coupled equations

$$b^P = S_{12}^{PP} a^P + S_{12}^{PQ} c^Q,$$
$$\sigma_x c^Q = S_{12}^{QP} a^P + S_{12}^{QQ} c^Q.$$

Eliminating $c^Q$ from this pair of equations, we obtain

$$b^P = S_{12}^{PP} a^P + S_{12}^{PQ} \frac{1}{\sigma_x - S_{12}^{QQ}} S_{12}^{QP} a^P.$$

The resulting $S$ matrix satisfies the relation

$$b^P = S a^P,$$

and is thus given by

$$S = S_{12}^{PP} + S_{12}^{PQ} \frac{1}{\sigma_x - S_{12}^{QQ}} S_{12}^{QP}. \tag{2.200}$$

This result is identical to that of eqn (2.193).

### 2.1.7 Transformation of the scattering and the transfer matrices under a translation

Figure 2.6 shows a potential $\mathring{V}(x)$ and the same potential translated a distance $d$, that we denote by $V(x)$. The amplitudes $a^{(i)}$ and $b^{(i)}$ ($i = 1, 2$) indicated in the figure correspond to scattering by the potential $V(x)$; we denote by $\mathring{a}^{(i)}$ and $\mathring{b}^{(i)}$ ($i = 1, 2$) the amplitudes corresponding to scattering by the potential $\mathring{V}(x)$. We write the various wave functions shown in Fig. 2.6 as

$$
\begin{aligned}
a^{(1)}\mathrm{e}^{\mathrm{i}kx} &= a^{(1)}\mathrm{e}^{\mathrm{i}kd}\mathrm{e}^{\mathrm{i}k(x-d)} = \mathring{a}^{(1)}\mathrm{e}^{\mathrm{i}k(x-d)}, \\
\mathring{a}^{(1)} &= \mathrm{e}^{\mathrm{i}kd}a^{(1)}, \\
a^{(2)}\mathrm{e}^{-\mathrm{i}kx} &= a^{(2)}\mathrm{e}^{-\mathrm{i}kd}\mathrm{e}^{-\mathrm{i}k(x-d)} = \mathring{a}^{(2)}\mathrm{e}^{-\mathrm{i}k(x-d)}, \\
\mathring{a}^{(2)} &= \mathrm{e}^{-\mathrm{i}kd}a^{(2)}, \\
b^{(1)}\mathrm{e}^{-\mathrm{i}kx} &= b^{(1)}\mathrm{e}^{-\mathrm{i}kd}\mathrm{e}^{-\mathrm{i}k(x-d)} = \mathring{b}^{(1)}\mathrm{e}^{-\mathrm{i}k(x-d)}, \\
\mathring{b}^{(1)} &= \mathrm{e}^{-\mathrm{i}kd}b^{(1)}, \\
b^{(2)}\mathrm{e}^{\mathrm{i}kx} &= b^{(2)}\mathrm{e}^{\mathrm{i}kd}\mathrm{e}^{\mathrm{i}k(x-d)} = \mathring{b}^{(2)}\mathrm{e}^{\mathrm{i}k(x-d)}, \\
\mathring{b}^{(2)} &= \mathrm{e}^{\mathrm{i}kd}b^{(2)},
\end{aligned}
\quad (2.201)
$$

so that the two sets of amplitudes are related as follows:

$$
\begin{bmatrix} \mathring{b}^{(1)} \\ \mathring{b}^{(2)} \end{bmatrix} = \begin{bmatrix} \mathrm{e}^{-\mathrm{i}kd} & 0 \\ 0 & \mathrm{e}^{\mathrm{i}kd} \end{bmatrix} \begin{bmatrix} b^{(1)} \\ b^{(2)} \end{bmatrix},
$$
$$
\begin{bmatrix} \mathring{a}^{(1)} \\ \mathring{a}^{(2)} \end{bmatrix} = \begin{bmatrix} \mathrm{e}^{\mathrm{i}kd} & 0 \\ 0 & \mathrm{e}^{-\mathrm{i}kd} \end{bmatrix} \begin{bmatrix} a^{(1)} \\ a^{(2)} \end{bmatrix}.
\quad (2.202)
$$

The incoming and the outgoing amplitudes with a 'o' on top are related by the scattering matrix $\mathring{S}$ associated with the potential $\mathring{V}$, i.e.,

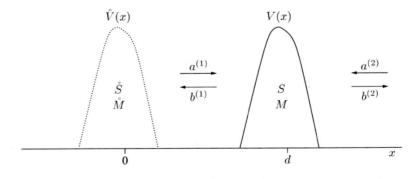

FIG. 2.6. The potential $\mathring{V}(x)$ and the same potential translated a distance $d$, denoted by $V(x)$. The amplitudes of the plane waves (2.2) in the most general solution of the Schrödinger equation for the potential $V(x)$ are indicated.

$$\begin{bmatrix} \mathring{b}^{(1)} \\ \mathring{b}^{(2)} \end{bmatrix} = \mathring{S} \begin{bmatrix} \mathring{a}^{(1)} \\ \mathring{a}^{(2)} \end{bmatrix}. \tag{2.203}$$

Substituting from eqn (2.202), we have

$$\begin{bmatrix} e^{-ikd} & 0 \\ 0 & e^{ikd} \end{bmatrix} \begin{bmatrix} b^{(1)} \\ b^{(2)} \end{bmatrix} = \mathring{S} \begin{bmatrix} e^{ikd} & 0 \\ 0 & e^{-ikd} \end{bmatrix} \begin{bmatrix} a^{(1)} \\ a^{(2)} \end{bmatrix}, \tag{2.204}$$

or

$$\begin{bmatrix} b^{(1)} \\ b^{(2)} \end{bmatrix} = \begin{bmatrix} e^{ikd} & 0 \\ 0 & e^{-ikd} \end{bmatrix} \mathring{S} \begin{bmatrix} e^{ikd} & 0 \\ 0 & e^{-ikd} \end{bmatrix} \begin{bmatrix} a^{(1)} \\ a^{(2)} \end{bmatrix}. \tag{2.205}$$

We thus identify the scattering matrix $S$ as

$$S = D(kd)\mathring{S}D(kd), \tag{2.206}$$

where the unitary displacement matrix $D(kd)$ is given by

$$D(kd) = \begin{bmatrix} e^{ikd} & 0 \\ 0 & e^{-ikd} \end{bmatrix} = \begin{bmatrix} T(kd) & 0 \\ 0 & T^{-1}(kd) \end{bmatrix}. \tag{2.207}$$

We have defined

$$T = e^{ikd}. \tag{2.208}$$

Using the structure (2.94), we thus find the following relations between the old and the new transmission and reflection amplitudes:

$$\begin{bmatrix} r & t' \\ t & r' \end{bmatrix} = \begin{bmatrix} e^{2ikd}\mathring{r} & \mathring{t}' \\ \mathring{t} & e^{-2ikd}\mathring{r}' \end{bmatrix}. \tag{2.209}$$

This result is physically clear. It describes an increase by $2kd$ in the optical path for the reflection process of waves incident from the left when the potential is shifted a distance $d$ to the right (see Fig. 2.6), and a decrease by the same amount for waves incident from the right, with no change for the transmission process.

Similarly, the transfer matrix $\mathring{M}$ associated with the potential $\mathring{V}$ satisfies

$$\begin{bmatrix} \mathring{b}^{(2)} \\ \mathring{a}^{(2)} \end{bmatrix} = \mathring{M} \begin{bmatrix} \mathring{a}^{(1)} \\ \mathring{b}^{(1)} \end{bmatrix}. \tag{2.210}$$

Substituting from eqn (2.202), we have

$$\begin{bmatrix} e^{ikd} & 0 \\ 0 & e^{-ikd} \end{bmatrix} \begin{bmatrix} b^{(2)} \\ a^{(2)} \end{bmatrix} = \mathring{M} \begin{bmatrix} e^{ikd} & 0 \\ 0 & e^{-ikd} \end{bmatrix} \begin{bmatrix} a^{(1)} \\ b^{(1)} \end{bmatrix}, \tag{2.211}$$

or

$$\begin{bmatrix} b^{(2)} \\ a^{(2)} \end{bmatrix} = \begin{bmatrix} e^{-ikd} & 0 \\ 0 & e^{ikd} \end{bmatrix} \mathring{M} \begin{bmatrix} e^{ikd} & 0 \\ 0 & e^{-ikd} \end{bmatrix} \begin{bmatrix} a^{(1)} \\ b^{(1)} \end{bmatrix}, \tag{2.212}$$

from which we identify the transfer matrix $M$ to be

$$M = D^{-1}(kd)\mathring{M}D(kd), \tag{2.213}$$

with $D(kd)$ defined in eqn (2.207). From eqn (2.170), the old and the new $M$-matrix elements are related by

$$\begin{bmatrix} \alpha & \beta \\ \gamma & \delta \end{bmatrix} = \begin{bmatrix} \mathring{\alpha} & e^{-2ikd}\mathring{\beta} \\ e^{2ikd}\mathring{\gamma} & \mathring{\delta} \end{bmatrix}. \tag{2.214}$$

### 2.1.8 An exactly soluble example

We consider again the spatial coordinate $x$ in the full interval $(-\infty, \infty)$ and study the scattering produced by the delta potential $U(x) = u_0\delta(x)$ centered at $x = 0$ (see Fig. 2.7). The Schrödinger equation can be written as

$$\left(\partial_x^2 + k^2\right)\psi(x) = u_0\delta(x)\psi(x). \tag{2.215}$$

Continuity of the wave function and discontinuity of its slope at $x = 0$ yield

$$a^{(2)} + b^{(2)} = a^{(1)} + b^{(1)}, \tag{2.216a}$$

$$ik(b^{(2)} - a^{(2)}) = ik(a^{(1)} - b^{(1)}) + u_0(a^{(1)} + b^{(1)}). \tag{2.216b}$$

*The scattering matrix $S$*

If we decide to relate incoming and outgoing waves, we are led, from eqn (2.95), to the $2 \times 2$ scattering matrix $S$. From eqns (2.216) we obtain

$$\begin{bmatrix} 1 & -1 \\ 1 - u_0/ik & 1 \end{bmatrix}\begin{bmatrix} b^{(1)} \\ b^{(2)} \end{bmatrix} = \begin{bmatrix} -1 & 1 \\ 1 + u_0/ik & 1 \end{bmatrix}\begin{bmatrix} a^{(1)} \\ a^{(2)} \end{bmatrix}. \tag{2.217}$$

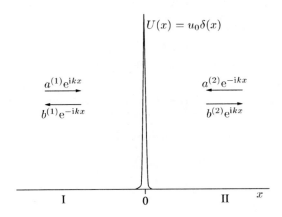

FIG. 2.7. Scattering by a one-dimensional delta potential: the most general solution. In the wave function, the factor $\sqrt{2\pi\hbar^2 k/m}$ of eqn (2.2) is common to all terms and is not indicated.

We then find the $S$ matrix to be

$$S = \begin{bmatrix} r & t' \\ t & r' \end{bmatrix} = \begin{bmatrix} (u_0/2ik)/[1 - u_0/2ik] & 1/[1 - u_0/2ik] \\ 1/[1 - u_0/2ik] & (u_0/2ik)/[1 - u_0/2ik] \end{bmatrix}. \quad (2.218)$$

A number of comments on this result are in order.

(i) We see explicitly that the $S$ matrix of eqn (2.218) is *unitary* and *symmetric*.
(ii) That we also have $r = r'$ reflects the fact that in the present case the potential is symmetric with respect to $x = 0$.
(iii) As $u_0/2ik \to 0$, we have $r \to 0$ and $t \to 1$, which is physically clear.
(iv) As $u_0/2ik \to \infty$, we have $r \to -1$ and $t \to 0$, and the particle is completely reflected from the potential.

*The transfer matrix $M$*

If we decide to relate the components on the two sides of the scatterer we are led, from the defining eqn (2.170), to the transfer matrix $M$. From eqns (2.216) we have

$$\begin{bmatrix} 1 & 1 \\ 1 & -1 \end{bmatrix} \begin{bmatrix} b^{(2)} \\ a^{(2)} \end{bmatrix} = \begin{bmatrix} 1 & 1 \\ 1 + u_0/ik & -1 + u_0/ik \end{bmatrix} \begin{bmatrix} a^{(1)} \\ b^{(1)} \end{bmatrix}. \quad (2.219)$$

From the relation (2.170) we find the transfer matrix to be

$$M = \begin{bmatrix} 1 + u_0/2ik & u_0/2ik \\ -u_0/2ik & 1 - u_0/2ik \end{bmatrix}. \quad (2.220)$$

We observe that $M$ given in (2.220) has the structure of eqn (2.181), and that $|\alpha|^2 - |\beta|^2 = 1$, as required by (2.182). We also see that the relations (2.183) between the parameters of the $S$ and $M$ matrices are satisfied in this example.

*The wave function*

We can write the wave function when we specify an incident plane wave and use outgoing or incoming wave boundary conditions, as shown in Table 2.1.

**Exercise 2.11** Using the wave functions of Table 2.1, with the amplitudes given in eqn (2.218), verify explicitly the general orthonormality relations (2.12).

We consider, in particular, the calculation of $\langle \psi_+^{(+)}(E') | \psi_+^{(+)}(E) \rangle$. We find

$$\left\langle \psi_+^{(+)}(E') | \psi_+^{(+)}(E) \right\rangle = \frac{m}{2\pi\hbar^2 \sqrt{kk'}} I, \quad (2.221)$$

where $I$ is the integral

$$I = \left\{ \int_{-\infty}^{0} \left[ e^{-ik'x} + r^*(k') e^{ik'x} \right] \left[ e^{ikx} + r(k) e^{-ikx} \right] dx \right. \\ \left. + \int_{0}^{\infty} \left[ t^*(k') e^{-ik'x} \right] \left[ t(k) e^{ikx} \right] dx \right\}. \quad (2.222)$$

TABLE 2.1. The wave function for the delta potential scattering problem discussed in the text. Here, $\phi_\pm(E;x)$ are the normalized plane waves of eqn (2.2) and $r$, $t$, $r'$ and $t'$ are given in eqn (2.218).

|  | $x < 0$ | $x > 0$ |
|---|---|---|
| $\psi_+^{(+)}(E;x)$ | $\phi_+(E;x) + r\phi_-(E;x)$ | $t\phi_+(E;x)$ |
| $\psi_-^{(+)}(E;x)$ | $t'\phi_-(E;x)$ | $\phi_-(E;x) + r'\phi_+(E;x)$ |
| $\psi_-^{(-)}(E;x) = [\psi_+^{(+)}(E;x)]^*$ | $\phi_-(E;x) + r^*\phi_+(E;x)$ | $t^*\phi_-(E;x)$ |
| $\psi_+^{(-)}(E;x) = [\psi_-^{(+)}(E;x)]^*$ | $(t')^*\phi_+(E;x)$ | $\phi_+(E;x) + (r')^*\phi_-(E;x)$ |

We need the identities

$$\int_{-\infty}^{0} e^{iKx}\,dx = \lim_{\epsilon \to 0} \int_{-\infty}^{0} e^{(iK+\epsilon)x}\,dx$$
$$= -i\lim_{\epsilon \to 0} \frac{1}{K - i\epsilon}$$
$$= -i\frac{\mathcal{P}}{K} + \pi\delta(K) \qquad (2.223)$$

and

$$\int_{0}^{\infty} e^{iKx}\,dx = \lim_{\epsilon \to 0} \int_{0}^{\infty} e^{(iK-\epsilon)x}\,dx$$
$$= i\lim_{\epsilon \to 0} \frac{1}{K + i\epsilon}$$
$$= i\frac{\mathcal{P}}{K} + \pi\delta(K). \qquad (2.224)$$

Here, $\mathcal{P}$ stands for 'principal value'. Adding (2.223) and (2.224), we obtain the well-known result

$$\int_{-\infty}^{\infty} e^{iKx}\,dx = 2\pi\delta(K). \qquad (2.225)$$

Substituting (2.223) and (2.224) into (2.222), we find

$$I = \left[-i\frac{\mathcal{P}}{k-k'} + \pi\delta(k-k')\right] + r(k)\left[i\frac{\mathcal{P}}{k+k'} + \pi\delta(k+k')\right]$$
$$+ r^*(k')\left[-i\frac{\mathcal{P}}{k+k'} + \pi\delta(k+k')\right] + r^*(k')r(k)\left[i\frac{\mathcal{P}}{k-k'} + \pi\delta(k-k')\right]$$
$$+ t^*(k')t(k)\left[i\frac{\mathcal{P}}{k-k'} + \pi\delta(k-k')\right]. \qquad (2.226)$$

Since $k, k' > 0$, we have $\delta(k+k') = 0$. Simplifying and combining terms, we write

$$I = i\frac{\mathcal{P}}{k-k'}\left[-1 + r^*(k')r(k) + t^*(k')t(k)\right] + i\frac{\mathcal{P}}{k+k'}\left[r(k) - r^*(k')\right] \qquad (2.227)$$
$$+ \pi\delta(k-k')\left[1 + r^*(k')r(k) + t^*(k')t(k)\right].$$

Using the amplitudes given in eqn (2.218), we find

$$r^*(k')r(k) + t^*(k')t(k) = \frac{1 + (u_0/2k)(u_0/2k')}{(1 - u_0/2ik)(1 + u_0/2ik')}, \qquad (2.228)$$

$$r(k) - r^*(k') = \frac{u_0/2ik + u_0/2ik'}{(1 - u_0/2ik)(1 + u_0/2ik')}. \qquad (2.229)$$

Substituting in (2.227), we have

$$I = \frac{u_0}{2ikk'}\left[-i\frac{\mathcal{P}}{k-k'}\frac{k-k'}{(1 - u_0/2ik)(1 + u_0/2ik')}\right.$$
$$\left. + i\frac{\mathcal{P}}{k+k'}\frac{k+k'}{(1 - u_0/2ik)(1 + u_0/2ik')}\right] + 2\pi\delta(k-k')$$
$$= 2\pi\delta(k-k'). \qquad (2.230)$$

Finally, we obtain for the overlap (2.221):

$$\left\langle \psi_+^{(+)}(E')|\psi_+^{(+)}(E)\right\rangle = \frac{m}{\hbar^2 k}\delta(k-k') = \delta(E-E'), \qquad (2.231)$$

which is the result we wanted to prove.

**Exercise 2.12** Using the wave functions of Table 2.1, with the amplitudes given in eqn (2.218), verify explicitly the relations (2.89) defining the $S$-matrix elements.

We consider, in particular, the calculation of $\langle \psi_+^{(-)}(E')|\psi_+^{(+)}(E)\rangle$. We find

$$\left\langle \psi_+^{(-)}(E')|\psi_+^{(+)}(E)\right\rangle = \frac{m}{2\pi\hbar^2\sqrt{kk'}}J, \qquad (2.232)$$

where $J$ is the integral

$$J = \left\{\int_{-\infty}^{0} t'(k')e^{-ik'x}\left[e^{ikx} + r(k)e^{-ikx}\right]dx \right.$$
$$\left. + \int_{0}^{\infty}\left[e^{-ik'x} + r'(k')e^{ik'x}\right]\left[t(k)e^{ikx}\right]dx\right\}. \qquad (2.233)$$

Using the identities (2.223) and (2.224), we write $J$ as

$$J = t'(k')\left[-i\frac{\mathcal{P}}{k-k'} + \pi\delta(k-k')\right] + t'(k')r(k)\left[i\frac{\mathcal{P}}{k+k'} + \pi\delta(k+k')\right]$$
$$+ t(k)\left[i\frac{\mathcal{P}}{k-k'} + \pi\delta(k-k')\right] + t(k)r'(k')\left[i\frac{\mathcal{P}}{k+k'} + \pi\delta(k+k')\right]. \qquad (2.234)$$

As in the previous exercise, we note that $k, k' > 0$ implies that $\delta(k+k') = 0$, so that

$$J = i\frac{\mathcal{P}}{k-k'}\left[t(k) - t'(k')\right] + i\frac{\mathcal{P}}{k+k'}\left[t'(k')r(k) + t(k)r'(k')\right]$$
$$+ \pi\delta(k-k')\left[t(k) + t'(k')\right]. \qquad (2.235)$$

Using the amplitudes given in eqn (2.218), we find

$$t(k) - t'(k') = \frac{u_0/2ik - u_0/2ik'}{(1 - u_0/2ik)(1 - u_0/2ik')},$$

$$t'(k')r(k) + t(k)r'(k') = \frac{u_0/2ik + u_0/2ik'}{(1 - u_0/2ik)(1 - u_0/2ik')}.$$

Substituting these results into (2.235), we have

$$J = \frac{u_0}{2ikk'}\left[-i\frac{\mathcal{P}}{k-k'}\frac{k-k'}{(1-u_0/2ik)(1-u_0/2ik')} + i\frac{\mathcal{P}}{k+k'}\frac{k+k'}{(1-u_0/2ik)(1-u_0/2ik')}\right] + 2\pi\delta(k-k')t(k).$$

Finally, we find

$$\left\langle \psi_+^{(-)}(E') | \psi_+^{(+)}(E) \right\rangle = \frac{m}{\hbar^2 k} t(k)\delta(k-k') = t(k)\delta(E-E'), \tag{2.236}$$

which coincides with eqn (2.114). Similarly, one can verify the relations for the remaining $S$-matrix elements.

### 2.1.9 Scattering by a step potential

Consider the scattering of a particle by the positive step potential $V(x) = V_0\theta(x)$ shown in Fig. 2.8, $\theta(x)$ being the step function. The Schrödinger equation can be written as

$$\left(\partial_x^2 + k^2\right)\psi(x) = K^2\theta(x)\psi(x), \tag{2.237}$$

with

$$K^2 = \frac{2mV_0}{\hbar^2}. \tag{2.238}$$

We consider two cases.

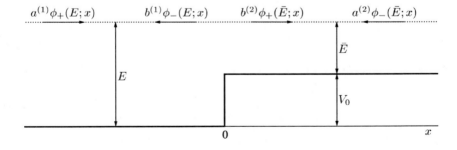

FIG. 2.8. A schematic representation of the most general solution to the problem of scattering by a step potential, when the energy $E$ is larger than the step $V_0$. The *kinetic* energy $\bar{E}$ of the particle on the right of the step is different from that on the left.

*Scattering above the step, $E > V_0$*

The Schrödinger equation reduces to

$$\left(\partial_x^2 + k^2\right)\psi(x) = 0, \quad x < 0, \tag{2.239a}$$

$$\left(\partial_x^2 + \bar{k}^2\right)\psi(x) = 0, \quad x > 0, \tag{2.239b}$$

where $k$ and $\bar{k}$ denote the wave numbers on the left and the right of the step, respectively, and they are given by

$$E = \frac{\hbar^2 k^2}{2m}, \tag{2.240a}$$

$$\bar{E} = \frac{\hbar^2 \bar{k}^2}{2m}, \tag{2.240b}$$

$$k^2 = \bar{k}^2 + K^2, \quad E = \bar{E} + V_0. \tag{2.240c}$$

The most general solution of these equations, indicated schematically in Fig. 2.8, is

$$\psi(E; x < 0) = a^{(1)}\phi_+(E; x) + b^{(1)}\phi_-(E; x), \tag{2.241a}$$

$$\psi(E; x > 0) = a^{(2)}\phi_-(\bar{E}; x) + b^{(2)}\phi_+(\bar{E}; x), \tag{2.241b}$$

where, just as in eqn (2.2), we have defined

$$\phi_s(\bar{E}; x) = \frac{e^{is\bar{k}x}}{\sqrt{2\pi\hbar^2\bar{k}/m}}, \quad \bar{k} > 0. \tag{2.242}$$

In contrast to the previous examples, this is a case where the asymptotic wave function has a different kinetic energy on the two sides of the scatterer. We shall comment more about this point later.

Continuity of the wave function (2.241) and its derivative at $x = 0$ yield

$$\frac{a^{(1)} + b^{(1)}}{\sqrt{k}} = \frac{a^{(2)} + b^{(2)}}{\sqrt{\bar{k}}}, \tag{2.243a}$$

$$i\sqrt{k}(a^{(1)} - b^{(1)}) = i\sqrt{\bar{k}}(b^{(2)} - a^{(2)}). \tag{2.243b}$$

From these equations we find the $S$ matrix, defined in the usual manner, to be

$$S = \begin{bmatrix} r & t' \\ t & r' \end{bmatrix} = \begin{bmatrix} (k - \bar{k})/(k + \bar{k}) & 2\sqrt{k\bar{k}}/(k + \bar{k}) \\ 2\sqrt{k\bar{k}}/(k + \bar{k}) & -(k - \bar{k})/(k + \bar{k}) \end{bmatrix}, \tag{2.244}$$

a unitary and symmetric matrix, and, similarly, the $M$ matrix is

$$M = \begin{bmatrix} (k + \bar{k})/2\sqrt{k\bar{k}} & -(k - \bar{k})/2\sqrt{k\bar{k}} \\ -(k - \bar{k})/2\sqrt{k\bar{k}} & (k + \bar{k})/2\sqrt{k\bar{k}} \end{bmatrix}, \tag{2.245}$$

a matrix with unit determinant.

POTENTIAL SCATTERING IN INFINITE ONE-DIMENSIONAL SPACE   59

The interaction potential contemplated in the present problem, given by the right-hand side of eqn (2.237) and illustrated in Fig. 2.8, is of infinite range. Thus we cannot apply the Lippmann–Schwinger theory, at least as it was developed earlier in this section, taking as the unperturbed Hamiltonian the kinetic energy and adding on to it the step potential. However, should we add to the present problem a finite-range potential, we could set up a Lippmann–Schwinger equation by taking the kinetic energy plus the step potential of Fig. 2.8 as the 'unperturbed' Hamiltonian; notice that the latter itself would then produce nonzero scattering, i.e., the 'unperturbed' solution would consist of an incident wave plus a reflected and a transmitted one. A similar situation will be encountered in Section 3.2, when we study the scattering of a wave in the interior of a structure consisting of an impenetrable cavity connected to the outside through (infinite) waveguides.

Going back to the step-potential problem, we denote by $\psi_+^{(+)}(E;x)$ the solution shown on the first row of Table 2.2. The reflection and the transmission amplitudes indicated there are given in eqn (2.244). Since the wave number is different on the two sides of the origin, i.e., $k$ on the left and $\bar{k}$ on the right, we cannot speak of an 'unperturbed' unit-amplitude plane wave traveling in the positive direction and extended all of the way from $-\infty$ to $+\infty$. The lower index '+' merely indicates that we have an incident unit-amplitude wave traveling with *positive* momentum $k$ on the *left* of the origin and the upper index '(+)' that, in addition, we have outgoing waves on the two sides. The label $E$ indicates the total energy, i.e., the eigenvalue of the Hamiltonian, which on the left of the origin is the kinetic energy, and on the right is the kinetic energy plus the height of the step $V_0$. Similarly, we have the solution $\psi_-^{(+)}(E;x)$ of the second row of

TABLE 2.2. The wave function for the step-potential scattering problem. The first row in each entry of the first column uses the notation $\psi_s^{(\pm)}(E;x)$ to denote the wave function, whereas the second row in each entry of the first column uses the 'lead index' notation discussed in the text.

|  | $x < 0$ | $x > 0$ |
|---|---|---|
| $\psi_+^{(+)}(E;x)$ <br> $\equiv \psi_1^{(+)}(E;x)$ | $\phi_+(E;x) + r\phi_-(E;x)$ | $t\phi_+(\bar{E};x)$ |
| $\psi_-^{(+)}(E;x)$ <br> $\equiv \psi_2^{(+)}(E;x)$ | $t'\phi_-(E;x)$ | $\phi_-(\bar{E};x) + r'\phi_+(\bar{E};x)$ |
| $\psi_-^{(-)}(E;x) = \theta\psi_+^{(+)}(E;x)$ <br> $\equiv \psi_1^{(-)}(E;x) = \theta\psi_1^{(+)}(E;x)$ | $\phi_-(E;x) + r^*\phi_+(E;x)$ | $t^*\phi_-(\bar{E};x)$ |
| $\psi_+^{(-)}(E;x) = \theta\psi_-^{(+)}(E;x)$ <br> $\equiv \psi_2^{(-)}(E;x) = \theta\psi_2^{(+)}(E;x)$ | $(t')^*\phi_+(E;x)$ | $\phi_+(\bar{E};x) + (r')^*\phi_-(\bar{E};x)$ |

Table 2.2, which consists of an incident wave traveling with *negative* momentum $-\bar{k}$ on the *right*, plus outgoing waves on the two sides. In an analogous way, the wave function $\psi_-^{(-)}(E;x)$ cannot be said to contain an 'unperturbed' unit-amplitude plane wave traveling in the negative direction and extended all of the way from $+\infty$ to $-\infty$. It is defined as the *time reverse* of the state $\psi_+^{(+)}(E;x)$ of the first row, i.e.,

$$\psi_-^{(-)}(E;x) = \theta \psi_+^{(+)}(E;x) = \left[\psi_+^{(+)}(E;x)\right]^*, \qquad (2.246)$$

and it is seen, from the third row of Table 2.2, to contain a unit-amplitude wave traveling with *negative* momentum $-k$ on the *left* of the origin, plus incoming waves on the two sides. Finally, $\psi_+^{(-)}(E;x)$ is defined as the *time reverse* of the state $\psi_-^{(+)}(E;x)$ of the second row, i.e.,

$$\psi_+^{(-)}(E;x) = \theta \psi_-^{(+)}(E;x) = \left[\psi_-^{(+)}(E;x)\right]^*. \qquad (2.247)$$

It is seen, from the last row of Table 2.2, to contain a unit-amplitude wave traveling with *positive* momentum $k$ on the *right* of the origin, plus incoming waves on the two sides. Equations (2.246) and (2.247) are the extension of eqn (2.21) to the present problem. The notation $\psi_s^{(\pm)}(E;x)$ just discussed is used in the first column of Table 2.2, in the first row of each entry, to designate the wave function.

There are several relations which, although usually proved from the Lippmann–Schwinger formalism for a finite-range potential, are valid more generally. For instance, the orthonormality relation of eqn (2.12) can be verified for the step-potential problem. The fundamental relation between the $S$ matrix and the scalar product of a $\psi^{(-)}$ and a $\psi^{(+)}$ state, although proved at the beginning of Section 2.1.4 using the Lippmann–Schwinger equations, was demonstrated in Exercise 2.6 using only the Schrödinger equation. The latter proof can be modified to account for the different values of the kinetic energy in the two asymptotic regions. One could further allow for the presence of a finite-range scatterer, although this is not contemplated in the example discussed in this subsection. The validity of eqn (2.114) for the present problem is verified in Exercise 2.13.

**Exercise 2.13** By direct integration, verify that the wave functions of Table 2.2, with the reflection and transmission amplitudes of eqn (2.244), satisfy the relation (2.114).

We calculate the overlap

$$\left\langle \psi_+^{(-)}(E') \middle| \psi_+^{(+)}(E) \right\rangle = \frac{m}{2\pi\hbar^2} L, \qquad (2.248)$$

where $L$ is the integral

# POTENTIAL SCATTERING IN INFINITE ONE-DIMENSIONAL SPACE

$$L = \left\{ \frac{1}{\sqrt{kk'}} \int_{-\infty}^{0} t'(E') e^{-ik'x} \left[ e^{ikx} + r(E) e^{-ikx} \right] dx \right.$$
$$\left. + \frac{1}{\sqrt{\bar{k}\bar{k}'}} \int_{0}^{\infty} \left[ e^{-i\bar{k}'x} + r'(\bar{k}') e^{i\bar{k}'x} \right] \left[ t(E) e^{i\bar{k}x} \right] dx \right\}. \tag{2.249}$$

Using the identities (2.223) and (2.224), we write $L$ as

$$L = \frac{t'(E')}{\sqrt{kk'}} \left[ -i\frac{\mathcal{P}}{k-k'} + \pi\delta(k-k') \right] + \frac{t'(E')r(E)}{\sqrt{kk'}} \left[ i\frac{\mathcal{P}}{k+k'} + \pi\delta(k+k') \right]$$
$$+ \frac{t(E)}{\sqrt{\bar{k}\bar{k}'}} \left[ i\frac{\mathcal{P}}{\bar{k}-\bar{k}'} + \pi\delta(\bar{k}-\bar{k}') \right] + \frac{r'(E')t(E)}{\sqrt{\bar{k}\bar{k}'}} \left[ i\frac{\mathcal{P}}{\bar{k}+\bar{k}'} + \pi\delta(\bar{k}+\bar{k}') \right]. \tag{2.250}$$

Since $k, k', \bar{k}, \bar{k}' > 0$, we have $\delta(k+k') = 0$ and $\delta(\bar{k}+\bar{k}') = 0$. Using the amplitudes given in eqn (2.244), we find

$$L = \sqrt{\frac{\bar{k}'}{k}} \frac{2}{k'+\bar{k}'} \left[ \pi\delta(k-k') - i\frac{\mathcal{P}}{k-k'} + i\frac{k-\bar{k}}{k+\bar{k}} \frac{1}{k+k'} \right]$$
$$+ \sqrt{\frac{k}{\bar{k}'}} \frac{2}{k+\bar{k}} \left[ \pi\delta(\bar{k}-\bar{k}') + i\frac{\mathcal{P}}{\bar{k}-\bar{k}'} - i\frac{k'-\bar{k}'}{k'+\bar{k}'} \frac{1}{\bar{k}+\bar{k}'} \right]. \tag{2.251}$$

Denoting by $Y$ the sum of the second, third, fifth and sixth terms in the above equation, we find

$$-\frac{i}{2}Y = \sqrt{\frac{\bar{k}'}{k}} \frac{1}{k'+\bar{k}'} \left[ -\frac{\mathcal{P}}{k-k'} + \frac{k-\bar{k}}{(k+\bar{k})(k+k')} \right]$$
$$+ \sqrt{\frac{k}{\bar{k}'}} \frac{1}{k+\bar{k}} \left[ \frac{\mathcal{P}}{\bar{k}-\bar{k}'} - \frac{k'-\bar{k}'}{(k'+\bar{k}')(\bar{k}+\bar{k}')} \right]$$
$$= -2\sqrt{\frac{\bar{k}'}{k}} \mathcal{P} \frac{k(\bar{k}+\bar{k}')}{(k'+\bar{k}')(k+\bar{k})(k+k')(k-k')}$$
$$+ 2\sqrt{\frac{k}{\bar{k}'}} \mathcal{P} \frac{\bar{k}'(\bar{k}+k')}{(k+\bar{k})(k'+\bar{k}')(\bar{k}+\bar{k}')(\bar{k}-\bar{k}')}$$
$$= \frac{2\sqrt{k\bar{k}'}(\bar{k}+k')}{(k+\bar{k})(k'+\bar{k}')} \left[ -\mathcal{P}\frac{1}{k^2-k'^2} + \mathcal{P}\frac{1}{\bar{k}^2-\bar{k}'^2} \right]. \tag{2.252}$$

From eqn (2.240) we find

$$k^2 - \bar{k}^2 = k'^2 - \bar{k}'^2 = K^2, \tag{2.253a}$$

and so

$$k^2 - k'^2 = \bar{k}^2 - \bar{k}'^2,$$
$$E - E' = \bar{E} - \bar{E}'. \tag{2.253b}$$

We thus conclude that $Y = 0$.

We now turn to the first and the fourth terms in eqn (2.251). Recall that, by convention, $k, k', \bar{k}, \bar{k}' > 0$. The first delta function implies that $k = k'$, and eqn (2.253b) further gives $\bar{k} = \bar{k}'$. Thus

$$\frac{\delta(k-k')}{k'+\bar{k}'} = \frac{\delta(k-k')}{k+\bar{k}}$$
$$= \frac{\hbar^2 k}{m}\frac{\delta(E-E')}{k+\bar{k}}. \tag{2.254}$$

Similarly, the fourth term in eqn (2.251) contains the quantity

$$\frac{\delta(\bar{k}-\bar{k}')}{k+\bar{k}} = \frac{\hbar^2 \bar{k}}{m}\frac{\delta(\bar{E}-\bar{E}')}{k+\bar{k}}$$
$$= \frac{\hbar^2 \bar{k}}{m}\frac{\delta(E-E')}{k+\bar{k}}. \tag{2.255}$$

We finally find

$$\left\langle \psi_+^{(-)}(E')|\psi_+^{(+)}(E)\right\rangle = 2\frac{\sqrt{kk'}}{k+\bar{k}}\delta(E-E')$$
$$= 2\frac{\sqrt{k\bar{k}}}{k+\bar{k}}\delta(E-E')$$
$$= t(E)\delta(E-E'), \tag{2.256}$$

which coincides with eqn (2.114). Similarly, one can verify the relations for the remaining $S$-matrix elements.

---

*The 'lead' notation.* We shall find it advantageous to think of the region where the scattering takes place—in the present problem, the origin, where the discontinuity in the potential occurs, plus, in the presence of an additional short-range potential, the region where that potential is nonzero—as a 'system' connected to the outside by two 'leads'. Lead 1(2) represents the negative (positive) $x$-axis, and we introduce the *lead index* $l = 1, 2$, as opposed to the index $s = \pm$ used so far. The outgoing boundary condition wave function

$$\psi_l^{(+)}(E;x) \tag{2.257a}$$

is defined as having an *incoming wave in lead $l$ only and outgoing waves in the two leads*; thus $\psi_+^{(+)}(E;x) = \psi_1^{(+)}(E;x)$ and $\psi_-^{(+)}(E;x) = \psi_2^{(+)}(E;x)$. The incoming boundary condition function $\psi_l^{(-)}(E;x)$, defined as the time reverse of the state (2.257a), i.e.,

$$\psi_l^{(-)}(E;x) = \theta\psi_l^{(+)}(E;x) = \left\{\psi_l^{(+)}(E;x)\right\}^*, \tag{2.257b}$$

contains an *outgoing wave in lead $l$ only and incoming waves in the two leads*; thus $\psi_-^{(-)}(E;x) = \psi_1^{(-)}(E;x)$ and $\psi_+^{(-)}(E;x) = \psi_2^{(-)}(E;x)$. The lead index notation is used in the first column of Table 2.2, in the second row of each entry, to designate the wave function. Using this notation, relations (2.89) and (2.93) between the

$S$-matrix elements and the reflection and transmission amplitudes can be written as

$$\left\langle \psi_l^{(-)}(E') \mid \psi_l^{(+)}(E) \right\rangle = r^{ll}\delta(E'-E), \tag{2.258a}$$

with

$$r^{11} = r, \quad r^{22} = r', \tag{2.258b}$$

and

$$\left\langle \psi_{l'}^{(-)}(E') \mid \psi_l^{(+)}(E) \right\rangle = t^{l'l}\delta(E'-E), \quad l' \neq l, \tag{2.259a}$$

with

$$t^{21} = t, \quad t^{12} = t'. \tag{2.259b}$$

Equation (2.94) can thus be written, alternatively, as

$$S = \begin{bmatrix} r^{11} & t^{12} \\ t^{21} & r^{22} \end{bmatrix}. \tag{2.260}$$

We shall find the 'lead notation' particularly useful in Section 3.2, on open cavities with leads.

*Scattering by a barrier.* We now use the above results to solve, in a simple way, the problem of scattering by the barrier shown in Fig. 2.9. First, the $M$ matrix for the step potential shown with a dashed line in Fig. 2.10 can be obtained from eqn (2.245) by simply interchanging $k$ and $\bar{k}$, i.e.,

$$\overset{\circ}{M} = \begin{bmatrix} (k+\bar{k})/2\sqrt{k\bar{k}} & (k-\bar{k})/2\sqrt{k\bar{k}} \\ (k-\bar{k})/2\sqrt{k\bar{k}} & (k+\bar{k})/2\sqrt{k\bar{k}} \end{bmatrix}. \tag{2.261}$$

The $M$ matrix for the step shown with a full line in Fig. 2.10 can now be obtained from $\overset{\circ}{M}$ by a translation. A simple extension of the argument of Section 2.1.7

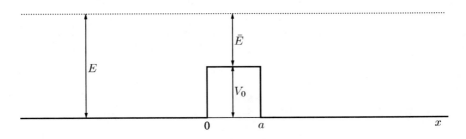

FIG. 2.9. The potential whose $M$ matrix is given in eqn (2.263).

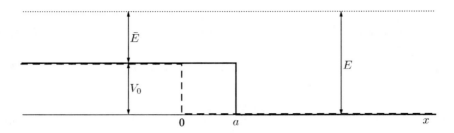

FIG. 2.10. The auxiliary potentials whose $M$ matrices are given in eqns (2.261) and (2.262).

(the wave numbers $k$ and $\bar{k}$ are now different on the two sides of the step) shows that

$$M = \begin{bmatrix} e^{-ika} & 0 \\ 0 & e^{ika} \end{bmatrix} \overset{\circ}{M} \begin{bmatrix} e^{i\bar{k}a} & 0 \\ 0 & e^{-i\bar{k}a} \end{bmatrix}$$
$$= \begin{bmatrix} \left((k+\bar{k})/2\sqrt{k\bar{k}}\right) e^{-i(k-\bar{k})a} & \left((k-\bar{k})/2\sqrt{k\bar{k}}\right) e^{-i(k+\bar{k})a} \\ \left((k-\bar{k})/2\sqrt{k\bar{k}}\right) e^{i(k+\bar{k})a} & \left((k+\bar{k})/2\sqrt{k\bar{k}}\right) e^{i(k-\bar{k})a} \end{bmatrix}. \quad (2.262)$$

With this information we can now solve the scattering problem for the potential shown in Fig. 2.9, by simply multiplying the transfer matrices of eqns (2.262) and (2.245) (in that order), to obtain

$$M = \begin{bmatrix} e^{-ika}\left\{\cos\bar{k}a + i\left((k^2+\bar{k}^2)/2k\bar{k}\right)\sin\bar{k}a\right\} & -ie^{-ika}\left((k^2-\bar{k}^2)/2k\bar{k}\right)\sin\bar{k}a \\ ie^{ika}\left((k^2-\bar{k}^2)/2k\bar{k}\right)\sin\bar{k}a & e^{ika}\left\{\cos\bar{k}a - i\left((k^2+\bar{k}^2)/2k\bar{k}\right)\sin\bar{k}a\right\} \end{bmatrix}. \quad (2.263)$$

This $M$ matrix satisfies the requirements of eqns (2.180) and (2.182). The reflection and transmission amplitudes can be found from eqns (2.184). This last expression, eqn (2.263), can be used to describe and discuss a number of physically interesting situations, like that corresponding to the so-called *barrier-top resonances*, which occur at energies $E$ above and close to the barrier.

*Scattering below the step, $E < V_0$*

Now the incident energy $E$ is an amount $\varepsilon$ below the step of Fig. 2.8. The Schrödinger equation reduces to

$$\left(\partial_x^2 + k^2\right)\psi(x) = 0, \quad x < 0, \quad (2.264a)$$
$$\left(\partial_x^2 - \kappa^2\right)\psi(x) = 0, \quad x > 0, \quad (2.264b)$$

where $k$ and $\kappa$ are related by
$$k^2 = -\kappa^2 + K^2, \quad E = -\varepsilon + V_0, \tag{2.265}$$
with
$$\varepsilon = \frac{\hbar^2 \kappa^2}{2m}. \tag{2.266}$$

In this case there is only *one independent solution* (normalizable in the sense of Dirac delta functions) at every energy, in contrast with the two solutions we had in the previous example, which was for an energy above the step. For an incident plane wave from the left, we have a reflected one on the same side, while on the right the wave function decays exponentially.

We shall find it convenient to define, by analogy with the plane-wave solutions (2.242), the solutions (certainly non-normalizable in the interval $-\infty < x < +\infty$) of eqn (2.264b) to be
$$\phi_s(-\varepsilon; x) = \frac{e^{-s\kappa x}}{e^{i\pi/4}\sqrt{2\pi\hbar^2\kappa/m}}, \quad \kappa > 0. \tag{2.267}$$

The usefulness of these functions will become clearer later on (see eqn (2.276)). Notice that $\phi_s(-\varepsilon; x)$ is the *analytic continuation* of the wave function of eqn (2.242) when
$$\bar{k} \Longrightarrow +i\kappa, \tag{2.268a}$$
$$\sqrt{\bar{k}} \Longrightarrow e^{i\pi/4}\kappa. \tag{2.268b}$$

The wave function described after eqn (2.266) can be written as
$$\psi(E; x < 0) = \phi_+(E; x) + r\phi_-(E; x), \tag{2.269a}$$
$$\psi(E; x > 0) = t\phi_+(-\varepsilon; x). \tag{2.269b}$$

Notice that the wave function $\phi_+(-\varepsilon; x)$ decreases exponentially for $x > 0$. We now have 'running waves' for $x < 0$ (an open channel in 'lead' 1, using the nomenclature introduced above) and an 'evanescent wave' for $x > 0$ (a closed channel in 'lead' 2). Continuity of the wave function $\psi(E; x)$ and its derivative at $x = 0$ yields
$$\frac{1+r}{\sqrt{k}} = \frac{t}{e^{i\pi/4}\sqrt{\kappa}}, \tag{2.270a}$$
$$ik\frac{1-r}{\sqrt{k}} = \frac{-\kappa t}{e^{i\pi/4}\sqrt{\kappa}}. \tag{2.270b}$$

From these equations we find
$$r = \frac{k - i\kappa}{k + i\kappa}, \tag{2.271}$$
$$t = 2e^{i\pi/4}\frac{\sqrt{k\kappa}}{k + i\kappa}, \tag{2.272}$$

which are the *analytic continuation* (2.268) *of the reflection and transmission amplitudes of eqn* (2.244). Notice that *the S matrix is now one-dimensional* and is given by the reflection amplitude $r$, a complex number of modulus 1, as follows:

$$S = r = \frac{k - i\kappa}{k + i\kappa}, \quad |S|^2 = 1, \tag{2.273}$$

implying total reflection. Notice therefore that a non-vanishing $t$ is merely the amplitude of the evanescent wave for $x > 0$ and does not imply any transmitted current.

Since the potential is real, $\psi(E; x)$ being a solution implies that its complex conjugate $\psi^*(E; x)$ is also a solution, for the same energy. But since we have only one independent solution (normalizable, in the sense of Dirac delta functions), $\psi(E; x)$ and $\psi^*(E; x)$ must be proportional. In fact, from the above form (2.269) of the wave function, we find, for all $x$, that

$$\psi^*(E; x) = r^* \psi(E; x). \tag{2.274}$$

We can thus write

$$(\psi^*(E'; x), \psi(E; x)) = r\delta(E' - E), \tag{2.275}$$

a relation analogous to that of eqn (2.258a) with (2.257b), that was deduced for scattering above the step.

Suppose now that we seek the $M$ matrix for the potential of Fig. 2.9 for the present case $E < V_0$, insisting on performing the calculation by multiplying two appropriate $M$ matrices for the left and the right portions of the potential, just as we did for $E > V_0$ in Section 2.1.9. Since in the region of the potential the wave function is, in general, *a linear combination of increasing and decreasing real exponentials*, it is clear that the individual transfer matrices that we need to multiply must connect the most general linear combination of waves on either side, just as happens with running waves.

We first consider the most general solution—not necessarily normalizable—of the step problem of Fig. 2.8 when $E < V_0$, i.e., of eqns (2.264). We write it as

$$\psi(E; x < 0) = a^{(1)} \phi_+(E; x) + b^{(1)} \phi_-(E; x), \tag{2.276a}$$

$$\psi(E; x > 0) = a^{(2)} \phi_-(-\varepsilon; x) + b^{(2)} \phi_+(-\varepsilon; x). \tag{2.276b}$$

This wave function is seen to be the analytic continuation (2.268) of the one given in eqn (2.241) for $E > V_0$.

We define the *extended* $\tilde{S}$ and $\tilde{M}$ matrices for this problem as satisfying the same eqns (2.95) and (2.170) as for the conventional $S$ and $M$ matrices. These extended matrices are then just the analytic continuation (2.268) of eqns (2.244) and (2.245). We thus have the extended $\tilde{S}$ matrix given by

$$\tilde{S} = \begin{bmatrix} (k - i\kappa)/(k + i\kappa) & 2e^{i\pi/4}\sqrt{k\kappa}/(k + i\kappa) \\ 2e^{i\pi/4}\sqrt{k\kappa}/(k + i\kappa) & -(k - i\kappa)/(k + i\kappa) \end{bmatrix}, \tag{2.277}$$

POTENTIAL SCATTERING IN INFINITE ONE-DIMENSIONAL SPACE 67

(notice that it is *only the running wave part*, i.e., the 1,1 matrix element, that is *unitary* and coincides with the standard, open-channel $S$ of eqn (2.273)), and the extended $\tilde{M}$ matrix given by

$$\tilde{M} = \begin{bmatrix} (k+i\kappa)/2e^{i\pi/4}\sqrt{k\kappa} & -(k-i\kappa)/2e^{i\pi/4}\sqrt{k\kappa} \\ -(k-i\kappa)/2e^{i\pi/4}\sqrt{k\kappa} & (k+i\kappa)/2e^{i\pi/4}\sqrt{k\kappa} \end{bmatrix}. \tag{2.278}$$

Similar considerations can be made for the step potentials of Fig. 2.10. The $M$ matrix for the potential barrier of Fig. 2.9 when $E < V_0$ can thus be obtained by multiplying the extended $\tilde{M}$ matrices for the two steps, which are given by the analytic continuation (2.268) of (2.262) and by eqn (2.278). Their matrix multiplication can be simply read off from eqn (2.263) as

$$M = \begin{bmatrix} e^{-ika}\left\{\cosh\kappa a + i\left((k^2-\kappa^2)/2k\kappa\right)\sinh\kappa a\right\} & -ie^{-ika}\left((k^2+\kappa^2)/2k\kappa\right)\sinh\kappa a \\ ie^{ika}\left((k^2+\kappa^2)/2k\kappa\right)\sinh\kappa a & e^{ika}\left\{\cosh\kappa a - i\left((k^2-\kappa^2)/2k\kappa\right)\sinh\kappa a\right\} \end{bmatrix}. \tag{2.279}$$

This $M$ matrix satisfies the requirements of eqns (2.180) and (2.182). The reflection and transmission amplitudes can be found from eqns (2.184).

As a check, we can easily verify that the $M$ matrix of eqn (2.279) goes over into that of eqn (2.220) for the delta potential when we take the limit $a \to 0$, $K^2 \to \infty$, $K^2 a = u_0$.

It should be clear that the use of the rather unconventional extended $\tilde{M}$ matrices is essential in the present case to obtain the right answer when composing two transfer matrices by multiplication. To stress this point further we remark that, although the decaying exponential in the wave function (2.269b) carries no current, the wave function in the region below the barrier in Fig. 2.9 in general does carry a nonzero current, even for $E < V_0$. This is the case when the current is nonzero on the left and on the right of the barrier since, in the stationary state, the current must be the same in the three regions. An elementary calculation shows that the most general solution in that region, i.e.,

$$\psi(x) = c\phi_+(-\varepsilon;x) + d\phi_-(-\varepsilon;x) \tag{2.280a}$$
$$= Ce^{-\kappa x} + De^{\kappa x}, \tag{2.280b}$$

carries a current

$$I = \frac{\hbar}{m}i\kappa\left(CD^* - C^*D\right) = \frac{i}{\hbar}\left(cd^* - c^*d\right). \tag{2.281}$$

If either $c$ or $d$ vanishes—as occurs below the barrier in the step problem—or if the product $cd^*$ is real, then the wave function (2.280) carries no current. However, in the region below the barrier in Fig. 2.9 there is, in general, a non-vanishing current.

2.1.10 *Combination of reflection and transmission amplitudes for a one-dimensional disordered conductor: invariant imbedding equations*

In this subsection we will digress a little to pose and solve the problem of finding, in the continuum limit, the equations governing the evolution in length $L$ of the reflection and transmission amplitudes for a one-dimensional disordered conductor having a scattering potential $V(x)$. We will approach the problem via the method of 'invariant imbedding' [85, 109, 110, 154]. Our motivation for doing this is twofold. Firstly, it allows us to introduce the method which is of considerable interest in itself, and is non-perturbative. Secondly, it enables one to deal with a potential of arbitrary shape. Moreover, it is microscopic in that it essentially reformulates the time-independent Schrödinger equation with the given sample-specific potential $V(x)$ as an equation directly giving the evolution of the physically important emergent quantities, namely the elements ($r$, $r'$, $t$, $t'$) of the 2×2 $S$ matrix, see eqn (2.94), with the length $L$ of the sample.

Thus motivated, consider the scattering of a plane wave of unit amplitude and wave number $k$ by a disordered conductor of length $L + \delta L$ located in the interval $0 \leqslant x \leqslant L + \delta L$. The sample is assumed to be connected to two infinitely long ideal leads (i.e., without scatterers) at the two ends, providing for free propagation of the plane waves.

We can now imagine a subsample of length $L$, located in the interval $0 \leqslant x \leqslant L$, as 'imbedded' in the given sample of length $L + \delta L$, and consider the total scattering to arise from the serial composition of scattering from the subsample $L$ and that from the element $\delta L$.

The imbedding is said to be *invariant* in the sense that the laws describing the behavior of the sample and the subsample are one and the same. For $k\delta L \ll 1$ (eventually we will let $\delta L \to 0$), we can lump the distributed scatterers in the interval $\delta L$ as a single delta-function scatterer of strength

$$u_0 \equiv \frac{2m}{\hbar^2} V(L) \delta L \qquad (2.282)$$

(in the notation of Section 2.1.8), placed at $x = L + \delta L$. It is readily verified that the location of this delta-function scatterer elsewhere in the interval $\delta L$ does not affect the end result as $\delta L \to 0$.

Thus, we now have merely to compose the subsample-$L$ scattering and the scattering by the delta potential at $x = L + \delta L$ in series, using the composition law provided by eqn (2.194) as specialized to the present case. The quantities with subscript 1 in eqn (2.194) correspond to the subsample and will be denoted by

$$r_1 = r(L), \qquad (2.283a)$$
$$t_1 = t(L), \qquad (2.283b)$$
$$r'_1 = r'(L), \qquad (2.283c)$$

while those with subscript 2 correspond to the delta potential and will be denoted by

$$r_2 = e^{2ik(L+\delta L)}\delta r, \qquad (2.284a)$$

$$t_2 = \delta t, \qquad (2.284b)$$

$$r_2' = e^{-2ik(L+\delta L)}\delta r', \qquad (2.284c)$$

where we have used the translation relations (2.209). Here

$$\delta r = \delta r' = \frac{u_0/2ik}{1 - u_0/2ik}, \qquad (2.285a)$$

$$\delta t = \frac{1}{1 - u_0/2ik} \qquad (2.285b)$$

are the elements of the $S$ matrix for the delta-function scatterer located at the origin and taken from eqn (2.218). Equation (2.194) thus gives

$$r(L + \delta L) = r(L) + \frac{[t(L)]^2 e^{2ik(L+\delta L)}\delta r}{1 - r'(L)e^{2ik(L+\delta L)}\delta r}, \qquad (2.286a)$$

$$r'(L + \delta L) = e^{-2ik(L+\delta L)}\delta r' + \frac{(\delta t)^2 r'(L)}{1 - r'(L)e^{2ik(L+\delta L)}\delta r}, \qquad (2.286b)$$

$$t(L + \delta L) = \frac{\delta t \cdot t(L)}{1 - r'(L)e^{2ik(L+\delta L)}\delta r}. \qquad (2.286c)$$

Expanding the right-hand side of the above equations in powers of $k\delta L$ ($\ll 1$) and writing $[r(L + \delta L) - r(L)]/\delta L$ as $\partial r/\partial L$ in the limit as $\delta L \to 0$, we obtain

$$\frac{\partial r(L)}{\partial L} = \frac{ik\eta(L)}{2}e^{2ikL}[t(L)]^2, \qquad (2.287a)$$

$$\frac{\partial r'(L)}{\partial L} = \frac{ik\eta(L)}{2}e^{2ikL}\left[e^{-2ikL} + r'(L)\right]^2, \qquad (2.287b)$$

$$\frac{\partial t(L)}{\partial L} = \frac{ik\eta(L)}{2}\left[1 + r'(L)e^{2ikL}\right]t(L), \qquad (2.287c)$$

where $\eta(L) = -2mV(L)/\hbar^2 k^2$.

Equations (2.287) are the *invariant imbedding equations*. They give the variation of the reflection and transmission amplitudes with the sample length. The point to note is that we have here a set of first-order, albeit nonlinear, differential equations directly giving the evolution with the length $L$ of the emergent complex quantities $r(L)$, $r'(L)$ and $t(L)$, which are the quantities of physical interest in a scattering problem. The *initial* value problem in $L$ has the natural *initial* conditions, namely $r(L) \to 0$, $r'(L) \to 0$ and $t(L) \to 1$ as $L \to 0$. Of particular interest is the Riccati equation (2.287b) for $r'(L)$, which is *not* coupled to $t(L)$, and can be solved for a given $V(x)$. For example, for $V(x)$ random in $L$, the complex $r'(L)$ performs a two-dimensional random walk in the *unit* disc $|r(L)| \leq 1$. Indeed, for a weak, delta-correlated Gaussian random potential $V(x)$, the Fokker–Planck equation associated with the stochastic differential equation

(2.287b) turns out to be [109] the Mel'nikov equation [139, 140] for the probability distribution of the four-probe sample resistance. This microscopically derived result serves as a check on our MEA-based treatment of the $N$-channel problem specialized to the $N = 1$ case (see Chapter 7).

Consider again our sample of length $L$, located along the $x$-axis between $x = 0$ and $x = L$. We adopt the 'lead notation' described at the end of Section 2.1.9, and consider the negative axis $x < 0$ as lead 1 and the region $x > L$ as lead 2. Following a convention that we shall introduce in Section 3.2, we define the axis $x_1$ as running *away from the lead* from $x = 0$ to $-\infty$, with the origin at $x = 0$, and the axis $x_2$ as running from $x = L$ to $+\infty$, also away from the lead, with the origin at $x = L$. The plane waves in leads 1(2) will be taken to have the form $\exp(\pm ikx_{1(2)})$, and the associated scattering amplitudes will be indicated by a tilde.

**Exercise 2.14** Show that the relation between the old and the new sets of amplitudes is
$$\widetilde{r} = r, \qquad \widetilde{t'} = e^{ikL}t', \qquad (2.288)$$
$$\widetilde{t} = e^{ikL}t, \qquad \widetilde{r'} = e^{2ikL}r'.$$

**Exercise 2.15** Show that the imbedding equations for the new scattering amplitudes are
$$\frac{\partial \widetilde{r}(L)}{\partial L} = \frac{ik\eta(L)}{2}\left[\widetilde{t}(L)\right]^2, \qquad (2.289a)$$
$$\frac{\partial \widetilde{r'}(L)}{\partial L} = 2ik\widetilde{r'}(L) + \frac{ik\eta(L)}{2}\left[1 + \widetilde{r'}(L)\right]^2, \qquad (2.289b)$$
$$\frac{\partial \widetilde{t}(L)}{\partial L} = ik\widetilde{t}(L) + \frac{ik\eta(L)}{2}\left[1 + \widetilde{r'}(L)\right]\widetilde{t}(L). \qquad (2.289c)$$

Some observations are now in order at this point. It is readily verified that the imbedding equation (2.289b) for $\widetilde{r'}(L)$ is form-invariant under $\widetilde{r'}(L) \to 1/(\widetilde{r'}(L))^*$. Similar *duality* holds for eqn (2.289c) for $\widetilde{t}(L) \to 1/(\widetilde{t}(L))^*$, simultaneous with $\widetilde{r'}(L) \to (\widetilde{r'}(L))^*$. (In the case of complex $V(x)$, one has to complex-conjugate $V(x)$ as well. This has obvious relevance to the (bosonic) case of a light wave propagating through a coherently amplifying/absorbing medium. However, this is clearly not quite relevant to the (fermionic) case of electron waves.) Finally, the above treatment can be generalized to the case of more than one channel, i.e., a physical wire [154].

## 2.2 Potential scattering in semi-infinite one-dimensional space: resonance theory

Resonant scattering is one of the most interesting phenomena occurring in scattering theory. Its study is taken up in the present section, where, for simplicity, the analysis is carried out in the context of scattering in a semi-infinite one-dimensional space. We first illustrate the main concepts using a simple soluble model, based on the discussion by McVoy [123]. The general theory is the topic of Section 2.2.7.

The notions learned from resonance theory will be very relevant to Chapter 6, which is devoted to the study of scattering inside waveguides and cavities.

### 2.2.1 *A soluble model for the study of resonances*

Consider a particle which, due to an impenetrable 'wall' at $x = 0$, is allowed to move on the positive $x$-axis only, where it feels a 'barrier' in the form of a repulsive delta potential at $x = a$, i.e.,

$$U(x) = u_0 \delta(x - a), \quad u_0 > 0. \tag{2.290}$$

Just as in Section 2.1.8, $u_0$ has dimensions of inverse length, so that the ratios $u_0/k$ and $u_0 a$ which we shall find again below are dimensionless. The situation is illustrated in Fig. 2.11. The particle is incident from the right and suffers multiple scattering at the delta potential at $x = a$ and the hard wall at $x = 0$, and is finally totally reflected back to the region $x > a$. The interesting feature of this problem is that there are particular energies at which the particle stays inside the region $0 < x < a$ for an unusually long time; we shall find that these *resonance energies* are close to those of the bound states that would occur in that region should the delta potential at $x = a$ be an impenetrable barrier. On the other hand, in between the resonances the particle is reflected back rather promptly.

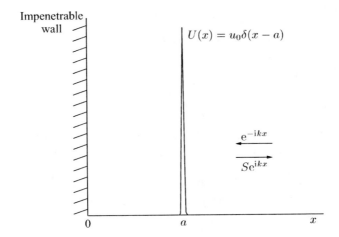

FIG. 2.11. Scattering in a semi-infinite one-dimensional space. There is an impenetrable wall at $x = 0$ and a delta potential barrier, eqn (2.290), at $x = a$.

The Schrödinger equation for this problem is

$$\left(\partial_x^2 + k^2\right)\psi(x) = u_0 \delta(x-a)\psi(x), \qquad (2.291\text{a})$$

with the boundary condition

$$\psi(0) = 0. \qquad (2.291\text{b})$$

The solution must then have the structure

$$\psi(x < a) = A\sin kx, \qquad (2.292\text{a})$$

$$\psi(x > a) = e^{-ik(x-a)} + Se^{ik(x-a)}. \qquad (2.292\text{b})$$

Equations (2.292a) and (2.292b) represent the internal and the external wave functions, respectively. For a given energy, the solution is non-degenerate. In our original definition (2.95), $S$ is a $2\times 2$ matrix which relates the coefficients of plane waves of the type $e^{\pm ikx}$. In eqn (2.292), $S$ is a $1\times 1$ matrix and it relates the outgoing wave, written as $e^{ik(x-a)}$, to the incoming wave, written as $e^{-ik(x-a)}$. It is as if we had shifted the origin to $x = a$, where the delta potential is located; this choice has certain advantages which we shall see during the discussion.

Continuity of the wave function at $x = a$ and discontinuity of its slope, obtained from eqn (2.291) as

$$[\psi'(x)]_{a^-}^{a^+} = u_0 \psi(a), \qquad (2.293)$$

give the pair of equations

$$1 + S = A\sin ka, \qquad (2.294\text{a})$$

$$ik(S-1) - kA\cos ka = u_0(1+S), \qquad (2.294\text{b})$$

whose solution is

$$A = -\frac{2ik/u_0}{((k/u_0)\cos ka + \sin ka) - i(k/u_0)\sin ka}, \qquad (2.295)$$

$$S = -\frac{(\sin ka + (k/u_0)\cos ka) + i(k/u_0)\sin ka}{(\sin ka + (k/u_0)\cos ka) - i(k/u_0)\sin ka}. \qquad (2.296)$$

We observe that $S$ is explicitly *unitary*, i.e., a complex number of unit modulus. Denoting its phase by $\theta$, we write

$$S = e^{i\theta} = -e^{2i\delta}, \qquad (2.297)$$

where we have also introduced the quantity $\delta$, known as the *phase shift* for reasons that will become clear below. As a check, we observe that, if $u_0 = 0$, i.e., in the absence of the potential, then $A = -2ie^{ika}$ and $S = -e^{2ika}$, which correctly imply, for the wave function, that

$$\psi(x) = -2ie^{ika} \sin kx, \tag{2.298}$$

for $0 \leqslant x < \infty$. On the other hand, if $ka \neq n\pi$ and we let $u_0 \to \infty$, then $A \to 0$ and $S \to -1$, so that

$$\psi(x < a) = 0, \tag{2.299a}$$
$$\psi(x > a) = -2i \sin k(x - a), \tag{2.299b}$$

and the scattering wave function vanishes for $x \leqslant a$, as it should.

For arbitrary $u_0$ we can write $\psi(x > a)$ in terms of the phase shift $\delta$ as

$$\psi(x > a) = -2ie^{i\delta} \sin\left[k(x - a) + \delta\right]. \tag{2.300}$$

We notice that the wave function is shifted, spatially, by a phase precisely equal to $\delta$, with respect to the case of eqn (2.299b) when the potential is impenetrable. For instance, a positive $\delta$ would mean that the wave function has been 'pulled in' due to the penetrable potential. Notice that our definition of phase shift is, at this point, slightly different from the standard one, which is defined using plane waves of the type $e^{\pm ikx}$, with a vanishing phase shift corresponding to the absence of a potential and complete reflection at $x = 0$; here, $\delta = 0$ means an impenetrable potential at $x = a$.

Since the system is invariant under time reversal ($U(x)$ is *real*), we should be able to write the wave function as real. This we obtain by dividing the wave function throughout by $-2ie^{i\delta}$, to find

$$\psi(x \leqslant a) = B \sin kx, \tag{2.301a}$$

with

$$B = \frac{1}{\left[\left((u_0/k) \sin ka + \cos ka\right)^2 + \sin^2 ka\right]^{1/2}}, \tag{2.301b}$$

and

$$\psi(x \geqslant a) = \sin\left[k(x - a) + \delta\right]. \tag{2.301c}$$

### 2.2.2 Behavior of the phase shift

From eqn (2.296) we find

$$\tan \delta = \frac{\tan ka}{1 + (u_0/k) \tan ka}. \tag{2.302}$$

As a check, we see that, for $u_0 = 0$, i.e., in the absence of the potential,

$$\delta = ka, \quad \theta = 2ka + \pi, \tag{2.303}$$

so that the wave function of eqn (2.301) is $\psi(x) = \sin kx$, for $0 \leqslant x < \infty$. This behavior of $\delta$ is shown in Fig. 2.12. The slope of $\delta(k)$ as a function of $k$ has

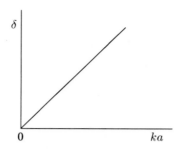

FIG. 2.12. The phase shift $\delta$ as a function of $ka$ in the absence of the delta potential. The positive slope indicates a *time delay* with respect to the situation of an impenetrable potential at $x = a$. In fact, in the absence of the delta potential the particle has to travel all the way to the origin and back.

a very appealing physical meaning, as it is related to the concept of *Wigner's time delay*, which measures the delay suffered by the centroid of a wave packet because of the scattering process [20, 163, 175]. Wigner's time delay $\tau$ is given by

$$\tau = \hbar \frac{\partial \theta}{\partial E} = \frac{2}{v} \frac{\partial \delta}{\partial k}, \qquad (2.304)$$

with $v = \hbar k/m$ representing the velocity. Applying this concept to the case described by eqn (2.303), we obtain

$$\tau = \frac{2a}{v}, \qquad (2.305)$$

which is precisely the extra time taken by the particle to go from $x = a$ to $x = 0$ and back. Our choice of $e^{\pm ik(x-a)}$ instead of $e^{\pm ikx}$ in eqn (2.292) has the effect that the resulting $\tau$ measures *the time delay suffered by the centroid of a wave packet with respect to the situation of complete reflection at $x = a$*.

Now consider the other extreme case of a strong barrier. We shall be interested in the scattering phenomena occurring when the incident energy is in the vicinity of $\hbar^2(n\pi)^2/2ma^2$, the energy where the $n$th bound state would occur should the barrier be impenetrable; in other words, we consider

$$ka \approx n\pi. \qquad (2.306)$$

We speak of a strong barrier if

$$u_0 \gg k. \qquad (2.307a)$$

Then

$$u_0 a \gg ka \approx n\pi, \qquad (2.307b)$$

and

$$\alpha \equiv \frac{1}{u_0 a} \ll 1. \qquad (2.307c)$$

Figure 2.13 shows $\tan\delta$ calculated from eqn (2.302) as a function of $ka$, for $\alpha = 0.1$. For this same case, Fig. 2.14 shows $\delta$ itself as a function of $ka$. We first see that $\delta$ is an ever increasing function of $ka$, thus indicating a *time delay for every energy*. Due to the large value of $u_0/k$ for the $k$s shown, $\delta$ is almost flat everywhere. This indicates a small time delay, except in the vicinity of a zero of the denominator of eqn (2.302), where $\delta$ suffers *abrupt increases* of about $\pi$, thus signalling a *large time delay* and hence the presence of what we shall call a *resonance*. The zeros of the denominator of eqn (2.302) occur at the intersection of the curves $\tan ka$ and $-ka/u_0a$, also shown in Fig. 2.13; at these points, which we shall call $k_n^\delta a$, we have $\tan\delta = \infty$ and

$$\delta = \frac{\pi}{2}, \frac{3\pi}{2}, \text{etc.} \tag{2.308}$$

We remark though that it is really the sharp increase of $\delta$, not its value, that signals the presence of a resonance, as one can see in more general examples. The quantities $k_n^\delta a$ satisfy the equation

$$\tan k_n^\delta a = -\alpha k_n^\delta a, \tag{2.309}$$

where $\alpha$ is defined in eqn (2.307c). For a strong barrier, $k_n^\delta a$ must be in the vicinity of the bound-state value $k_n^{BS}a = n\pi$ for a box of length $a$. In fact,

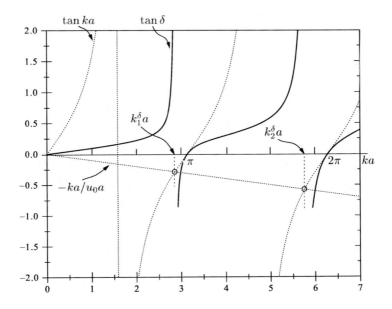

FIG. 2.13. The behavior of $\tan\delta$ of eqn (2.302) as a function of $ka$, for $\alpha = 1/u_0a = 0.1$.

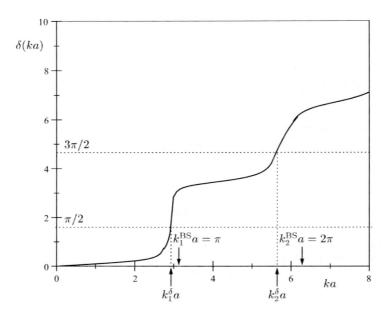

FIG. 2.14. The behavior of the phase shift $\delta$ as a function of $ka$, for $u_0 a = 10$.

the following exercise finds the result (2.314) for $k_n^\delta a$, up to third order in $\alpha$. Including only the correction to first order in $\alpha$, eqn (2.314) gives

$$k_n^\delta a = n\pi(1 - \alpha) + O(\alpha^2). \tag{2.310}$$

**Exercise 2.16** Find, to $O(\alpha^3)$, the values $k_n^\delta$ of $k$ for which the phase shift $\delta = \pi/2$, $3\pi/2$, etc., for the problem of an infinite wall and the highly reflecting delta potential barrier illustrated in Fig. 2.11.

The denominator of eqn (2.302) vanishes at values of $k$ that we call $k_n^\delta$, where the phase shift $\delta = \pi/2, 3\pi/2, \ldots$. We shall be interested in the case of a strong barrier, or $\alpha \ll 1$. The $k_n^\delta a$ satisfy the transcendental equation (2.309). Since, evidently, $k_n^\delta a$ must be in the vicinity of $n\pi$, i.e., where we would have bound states for an impenetrable barrier, we write

$$k_n^\delta a = n\pi + \eta, \tag{2.311}$$

with $\eta \ll 1$, and assume that $\eta = O(\alpha)$. Substituting in eqn (2.309), using $\tan k_n a = \tan \eta$ and expanding the tangent, we have

$$\eta_n = -n\pi\alpha - \alpha\eta_n - \frac{1}{3}\eta_n^3 + O(\alpha^4). \tag{2.312}$$

We solve this equation by successive iterations, to obtain

# RESONANCE THEORY

$$\eta_n^{(0)} = 0, \tag{2.313a}$$

$$\eta_n^{(1)} = -n\pi\alpha, \tag{2.313b}$$

$$\eta_n^{(2)} = n\pi(-\alpha + \alpha^2), \tag{2.313c}$$

$$\eta_n^{(3)} = n\pi(-\alpha + \alpha^2 - \alpha^3) + \frac{1}{3}(n\pi\alpha)^3, \tag{2.313d}$$

etc., so that

$$k_n^\delta a = n\pi\left[1 - \alpha + \alpha^2 - \alpha^3 + \frac{1}{3}(n\pi)^2\alpha^3\right] + O(\alpha^4). \tag{2.314}$$

In the following exercise the values $k_n^\tau$ of the momentum $k$ for which *the time delay is a maximum* are found up to third order in $\alpha$ (eqn (2.320)). Retaining only the first-order correction, we have

$$k_n^\tau a = n\pi(1 - \alpha) + O(\alpha)^2. \tag{2.315}$$

**Exercise 2.17** For the problem of an impenetrable wall plus a very reflecting delta potential barrier shown in Fig. 2.11, find, to $O(\alpha^3)$, the values $k_n^\tau$ of the momentum $k$ for which the time delay is a maximum.

From eqns (2.302) and (2.304) we find the time delay to be

$$\tau(ka) = \frac{2ma^2}{\hbar}\alpha \frac{1 - \cos^2 ka + \alpha(ka)^2}{ka\left[2\alpha ka \cos ka \sin ka + 1 - \cos^2 ka + \alpha^2(ka)^2\right]}. \tag{2.316}$$

The condition for a maximum is found from $\partial\tau(ka)/\partial(ka) = 0$. Writing the solutions $k_n^\tau a$ as

$$k_n^\tau a = n\pi + \epsilon_n, \tag{2.317}$$

and expanding in powers of $\epsilon$, we find

$$\epsilon_n = -n\pi\alpha - 3\alpha\epsilon_n - \frac{n\pi}{2}\alpha^2 - \frac{3}{2n\pi}\epsilon_n^2 - 2\alpha^2\epsilon_n \\ + \left(2n\pi - \frac{9}{2n\pi}\right)\alpha\epsilon_n^2 + \left[\frac{2}{3} - \frac{2}{(n\pi)^2}\right]\epsilon_n^3 + \cdots. \tag{2.318}$$

We solve this equation by successive iterations, to find

$$\epsilon_n^{(0)} = 0, \tag{2.319a}$$

$$\epsilon_n^{(1)} = -n\pi\alpha, \tag{2.319b}$$

$$\epsilon_n^{(2)} = n\pi\left(-\alpha + \alpha^2\right), \tag{2.319c}$$

$$\epsilon_n^{(3)} = n\pi\left(-\alpha + \alpha^2 - \frac{1}{2}\alpha^3\right) + \frac{4}{3}(n\pi\alpha)^3, \tag{2.319d}$$

etc., so that

$$k_n^r a = n\pi \left(1 - \alpha + \alpha^2 - \frac{\alpha^3}{2}\right) + \frac{4}{3}(n\pi\alpha)^3 + O(\alpha^4). \tag{2.320}$$

Thus, *resonances appear at energies slightly below the bound states that would occur should the delta potential be an impenetrable barrier.* This correction increases with $n$, as is also apparent from Figs 2.13 and 2.14.

### 2.2.3  Behavior of the wave function

We study the behavior of the wave function in the strong-potential limit discussed above. The detailed analysis is performed in Exercise 2.18, and so only the results are summarized here.

Figure 2.15 shows that for $ka = \pi/2$ the amplitude of the internal wave function is smaller than the amplitude of the external one—we are *off-resonance*. The situation is very different when $ka$ is in the vicinity of $\pi$. For $ka = \pi(1-2\alpha)$ we see, in Fig. 2.16, that the internal and the external amplitudes are equal. We are approaching a *resonance*. For $ka = \pi(1-\alpha)$, i.e., *close to a resonance* according to eqn (2.310), *the internal amplitude is larger than the external one* (Fig. 2.17). Notice also that in this case the wave function emerges from the potential, i.e., at $x = a^+$, almost with zero slope. At $ka = k_1^\delta a$, i.e., when $\delta = \pi/2$, *the wave function emerges from the potential with exactly zero slope.* For $ka = \pi$ the wave function has a node precisely at $x = a$ and it does not feel

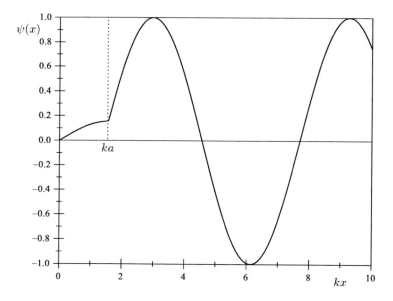

FIG. 2.15. Behavior of the wave function for the delta potential of Fig. 2.11, for $ka = \pi/2$ and $u_0 a = 10$. Notice that the wave function in the internal region is smaller than the one outside. We are *off-resonance*.

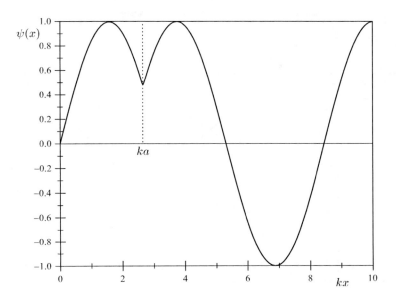

FIG. 2.16. The wave function for the same system as in Fig. 2.15, for $ka = \pi(1 - 2\alpha)$. Notice that the amplitude in the internal and the external regions are equal. We are approaching a *resonance*.

the potential; the behavior is indicated in Fig. 2.18. Finally, at $ka = 3\pi/2$ we have the situation indicated in Fig. 2.19, where, again, we are off-resonance and the internal wave function is smaller than the external one.

**Exercise 2.18** Analyze, in the region of the first resonance, the behavior of the wave function for the problem of an infinite wall and a highly reflecting delta potential barrier illustrated in Fig. 2.11.

The behavior of the wave function at various energies can be better understood if we calculate its value at $x = a$ and its slope on the two sides of $x = a$. From eqn (2.301c), the value of the wave function at $x = a$ is given by

$$\psi(a) = \sin \delta, \tag{2.321}$$

with $\tan \delta$ being given in eqn (2.302). The slope of the wave function just to the right of $x = a$ is

$$\psi'(a^+) = k \cos \delta,$$
$$\left(\frac{\partial \psi}{\partial (kx)}\right)_{a^+} = \cos \delta, \tag{2.322}$$

and the slope just to the left (see eqn (2.293)) is

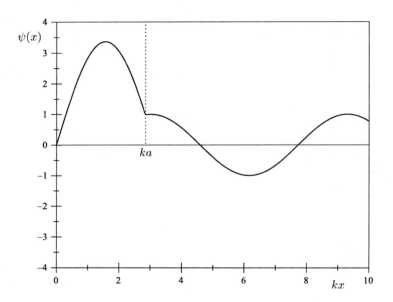

FIG. 2.17. The wave function for the same system as in Fig. 2.15, for $ka = \pi(1 - \alpha)$, i.e., almost at resonance. Notice that *the internal amplitude is larger than the external one.* Exactly at $k = k_1^\delta$, when $\delta = \pi/2$, the wave function emerges from the potential, i.e., at $x = a^+$, with zero slope.

$$\psi'(a^-) = \psi'(a^+) - u_0\psi(a),$$
$$\left(\frac{\partial \psi}{\partial(kx)}\right)_{a^-} = \cos\delta - \frac{\sin\delta}{\alpha k a}. \quad (2.323)$$

We now analyze $\psi(a)$, $(\partial\psi/\partial(kx))_{a^+}$ and $(\partial\psi/\partial(kx))_{a^-}$ at various energies, assuming a strong barrier, as defined in eqns (2.307).

1. Let $ka = \pi/2$. We find

$$\tan\delta = \alpha ka = \frac{\pi}{2}\alpha \ll 1, \quad (2.324a)$$
$$\sin\delta \approx \frac{\pi}{2}\alpha \ll 1, \quad (2.324b)$$
$$\cos\delta \approx 1. \quad (2.324c)$$

Hence

$$\psi(a) \approx \frac{\pi}{2}\alpha \ll 1, \quad (2.325a)$$
$$\left(\frac{\partial\psi}{\partial(kx)}\right)_{a^-} = 0, \quad (2.325b)$$
$$\left(\frac{\partial\psi}{\partial(kx)}\right)_{a^+} \approx 1. \quad (2.325c)$$

The actual behavior of the wave function is shown in Fig. 2.15.

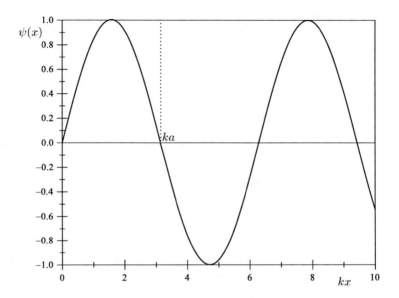

FIG. 2.18. The wave function for the same system as in Fig. 2.15, for $ka = \pi$. The wave function has a node precisely at $x = a$, so that it behaves as if the delta potential were not there.

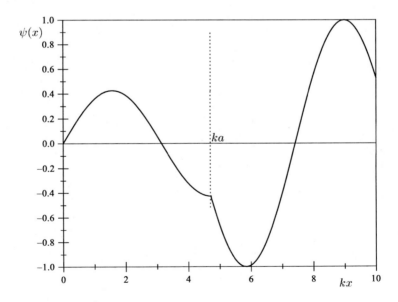

FIG. 2.19. The wave function for the same system as in Fig. 2.15, for $ka = 3\pi/2$. We are again off-resonance and the internal amplitude is smaller than the external one.

2. Let $ka \approx \pi$. We write
$$ka = \pi + \eta,$$
with $\eta \ll 1$. We find
$$\tan \delta = \alpha ka \frac{\tan \eta}{\alpha ka + \tan \eta}. \qquad (2.326)$$

(a) For $\eta = -2\pi\alpha$, we have
$$\tan \delta \approx \pi \frac{-2\pi\alpha}{\pi - 2\pi} = 2\pi\alpha \ll 1, \qquad (2.327a)$$
$$\sin \delta \approx 2\pi\alpha \ll 1, \qquad (2.327b)$$
$$\cos \delta \approx 1. \qquad (2.327c)$$

Then
$$\psi(a) \approx 2\pi\alpha \ll 1, \qquad (2.328a)$$
$$\left(\frac{\partial \psi}{\partial (kx)}\right)_{a^-} \approx 1 - 2 = -1, \qquad (2.328b)$$
$$\left(\frac{\partial \psi}{\partial (kx)}\right)_{a^+} \approx 1. \qquad (2.328c)$$

The wave function is shown in Fig. 2.16.

(b) For $\eta = -\pi\alpha$, we have
$$\alpha ka + \tan \eta = \alpha\pi(1-\alpha) - \pi\alpha - \frac{(\pi\alpha)^3}{3} + \cdots$$
$$= -\pi\alpha^2 - \frac{(\pi\alpha)^3}{3} + \cdots,$$

and so
$$\tan \delta = \pi(1-\alpha)\frac{1 + (\pi\alpha)^2/3 + \cdots}{1 + (\pi^2\alpha)/3 + \cdots} = \pi + O(\alpha), \qquad (2.329a)$$
$$\sin \delta \approx \frac{\pi}{\sqrt{1+\pi^2}}, \qquad (2.329b)$$
$$\cos \delta \approx \frac{1}{\sqrt{1+\pi^2}}. \qquad (2.329c)$$

Thus
$$\psi(a) \approx \frac{\pi}{\sqrt{1+\pi^2}}, \qquad (2.330a)$$
$$\left(\frac{\partial \psi}{\partial (kx)}\right)_{a^-} \approx -\frac{1}{\alpha\sqrt{1+\pi^2}}, \qquad (2.330b)$$
$$\left(\frac{\partial \psi}{\partial (kx)}\right)_{a^+} \approx \frac{1}{\sqrt{1+\pi^2}}. \qquad (2.330c)$$

The wave function is shown in Fig. 2.17.

(c) For $\eta = -\pi\alpha + \pi\alpha^2$ (see eqn (2.314); $ka$ is still smaller than $k_1^\delta a$), we have

$$\alpha ka + \tan\eta = \alpha\pi\left(1 - \alpha + \alpha^2\right) + \left(-\pi\alpha + \pi\alpha^2\right) + \frac{1}{3}\left[-(\pi\alpha)^3 + \cdots\right]$$
$$= \pi\left(1 - \frac{\pi^2}{3}\right)\alpha^3 + \cdots,$$

and so

$$\tan\delta = \pi\left(1 - \alpha + \alpha^2\right)\frac{1 - \alpha + \cdots}{(\pi^2/3 - 1)\alpha + \cdots}$$
$$\approx \frac{\pi}{\pi^2/3 - 1}\frac{1}{\alpha} + \cdots \gg 1, \qquad (2.331\text{a})$$
$$\sin\delta \approx 1, \qquad (2.331\text{b})$$
$$\cos\delta = \left[1 + \left(\frac{\pi}{\pi^2/3 - 1}\frac{1}{\alpha} + \cdots\right)^2\right]^{-1/2}$$
$$\approx \frac{\pi^2/3 - 1}{\pi}\alpha + \cdots \ll 1. \qquad (2.331\text{c})$$

Thus

$$\psi(a) \approx 1, \qquad (2.332\text{a})$$
$$\left(\frac{\partial\psi}{\partial(kx)}\right)_{a-} \approx -\frac{1}{\pi\alpha}, \qquad (2.332\text{b})$$
$$\left(\frac{\partial\psi}{\partial(kx)}\right)_{a+} \approx \frac{\pi^2/3 - 1}{\pi}\alpha \ll 1. \qquad (2.332\text{c})$$

(d) Exactly at the pole of $\tan\delta$, we have

$$\sin\delta = 1, \quad \cos\delta = 0, \qquad (2.333)$$

and so

$$\psi(a) = 1, \qquad (2.334\text{a})$$
$$\left(\frac{\partial\psi}{\partial(kx)}\right)_{a-} = 0 - \frac{1}{\alpha ka} \approx -\frac{1}{\pi\alpha}, \qquad (2.334\text{b})$$
$$\left(\frac{\partial\psi}{\partial(kx)}\right)_{a+} = 0, \qquad (2.334\text{c})$$

and the wave function 'emerges' from the delta potential with exactly zero slope.

(e) For $\eta = 0$, we have

$$\sin\delta = 0, \quad \cos\delta = -1, \qquad (2.335)$$

and so

$$\psi(a) = 0, \qquad (2.336a)$$

$$\left(\frac{\partial \psi}{\partial (kx)}\right)_{a^-} = -1, \qquad (2.336b)$$

$$\left(\frac{\partial \psi}{\partial (kx)}\right)_{a^+} = -1. \qquad (2.336c)$$

The wave function is shown in Fig. 2.18.

3. For $ka = 3\pi/2$, we have

$$\tan \delta = \frac{3\pi}{2}\alpha \ll 1, \qquad (2.337a)$$

$$\sin \delta \approx -\frac{3\pi}{2}\alpha, \qquad (2.337b)$$

$$\cos \delta \approx -1, \qquad (2.337c)$$

and so

$$\psi(a) \approx -\frac{3\pi}{2}\alpha, \qquad (2.338a)$$

$$\left(\frac{\partial \psi}{\partial (kx)}\right)_{a^-} \approx -1 + 1 = 0, \qquad (2.338b)$$

$$\left(\frac{\partial \psi}{\partial (kx)}\right)_{a^+} \approx -1. \qquad (2.338c)$$

The wave function is shown in Fig. 2.19.

### 2.2.4 Analytical study of the internal amplitude of the wave function near resonance

The analysis of the behavior of the wave function presented above is now complemented by studying, in the vicinity of a resonance, the amplitude $B$ of the internal wave function, eqn (2.301b). The following exercise finds, up to third order in $\alpha$, the result given in eqn (2.344) for the values $k_n^B$ of $k$ at which $|B|^2$ is a maximum. Including only the first-order correction, the result is

$$k_n^B a = n\pi(1 - \alpha) + O(\alpha^2). \qquad (2.339)$$

**Exercise 2.19** For the problem of the infinite wall and a highly reflecting delta potential barrier, find, to $O(\alpha^3)$, the values $k_n^B$ of the momentum $k$ for which the square of the amplitude, $|B|^2$, of the internal wave function is a maximum.

Differentiating $|B|^2$ of eqn (2.347) with respect to $ka$, we find the condition for a maximum to be

$$\left(\tan k_n^B a - k_n^B a\right)\left(\tan k_n^B a + \alpha k_n^B a\right) = -\alpha \left(k_n^B a\right)^2 \tan^2 k_n^B a. \qquad (2.340)$$

We set
$$k_n^B a = n\pi + \zeta_n, \tag{2.341}$$
with $\zeta_n = O(\alpha)$, and expand $\tan \zeta_n$ in powers of $\zeta_n$ to get
$$\zeta_n = -n\pi\alpha - \alpha\zeta_n - \frac{1}{3}\zeta_n^3 + n\pi\alpha\zeta_n^2 + O(\alpha^4), \tag{2.342}$$
which can be solved iteratively to give
$$\zeta_n^{(0)} = 0, \tag{2.343a}$$
$$\zeta_n^{(1)} = -n\pi\alpha, \tag{2.343b}$$
$$\zeta_n^{(2)} = -n\pi\alpha + n\pi\alpha^2, \tag{2.343c}$$
$$\zeta_n^{(3)} = n\pi\left(-\alpha + \alpha^2 - \alpha^3\right) + \frac{4}{3}(n\pi\alpha)^3. \tag{2.343d}$$
For $k_n^B a$ we finally obtain
$$k_n^B a = n\pi(1 - \alpha + \alpha^2 - \alpha^3) + \frac{4}{3}(n\pi\alpha)^3 + O(\alpha^4). \tag{2.344}$$

Exercise 2.20 shows that, in the vicinity of its maximum, $|B|^2$ has the behavior
$$|B|^2 \approx \frac{\Gamma_n^{(k)} a/2}{(ka - k_n^\delta a)^2 + \left(\Gamma_n^{(k)} a/2\right)^2}, \tag{2.345}$$
where we have introduced the *half-width at half-maximum*
$$\Gamma_n^{(k)} \frac{a}{2} = (n\pi\alpha)^2. \tag{2.346}$$
We notice that the width of the resonance increases with energy, i.e., with $n$, indicating that the delta potential becomes more transparent at higher energies.

**Exercise 2.20** Find the magnitude of the internal wave function in the vicinity of a resonance, for the problem of an impenetrable wall and a highly reflecting delta potential barrier illustrated in Fig. 2.11.

From eqn (2.301b), we can write
$$|B|^2 = \frac{\alpha^2 (ka)^2}{(\sin ka + \alpha ka \cos ka)^2 + (\alpha ka \sin ka)^2}. \tag{2.347}$$
It is convenient to define the quantities
$$f(ka) = \sin ka + \alpha ka \cos ka, \tag{2.348}$$
$$g(ka) = \alpha ka \sin ka. \tag{2.349}$$

We should study $|B|^2$ in the vicinity of one of its maxima $k_n^B a$, eqn (2.344). Instead, we find it easier to study $|B|^2$ in the vicinity of the points $k_n^\delta a$, eqn (2.314) (where $\delta = (2n+1)\pi/2$), which agree with $k_n^B a$ up to order $\alpha^2$.

We find, from eqn (2.309),
$$f(k_n^\delta a) = 0, \tag{2.350}$$
so that, close enough to a resonance, we have
$$f(ka) = 0 + f'(k_n^\delta a)\left(ka - k_n^\delta a\right) + \cdots. \tag{2.351}$$

We calculate the derivative $f'(k_n^\delta a)$ (the prime indicating differentiation with respect to the argument $ka$, evaluated at $k_n^\delta a$) to be

$$\begin{aligned} f'(k_n^\delta a) &= (1+\alpha)\cos k_n^\delta a - \alpha k_n^\delta a \sin k_n^\delta a \\ &= \left[1 + \alpha + \left(\alpha k_n^\delta a\right)^2\right]\cos k_n^\delta a, \end{aligned} \tag{2.352}$$

where we have used eqn (2.309). Now using eqn (2.310) for the approximate value of the quantity $k_n^\delta a$, we have

$$f'(k_n^\delta a) = (-)^n \left[1 + \alpha + O(\alpha)^2\right]. \tag{2.353}$$

The Taylor expansion of $f(ka)$ around $k_n^\delta a$ is then, according to eqns (2.351) and (2.353),
$$f(ka) = (-)^n \left[1 + \alpha + O(\alpha^2)\right]\left(ka - k_n^\delta a\right) + \cdots. \tag{2.354}$$

Then, finally,
$$[f(ka)]^2 = \left[1 + 2\alpha + O(\alpha^2)\right]\left(ka - k_n^\delta a\right)^2 + \cdots. \tag{2.355}$$

A similar expansion can be carried out for $g(ka)$, eqn (2.349), i.e.,
$$g(ka) = g(k_n^\delta a) + g'(k_n^\delta a)\left(ka - k_n^\delta a\right) + \cdots. \tag{2.356}$$

From eqn (2.309), we write

$$\begin{aligned} g(k_n^\delta a) &= -\left(\alpha k_n^\delta a\right)^2 \cos k_n^\delta a \\ &\approx (-)^{n+1}(n\pi\alpha)^2 + O(\alpha^3), \\ g'(k_n^\delta a) &= \alpha \sin k_n^\delta a + \alpha k_n^\delta a \cos k_n^\delta a \\ &= \alpha k_n^\delta a\,(1-\alpha)\cos k_n^\delta a \\ &\approx (-)^n (n\pi\alpha) + O(\alpha^2), \end{aligned} \tag{2.357}$$
$$\tag{2.358}$$

so that, from eqn (2.356), $g(ka)$ and $[g(ka)]^2$ take the form

$$g(ka) = (-)^n \left\{-(n\pi\alpha)^2 + O(\alpha^3) + \left[(n\pi\alpha) + O(\alpha^2)\right](ka - k_n^\delta a)\right\}, \tag{2.359a}$$

$$[g(ka)]^2 = (n\pi\alpha)^2 \left\{(n\pi\alpha) + O(\alpha^2) - [1 + O(\alpha)](ka - k_n^\delta a)\right\}^2. \tag{2.359b}$$

Suppose now that we are interested in a neighborhood of order $\alpha^2$ around $k_n^\delta a$, namely

$$\left|ka - k_n^\delta a\right| \lesssim O\left(\alpha^2\right) \ll 1; \qquad (2.360)$$

this neighborhood will turn out to be a *fixed* (i.e., independent of $\alpha$) multiple of the level width, eqn (2.346). Then, eqns (2.355) and (2.359b) give

$$[f(ka)]^2 \approx \left(ka - k_n^\delta a\right)^2 = O\left(\alpha^4\right), \qquad (2.361)$$

$$[g(ka)]^2 \approx (n\pi\alpha)^4 = O\left(\alpha^4\right), \qquad (2.362)$$

to lowest order in $\alpha$, so that the two terms in the denominator of eqn (2.347) are of the same order of magnitude. We then obtain, for $|B|^2$, eqn (2.345) in the text.

Expression (2.345) shows that the intensity of the wave function in the internal region $0 < x < a$ has, near a resonance, a *Breit–Wigner form* centered at $k \approx k_n^\delta$, with a half-width at half-maximum $\Gamma_n^{(k)}/2$. At $k = k_n^\delta$, we have

$$\left|B_{\text{res}}\right|^2 \approx \frac{2}{\Gamma_n^{(k)} a} = \frac{1}{(n\pi\alpha)^2} \gg 1, \qquad (2.363)$$

which is to be compared with Fig. 2.17. At $ka = \pi$ the exact expression (2.301b) gives

$$|B|^2 = 1 \ll |B_{\text{res}}|^2, \qquad (2.364)$$

which is to be compared with Fig. 2.18.

### 2.2.5 *The analytic structure of $S(k)$ in the complex-momentum plane*

We first look for the *poles* of $S(k)$ in the complex-$k$ plane, i.e., for the zeros $k_n$ of the denominator of $S$, eqn (2.296), which satisfy

$$\sin k_n a + \alpha k_n a \cos k_n a = i\alpha k_n a \sin k_n a, \qquad (2.365)$$

with $\alpha$ defined in eqn (2.307c). For every solution $k_n$ of this equation there is a *zero* of the numerator of eqn (2.296), precisely at *the complex conjugate position* $k_n^*$; this actually ensures unitarity. Taking the complex conjugate of eqn (2.365), we also find that for every pole $k_n$ there is a pole at $-k_n^*$, i.e., at the reflected position with respect to the imaginary axis.

As a check, we see from eqn (2.365) that, for an impenetrable barrier ($u_0 \to \infty$), we have

$$k_n a \to n\pi = k_n^{\text{BS}} a, \qquad (2.366)$$

which are precisely the bound states of a one-dimensional box of length $a$. This does not mean that $S$ has poles *on* the real axis, since unitarity would be violated. In fact, every pole is cancelled by the complex conjugate zero, which now coincides with the pole.

We rewrite eqn (2.365) as

$$(1 - i\alpha k_n a)\tan k_n a = -\alpha k_n a. \qquad (2.367)$$

We investigate the situation when the barrier is not impenetrable, but still very strong. We then expect the poles of $S$ to differ only slightly from the values given by eqn (2.366). In fact, Exercise 2.21 below finds, for $k_n a$, the result (2.377), up to third order in $\alpha$. Including only the leading correction in $\alpha$ in the real and the imaginary parts of the pole, eqn (2.377) gives

$$k_n a = n\pi(1 - \alpha + \cdots) - i(n\pi)^2 (\alpha^2 + \cdots). \tag{2.368}$$

We verify explicitly from this expression that for every resonance pole associated with positive $n$ there is another pole with negative $n$, the two poles being reflected images of each other with respect to the imaginary axis. The leading correction to the real part of the pole $k_n$ is first order in $\alpha$ and coincides with that for $k_n^\delta$, eqn (2.310) (in fact, the coincidence occurs up to second order in $\alpha$, as seen from eqns (2.377) and (2.314)). The imaginary part is negative and, again to leading order, is quadratic in $\alpha$ and coincides precisely with $-\Gamma_n^{(k)} a/2$, the half-width at half-maximum defined earlier, in eqn (2.346). Thus the poles of the $S$ matrix have the form

$$k_n = k_n' + i k_n'' \approx k_n^\delta - \frac{i\Gamma_n^{(k)}}{2}. \tag{2.369}$$

As $n$ increases, the poles of $S$ move further away from the real axis, again indicating the better transparency of the barrier for higher energies. The poles of $S(k)$ given by eqn (2.369) are referred to as *resonance poles*.

**Exercise 2.21** Find, to $O(\alpha^3)$, the poles $k_n$ of $S(k)$ in the complex-momentum plane for the problem of an infinite wall and a highly reflecting delta potential barrier illustrated in Fig. 2.11.

Since we expect the poles of $S$ to differ only slightly from the bound-state values given by eqn (2.366), we write

$$k_n a = n\pi + \xi_n, \tag{2.370}$$

with $\xi = O(\alpha)$. The analysis of eqn (2.367) is then similar to the one performed above for eqn (2.309). Substituting eqn (2.370) into (2.367) and expanding the tangent, we have

$$[1 - i\alpha(n\pi + \xi_n)]\left[\xi_n + \frac{1}{3}\xi_n^3 + O(\alpha^4)\right] = -\alpha(n\pi + \xi_n), \tag{2.371}$$

or

$$\xi_n = -n\pi\alpha - (1 - in\pi)\alpha\xi_n + \left(-\frac{1}{3}\xi_n^3 + i\alpha\xi_n^2\right) + O(\alpha^4). \tag{2.372}$$

We can solve this equation iteratively, by successive approximations. To zeroth order we have

$$\xi_n^{(0)} = 0; \tag{2.373}$$

to first order in $\alpha$ we have

$$\xi_n^{(1)} = -n\pi\alpha; \tag{2.374}$$

to second order we have

$$\xi_n^{(2)} = -n\pi\alpha + n\pi(1 - in\pi)\alpha^2; \tag{2.375}$$

and to third order we have

$$\xi_n^{(3)} = n\pi\left(-\alpha + \alpha^2 - \alpha^3\right) + \frac{4}{3}(n\pi\alpha)^3 + i[-(n\pi\alpha)^2 + 3(n\pi)^2\alpha^3]. \tag{2.376}$$

Finally, from (2.370), we have

$$k_n a = n\pi\left[1 - \alpha + \alpha^2 - \alpha^3 + \frac{4}{3}(n\pi)^2\alpha^3 + O(\alpha^4)\right] - i(n\pi)^2[\alpha^2 - 3\alpha^3 + O(\alpha^4)]. \tag{2.377}$$

This is a good place to summarize the various physical phenomena that signal the occurrence of a resonance, as we have seen in the model problem of an impenetrable wall and a strong delta potential barrier discussed here and in the previous sections. Equation (2.344) gives the values of the momentum $k$, denoted by $k_n^B$, at which *the intensity of the internal wave function is a maximum*, and eqn (2.320) gives those values, $k_n^\tau$, at which *the time delay is a maximum*. The two sets of $k$s are not identical, but are very close for small $\alpha$ (in fact, they coincide up to order $\alpha^2$), i.e., when the delta potential barrier is highly reflecting and hence the resonances are very narrow. In the present example, these values of $k$ occur very near to the $k_n^\delta$ given in eqn (2.314), for which *the phase shift takes on the values* $\pi/2, 3\pi/2, \ldots$ and *the wave function emerges from the delta potential with zero slope*. The $k_n^\delta$s, in turn, are also close (again to $O(\alpha^2)$) to *the real part of the resonance poles $k_n$ of the $S$ matrix in the complex-momentum plane* given by eqn (2.377). It is remarkable that the various values of $k$ described above are so close to each other for small $\alpha$, at least within the model we have analyzed. For a more transparent barrier—giving rise to wider resonances—or presumably for a more general potential, such a close coincidence deteriorates.

The analytic structure of $S(k)$ in the complex-$ka$ plane is illustrated schematically in Fig. 2.20. We observe that the $S$ matrix is *analytic* in the *upper half of the complex-ka plane*. For a more general potential one may also find poles *on the positive imaginary axis*, corresponding to the *bound states* of the problem [123], which are not there in the present example. In the *lower half of the complex-ka plane* the $S$ matrix has the poles that we calculated above, corresponding to *resonances*. In a more general case one may find poles *on the negative imaginary axis*, corresponding to what are called *virtual states* [123]. Also indicated in Fig. 2.20 are the pairs of poles at reflected positions with respect to the imaginary axis, as well as the zero-pole pairs at complex conjugate positions that ensure unitarity.

In a more general case, if the pole closest to the origin in the complex-$k$ plane is that associated with a bound state at $k = ik_0$, or with a virtual state at $k = -ik_0$, with $k_0 > 0$, we can approximate $S$ for $k \approx 0$ by

$$S \approx -\frac{k + ik_0}{k - ik_0}, \tag{2.378a}$$

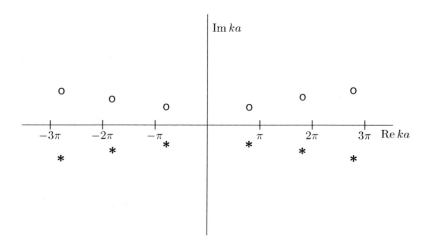

FIG. 2.20. Analytic structure (schematic) of $S(k)$ in the complex-$ka$ plane for the problem illustrated in Fig. 2.11. The only poles occurring for this potential are due to resonances.

or
$$S \approx -\frac{k - ik_0}{k + ik_0}, \tag{2.378b}$$

respectively. Notice that in the first case the phase shift $\delta$ decreases with increasing $k$, with the opposite behavior for a virtual state. The one-dimensional delta potential model whose $2 \times 2$ $S$ matrix is given in eqn (2.218) admits a bound state for an attractive potential, $u_0 < 0$, while it gives rise to a virtual state for a repulsive one, $u_0 > 0$. The determinant of that $S$ matrix has precisely the behavior indicated above, in eqns (2.378).

Finally, we would like to remark that a complete set of functions in the Hilbert space comprises only the negative energy bound states and the positive energy scattering states. The latter, for instance, do not include resonances or virtual states. Nevertheless, these non-normalizable states are physically important, as we discussed earlier.

2.2.6 *Analytic structure of $S(E)$ in the complex-energy plane*

Here we consider the $S$ matrix as a function of the energy variable $E$, and its analytic continuation in the complex-$E$ plane. In some of the equations we shall find it convenient to use the dimensionless variable $e$, which differs from $E$ by a factor, so that
$$E = \frac{\hbar^2 e^2}{2ma^2}, \quad e = (ka)^2. \tag{2.379}$$

We can write the $S$ matrix of eqn (2.296) as

$$S(E) = -\frac{(\sin\sqrt{e} + \alpha\sqrt{e}\cos\sqrt{e}) + i\alpha\sqrt{e}\sin\sqrt{e}}{(\sin\sqrt{e} + \alpha\sqrt{e}\cos\sqrt{e}) - i\alpha\sqrt{e}\sin\sqrt{e}}. \quad (2.380)$$

Notice that we will use $e$ and $E$ interchangeably. We first observe that $S(E)$ has a *branch point* at $E = 0$ and that we need two Riemann sheets for its analytic description. For convenience, we choose the cut along the positive real-$E$-axis and obtain the analytic structure shown in Fig. 2.21. The edges of the cut that have to be identified have been labeled by the same letter. The zeros in the upper half of the complex-$k$ plane appear in the first $E$-sheet, or $E^{(1)}$, while the resonance poles appear in the second $E$-sheet, or $E^{(2)}$. Should the problem admit bound states, they would appear on the positive imaginary axis of the complex-$k$ plane, and thus on the real negative axis of $E^{(1)}$, also called the 'physical sheet'. If we consider the analytic continuation of $S(E)$ from the first quadrant of $E^{(1)}$ down to the fourth quadrant of $E^{(2)}$ across the edge A of the cut in Fig. 2.21, then we obtain the structure shown in Fig. 2.22 on the right of the imaginary axis. In that region, $S(E)$ can be considered to be a meromorphic function if we exclude the origin; it has poles in the lower half-plane and zeros at the complex conjugate positions. From (2.368), we obtain, keeping only the leading correction to the real and imaginary parts, the following complex poles of $S(E)$ in the variable $e$:

$$(k_n a)^2 = (1 - 2\alpha + \cdots)(n\pi)^2 - 2i(n\pi)^3\alpha^2 + \cdots, \quad (2.381)$$

and, in actual energy units, we have

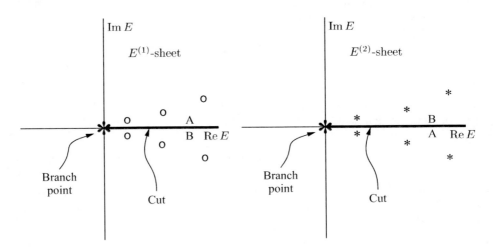

FIG. 2.21. Analytic structure of $S(E)$, eqn (2.380), in the complex-energy plane for the problem illustrated in Fig. 2.11. We have a branch point at the origin and we need two Riemann sheets, $E^{(1)}$ and $E^{(2)}$, for its analytic description. The only poles occurring for this potential are due to resonances; they appear in the second energy sheet.

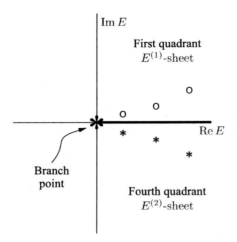

FIG. 2.22. Analytic continuation of $S(E)$ from the first quadrant of $E^{(1)}$ to the fourth quadrant of $E^{(2)}$. We have poles in the lower half-plane and zeros at the complex conjugate positions.

$$\mathcal{E}_n = \frac{\hbar^2 k_n^2}{2m} \equiv E_n - \frac{\mathrm{i}\Gamma_n}{2}. \tag{2.382}$$

The real part $E_n$ of the pole is seen to coincide, to leading order, with $E_n^\delta = \hbar^2(k_n^\delta)^2/2m$ ($k_n^\delta$ being given by eqn (2.310)), i.e.,

$$E_n \approx E_n^\delta = \frac{\hbar^2}{2ma^2}(1 - 2\alpha)(n\pi)^2, \tag{2.383}$$

which is the energy at which the phase shift $\delta$ is an odd multiple of $\pi/2$, and the wave function emerges from the delta potential with zero slope. The imaginary part of the pole has been written as $-\Gamma_n/2$, with

$$\frac{\Gamma_n}{2} = 2\frac{\hbar^2}{2ma^2}(n\pi)^3\alpha^2 + \cdots. \tag{2.384}$$

In Exercise 2.22 we find the structure of $S$ as a function of energy, near a resonance. In terms of the actual energy $E$, eqn (2.392) below provides the *one-pole approximation*

$$S(E) \approx -\frac{E - E_n - \mathrm{i}\Gamma_n/2}{E - E_n + \mathrm{i}\Gamma_n/2}, \tag{2.385}$$

where $E_n$ and $\Gamma_n/2$ are the quantities given above.

---

**Exercise 2.22** Find the structure of $S(E)$ near a resonance energy for the problem of an infinite wall and a highly reflecting delta potential barrier illustrated in Fig. 2.11.

Using eqns (2.348) and (2.349), we write $S(E)$ of eqn (2.380) as

$$S(E) = -\frac{f(e) + ig(e)}{f(e) - ig(e)}. \qquad (2.386)$$

We now use the expansions (2.354) and (2.359a) to write

$$f(e) - ig(e) \approx (-)^n \left(\sqrt{e} - \sqrt{e_n^\delta}\right)[1 + O(\alpha)] + i\left[(-)^n(n\pi\alpha)^2 + O(\alpha^3)\right], \qquad (2.387)$$

where, from eqn (2.310), we have

$$e_n^\delta = (k_n^\delta a)^2 \approx [n\pi(1 - \alpha)]^2 \approx (1 - 2\alpha)(n\pi)^2. \qquad (2.388)$$

We choose $\sqrt{e} - \sqrt{e_n^\delta} \lesssim O(\alpha^2)$, as in eqn (2.360). Hence both $f(e)$ and $g(e)$ are $O(\alpha^2)$. Thus $S(E)$ becomes

$$S(E) \approx -\frac{\sqrt{e} - \sqrt{e_n^\delta} - i(n\pi\alpha)^2}{\sqrt{e} - \sqrt{e_n^\delta} + i(n\pi\alpha)^2}. \qquad (2.389)$$

We approximate $\sqrt{e} - \sqrt{e_n^\delta}$ by

$$\sqrt{e} - \sqrt{e_n^\delta} = \frac{\sqrt{e} - \sqrt{e_n^\delta}}{\sqrt{e} + \sqrt{e_n^\delta}}\left(\sqrt{e} + \sqrt{e_n^\delta}\right) \approx \frac{e - e_n^\delta}{2\sqrt{e_n^\delta}}, \qquad (2.390)$$

so that

$$S(E) \approx -\frac{e - e_n^\delta - 2i(n\pi)^3\alpha^2}{e - e_n^\delta + 2i(n\pi)^3\alpha^2}, \qquad (2.391)$$

where use has been made of eqn (2.388).

In terms of the actual energy $E$, we then write the last equation as (using eqns (2.383) and (2.384))

$$S(E) \approx -\frac{E - E_n - i\Gamma_n/2}{E - E_n + i\Gamma_n/2}. \qquad (2.392)$$

### 2.2.7 The R-matrix theory of scattering

Just as in the last discussion, we again consider scattering in a semi-infinite one-dimensional space and take advantage of the relative simplicity of this geometry to illustrate the Wigner–Eisenbud $R$-matrix theory of scattering [177]. This theory, which was originally introduced to study nuclear reactions, will be found to be very important in many of our future discussions.

When Bohr proposed the 'compound nucleus' mechanism [33, 34], there was no mathematical framework in quantum mechanics for its description. Breit and Wigner [38, 39] treated nuclear reactions in perturbation theory and obtained the famous formula that bears their names.

The unsatisfactory application of perturbation theory in the nuclear domain was overcome by Kapur and Peierls [101, 149] in 1938, who presented a rigorous theory of reactions. Its essential feature is the introduction of a complete set of formal *discrete* states—further identified with the states of the compound nucleus—defined in a volume of about the nuclear size by imposing, on the surface of such a volume, boundary conditions describing outgoing waves at the incident energy of the scattering problem. These states form a complete set and the boundary conditions defining them are *complex* and *energy dependent*. As a result, almost all of the quantities occurring in the Kapur–Peierls formalism are energy dependent. The actual scattering wave function inside the nucleus is expanded in terms of this complete set of states and is then connected to the outside world in order to find the $S$ matrix and cross-sections.

Since the appearance of the Kapur–Peierls formalism, Breit [36, 37], Wigner [172, 173] and co-workers studied the same problem of providing a rigorous theory of nuclear reactions. In the '$R$-matrix theory' of Wigner and Eisenbud [177], a complete set of states in the inner region is again introduced, but now with *energy-independent boundary conditions*. A great advantage of this theory is that the energy dependence of all the quantities is made as explicit as possible. We emphasize that this theory is rigorous and that, in principle, it can describe processes that proceed by the compound nucleus mechanism, as well as processes whose main contribution comes from direct mechanisms, with a rather low probability for the formation of a compound nucleus. These remarks from nuclear physics have been made in anticipation of their relevance to the treatment of chaotic scattering, studied in Chapter 6.

An excellent review on $R$-matrix theory was written by Lane and Thomas in 1958 [112]. The reader is strongly advised to consult this article, which also contains an invaluable list of references.

We now proceed to describe the $R$-matrix theory for the simplified situation of a spinless particle moving in a semi-infinite one-dimensional space, as was mentioned above. Here, *TRI is assumed throughout*. At $x = -a$ there is an impenetrable wall, and for $x > -a$ there is an arbitrary but finite-range potential, confined to $-a < x < 0$, as illustrated in Fig. 2.23. In eqn (2.292b) above, the choice of plane waves was equivalent to having shifted the origin to $x = a$. At this point we implement that choice explicitly, placing $x = 0$ *outside* the potential. The wave function satisfies Schrödinger's equation

$$-\frac{\hbar^2}{2m}\frac{\partial^2 \psi(E;x)}{\partial x^2} + V(x)\psi(E;x) = E\psi(E;x), \quad -a < x < \infty, \quad (2.393a)$$

or

$$\left(\partial_x^2 + k^2\right)\psi(E;x) = U(x)\psi(E;x), \quad (2.393b)$$

with the boundary condition

$$\psi(x = -a) = 0. \quad (2.393c)$$

For $x > 0$ the particle is thus free and its wave function has the structure

$$\psi(E; x > 0) = e^{-ikx} + S(E)e^{ikx}. \quad (2.394)$$

As usual, $E = \hbar^2 k^2/2m$ and $U(x) = 2mV(x)/\hbar^2$.

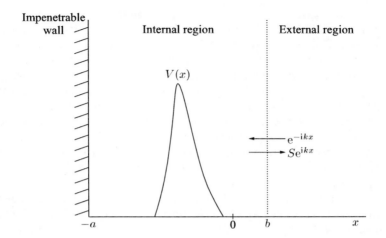

FIG. 2.23. Scattering in a semi-infinite one-dimensional space. There is an impenetrable wall at $x = -a$ and, in addition, an arbitrary but finite-range potential confined to $-a < x < 0$. For $x > 0$ the particle is free to move and its wave function is a linear combination of an incoming and an outgoing plane wave as shown. At $x = b$ we have introduced the fictitious '$R$-matrix boundary', which separates the *internal* from the *external* regions.

Now comes the basic feature of $R$-matrix theory. We first consider an arbitrarily chosen point $x = b \geqslant 0$, to be called the '$R$-matrix boundary', which separates the *internal* $(-a < x < b)$ from the *external* $(x > b)$ regions, as shown in Fig. 2.23. Next, we introduce, *in the internal region* $-a < x < b$, the solutions of the Schrödinger equation

$$-\frac{\hbar^2}{2m}\frac{\partial^2 \psi_\lambda(x)}{\partial x^2} + V(x)\psi_\lambda(x) = E\psi_\lambda(x), \quad -a < x < b, \qquad (2.395a)$$

or

$$\left(\partial_x^2 + K_\lambda^2\right)\psi_\lambda(x) = U(x)\psi_\lambda(x), \qquad (2.395b)$$

with the boundary conditions

$$\psi_\lambda(x = -a) = 0, \qquad (2.395c)$$

$$\left[\frac{\mathrm{d}\psi_\lambda(x)}{\mathrm{d}x}\right]_{x=b} = 0. \qquad (2.395d)$$

These mixed boundary conditions maintain the Hermiticity of the problem in the internal region. The problem so defined in the internal region has the *discrete spectrum*

$$E_\lambda = \frac{\hbar^2 K_\lambda^2}{2m}. \qquad (2.396)$$

The condition of (2.395c) is the conventional one associated with an impenetrable wall at $x = -a$. The boundary condition (2.395d) states that the wave function reaches the '$R$-matrix boundary' $x = b$ with zero slope. In the problem we are describing, there is no degeneracy in the $R$-matrix $\lambda$ spectrum. Similarly, in the scattering problem (2.393) we have only one state at every energy $E$. Thus, for $E = E_\lambda$ the scattering wave function $\psi(E = E_\lambda; x)$ is proportional to $\psi_\lambda(x)$ in the internal region $-a < x < b$ and reaches $x = b$ with zero slope. In the case of the delta potential of Fig. 2.11 studied above, we saw in Fig. 2.17 that, precisely at resonance, the scattering wave function just outside the potential has zero slope. For that case, and if the $R$-matrix boundary $x = b$ is chosen right outside the potential, the $R$-matrix eigenvalues $E_\lambda$ are precisely the resonance energies. For a more general potential, like the one illustrated in Fig. 2.23, we still expect the $E_\lambda$s to have some connection with the resonance energies. However, whatever the physical properties of these states, they serve to parametrize the basic object of the present formulation, the $R$ function, to be described below.

In specific applications in which the wave functions are sufficiently simple, it may be possible to calculate from a suitable model, using eqns (2.395), at least some of the $R$-matrix states and energies, and make predictions for the scattering problem (in the present problem the only scattering quantity is the phase shift; in the situations described in the next chapter we also have the various transmission and reflection coefficients; in three-dimensional problems in infinite space described in most textbooks one has the various cross-sections) which can then be compared with experiments. In more complex situations in which the analytical solution of the $R$-matrix equations is out of the question, one can set up a statistical model for the various $R$-matrix parameters and thereby make statistical predictions for the resulting scattering quantities—this is the route followed by a number of authors. Under these circumstances it has even been possible to construct statistical models directly for the $S$ matrix of the problem, without resorting to the $R$-matrix description; however, we defer the discussion of this point until later chapters.

From an experimental point of view, given the results of a scattering experiment over a certain energy interval, one can think of extracting the parameters associated with a finite number of resonances, making some suitable approximation to describe the background of the remaining ones.

In some instances within the discussion that follows, we shall find it useful to have at hand a soluble example in order to illustrate the various quantities that appear in the $R$-matrix formulation of the problem, as well as to check the validity of certain approximations that we shall introduce later, such as the single-level and the single-level-plus-background approximations. The problem consisting of an impenetrable barrier plus a delta potential, that has been studied earlier in detail, will serve this purpose very well, thus playing the role of 'experimental data' against which comparisons can be made.

The point of view taken in the formulation of $R$-matrix theory is that the states $\psi_\lambda(x)$, being the solutions of the Hermitian problem defined by eqns (2.395), form a complete set of orthonormal states in the interval $-a < x < b$, in terms of which we can *expand, in that interval, the scattering wave function* $\psi(E;x)$ of eqn (2.393).

To prove the orthogonality of the $R$-matrix states, we first write the complex conjugate of eqn (2.395b) for $\psi_{\lambda'}(x)$, say (with $E_{\lambda'} = \hbar^2 K_{\lambda'}^2/2m$), as follows, remembering that $U(x)$ is real:

$$\left(\partial_x^2 + K_{\lambda'}^2\right)\psi_{\lambda'}^*(x) = U(x)\psi_{\lambda'}^*(x), \tag{2.397a}$$

with the corresponding boundary conditions

$$\psi_{\lambda'}(-a) = 0, \tag{2.397b}$$

$$\left[\frac{d\psi_{\lambda'}(x)}{dx}\right]_{x=b} = 0. \tag{2.397c}$$

We write eqn (2.397a) as shown, although we know that when $U(x)$ is real (we assume TRI) the $\psi_\lambda(x)$s can be written as real functions. We multiply eqn (2.395b) by $\psi_{\lambda'}^*(x)$, eqn (2.397a) by $\psi_\lambda(x)$, subtract the two equations and integrate from $x = -a$ to $x = b$, to obtain the Green identity

$$\left[\psi_{\lambda'}^*(x)\frac{\partial\psi_\lambda(x)}{\partial x} - \psi_\lambda(x)\frac{\partial\psi_{\lambda'}^*(x)}{\partial x}\right]_{x=-a}^{b}$$
$$= \left(K_{\lambda'}^2 - K_\lambda^2\right)\int_{-a}^{b}\psi_{\lambda'}^*(x)\psi_\lambda(x)\,dx. \tag{2.398}$$

Using the above boundary conditions of $\psi_\lambda(x)$ and $\psi_{\lambda'}(x)$ at $x=-a$ and $x=b$, we see that the left-hand side of eqn (2.398) vanishes, so that

$$\left(K_{\lambda'}^2 - K_\lambda^2\right)(\psi_{\lambda'}(x),\psi_\lambda(x)) = 0, \tag{2.399}$$

where we have defined the scalar product

$$(\psi_{\lambda'}(x),\psi_\lambda(x)) = \int_{-a}^{b}\psi_{\lambda'}^*(x)\psi_\lambda(x)\,dx. \tag{2.400}$$

For $K_{\lambda'}^2 \neq K_\lambda^2$, and hence $E_\lambda' \neq E_\lambda$, which is what occurs for $\lambda' \neq \lambda$ in a one-dimensional problem (i.e., non-degeneracy), we thus have the orthogonality relation

$$(\psi_{\lambda'}(x),\psi_\lambda(x)) = 0, \quad \lambda' \neq \lambda. \tag{2.401}$$

For $\lambda' = \lambda$ we can normalize the wave functions to 1, so that

$$(\psi_{\lambda'}(x),\psi_\lambda(x)) = \delta_{\lambda\lambda'}. \tag{2.402}$$

One can also show completeness of the orthonormal states in the same interval.

Although TRI has been assumed throughout, in that the Hermitian potential in eqn (2.395) is local and real, the analysis can be extended to a potential not obeying TRI. For instance, one can show that a non-TRI potential of the type (2.133) with $v(x) \equiv 0$ for $x \geqslant 0$, or a non-TRI Hamiltonian as in eqn (2.156), even with an extra TRI potential added to it (all potentials vanishing for $x \geqslant 0$), define Hermitian problems as well. The $R$-matrix formalism for these Hermitian problems can be set up in a similar fashion.

As indicated above, we now expand, in the internal region, the scattering wave function $\psi(E;x)$ of eqn (2.393) as a linear combination of the $\psi_\lambda(x)$s, i.e.,

$$\psi(-a \leqslant x \leqslant b) = \sum_\lambda A_\lambda \psi_\lambda(x). \qquad (2.403)$$

We multiply eqn (2.393b) by $\psi_\lambda^*(x)$, the complex conjugate of eqn (2.395b) by $\psi(x)$, subtract the two resulting equations and integrate from $x = -a$ to $x = b$, to obtain

$$\left[\psi_\lambda^*(x)\frac{\partial \psi(x)}{\partial x} - \psi(x)\frac{\partial \psi_\lambda^*(x)}{\partial x}\right]_{x=-a}^{b} = \left(K_\lambda^2 - k^2\right)(\psi_\lambda(x), \psi(x))$$
$$= \left(K_\lambda^2 - k^2\right) A_\lambda, \qquad (2.404)$$

where we have used the expansion (2.403) for $\psi(x)$ and the orthonormality relation (2.402). Using the boundary conditions (2.393c) and (2.395c,d), we then find the following expression for the expansion coefficients $A_\lambda$:

$$A_\lambda = \frac{\hbar^2}{2m} \frac{\psi_\lambda^*(b)}{E_\lambda - E} \psi'(b), \qquad (2.405)$$

where $\psi'(b) = [\partial \psi(x)/\partial x]_{x=b}$. Using this result in the expansion (2.403), we thus find

$$\psi(-a \leqslant x \leqslant b) = \frac{\hbar^2}{2m} \left[\sum_\lambda \frac{\psi_\lambda(x)\psi_\lambda^*(b)}{E_\lambda - E}\right] \psi'(b). \qquad (2.406)$$

This expression can be evaluated at $x = b^-$, to give

$$\psi(b^-) = \frac{\hbar^2}{2m} \left[\sum_\lambda \frac{[\psi_\lambda(b)]^2}{E_\lambda - E}\right] \psi'(b); \qquad (2.407)$$

we have now used explicitly the reality of the wave functions $\psi_\lambda(x)$.

If we were to differentiate eqn (2.406) at $x = b$ term by term, we would obtain

$$\psi'(b^-) = \frac{\hbar^2}{2m} \left[\sum_\lambda \frac{\psi_\lambda'(b)\psi_\lambda^*(b)}{E_\lambda - E}\right] \psi'(b). \qquad (2.408)$$

This expression vanishes, by virtue of the boundary condition (2.395d), which is clearly a contradiction, since $\psi'(x)$ does not have to vanish at $x = b$. The reason for the difficulty is that we are expanding $\psi(x)$ in terms of a set of functions $\psi_\lambda(x)$, all of which have zero slope at $x = b$. The resulting series, although convergent for $-a \leqslant x \leqslant b$, is not uniformly convergent at $x = b$, and hence cannot be differentiated term by term at that point. To see this more clearly, suppose that we have zero potential, $U(x) = 0$, that we set momentarily the boundary $b$ at $x = 0$ for simplicity and that, at some energy, the internal wave function $\psi(x \leqslant 0)$ has the form shown in Fig. 2.24. Two of the functions $\psi_\lambda(x)$ with zero slope at $x = 0$ are also shown in the figure; an expansion in terms of them is then clearly a standard Fourier expansion. The extension of the various functions beyond $x = 0$ (not the actual wave functions) is also shown in the figure using dotted lines. We see that the $\psi_\lambda(x)$s are symmetric with respect to $x = 0$, and so will be the resulting expansion of $\psi(x)$, also indicated by a dotted line. A finite series with a large number of terms will be very similar to the $\psi(x)$ and its extension to $x > 0$ shown in the figure, except that the cusp at $x = 0$ will certainly be rounded, thus giving zero slope, just as we found above! In conclusion, we have to be careful and *not* differentiate the expansion (2.406) term by term at $x = b$.

Following Wigner, we define the $R$ function as the inverse logarithmic derivative of the actual scattering wave function $\psi(x)$ at $x = b$ (where $\psi(x)$ and $\psi'(x)$ are certainly continuous), i.e.,

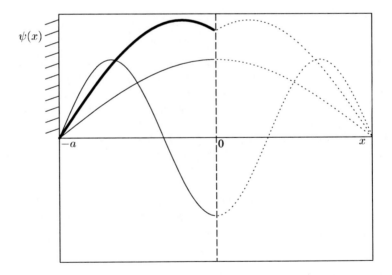

FIG. 2.24. Scattering in a semi-infinite one-dimensional space with zero potential; the $R$-matrix boundary is set at $x = 0$. The actual wave function for $-a < x < 0$ and two of the $R$-matrix functions with zero slope at $x = 0$ are represented by the solid heavy and light lines, respectively. The continuation provided by the Fourier series for $x > 0$ is indicated by the dotted lines.

$$R(E) = \frac{\psi(b)}{\psi'(b)}, \qquad (2.409)$$

which has *dimensions of length*. From eqn (2.407) we find $R(E)$ to be

$$R(E) = \sum_\lambda \frac{\gamma_\lambda^2}{E_\lambda - E}, \qquad (2.410)$$

where

$$\gamma_\lambda = \sqrt{\frac{\hbar^2}{2m}} \psi_\lambda(b). \qquad (2.411)$$

The real quantity $\gamma_\lambda$, called the *reduced amplitude*, is proportional to the *value of the $\lambda$th eigenfunction at the boundary* $x = b^-$. Its square, $\gamma_\lambda^2$, which has dimensions of energy · length, is referred to as the *reduced width*.

Across $x = b$ we must have continuity of the logarithmic derivative of the scattering wave function. From eqn (2.394) we have, for $x = b$,

$$\frac{\psi(b)}{\psi'(b)} = \frac{e^{2ikb}S + 1}{ik(e^{2ikb}S - 1)}. \qquad (2.412)$$

Equating this result with (2.409), we can express the $S$ function in terms of the $R$ function as

$$S(E) = -e^{-2ikb}\frac{1 + ikR(E)}{1 - ikR(E)} = -e^{2i\delta}. \qquad (2.413)$$

At this point we could say that the main achievement of the present formulation of scattering theory is Wigner's observation that the inverse logarithmic derivative of the scattering wave function at $x = b$, or, equivalently, the combination involving the $S$ function shown on the right-hand side of eqn (2.412), is a *meromorphic function of the energy*, called $R(E)$, which is real on the real axis, whose *poles occur only on the real axis* and whose *residues are real and negative*, as shown in eqn (2.410). Such poles and residues could, in principle, be calculated as indicated above; in any case, they have proved to be a convenient way to parametrize the scattering data.

Since the Wigner $R$ function is real for real $E$, the function $S$ of eqn (2.413) is explicitly *unitary*. The phase shift $\delta$, defined by $S = -e^{2i\delta}$, is given in terms of the $R$ function as

$$\tan(\delta + kb) = kR. \qquad (2.414)$$

We can thus make the following statements.

(a) The poles of the $R$ function occur at those energies for which the phase $\delta + kb$ takes on the values $\pi/2$, $3\pi/2$, etc.

(b) As already noted, exactly at these energies the scattering wave function coincides, in the region $x < b$, with an $R$-matrix state (up to a normalization factor) and thus has zero slope at the $R$-matrix boundary.

For the particular case in which the $R$-matrix boundary coincides with the origin, i.e., $b = 0$, eqn (2.413) takes the simpler form

$$S(E) = -\frac{1 + ikR(E)}{1 - ikR(E)}, \quad b = 0, \tag{2.415}$$

implying that
$$\tan \delta = kR, \quad b = 0. \tag{2.416}$$

As an illustration, consider the delta potential model studied in detail in Section 2.2.1 and subsequent sections. The $S$ matrix, given in eqn (2.296), is reproduced here for convenience, i.e.,

$$S = -\frac{(\sin ka + \alpha ka \cos ka) + i\alpha ka \sin ka}{(\sin ka + \alpha ka \cos ka) - i\alpha ka \sin ka}. \tag{2.417}$$

Figures 2.25 and 2.26 are adaptations of Fig. 2.11 to the notation introduced here in relation to the $R$-matrix formalism. In the situation illustrated in Fig. 2.25, the $R$-matrix boundary is chosen right outside the delta potential, i.e., $b = 0^+$. The corresponding $R$ function is calculated in the following exercise.

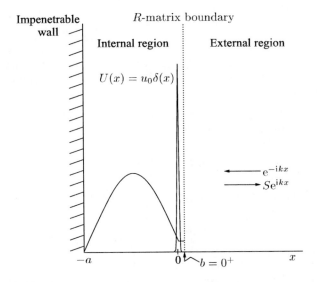

FIG. 2.25. The situation in which the $R$-matrix boundary is chosen just outside the delta potential, i.e., $b = 0^+$. The first $R$-matrix state $\psi_{\lambda=1}(x)$ is shown in its region of definition, i.e., the internal region; this state is seen to reach the $R$-matrix boundary with zero slope.

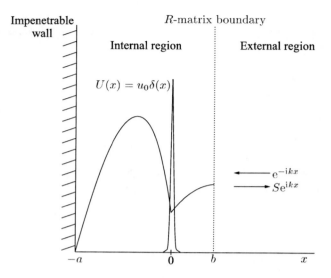

FIG. 2.26. The situation in which the $R$-matrix boundary ($x = b$) is chosen to be a distance $b$ from the delta potential. The first $R$-matrix state $\psi_{\lambda=1}(x)$ is shown in the internal region, i.e., $-a \leqslant x \leqslant b$. This state is seen to reach the $R$-matrix boundary with zero slope. The distance from the perfectly reflecting wall to the delta potential is $a$, just as in Fig. 2.25.

**Exercise 2.23** Find the $R$ function for the delta potential model studied in Section 2.2.1, when the $R$-matrix boundary is chosen right outside the delta potential, i.e., $b = 0^+$, as illustrated in Fig. 2.25.

We solve eqn (2.415) for the $R$ function, which we now call $R_0$, to remind us of the position, $b = 0^+$, of the $R$-matrix boundary, to give

$$kR_0 = -i\frac{S+1}{S-1}, \qquad (2.418)$$

which, for the $S$ matrix of eqn (2.417), gives the $R$ function

$$kR_0 = \tan \delta = \alpha k a \frac{\tan ka}{\alpha ka + \tan ka} = \frac{g(ka)}{f(ka)}, \qquad (2.419)$$

where we have used the definitions (2.348) and (2.349) of $f(ka)$ and $g(ka)$, respectively. This result could also have been obtained from eqn (2.416) and the explicit expression, eqn (2.302), for the phase shift for the delta potential model. The $R$-function poles occur at the zeros of the denominator in eqn (2.419) (at which $\delta = \pi/2, \pi/2, \ldots$) and were discussed in detail in Section 2.2.2. Also, as discussed in eqn (2.334c), Section 2.2.3, the scattering wave function at these energies emerges from the delta potential with zero slope.

For the delta potential model and for this choice of $R$-matrix boundary, the $R$-matrix states, their energies and reduced widths are calculated in the following exercise.

**Exercise 2.24** The $R$-matrix boundary is chosen right outside the delta potential, as in the previous exercise (see Fig. 2.25), i.e., $b = 0^+$. Find the $R$-matrix states $\psi_\lambda(x)$ and the $R$-matrix poles and residues for the case of a strong delta potential, i.e., $\alpha \ll 1$.

In their region of definition, i.e., up to a point right outside the delta potential, the $R$-matrix states $\psi_\lambda(x)$ coincide, up to a normalization factor, with those scattering wave functions that emerge from the delta potential with precisely zero slope, and are associated with a phase shift $\delta = \pi/2, 3\pi/2, \ldots$. The corresponding wave numbers were calculated in Exercise 2.16.

Thus the $R$-matrix states are those with $K_\lambda = k_\lambda^\delta$, the latter being given in eqn (2.314), i.e.,

$$\psi_\lambda(x) = C_\lambda \sin[k_\lambda^\delta(x+a)], \quad -a \leqslant x \leqslant 0, \qquad (2.420a)$$

$$k_\lambda^\delta a = \lambda\pi(1 - \alpha + \alpha^2 + \cdots), \qquad (2.420b)$$

with $\lambda = 1, 2, \ldots$. The normalization coefficient $C_\lambda$ satisfies

$$1 = \frac{C_\lambda^2 a}{2}\left[1 - \frac{\sin 2k_\lambda^\delta a}{2k_\lambda^\delta a}\right] \qquad (2.421a)$$

$$= \frac{C_\lambda^2 a}{2}\left[1 - \frac{2\lambda\pi(-\alpha + \alpha^2) + O(\alpha^3)}{2\lambda\pi(1 - \alpha + \alpha^2) + O(\alpha^3)}\right] \qquad (2.421b)$$

$$= \frac{C_\lambda^2 a}{2}\frac{1 + O(\alpha^3)}{1 - \alpha + \alpha^2 + O(\alpha^3)}, \qquad (2.421c)$$

and is thus given by

$$C_\lambda^2 = \frac{2}{a}\left[1 - \alpha + \alpha^2 + O(\alpha^3)\right]. \qquad (2.422)$$

From eqns (2.420) and (2.422), we find the square of the wave function $\psi_\lambda(x)$ at the boundary $x = 0$ to be

$$[\psi_\lambda(0)]^2 = C_\lambda^2 \sin^2 k_\lambda^\delta a$$

$$= \frac{2}{a}(\lambda\pi)^2\left[\alpha^2 + O(\alpha^3)\right]. \qquad (2.423)$$

The $R$-matrix poles and residues, or reduced widths, are then given by (see eqns (2.420b) and (2.411))

$$E_\lambda = \frac{\hbar^2}{2ma^2}(\lambda\pi)^2[1 - 2\alpha + 3\alpha^2 + O(\alpha^3)], \qquad (2.424a)$$

$$\gamma_\lambda^2 = \frac{\hbar^2}{ma}(\lambda\pi)^2 \left[\alpha^2 + O(\alpha^3)\right].\tag{2.424b}$$

Another choice of the $R$-matrix boundary is discussed below. In Figs 2.25 and 2.26 we have exactly the same physical system, with the same $S$ matrix, the difference being just a shift in the $R$-matrix boundary; the corresponding $R$ functions will thus be different. This is discussed in the following exercise, first in general and then for the particular example of the delta potential model.

**Exercise 2.25** Denote by $R_0$ and $R_b$ the $R$ matrices for the cases in which the $R$-matrix boundary is at $x = 0$ and $x = b$, respectively. A particular case is shown in Figs 2.25 and 2.26 for the delta potential problem. Find the relation between $R_0$ and $R_b$ for an arbitrary potential and, as a particular case, for the delta potential problem.

From eqns (2.414) and (2.416) we have

$$kR_b = \tan(\delta + kb),\tag{2.425a}$$
$$kR_0 = \tan\delta,\tag{2.425b}$$

so that the relation between the two $R$ functions is

$$kR_b = \frac{kR_0 + \tan kb}{1 - kR_0 \tan kb}.\tag{2.426}$$

As was mentioned immediately after eqn (2.414), the poles of $R_b$ occur at those energies for which the phase $\delta + kb = \pi/2, 3\pi/2, \ldots$. At those energies the scattering wave function has zero slope at $x = b$ and coincides, for $x < b$ and up to a normalization constant, with one of the $R$-function states (Fig. 2.26). From eqn (2.426) we see that the $R$-function poles $E_\lambda^b$ satisfy the equation

$$1 = [kR_0(E)\tan kb]_{E=E_\lambda^b},\tag{2.427a}$$

or

$$\left[R_0(E) = \frac{1}{k\tan kb}\right]_{E=E_\lambda^b},\tag{2.427b}$$

written in terms of the $R$ function $R_0$, whose poles are $E_\lambda^0$; there is seen to be a pole of $R_b$ between two successive poles of $R_0$.

For the delta potential problem we use the result of Exercise 2.23 to find

$$kR_b = \frac{g(ka) + f(ka)\tan kb}{f(ka) - g(ka)\tan kb}.\tag{2.428}$$

The structure of the $R$-matrix states for the problem of a strong delta potential and the choice of the origin shown in Fig. 2.26 is discussed in Exercise 2.28 at the end of the chapter.

*The poles of $S(k)$ in R-matrix language*

The poles $k_\lambda$ of the scattering matrix considered as a function of momentum, i.e., $S(k)$, are the zeros of the denominator of eqn (2.413), i.e.,

$$1 = \mathrm{i}k_\lambda R(\mathcal{E}_\lambda), \qquad (2.429\mathrm{a})$$

where $k_\lambda$ and $\mathcal{E}_\lambda$ are related by

$$\mathcal{E}_\lambda = \frac{\hbar^2 k_\lambda^2}{2m}, \qquad (2.429\mathrm{b})$$

according to the notation introduced in Sections 2.2.5 and 2.2.6. Taking the complex conjugate of eqn (2.429a), we find that *the poles of $S$ occur in pairs*, i.e., $k_\lambda$ and $-k_\lambda^*$, as we already found in Section 2.2.5 for the example of the delta potential.

**Exercise 2.26** Write eqn (2.429) for the $S$-matrix poles for the special case of the delta potential model in terms of the $R$ function $R_0$, i.e., for the $R$-matrix boundary chosen right outside the delta potential, as in Fig. 2.25.

From eqns (2.429a) and (2.419) we find

$$\alpha k_\lambda a + \tan k_\lambda a = \mathrm{i}\alpha k_\lambda a \tan k_\lambda a, \qquad (2.430)$$

which coincides with eqn (2.367). Of course, exactly the same equation would have been obtained using the $R$ function $R_b$ of eqn (2.428).

*The case of isolated resonances.* This situation is realized when the internal states are only weakly coupled to the outside. For $E \approx E_\lambda$, and hence $k \approx \pm K_\lambda$ (with $E_\lambda = \hbar^2 K_\lambda^2/2m$), the summation in eqn (2.410) will be approximated by keeping only the single term $\lambda$, i.e.,

$$R(E) \approx \frac{\gamma_\lambda^2}{E_\lambda - E}. \qquad (2.431)$$

In the *single-level approximation* (2.431) we write the $S$ function of eqn (2.413) as

$$S(k) \approx -\mathrm{e}^{-2\mathrm{i}kb} \frac{E_\lambda - E + \mathrm{i}k\gamma_\lambda^2}{E_\lambda - E - \mathrm{i}k\gamma_\lambda^2}. \qquad (2.432)$$

This expression has poles where the denominator vanishes, i.e.,

$$\mathcal{E}_\lambda + \mathrm{i}k_\lambda \gamma_\lambda^2 - E_\lambda = 0, \qquad (2.433)$$

with the solution

$$k_\lambda = k_\lambda' + \mathrm{i}k_\lambda''$$
$$= \pm\sqrt{K_\lambda^2 - \frac{1}{4}\psi_\lambda^4(b)} - \frac{\mathrm{i}}{2}\psi_\lambda^2(b), \qquad (2.434)$$

an expression that exhibits explicitly the two poles at positions which are symmetric with respect to the imaginary axis.

We now illustrate the above expressions for the case of the *strong delta potential barrier* that was studied earlier in great detail. We compare the poles of $S$, as given by eqn (2.434) in $R$-matrix theory in the single-level approximation, with those discussed in Section 2.2.6.

For the choice of the $R$-matrix boundary shown in Fig. 2.25, i.e., $b = 0^+$, we introduce into eqn (2.434) the results of Exercise 2.24 to find the poles $k_\lambda$ of $S$, namely

$$k_\lambda a = \pm(\lambda\pi)\left[1 - \alpha + \alpha^2 + O(\alpha^3)\right] - i(\lambda\pi)^2\left[\alpha^2 + O(\alpha^3)\right], \quad (2.435)$$

just as we encountered in eqn (2.377) without using the $R$-matrix formalism.

As will be discussed in Exercise 2.28 below, a change in the $R$-matrix boundary, i.e., in the quantity $b$ shown in Fig. 2.26, entails a change in the $R$-matrix energies $E_\lambda$ and reduced widths $\gamma_\lambda^2$. Of course, as long as we use the exact $R$ function, the poles of the resulting $S$ matrix cannot change; differences may be expected though if we truncate the $R$ function, as in the single-level approximation indicated above. This we wish to investigate now, using our delta potential model as an illustration. When the $R$-matrix boundary is not right outside the delta potential, i.e., when $b \neq 0$ as in Fig. 2.26, instead of proposing a value for $b$ and finding the $R$-matrix parameters therefrom, we shall find it simpler, for the sake of illustration, to proceed in the opposite direction, as explained in the following exercise.

**Exercise 2.27** Writing the $R$-matrix poles for $R_b$ as $E_\lambda^b = \hbar^2(K_\lambda^b)^2/2m$, we *postulate*, for the *first pole*, that

$$K_{\lambda=1}^b a = \pi(1 - \alpha). \quad (2.436)$$

Find the location of the $R$-matrix boundary, i.e., the value of $b$ in Fig. 2.26, for which $K_{\lambda=1}^b a$ takes on the stated value. Using the results of Exercise 2.28, find the properties of the corresponding $R$-matrix state $\psi_{\lambda=1}(x)$ and, from eqn (2.434), find the $S$-matrix pole that would arise in the *single-level approximation*.

The difference between the proposed $K_{\lambda=1}^b a$ of eqn (2.436) and the value given in eqn (2.420b) is

$$K_{\lambda=1}^b a - k_{\lambda=1}^\delta a = \pi(1 - \alpha) - \pi(1 - \alpha + \alpha^2 + \cdots)$$
$$= -\pi\alpha^2 + \cdots. \quad (2.437)$$

From eqn (2.466a) of Exercise 2.28 we find

$$K_1^b b = \frac{\pi}{2} - \delta(K_1^b). \quad (2.438)$$

The phase shift for this case was calculated in eqns (2.329), so that

$$\cos K_1^b b = \sin\delta(K_1^b) \quad (2.439\text{a})$$

$$\approx \frac{\pi}{\sqrt{1+\pi^2}}, \quad (2.439\text{b})$$

from which we find
$$K_1^b b \approx 0.31, \quad \frac{b}{a} \approx 0.1. \tag{2.440}$$
This shows that in the vicinity of a sharp resonance the phase shift changes so rapidly that the difference of order $\alpha^2$, eqn (2.437), between the $K_{\lambda=1}^b a$ of eqn (2.436) and that of eqn (2.420b) produces a change in the position of the $R$-matrix boundary of the order of ten per cent!

Using eqn (2.437) in eqn (2.471) of Exercise 2.28, we find $B_1^2$ to be
$$B_1^2 = \frac{(\pi\alpha)^2}{(\pi\alpha^2)^2 + (\pi\alpha)^4} = \frac{1}{(\pi^2+1)\alpha^2}. \tag{2.441}$$

Using this result and the above values of $K_1^b a$ and $K_1^b b$ in eqn (2.468), we find
$$I = \frac{a}{2} \left\{ \frac{1}{(\pi^2+1)\alpha^2} \left[1 + \frac{\sin 2\pi\alpha}{2\pi(1-\alpha)}\right] + 0.1 \left[1 + \frac{2\pi}{0.62(1+\pi^2)}\right] \right\}$$
$$= \frac{a}{2} \frac{1}{(\pi^2+1)\alpha^2} + O\left(\frac{1}{\alpha}\right). \tag{2.442}$$

From eqns (2.469) we find the square of the $R$-matrix wave function at the $R$-matrix boundary and the reduced width $\gamma_{\lambda=1}^2$ to be
$$|\psi_{\lambda=1}(b)|^2 = \frac{2}{a}(\pi^2+1)\alpha^2 + O(\alpha^3), \tag{2.443a}$$
$$\gamma_{\lambda=1}^2 = \frac{\hbar^2}{2m}\frac{2}{a}(\pi^2+1)\alpha^2 + O(\alpha^3). \tag{2.443b}$$

Equation (2.434) now gives the following $S$-matrix pair of poles in the single-level approximation:
$$k_{\lambda=1}a = \pm\left[\pi(1-\alpha) + O(\alpha^4)\right] - \mathrm{i}(\pi^2+1)\left[\alpha^2 + O(\alpha^3)\right]. \tag{2.444}$$

Comparing the result of the last exercise, eqn (2.444), with that of eqn (2.377), obtained directly, i.e., without making use of the $R$-matrix formalism, we see that the $R$-matrix single-level approximation gives a shift in the real part of the pole of order $\alpha^2$, which is of the order of the width itself, while the imaginary part differs from the true one by the multiplicative factor $(\pi^2+1)/\pi^2$, which is independent of $\alpha$.

This discrepancy can only be due to the extreme single-level approximation, eqn (2.431). We can improve on this without much complication by including the effect of the other levels as a background term $R^B$ and writing
$$R(E) \approx \frac{\gamma_\lambda^2}{E_\lambda - E} + R^B. \tag{2.445}$$

The quantity $R^B$ will be approximated by a constant in the region of interest, i.e., for $E$ in the vicinity of $E_\lambda$. Notice that for $|E_\lambda - E| \sim k\gamma_\lambda^2$, i.e., about a width away from $E_\lambda$, the first term in eqn (2.445) is $O(\alpha^0)$. Thus, should we find that $R^B \sim O(\alpha^0)$, then the background correction would be of the same order as the resonance term. Since for our delta potential model we know the $R$ function exactly, we can calculate the background term $R^B$ explicitly. This is done in Exercises 2.29 and 2.30 at the end of the chapter. For the first choice of $R$-matrix boundary, i.e., right outside the delta potential (see Fig. 2.25), one finds the result given in eqn (2.479), which shows that the background term $R^B \sim O(\alpha)$ and can thus be made very small for a sufficiently strong delta barrier. In contrast, for the choice of $R$-matrix boundary shown in Fig. 2.26 the result is given in eqn (2.490). Now the background term $R^B \sim O(\alpha^0)$ and it cannot be made arbitrarily small by increasing the strength of the delta barrier.

Expression (2.434) gives the $S$-matrix poles based on the extreme $R$-matrix single-level approximation (2.431). We should now find an improved expression for the $S$-matrix poles using the single-level-plus-background approximation, eqn (2.445). The exact pole condition, eqn (2.429), now reads

$$1 = \mathrm{i}k_\lambda \left( \frac{\gamma_\lambda^2}{E_\lambda - \mathcal{E}_\lambda} + R^B \right), \tag{2.446a}$$

or

$$k_\lambda^2 = K_\lambda^2 - \mathrm{i}k_\lambda [\psi_\lambda(b)]^2 + \mathrm{i}k_\lambda R^B(k_\lambda^2 - K_\lambda^2). \tag{2.446b}$$

This is a third-degree equation in $k_\lambda$, for which we shall find an approximate solution. Writing

$$k_\lambda = K_\lambda + \zeta_\lambda, \tag{2.447}$$

eqn (2.446) can be rewritten as

$$\zeta_\lambda = \frac{1}{1 - \mathrm{i}K_\lambda R^B} \left[ -\frac{\mathrm{i}}{2} \psi_\lambda^2(b) - \frac{\mathrm{i}}{2K_\lambda} \psi_\lambda^2(b)\zeta_\lambda + \frac{3\mathrm{i}}{2} R^B \zeta_\lambda^2 - \frac{\zeta_\lambda^2}{2K_\lambda} + \frac{\mathrm{i}}{2K_\lambda} R^B \zeta_\lambda^3 \right]. \tag{2.448}$$

For the choice (2.436) of $K_\lambda$ that we shall be interested in (for $\lambda = 1$), a comparison with the pole expression, eqn (2.377), shows that $\zeta_\lambda = O(\alpha^2)$. Similarly, from eqn (2.443) we see that $\psi_\lambda^2(b) = O(\alpha^2)$ and, from eqn (2.490), that $R^B = O(\alpha^0)$. Thus the terms on the right-hand side of eqn (2.448) are arranged in order of increasing powers of $\alpha$. We find the solution of this equation by successive approximations to be

$$\zeta_\lambda^{(0)} = 0, \tag{2.449a}$$

$$\zeta_\lambda^{(2)} = -\frac{\mathrm{i}}{2} \frac{\psi_\lambda^2(b)}{1 - \mathrm{i}K_\lambda R^B}, \tag{2.449b}$$

etc., so that we have the expression for the pole, to order $\alpha^2$, as

$$k_\lambda a = K_\lambda a + \frac{1}{2} \frac{K_\lambda R^B}{1 + (K_\lambda R^B)^2} [\psi_\lambda^2(b)a] - \frac{\mathrm{i}}{2} \frac{[\psi_\lambda^2(b)a]}{1 + (K_\lambda R^B)^2} + \cdots . \tag{2.450}$$

The important result is that the $R^B$ correction renormalizes the imaginary part of the pole and introduces an additive correction to its real part. We apply this result to our case of interest, defined by eqns (2.436) and (2.443), and, using (2.490), we find

$$k_{\lambda=1}a = \pi(1-\alpha) + \frac{1}{2}\frac{(1/\pi)\left[1+O(\alpha)\right]}{1+(1/\pi^2)\left[1+O(\alpha)\right]}\left[2(1+\pi^2)\alpha^2 + \cdots\right]$$

$$- \frac{i}{2}\frac{2(1+\pi^2)\alpha^2 + \cdots}{1+(1/\pi^2)\left[1+O(\alpha)\right]}$$

$$= \pi(1-\alpha) + \pi\alpha^2 + \cdots - i(\pi^2\alpha^2 + \cdots) \qquad (2.451)$$

and the right pole position, given by eqn (2.377), is recovered.

### 2.2.8  *The 'motion' of the S matrix as a function of energy*

In the present section on potential scattering in semi-infinite one-dimensional space we have seen that the $S$ matrix is one-dimensional, so that from unitarity it must be a complex number of unit modulus at every energy (see eqn (2.297)). In the Argand diagram of Fig. 2.27, $S(E)$, for a given energy, is represented by *a point on the unitarity circle*; this point is defined by the angle $\theta(E)$. As the energy changes, so does the representative point; this is what we call, pictorially, the 'motion' of $S(E)$ as a function of energy. It resembles the motion, in phase space, of the representative point of a classical system as a function of time. This picture also applies to the $S(E)$ function that describes scattering by a cavity of the type discussed in Chapter 6, connected to the outside by one waveguide, which supports only one open channel.

As an illustration, consider the soluble model studied in detail in Section 2.2.1, consisting of an impenetrable barrier and a delta potential in front of it. In particular, take the case described by eqn (2.303), corresponding to no

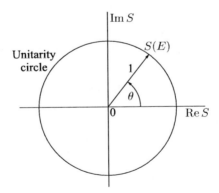

FIG. 2.27. The Argand diagram Re $S$, Im $S$. Due to unitarity, $S(E)$, for a given energy $E$, is represented by a point, defined by the angle $\theta(E)$, *on* the unit circle. As $E$ changes, $\theta(E)$ changes and the representative point moves along the circle.

barrier, i.e., $u_0 = 0$. Here $S(k)$ moves along the circumference—or equivalently $\theta$ increases—at uniform speed as a function of momentum. As discussed in detail in Section 2.2.1, in the opposite situation of a strong barrier, i.e., for $\alpha = 1/u_0 a \ll 1$, $S(k) \approx -1$ between resonances and makes a full turn around the unitarity circle every time a resonance is hit.

It is natural to ask the following question: as we move within an interval of the energy axes centered at $E_0$, *what fraction of the time do we find the phase* $\theta$ *of* $S = e^{i\theta}$ *lying inside* $(\theta, \theta + d\theta)$?

The property referred to in the last paragraph clearly has a *local* character, in the sense that it may change as we go over to a different energy region. Consider the $S(E)$ given by eqn (2.415) in terms of Wigner's $R(E)$ function. If, in that equation, the variation of $k = \sqrt{2mE}/\hbar$ with energy inside the interval in question is so much smaller than that of $R(E)$, then, in a local description, it is reasonable to consider it to be independent of $E$ and replace it by the *constant* value $k_0 = \sqrt{2mE_0}/\hbar$. The analytic properties of the $S(E)$ so constructed are, locally, similar to those of the original function. However, in the whole complex-$E$ plane they are much simpler. In fact, $S(E)$ no longer has a branch point at $E = 0$; it is a meromorphic function of $E$ (i.e., with no singularities other than poles) in the entire complex-$E$ plane, it is analytic in the upper half-plane (we call it a 'causal' function) and unitary on the full real axis, i.e., for $-\infty < E < +\infty$. In the lower half of the $E$ plane $S(E)$ has resonance poles.

We can easily verify this last property. We write $k_0 R(E) = K(E)$ and $k_0 \gamma_\lambda^2 = \Gamma_\lambda/2$. For a complex energy $\mathcal{E} = E' + iE''$, the $K$ function can be written as

$$K(E' + iE'') = \frac{1}{2} \sum_\lambda \frac{\Gamma_\lambda}{E_\lambda - E' - iE''} = K' + iK'', \qquad (2.452)$$

where

$$K' = \frac{1}{2} \sum_\lambda \Gamma_\lambda \frac{E_\lambda - E}{(E_\lambda - E')^2 + (E'')^2}, \qquad (2.453a)$$

$$K'' = \frac{1}{2} \sum_\lambda \Gamma_\lambda \frac{E''}{(E_\lambda - E')^2 + (E'')^2}. \qquad (2.453b)$$

The function $K$ is sometimes called a 'following function' [174], meaning that $K''$ is positive (negative) in the upper (lower) half of the energy plane.

The $S$ function evaluated at $\mathcal{E} = E' + iE''$ is thus given by

$$S(E' + iE'') = -\frac{1 - K'' + iK'}{1 + K'' - iK'}. \qquad (2.454)$$

The poles of $S$ occur when $K'' = -1$ and $K' = 0$. Since $K$ is a following function, the former relation can only be satisfied in the lower half of the energy plane.

From eqn (2.454) we find

$$|S|^2 = \frac{(1-K'')^2 + (K')^2}{(1+K'')^2 + (K')^2}. \qquad (2.455)$$

In the upper half-plane we have $K'' > 0$ and $|S|^2 < 1$, i.e., $S$ is 'subunitary'.

In the process of answering the above-posed question we shall need to calculate energy averages. For this purpose we define a weight function $W_{E_0,I}(E)$ that selects the energy region of interest. For convenience, the weight function is chosen as a Lorentzian centered at $E_0$ and with a width $2I$, i.e.,

$$W_{E_0,I}(E) = \frac{I/\pi}{(E-E_0)^2 + I^2}. \qquad (2.456)$$

The average of $S$ over an interval of size $2I$ centered at $E_0$ is then defined as

$$\overline{S}(E_0, I) \equiv \int_{-\infty}^{+\infty} S(E) \frac{I/\pi}{(E-E_0)^2 + I^2}\, dE. \qquad (2.457)$$

In order to evaluate this integral, we recall that $S(E)$ is analytic in the upper half of the energy plane, while the Lorentzian (2.456) has poles at $E_0 + iI$ and $E_0 - iI$. The integral (2.457) can then be performed using Cauchy's theorem of contour integration, closing the contour in the upper half-plane (where $S$ is bounded and hence the integrand of (2.457) tends to zero sufficiently fast as we approach the circle at infinity), with the result

$$\overline{S}(E_0, I) = S(E_0 + iI). \qquad (2.458)$$

Thus we have shown that the Lorentzian energy average of $S$ over an interval of size $2I$ centered at $E_0$ can be obtained by evaluating $S$ at the complex energy $E_0 + iI$. Since this complex energy is in the upper half-plane, $S$ evaluated there is subunitary, i.e., its modulus is less than unity, as expected for the average of a function which is unitary on the real axis. Pictorially, we could say that $S(E_0 + iI)$ 'sees' the poles from further away than $S(E)$ does, and is therefore subunitary.

In a similar way we can calculate the average of the $m$th power of $S$, $S^m(E)$, with the same Lorentzian weight function as above, i.e.,

$$\overline{S^m}(E_0, I) = \int_{-\infty}^{+\infty} [S(E)]^m\, W_{E_0,I}(E)\, dE, \qquad (2.459)$$

to find

$$\overline{S^m}(E_0, I) = S^m(E_0 + iI) = \left[\overline{S}(E_0, I)\right]^m. \qquad (2.460)$$

Thus, as a result of $S$ being a causal function, *the average of the $m$th power of $S$ coincides with the $m$th power of the average of $S$.* The average of $S$, i.e.,

$\overline{S}(E_0, I)$, is referred to in the literature as the 'optical $S$ matrix'. In the spirit of the usual frequentist viewpoint, we inquire whether it is possible to find a *measure* $dP(S) = p(\theta)\, d\theta$, independent of $k$, that can be used to evaluate the averages (2.460), i.e.,

$$\overline{S^m}(E_0, I) = \int S^m\, dP(S) = \int_0^{2\pi} e^{im\theta} p(\theta)\, d\theta. \tag{2.461}$$

We expand $p(\theta)$ in a Fourier series, i.e.,

$$p(\theta) = \sum_m a_k e^{im\theta}, \tag{2.462}$$

with $a_m = a_{-m}^*$ to ensure the reality of $p(\theta)$. We find the following expression for the expansion coefficients:

$$a_{-m} = \frac{1}{2\pi} \int p(\theta) e^{im\theta}\, d\theta = \frac{\overline{S^m}}{2\pi} = \frac{\overline{S}^m}{2\pi}. \tag{2.463}$$

We can sum the series, with the result [121]

$$p(\theta) = \frac{1}{2\pi} \frac{1 - |\overline{S}|^2}{|S - \overline{S}|^2}, \quad S = e^{i\theta}. \tag{2.464}$$

We have thus found that eqn (2.461) indeed holds, with the *unique* solution (2.464) for $p(\theta)$, also known as *Poisson's kernel* [121]. Note that its dependence on $E_0$ and $I$ is hidden in the value of $\overline{S}(E_0, I)$. Clearly, this result is true whether $S$ has a single resonance pole, some regular pattern of poles or a random distribution of them; the result also holds for arbitrary values of $E_0$ and $I$.

We have thus answered the question posed on p. 110 in the following circumstances: for a given $S(E)$, $dP(S) = p(\theta)\, d\theta$ is the fraction of time that $\theta$ falls inside $(\theta, \theta + d\theta)$, when the energy axis is visited from $-\infty$ to $+\infty$ and weighted with the Lorentzian $W_{E_0, I}(E)$ of eqn (2.456).

The delta potential model studied in great detail earlier was taken as an example to verify the above result [81]. An energy stretch containing 100 resonances starting from $ka = 10\,000$ (so that secular variations can be neglected) was sampled to find the fraction of time that the phase of $S$ falls within a small interval around $\theta$. The result is compared in Fig. 2.28 with Poisson's kernel (2.464), using a value for the optical $\overline{S}$ that was extracted from the numerical data inside the energy stretch mentioned above; in this sense we have a parameter-free fit. We observe that the agreement is excellent.

As we shall see in Chapter 6, of particular interest are those cases where $S(E)$ defines a *stationary random process*. In such cases we have an *ensemble* of $S$ matrices; ensemble averages will be indicated with angular brackets, to distinguish them from the *energy averages* considered above. If the correlation function $c(E - E') = \langle S(E) S(E') \rangle - \langle S(E) \rangle \langle S(E') \rangle$ vanishes in the limit $|E - E'| \to \infty$,

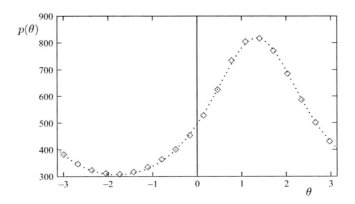

FIG. 2.28. In the delta potential model a stretch of energy containing 100 resonances, starting from $ka = 10\,000$, was sampled to find the fraction of time that the phase of $S$ falls within a small interval around $\theta$; the result is indicated by diamonds. The curve is a plot of Poisson's kernel (2.464), with the value of $\overline{S}$ extracted from the numerical data, in order to have a parameter-free fit. The agreement is excellent.

one can show that an ensemble average evaluated at a fixed energy coincides, up to a set of zero measure, with an energy average performed with a sufficiently smooth weight function $W_{E_0,I}(E)$, *in the limit when its width $I$ becomes infinitely large*. This property of *ergodicity* [181] thus implies that, in this limit, the various energy averages become independent of the weight function $W_{E_0,I}(E)$. In particular, under such circumstances the Lorentzian-weighted averages will also be equal to averages performed with a rectangular weight function—over sufficiently large intervals—which are more physical, and both types of energy averages will be equal to the average obtained with the ensemble measure (6.11), which contains the single parameter $\langle S \rangle$, now independent of $E_0$ and $I$. For a given realization $S(E)$ of the ensemble, $\mathrm{d}P(S) = p(\theta)\,\mathrm{d}\theta$ is now the fraction of the 'time'—inside a very large uniformly weighted energy interval—that $\theta$ falls inside $(\theta, \theta + \mathrm{d}\theta)$; it depends on the single parameter $\langle S \rangle$, often referred to as the *optical S matrix*. A few comments are in order here. The form of $S$ in eqn (2.415) corresponds to the particular choice $b = 0$ of the $R$-matrix boundary $b$ in eqn (2.413). This is most convenient in that the dominant variation of $S$ with energy is contained in $R(E)$ itself, as assumed in the foregoing treatment. For a one-dimensional potential, the variation of $k$ in the prefactor of eqn (2.413) would have to be explicitly taken into account. As we shall see later on, the scattering from a higher-dimensional cavity connected to a one-channel lead can be described by a $1 \times 1$ $S$ matrix. In such a case, the level density would be sufficiently high to justify neglecting the variation coming from the prefactor mentioned above.

**Exercise 2.28** Find the structure of the $R$-matrix states for the problem of a strong delta potential and the choice of the $R$-matrix boundary $b$ shown in Fig. 2.26.

From eqns (2.301) we write the scattering wave function at energy $E$ as

$$\psi(-a \leqslant x < 0) = B\sin[k(x+a)], \tag{2.465a}$$

$$\psi(x \geqslant 0) = \sin[kx + \delta(k)], \tag{2.465b}$$

with

$$B^2 = \frac{\alpha^2(ka)^2}{(\sin ka + \alpha ka \cos ka)^2 + (\alpha ka \sin ka)^2}, \tag{2.465c}$$

$$\tan\delta = \frac{\alpha ka \tan ka}{\alpha ka + \tan ka}. \tag{2.465d}$$

The $R$-matrix energies $E_\lambda^b = \hbar^2[K_\lambda^b]^2/2m$ satisfy eqn (2.427). Alternatively, we notice that at these energies the scattering wave function at $x = b$ has zero slope, $\psi'(b) = 0$. Thus, from eqn (2.465b) (see also eqn (2.425a)), $K_\lambda^b$ must satisfy

$$\delta(K_\lambda^b) + K_\lambda^b b = (2\lambda - 1)\frac{\pi}{2}, \tag{2.466a}$$

$$\cos\left[\delta(K_\lambda^b) + K_\lambda^b b\right] = 0, \tag{2.466b}$$

$$\sin\left[\delta(K_\lambda^b) + K_\lambda^b b\right] = (-1)^{\lambda+1}. \tag{2.466c}$$

We thus have, in the internal region, the $R$-matrix states ($B_\lambda$ denoting $B$ evaluated at $k = K_\lambda^b$)

$$\psi_\lambda(-a \leqslant x < 0) = C_\lambda \sin[K_\lambda^b(x+a)], \quad C_\lambda = \frac{B_\lambda}{\sqrt{I}}, \tag{2.467a}$$

$$\psi_\lambda(0 \leqslant x \leqslant b) = \frac{1}{\sqrt{I}}\sin\left[K_\lambda^b x + \delta(K_\lambda^b)\right], \tag{2.467b}$$

where $I$ is the normalization integral

$$I = B_\lambda^2 \int_{-a}^{0} \sin^2[K_\lambda^b(x+a)]\,dx + \int_{0}^{b} \sin^2\left[K_\lambda^b x + \delta(K_\lambda^b)\right]\,dx$$

$$= \frac{B_\lambda^2 a}{2}\left[1 - \frac{\sin 2K_\lambda^b a}{2K_\lambda^b a}\right] + \frac{b}{2}\left[1 + \frac{\sin 2K_\lambda^b b}{2K_\lambda^b b}\right], \tag{2.468}$$

where use has been made of eqn (2.466a) to write $\sin[2\delta(K_\lambda^b)] = \sin 2K_\lambda^b b$. The square of the $R$-matrix wave function $\psi_\lambda(x)$ at the boundary $x = b$ and the corresponding reduced width are thus

$$[\psi_\lambda(b)]^2 = \frac{1}{I}, \tag{2.469a}$$

$$\gamma_\lambda^2 = \frac{\hbar^2}{2mI}. \tag{2.469b}$$

If $K_\lambda^b a$ is not further away from the $k_\lambda^\delta a$ of eqn (2.420b) than $O(\alpha^2)$, i.e.,

$$|K_\lambda^b a - k_\lambda^\delta a| \lesssim O(\alpha^2), \tag{2.470}$$

we can use, for $B_\lambda^2$, the approximation given in eqn (2.345), i.e.,

$$[B_\lambda]^2 \approx \frac{\Gamma_\lambda^{(k)} a/2}{\left(K_\lambda^b a - k_\lambda^\delta a\right)^2 + \left(\Gamma_\lambda^{(k)} a/2\right)^2}, \tag{2.471}$$

with $\Gamma_\lambda^{(k)}$ given in eqn (2.346) as

$$\Gamma_\lambda^{(k)} \frac{a}{2} = (\lambda \pi \alpha)^2. \tag{2.472}$$

The above expressions should reduce, in the particular case $b = 0$, to the results obtained in Exercise 2.24. Since for $b = 0$, we have $K_\lambda^b = k_\lambda^\delta$, eqns (2.471), (2.468) and (2.467a) give, for $B_\lambda^2$, $I$ and $C_\lambda$, to leading order in the small quantity $\alpha \ll 1$,

$$B_\lambda^2 \approx \frac{2}{\Gamma_\lambda^{(k)} a} \approx \frac{1}{(\lambda \pi \alpha)^2}, \tag{2.473a}$$

$$I = \frac{B_\lambda^2 a}{2}\left[1 - \frac{\sin 2k_\lambda^\delta a}{2k_\lambda^\delta a}\right] \approx \frac{B_\lambda^2 a}{2} \approx \frac{a}{2(\lambda \pi \alpha)^2}, \tag{2.473b}$$

$$C_\lambda = \frac{B_\lambda}{\sqrt{I}} \approx \sqrt{\frac{2}{a}}. \tag{2.473c}$$

The normalization constant $C_\lambda$ coincides, to leading order $\alpha$, with that given in eqn (2.422). From eqn (2.469) we find

$$[\psi_\lambda(0)]^2 = \frac{1}{I} \approx \frac{2}{a}(\lambda \pi \alpha)^2, \tag{2.474}$$

to be compared with the result (2.423).

**Exercise 2.29** Equation (2.419) gives the $R$ function for the delta potential model, with the choice of the $R$-matrix boundary shown in Fig. 2.25. Expand that expression around one of the $R$-matrix levels, in order to express it in the single-level-plus-background approximation of eqn (2.445). Extract the reduced width for that level and the background term $R^B$.

Equation (2.419) gives the $R$ function

$$kR_0 = \frac{g(ka)}{f(ka)}. \tag{2.475}$$

As in Exercise 2.20, we expand the functions $g(ka)$ and $f(ka)$ around the zeros of $f(ka)$, which are the poles of $R_0$, to find

$$kR_0 = \frac{g(k_\lambda^\delta a) + g'(k_\lambda^\delta a)(ka - k_\lambda^\delta a) + \frac{1}{2}g''(k_\lambda^\delta a)(ka - k_\lambda^\delta a)^2 + \cdots}{f'(k_\lambda^\delta a)(ka - k_\lambda^\delta a) + \frac{1}{2}f''(k_\lambda^\delta a)(ka - k_\lambda^\delta a)^2 + \cdots}$$

$$= \frac{1}{f'(k_\lambda^\delta a)(ka - k_\lambda^\delta a)} \frac{g(k_\lambda^\delta a) + g'(k_\lambda^\delta a)(ka - k_\lambda^\delta a) + \cdots}{1 + \frac{1}{2}[f''(k_\lambda^\delta a)/f'(k_\lambda^\delta a)](ka - k_\lambda^\delta a) + \cdots}, \quad (2.476a)$$

and, expanding the denominator,

$$kR_0 = \frac{1}{f'(k_\lambda^\delta a)(ka - k_\lambda^\delta a)} \left\{ g(k_\lambda^\delta a) + \left[ g'(k_\lambda^\delta a) - \frac{1}{2} g(k_\lambda^\delta a) \frac{f''(k_\lambda^\delta a)}{f'(k_\lambda^\delta a)} \right] (ka - k_\lambda^\delta a) + \cdots \right\}$$

$$= \frac{g(k_\lambda^\delta a)}{f'(k_\lambda^\delta a)(ka - k_\lambda^\delta a)} + \left\{ \frac{g'(k_\lambda^\delta a)}{f'(k_\lambda^\delta a)} - \frac{1}{2} \frac{g(k_\lambda^\delta a) f''(k_\lambda^\delta a)}{[f'(k_\lambda^\delta a)]^2} \right\} + \cdots. \quad (2.476b)$$

Using the expansions in powers of $\alpha$ found in Exercise 2.20 for $g(k_\lambda^\delta a)$, $f(k_\lambda^\delta a)$ and their derivatives, we find

$$\frac{g(k_\lambda^\delta a)}{f'(k_\lambda^\delta a)} = -\frac{(\lambda \pi \alpha)^2 [1 + O(\alpha)]}{1 + \alpha + O(\alpha^2)} = -(\lambda \pi \alpha)^2 [1 + O(\alpha)], \quad (2.477)$$

$$\frac{g'(k_\lambda^\delta a)}{f'(k_\lambda^\delta a)} = \frac{\lambda \pi \alpha + O(\alpha^2)}{1 + \alpha + O(\alpha^2)} = \lambda \pi \alpha [1 + O(\alpha)]. \quad (2.478)$$

In the second term in the curly bracket of eqn (2.476b), we have $g(k_\lambda^\delta a) \sim O(\alpha^2)$ and $f''(k_\lambda^\delta a) = 2\alpha^2 (k_\lambda^\delta a) \cos k_\lambda^\delta a \sim O(\alpha^2)$. Substituting these expressions into eqn (2.476), we thus find

$$kR_0 = \frac{(\lambda \pi \alpha)^2 + \cdots}{(k_\lambda^\delta - k) a} + (\lambda \pi \alpha)[1 + O(\alpha)] + \cdots$$

$$= \frac{\frac{1}{2}(k + k_\lambda^\delta)(\hbar^2/ma)(\lambda \pi \alpha)^2}{E_\lambda^\delta - E} + (\lambda \pi \alpha)[1 + O(\alpha)] + \cdots$$

$$\approx \frac{k \gamma_\lambda^2}{E_\lambda - E} + (\lambda \pi \alpha)[1 + O(\alpha)] + \cdots, \quad (2.479)$$

where we have used the expression for the $R$-matrix poles and residues found for the present case in eqn (2.424). Result (2.479) has the form of the single-level-plus-background approximation, eqn (2.445). We see that the background term $R^B$ is $O(\alpha)$ and, for a fixed $\lambda$, can thus be made very small for a sufficiently strong delta barrier.

**Exercise 2.30** Equation (2.428) gives the $R$ function $R_b$ for the delta potential model, with the choice of the $R$-matrix boundary shown schematically in Fig. 2.26. Expand that expression for $R_b$ around the lowest $R$-matrix level, given in eqn (2.436), in order to write it in the single-level-plus-background approximation, eqn (2.445). Extract the reduced width for that level and the background term $R_b^B$.

Equation (2.428) gives $R_b$ as

$$kR_b = \frac{g(ka) + f(ka) \tan kb}{f(ka) - g(ka) \tan kb} \equiv \frac{G(ka)}{F(ka)}, \quad (2.480)$$

where, for convenience, we have defined the functions $G(ka)$ and $F(ka)$ as

$$G(ka) = g(ka) + f(ka) \tan \left( ka \frac{b}{a} \right), \quad (2.481a)$$

## RESONANCE THEORY

$$F(ka) = f(ka) - g(ka)\tan\left(ka\frac{b}{a}\right).\qquad(2.481\text{b})$$

We expand these functions around the first $R$-matrix level, i.e., around $K_1 a$, the zero of $F(ka)$ given by eqn (2.436), namely

$$K_{\lambda=1}a = \pi(1-\alpha).\qquad(2.482)$$

(The superscript $b$ in $K_{\lambda=1}^b$ is omitted for simplicity.) Proceeding as in the previous exercise, we find

$$kR_b = \frac{G(K_1 a)}{F'(K_1 a)(ka - K_1 a)} + \left\{\frac{G'(K_1 a)}{F'(K_1 a)} - \frac{1}{2}\frac{G(K_1 a)F''(K_1 a)}{[F'(K_1 a)]^2}\right\} + \cdots,\qquad(2.483)$$

where

$$F(K_1 a) = 0,\qquad(2.484\text{a})$$
$$F'(K_1 a) = -1 + O(\alpha^2),\qquad(2.484\text{b})$$
$$G(K_1 a) = (1 + \pi^2)\alpha^2 + O(\alpha^3),\qquad(2.484\text{c})$$
$$G'(K_1 a) = -\frac{1}{\pi}\left[1 + \left(2 + \frac{4}{3}\pi^2\right)\alpha + O(\alpha^2)\right].\qquad(2.484\text{d})$$

We have used

$$\tan K_1 b = \frac{1}{\pi}\left[1 + \left(\frac{1}{3}\pi^2 + 1\right)\alpha + O(\alpha^2)\right],\qquad(2.484\text{e})$$

obtained from eqns (2.484a) and (2.481b), i.e.,

$$\begin{aligned}\tan K_1 b &= \frac{f(K_1 a)}{g(K_1 a)}\\ &= \frac{\alpha K_1 a \cos K_1 a + \sin K_1 a}{\alpha K_1 a \sin K_1 a}\\ &= \frac{-\pi\alpha(1-\alpha)\cos\pi\alpha + \sin\pi\alpha}{\pi\alpha(1-\alpha)\sin\pi\alpha}\\ &= \frac{-\pi\alpha(1-\alpha)\left[1 - \pi^2\alpha^2/2 + O(\alpha^4)\right] + \left[\pi\alpha - \pi^3\alpha^3/6 + O(\alpha^5)\right]}{\pi\alpha(1-\alpha)\left[\pi\alpha - \pi^3\alpha^3/6 + O(\alpha^5)\right]}\\ &= \frac{\pi\alpha^2 + \frac{1}{3}\pi^3\alpha^3 + O(\alpha^4)}{\pi^2\alpha^2 - \pi^2\alpha^3 + O(\alpha^4)},\end{aligned}\qquad(2.485)$$

which gives the result (2.484e). Equation (2.484b) is obtained from

$$\begin{aligned}F'(K_1 a) &= (1+\alpha)\cos K_1 a - \alpha K_1 a \sin K_1 a\left[1 + \frac{b}{a}\sec^2 K_1 b\right]\\ &\quad - \alpha\sin K_1 a\tan K_1 b - \alpha K_1 a \cos K_1 a\tan K_1 b\\ &= -(1+\alpha)\cos\pi\alpha - \pi\alpha(1-\alpha)\sin\pi\alpha\left[1 + \frac{b}{a}\sec^2 K_1 b\right]\\ &\quad - \alpha\sin\pi\alpha\tan K_1 b + \pi\alpha(1-\alpha)\cos\pi\alpha\tan K_1 b\end{aligned}$$

$$= -(1+\alpha)\left[1+O(\alpha^2)\right] + O(\alpha^2)$$
$$+ \pi\alpha(1-\alpha)\left[1+O(\alpha^2)\right]\frac{1}{\pi}\left[1+\left(\frac{1}{3}\pi^2+1\right)\alpha+O(\alpha^2)\right]$$
$$= -1 + O(\alpha^2). \tag{2.486}$$

Equation (2.484c) is obtained from

$$\begin{aligned}G(K_1 a) &= \alpha K_1 a \sin K_1 a + [\alpha K_1 a \cos K_1 a + \sin K_1 a] \tan K_1 b \\ &= \pi\alpha(1-\alpha)\sin\pi\alpha + [-\pi\alpha(1-\alpha)\cos\pi\alpha + \sin\pi\alpha]\tan K_1 b \\ &= \pi\alpha(1-\alpha)\left[\pi\alpha+O(\alpha^3)\right] \\ &\quad + \left\{-\pi\alpha(1-\alpha)\left[1+O(\alpha^2)\right] + \pi\alpha + O(\alpha^3)\right\}\tan K_1 b \\ &= (1+\pi^2)\alpha^2 + O(\alpha^3). \end{aligned} \tag{2.487}$$

Equation (2.484d) is obtained from

$$\begin{aligned}G'(K_1 a) &= \alpha \sin K_1 a + \alpha K_1 a \cos K_1 a + [\alpha K_1 a \cos K_1 a + \sin K_1 a]\frac{b}{a}\sec^2 K_1 b \\ &\quad + [\alpha \cos K_1 a - \alpha K_1 a \sin K_1 a + \cos K_1 a]\tan K_1 b \\ &= \alpha \sin\pi\alpha - \pi\alpha(1-\alpha)\cos\pi\alpha \\ &\quad + [-\pi\alpha(1-\alpha)\cos\pi\alpha + \sin\pi\alpha]\frac{b}{a}(1+\tan^2 K_1 b) \\ &\quad + [-\alpha\cos\pi\alpha - \pi\alpha(1-\alpha)\sin\pi\alpha - \cos\pi\alpha]\tan K_1 b \\ &= \alpha\left[\pi\alpha + O(\alpha^3)\right] - \pi\alpha(1-\alpha)\left[1+O(\alpha^2)\right] \\ &\quad + \left\{\pi\alpha + O(\alpha^3) - \pi\alpha(1-\alpha)\left[1+O(\alpha^2)\right]\right\}\frac{b}{a}(1+\tan^2 K_1 b) \\ &\quad - \left\{\alpha\left[1+O(\alpha^2)\right] + \pi\alpha(1-\alpha)\left[\pi\alpha + O(\alpha^3)\right]\right. \\ &\quad \left. + 1 - \frac{\pi^2\alpha^2}{2} + O(\alpha^4)\right\}\tan K_1 b \\ &= -\frac{1}{\pi}\left[1+\left(2+\frac{4}{3}\pi^2\right)\alpha + O(\alpha^2)\right]. \end{aligned} \tag{2.488}$$

Substituting eqns (2.484) into (2.483), we find, for $kR_b$, (notice that in the last term of eqn (2.483), $G \sim \alpha^2$) that

$$\begin{aligned} kR_b &= \frac{(1+\pi^2)\alpha^2 + O(\alpha^3)}{[1+O(\alpha^2)](K_1-k)a} + \frac{\frac{1}{\pi}\left[1+\left(2+\frac{4}{3}\pi^2\right)\alpha + O(\alpha^2)\right]}{1+O(\alpha^2)} + \cdots \\ &= \frac{(1+\pi^2)\alpha^2}{(K_1-k)a} + \frac{1}{\pi}\left[1+\left(2+\frac{4}{3}\pi^2\right)\alpha + O(\alpha^2)\right] + \cdots \\ &\approx \frac{k\left[(\hbar^2/2m)(2/a)(1+\pi^2)\alpha^2\right]}{E_1 - E} + \frac{1}{\pi}\left[1+\left(2+\frac{4}{3}\pi^2\right)\alpha + O(\alpha^2)\right] + \cdots \\ &= \frac{k\gamma_1^2}{E_1 - E} + \frac{1}{\pi}\left[1+\left(2+\frac{4}{3}\pi^2\right)\alpha + O(\alpha^2)\right] + \cdots. \end{aligned} \tag{2.489}$$

We are assuming that we are in the region $|ka - K_1 a| \lesssim O(\alpha^2)$. Result (2.489) has the form of the single-level-plus-background approximation, eqn (2.445). The pole and residue of the resonant term coincide with the expressions given for the present case in eqns (2.436) and (2.443). The background term is given by

$$kR_b^B = K_1 R_b^B + O(\alpha^2) = \frac{1}{\pi}\left[1 + \left(2 + \frac{4}{3}\pi^2\right)\alpha + O(\alpha^2)\right] + \cdots; \qquad (2.490)$$

it is $O(\alpha^0)$ and hence *cannot be made arbitrarily small by increasing the strength of the delta barrier.*

# 3

# INTRODUCTION TO THE QUANTUM MECHANICAL TIME-INDEPENDENT SCATTERING THEORY II: SCATTERING INSIDE WAVEGUIDES AND CAVITIES

In this chapter we extend the analysis of Chapter 2 to the study of quantum mechanical scattering by systems in more than one dimension. In all of the problems to be considered here, the scattering takes place inside a structure that possesses impenetrable walls. It should be noted that the existence of the hard-wall boundaries makes the scattering problem markedly different from those treated in standard texts. Although we refer specifically to two-dimensional systems, the analysis can be easily generalized to higher dimensions. In Section 3.1 we study the problem of a waveguide with uniform cross-section containing scatterers in its interior; this is referred to as a *quasi-one-dimensional* system. We generalize to this case the Lippmann–Schwinger equation and the concepts of scattering matrix $S$ and transfer matrix $M$ that were introduced earlier in Chapter 2 for the strictly one-dimensional case. In Section 3.2 we deal with a cavity with an arbitrary shape connected to the outside by a number of waveguides and study the related scattering problem. Wigner's $R$-matrix theory is generalized to this situation in Section 3.3.

## 3.1 Quasi-one-dimensional scattering theory

### 3.1.1 *The reflection and transmission amplitudes; the Lippmann–Schwinger coupled equations*

Consider the one-particle scattering problem defined inside the two-dimensional structure shown in Fig. 3.1. The system consists of a waveguide of infinite length, constant width $W$ and *impenetrable walls*; the scattering potential is real and nonzero within a portion of the waveguide of length $L$. Only the case with TRI will be considered in detail. The generalization to non-TRI cases will be commented upon following eqn (3.85) and in the last paragraph of this chapter. If $H_0$ denotes the kinetic energy operator

$$H_0 = -\frac{\hbar^2}{2m}\left(\frac{\partial^2}{\partial x^2} + \frac{\partial^2}{\partial y^2}\right), \qquad (3.1)$$

then the problem inside the waveguide is described by the Schrödinger equation

$$(E - H_0)\,\psi(x,y) = V(x,y)\psi(x,y), \qquad (3.2\mathrm{a})$$

with the lateral boundary conditions

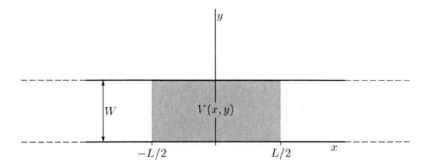

FIG. 3.1. The quasi-one-dimensional system studied in the text. The waveguide has a uniform width $W$ and the scattering potential $V(x,y)$ is nonzero only within a portion of length $L$.

$$\psi(x,0) = \psi(x,W) = 0. \tag{3.2b}$$

Equation (3.2a) can also be written as

$$\left(\frac{\partial^2}{\partial x^2} + \frac{\partial^2}{\partial y^2} + k^2\right)\psi(x,y) = U(x,y)\psi(x,y), \tag{3.2c}$$

where we have defined

$$k^2 = \frac{2mE}{\hbar^2}, \tag{3.3}$$

$$U = \frac{2mV}{\hbar^2}. \tag{3.4}$$

Equations (3.2) are the extension of eqns (2.6) to a quasi-one-dimensional system. Consider the auxiliary differential equation

$$\left(\frac{\partial^2}{\partial y^2} + K_{\perp n}^2\right)\chi_n(y) = 0, \tag{3.5a}$$

with the boundary conditions

$$\chi_n(0) = \chi_n(W) = 0. \tag{3.5b}$$

The solutions of (3.5), i.e.,

$$\chi_n(y) = \sqrt{\frac{2}{W}}\sin K_{\perp n} y, \quad K_{\perp n} = \frac{n\pi}{W}, \tag{3.6}$$

with energies

$$\mathcal{E}_{\perp n} = \frac{\hbar^2 K_{\perp n}^2}{2m}, \tag{3.7}$$

(the subscript '$\perp$' stands for 'transverse') are orthonormal, i.e.,

$$(\chi_n, \chi_m) = \int_0^W \chi_n(y)\chi_m(y)\, \mathrm{d}y = \delta_{nm}, \tag{3.8}$$

and form a complete set of functions in the transverse variable $y$, satisfying the boundary condition (3.2b). These solutions will be used extensively in what follows.

In the absence of a scattering potential inside the waveguide, the functions

$$\phi_s\left(E_\parallel; x\right) \chi_n(y) \tag{3.9}$$

are solutions of the unperturbed Schrödinger equation, with energy

$$E = E_\parallel + \mathcal{E}_{\perp n}. \tag{3.10}$$

Here, the subscript '$\parallel$' stands for 'longitudinal'. The wave function $\phi_s(E_\parallel; x)$ is the one-dimensional plane wave of eqn (2.2), i.e.,

$$\phi_s(E_\parallel; x) = \frac{e^{isk_\parallel x}}{\sqrt{2\pi\hbar^2 k_\parallel/m}}, \tag{3.11}$$

associated with the 'longitudinal energy'

$$E_\parallel = \frac{\hbar k_\parallel^2}{2m} \tag{3.12}$$

and the corresponding wave number $k_\parallel$. Thus the functions (3.9) satisfy the equation

$$\left[\frac{\partial^2}{\partial x^2} + \frac{\partial^2}{\partial y^2} + k_\parallel^2 + K_{\perp n}^2\right] \phi_s\left(E_\parallel; x\right) \chi_n(y) = 0 \tag{3.13}$$

and, for $E_\parallel \in (0, \infty)$ and $n \in (1, \infty)$, form a *complete set of orthonormal functions inside the waveguide*. The index $n$ in the wave function (3.9) denotes the so-called (transverse) *modes* or *channels*. Suppose now that *the total energy $E$ is given*. Then only those channels $n$ are allowed for which $\mathcal{E}_{\perp n} \leqslant E$; we say that these channels are *open* at that energy. If

$$N < \frac{kW}{\pi} < N + 1, \tag{3.14}$$

then there are precisely $N$ open channels. Thus, at the energy $E$ and for the channel $n$, the longitudinal energy $E_\parallel$ takes the value $E_{\parallel n}$ which satisfies

$$E_{\parallel n} + \mathcal{E}_{\perp n} = E, \tag{3.15a}$$

or, in terms of wave numbers,

$$k_{\|n}^2 + K_{\perp n}^2 = k^2, \tag{3.15b}$$

$E_{\|n}$ and $k_{\|n}$ being related as in eqn (3.12), i.e.,

$$E_{\|n} = \frac{\hbar^2 k_{\|n}^2}{2m}. \tag{3.15c}$$

From now on we shall generally omit, for simplicity of notation, the subscript '$\|$' from $k_{\|n}$ and $E_{\|n}$, and also the subscript '$\perp$' from $K_{\perp n}$ and $\mathcal{E}_{\perp n}$. Thus, $k_n$ will now denote the longitudinal wave number in channel $n$, corresponding to the longitudinal energy $E_n$, while $K_n$ will denote the transverse wave number for channel $n$, corresponding to the transverse channel energy $\mathcal{E}_n$.

We now turn to the complete Schrödinger equation (3.2), which admits the presence of a non-vanishing potential inside the waveguide. The wave function $\psi(x,y)$ can be expanded as a series in the functions $\chi_n(y)$, with $x$-dependent coefficients, i.e.,

$$\psi(x,y) = \sum_{n=1}^{\infty} [\psi(x)]_n \chi_n(y). \tag{3.16}$$

We substitute this expansion into the Schrödinger equation (3.2c), take the scalar product of both sides with $\chi_n(y)$ and, defining the matrix element

$$U_{nm}(x) = \int_0^W \chi_n(y) U(x,y) \chi_m(y) \, dy, \tag{3.17}$$

we obtain the coupled equations

$$\left(\frac{\partial^2}{\partial x^2} + k^2 - K_n^2\right) [\psi(x)]_n = \sum_{m=1}^{\infty} U_{nm}(x) [\psi(x)]_m. \tag{3.18}$$

Outside the range of the scattering potential, the components $[\psi(x)]_n$ satisfy the homogeneous equation

$$\left(\frac{\partial^2}{\partial x^2} + k^2 - K_n^2\right) [\psi(x)]_n = 0, \quad |x| > \frac{L}{2}. \tag{3.19}$$

For the open channels, i.e., $1 \leqslant n \leqslant N$, we have $K_n^2 < k^2$, so that $k_n^2 = k^2 - K_n^2 > 0$ and $k_n$ is real. Thus the $[\psi(x)]_n$ are linear combinations of the running waves of eqn (3.11) with $E_\| \equiv E_n$; for this reason, open channels are also referred to as *running modes*. On the other hand, for the *closed channels*, i.e., $n > N$, we have $K_n^2 > k^2$, so that now $k_n^2 = k^2 - K_n^2 < 0$, and hence $k_n = i\kappa_n$ is purely imaginary, giving rise to real exponentials. Since, asymptotically, i.e., for $x \to \pm\infty$, these modes can only give rise to exponentially decaying waves, they are also called *evanescent modes*. For closed channels we shall write the basic asymptotic solutions as

$$\phi_s(-\epsilon_n; x)\chi_n(y), \tag{3.20}$$

where

$$\phi_s(-\epsilon_n; x) = \frac{e^{-s\kappa_n x}}{e^{i\pi/4}\sqrt{2\pi\hbar^2\kappa_n/m}}. \tag{3.21}$$

We have defined

$$\epsilon_n = \frac{\hbar^2\kappa_n^2}{2m}, \tag{3.22}$$

which are related to the total and the transverse energies by

$$-\epsilon_n + \mathcal{E}_n = E, \tag{3.23a}$$

$$-\kappa_n^2 + K_n^2 = k^2. \tag{3.23b}$$

The $\phi_s(-\epsilon_n; x)$ of eqn (3.21) has been written as the *analytic continuation* of the plane wave $\phi_s(E_n; x)$ of eqn (3.11) ($E_\parallel$ having been replaced by $E_n$), when

$$k_n \Longrightarrow i\kappa_n, \tag{3.24a}$$

$$\sqrt{k_n} \Longrightarrow e^{i\pi/4}\sqrt{\kappa_n}. \tag{3.24b}$$

The above analytic continuation corresponds to the branch cut in the complex-$k$ plane being taken along the positive real axis from 0 to $\infty$, and staying on the first Riemann sheet.

The coupled eqns (3.18) thus split into the two sets

$$\left(\frac{\partial^2}{\partial x^2} + k_n^2\right)[\psi(x)]_n = \sum_{m=1}^{\infty} U_{nm}(x)[\psi(x)]_m, \tag{3.25a}$$

for the $N$ open channels, $n = 1, \ldots, N$, and

$$\left(\frac{\partial^2}{\partial x^2} - \kappa_n^2\right)[\psi(x)]_n = \sum_{m=1}^{\infty} U_{nm}(x)[\psi(x)]_m, \tag{3.25b}$$

for the closed channels, $n = N+1, \ldots, \infty$. Asymptotically, i.e., as $|x| \to \infty$, only the open channels contribute to the wave function, the contribution of the closed ones decreasing exponentially. Equations (3.25) represent an infinite set of coupled differential equations for the $[\psi(x)]_n$s, which have to be solved subject to the appropriate physical boundary conditions: an incident wave in one of the *open* channels, plus outgoing wave boundary conditions for $[\psi(x)]_n$, $n = 1, \ldots, N$, and, for $n \geqslant N+1$, asymptotically decaying waves, i.e., $[\psi(x)]_n \to 0$ as $|x| \to \infty$.

Just as in eqns (2.8) and (2.18) of the previous chapter, we introduce, for the $n$th open channel ($n = 1, \ldots, N$), the unperturbed Green function $g_0^{(\pm)}(E_n; x, x')$ for incoming or outgoing wave boundary conditions. It satisfies the equation

$$\left(\frac{\partial^2}{\partial x^2} + k_n^2\right) g_0^{(\pm)}(E_n; x, x') = \delta(x - x') \tag{3.26}$$

and is given by

$$g_0^{(\pm)}(E_n; x, x') = \pm \frac{e^{\pm ik_n|x-x'|}}{2ik_n}. \tag{3.27}$$

For the $n$th closed channel ($n = N+1, \ldots, \infty$) the corresponding Green function satisfies the equation

$$\left(\frac{\partial^2}{\partial x^2} - \kappa_n^2\right) g_0^{(\pm)}(-\epsilon_n; x, x') = \delta(x - x'). \tag{3.28}$$

For both the incoming and the outgoing wave boundary conditions the asymptotic closed-channel components of the wave function must decrease exponentially. We thus have the solution

$$g_0^{(\pm)}(-\epsilon_n; x, x') = -\frac{e^{-\kappa_n|x-x'|}}{2\kappa_n}. \tag{3.29}$$

Notice that a closed-channel Green function is related to an open-channel one through the relation

$$g_0^{(\pm)}(-\epsilon_n; x, x') = \left[g_0^{(\pm)}(k_n^2; x, x')\right]_{k_n = \pm i\kappa_n}. \tag{3.30}$$

Considering the right-hand side of eqns (3.25) as the inhomogeneous part of a set of otherwise homogeneous equations, we can use the unperturbed Green function to write down a set of integral equations for the perturbed wave function, just as we did in the previous chapter to obtain the Lippmann–Schwinger equation (2.10). We write the wave function of eqns (3.2) and (3.16) more explicitly by specifying the direction of propagation $s = \pm$ of the incident wave in channel $n_0$ (*an open channel*, of course) and the energy $E$, so that

$$\psi_{sn_0}^{(\pm)}(E; x, y) = \sum_{n=1}^{\infty} \left[\psi_{sn_0}^{(\pm)}(E; x)\right]_n \chi_n(y). \tag{3.31}$$

The generalization of the one-dimensional Lippmann–Schwinger equation (2.10) is now the set of coupled integral equations

$$\left[\psi_{sn_0}^{(\pm)}(E; x)\right]_n = \phi_s(E_{n_0}; x)\delta_{nn_0}$$
$$+ \sum_{m=1}^{\infty} \int g_0^{(\pm)}(E_n; x, x') U_{nm}(x') \left[\psi_{sn_0}^{(\pm)}(E; x')\right]_m dx', \tag{3.32a}$$

when $n$ is an open channel, and

$$\left[\psi_{sn_0}^{(\pm)}(E;x)\right]_n = \sum_{m=1}^{\infty}\int g_0^{(\pm)}(-\epsilon_n;x,x')U_{nm}(x')\left[\psi_{sn_0}^{(\pm)}(E;x')\right]_m dx', \quad (3.32b)$$

when $n$ is a closed channel. We have indicated explicitly that incidence is in the $s = \pm$ direction in the *open* channel $n_0 \leqslant N$, the total energy being $E$.

Just as in the one-channel case, here too the states $|\psi_{sn_0}^{(+)}(E)\rangle$ form a complete set of orthonormal states; this is also the case with the states $|\psi_{sn_0}^{(-)}(E)\rangle$. Should there be negative-energy bound states, they would have to be included.

Introducing the explicit form of eqns (3.27) and (3.29) for the Green function, we find the generalization of eqn (2.20) to be

$$\left[\psi_{sn_0}^{(\pm)}(E;x)\right]_n = \phi_s(E_{n_0};x)\delta_{nn_0}$$
$$\pm \frac{1}{2ik_n}\sum_{m=1}^{\infty}\int e^{\pm ik_n|x-x'|}U_{nm}(x')\left[\psi_{sn_0}^{(\pm)}(E;x')\right]_m dx', \quad (3.33a)$$

for open channels $n$, and

$$\left[\psi_{sn_0}^{(\pm)}(E;x)\right]_n = -\frac{1}{2\kappa_n}\sum_{m=1}^{\infty}\int e^{-\kappa_n|x-x'|}U_{nm}(x')\left[\psi_{sn_0}^{(\pm)}(E;x')\right]_m dx', \quad (3.33b)$$

for closed channels $n$.

*The unperturbed Green function in operator form*

Just as in eqn (2.44), we write the unperturbed Green function in operator form as

$$G_0^{(\pm)}(E) = \lim_{\eta\to 0^+}\frac{1}{E\pm i\eta - H_0}, \quad (3.34)$$

where $H_0$ is the two-dimensional kinetic energy operator given in eqn (3.1).

We now show that the matrix elements of the operator $G_0^{(\pm)}(E)$ give precisely the results (3.27) and (3.29). Consider the matrix element

$$\left[G_0^{(\pm)}(E;x,x')\right]_{nn'} = \lim_{\eta\to 0^+}\left\langle x,\chi_n\left|\frac{1}{E\pm i\eta - H_0}\right|x',\chi_{n'}\right\rangle. \quad (3.35)$$

We introduce, inside the matrix element, the complete set of states (3.9) to find

$$\left[G_0^{(\pm)}(E;x,x')\right]_{nn'}$$
$$= \lim_{\eta\to 0^+}\langle x,\chi_n|\frac{1}{E\pm i\eta - H_0}\sum_s\sum_{m=1}^{\infty}\int_0^{\infty}dE'\,|\phi_s(E')\chi_m\rangle\langle\phi_s(E')\chi_m|x',\chi_{n'}\rangle$$

$$= \lim_{\eta \to 0^+} \sum_s \int_0^\infty dE' \frac{\langle x, \chi_n | \phi_s(E')\chi_{n'}\rangle \langle \phi_s(E')\chi_{n'} | x', \chi_{n'}\rangle}{E \pm i\eta - (E' + \mathcal{E}_{n'})}$$

$$= \delta_{nn'} \lim_{\eta \to 0^+} \sum_s \int_0^\infty dE' \frac{\phi_s(E'; x)\phi_s^*(E'; x')}{(E - \mathcal{E}_{n'}) \pm i\eta - E'}. \tag{3.36}$$

In the evaluation of (3.36) we have two possibilities.

(i) $E > \mathcal{E}_n$

The calculation proceeds just as in Section 2.1.2, with the result

$$\left[G_0^{(\pm)}(E; x, x')\right]_{nn'} = \pm \frac{2m}{\hbar^2} \frac{e^{\pm i k_n |x - x'|}}{2ik_n} \delta_{nn'}$$

$$= \frac{2m}{\hbar^2} g_0^{(\pm)}(E_n; x, x')\delta_{nn'}, \quad E > E_n^t, \tag{3.37}$$

where $g_0^{(\pm)}(E_n; x, x')$ is given in eqn (3.27) and $E_n = E - \mathcal{E}_n$, as in eqn (3.15a). Notice that, whereas the first argument, $E$, of $G_0^{(\pm)}(E; x, x')$ indicates the total energy, the first argument, $E_n$, of $g_0^{(\pm)}(E_n; x, x')$ indicates the 'longitudinal' energy.

(ii) $E < \mathcal{E}_n$

We use eqn (3.23) to write

$$\left[G_0^{(\pm)}(E; x, x')\right]_{nn'} = -\frac{m}{\pi\hbar^2} \delta_{nn'} \lim_{\eta' \to 0^+} \int_{-\infty}^\infty \frac{e^{ik'(x-x')}}{(k')^2 + \kappa_n^2 \mp i\eta'} dk'. \tag{3.38}$$

The poles of the integrand in (3.38) are located at

$$k' \approx \begin{cases} i\kappa_n \left[1 \mp i\frac{\eta'}{2\kappa_n^2}\right] \equiv i\kappa_n \pm \zeta, \\ -i\kappa_n \left[1 \mp i\frac{\eta'}{2\kappa_n^2}\right] \equiv -(i\kappa_n \pm \zeta), \end{cases} \tag{3.39}$$

where $\zeta \to 0^+$ as $\eta' \to 0^+$. Fig. 3.2 shows the location of these poles in the complex-$k'$ plane for $G_0^{(+)}$ and $G_0^{(-)}$, as well as the appropriate paths of integration for $x - x' > 0$ and $x - x' < 0$. To show the poles explicitly, we write (3.38) as

$$\left[G_0^{(+)}(E; x, x')\right]_{nn'}$$
$$= -\frac{m}{\pi\hbar^2} \delta_{nn'} \lim_{\zeta \to 0^+} \int_{-\infty}^\infty \frac{e^{ik'(x-x')}}{[k' - (i\kappa_n + \zeta)][k' + (i\kappa_n + \zeta)]} dk', \tag{3.40}$$

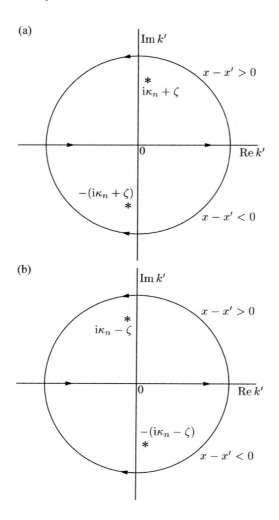

FIG. 3.2. Analytic properties of the integrand in eqn (3.38) for the Green function (a) $G^{(+)}$, and (b) $G^{(-)}$.

$$\left[G_0^{(-)}(E;x,x')\right]_{nn'}$$
$$= -\frac{m}{\pi\hbar^2}\delta_{nn'}\lim_{\zeta\to 0^+}\int_{-\infty}^{\infty}\frac{e^{ik'(x-x')}}{[k'-(i\kappa_n-\zeta)][k'+(i\kappa_n-\zeta)]}\,dk'. \tag{3.41}$$

Cauchy's theorem of contour integration then gives

$$\left[G_0^{(\pm)}(E; x > x')\right]_{nn'} = -\frac{2m}{\hbar^2}\frac{e^{-\kappa_n(x-x')}}{2\kappa_n}\delta_{nn'},$$
$$\left[G_0^{(\pm)}(E; x < x')\right]_{nn'} = -\frac{2m}{\hbar^2}\frac{e^{\kappa_n(x-x')}}{2\kappa_n}\delta_{nn'}.$$
(3.42)

These two equations can be condensed as

$$\left[G_0^{(\pm)}(E; x, x')\right]_{nn'} = -\frac{2m}{\hbar^2}\frac{e^{-\kappa_n|x-x'|}}{2\kappa_n}\delta_{nn'}$$
$$= \frac{2m}{\hbar^2}g_0^{(\pm)}(-\epsilon_n; x, x')\delta_{nn'}, \quad E < \mathcal{E}_n, \quad (3.43)$$

where $g_0^{(\pm)}(-\epsilon_n; x, x')$ is given by eqn (3.29) and $\epsilon_n = \mathcal{E}_n - E$, as in eqn (3.23a). This completes the proof.

Just as in eqn (2.46), we can now write the Lippmann–Schwinger equation in operator form as

$$\left|\psi_{sn_0}^{(\pm)}(E)\right\rangle = |\phi_s(E_{n_0})\chi_{n_0}\rangle + \frac{1}{E - H_0 \pm i\eta}V\left|\psi_{sn_0}^{(\pm)}(E)\right\rangle, \quad (3.44)$$

with the understanding that $\eta \to 0^+$. The overlap of this equation with the transverse wave functions $\chi_n$ gives the coupled eqns (3.32).

*The asymptotic form of the wave function*

From the above equations we can now write down the wave function outside the region of the potential. We consider outgoing wave boundary conditions and assume incidence from the left ($s = +$). For $x < -L/2$ we have, from eqns (3.33),

$$\left[\psi_{+n_0}^{(+)}(E; x)\right]_n = \phi_+(E_{n_0}; x)\delta_{nn_0}$$
$$- 2\pi i\phi_-(E_n; x)\sum_{m=1}^{\infty}\int \phi_+(E_n; x')V_{nm}(x')\left[\psi_{+n_0}^{(+)}(E; x')\right]_m dx',$$
(3.45a)

for $1 \leqslant n \leqslant N$, and

$$\left[\psi_{+n_0}^{(+)}(E; x)\right]_n$$
$$= -2\pi i\phi_-(-\epsilon_n; x)\sum_{m=1}^{\infty}\int \phi_+(-\epsilon_n; x')V_{nm}(x')\left[\psi_{+n_0}^{(+)}(E; x')\right]_m dx',$$
(3.45b)

for $N+1 \leqslant n$. The amplitude $r_{nn_0}$ of the reflected wave in channel $n$ when incidence is from channel $n_0$ is thus

$$r_{nn_0} = -2\pi i \sum_{m=1}^{\infty} \int \phi_+(E_n;x) V_{nm}(x) \left[\psi_{+n_0}^{(+)}(E;x)\right]_m dx$$

$$= -2\pi i \sum_{m=1}^{\infty} \left(\phi_-(E_n;x), V_{nm}(x) \left[\psi_{+n_0}^{(+)}(E;x)\right]_m\right), \quad 1 \leqslant n \leqslant N, \quad (3.46a)$$

$$r_{nn_0} = -2\pi i \sum_{m=1}^{\infty} \int \phi_+(-\epsilon_n;x) V_{nm}(x) \left[\psi_{+n_0}^{(+)}(E;x)\right]_m dx', \quad N+1 \leqslant n. \quad (3.46b)$$

The integral in eqn (3.46b) exists, since we have assumed our potential to be of finite support.

Similarly, for $x > L/2$ we have

$$\left[\psi_{+n_0}^{(+)}(E;x)\right]_n = \phi_+(E_{n_0};x)\delta_{nn_0}$$
$$- 2\pi i \phi_+(E_n;x) \sum_{m=1}^{\infty} \int \phi_-(E_n;x') V_{nm}(x') \left[\psi_{+n_0}^{(+)}(E;x')\right]_m dx', \quad (3.47a)$$

for $1 \leqslant n \leqslant N$, and

$$\left[\psi_{+n_0}^{(+)}(E;x)\right]_n$$
$$= -2\pi i \phi_+(-\epsilon_n;x) \sum_{m=1}^{\infty} \int \phi_-(-\epsilon_n;x') V_{nm}(x') \left[\psi_{+n_0}^{(+)}(E;x')\right]_m dx', \quad (3.47b)$$

for $N+1 \leqslant n$. The amplitude $t_{nn_0}$ of the transmitted wave in channel $n$ when incidence is from channel $n_0$ is thus

$$t_{nn_0} = \delta_{nn_0} - 2\pi i \sum_{m=1}^{\infty} \int \phi_-(E_n;x) V_{nm}(x) \left[\psi_{+n_0}^{(+)}(E;x)\right]_m dx$$

$$= \delta_{nn_0} - 2\pi i \sum_{m=1}^{\infty} \left(\phi_+(E_n;x), V_{nm}(x) \left[\psi_{+n_0}^{(+)}(E;x)\right]_m\right), \quad 1 \leqslant n \leqslant N, \quad (3.48a)$$

$$t_{nn_0} = -2\pi i \sum_{m=1}^{\infty} \int \phi_-(-\epsilon_n;x) V_{nm}(x) \left[\psi_{+n_0}^{(+)}(E;x)\right]_m dx, \quad N+1 \leqslant n. \quad (3.48b)$$

We notice that, using the expansion of eqn (3.16) and the definition (3.17) of the potential matrix elements, we can write the reflection amplitude of eqn (3.46) and the transmission amplitude of eqn (3.48) in terms of the full two-variable (i.e., $x$ and $y$) wave function and the potential $V(x, y)$ as

$$r_{nn_0} = -2\pi i \sum_{m=1}^{\infty} \left( \phi_-(E_n;x)\chi_n(y), V(x,y)\chi_m(y) \left[ \psi_{+n_0}^{(+)}(E;x) \right]_m \right)$$

$$= -2\pi i \left( \phi_-(E_n;x)\chi_n(y), V(x,y)\psi_{+n_0}^{(+)}(E;x,y) \right)$$

$$= -2\pi i \left\langle \phi_-(E_n)\chi_n \left| V \right| \psi_{+n_0}^{(+)}(E) \right\rangle, \quad 1 \leqslant n \leqslant N, \tag{3.49}$$

and

$$t_{nn_0} = \delta_{nn_0} - 2\pi i \left( \phi_+(E_n;x)\chi_n(y), V(x,y)\psi_{+n_0}^{(+)}(E;x,y) \right)$$

$$= \delta_{nn_0} - 2\pi i \left\langle \phi_+(E_n)\chi_n \left| V \right| \psi_{+n_0}^{(+)}(E) \right\rangle, \quad 1 \leqslant n \leqslant N, \tag{3.50}$$

respectively. These last two equations have a structure similar to the one-dimensional equations (2.32) and (2.33) which we derived in the previous chapter.

**Exercise 3.1** Write the coupled Lippmann–Schwinger equations for outgoing wave boundary conditions and the expressions for the reflection and the transmission amplitudes for a problem consisting of (i) two open channels, and (ii) one open and one closed channel.

(i) *Two open channels*

Firstly, suppose the two channels to be open, i.e.,

$$K_1^2 < K_2^2 < k^2, \tag{3.51a}$$

$$k_1^2 = k^2 - K_1^2 > 0, \quad k_2^2 = k^2 - K_2^2 > 0. \tag{3.51b}$$

Assuming also that *incidence is from the left on channel 1*, the two components of the wave function satisfy the coupled Lippmann–Schwinger equations

$$\left[ \psi_{+,1}^{(+)}(E;x) \right]_1 = \phi_+(E_1;x) + \sum_{m=1}^{2} \int \frac{e^{ik_1|x-x'|}}{2ik_1} U_{1m}(x') \left[ \psi_{+,1}^{(+)}(E;x') \right]_m dx', \tag{3.52a}$$

$$\left[ \psi_{+,1}^{(+)}(E;x) \right]_2 = \sum_{m=1}^{2} \int \frac{e^{ik_2|x-x'|}}{2ik_2} U_{2m}(x') \left[ \psi_{+,1}^{(+)}(E;x') \right]_m dx'. \tag{3.52b}$$

The $n$th component $[\psi_{sn_0}^{(+)}(E;x)]_n$ of the wave function was defined in eqn (3.31); we recall that the indices $s$ and $n_0$ represent the direction of propagation and the channel, respectively, of the incident wave.

Outside the potential region, using the definition (3.11) we have, for $x < -L/2$ (and outgoing wave boundary conditions),

$$\left[ \psi_{+,1}^{(+)}(E;x) \right]_1 = \phi_+(E_1;x)$$

$$- 2\pi i \phi_-(E_1;x) \sum_{m=1}^{2} \int \phi_+(E_1;x') V_{1m}(x') \left[ \psi_{+,1}^{(+)}(E;x') \right]_m dx', \tag{3.53a}$$

$$\left[\psi_{+,1}^{(+)}(E;x)\right]_2 = -2\pi i \phi_-(E_2;x) \sum_{m=1}^{2} \int \phi_+(E_2;x')V_{2m}(x')\left[\psi_{+,1}^{(+)}(E;x')\right]_m dx', \tag{3.53b}$$

giving the reflection amplitudes

$$r_{11} = -2\pi i \sum_{m=1}^{2} \int \phi_+(E_1;x)V_{1m}(x)\left[\psi_{+,1}^{(+)}(E;x)\right]_m dx, \tag{3.54a}$$

$$r_{21} = -2\pi i \sum_{m=1}^{2} \int \phi_+(E_2;x)V_{2m}(x)\left[\psi_{+,1}^{(+)}(E;x)\right]_m dx, \tag{3.54b}$$

with similar expressions for $r_{12}$ and $r_{22}$, which correspond to incidence on channel 2. For $x > L/2$, we have

$$\left[\psi_{+,1}^{(+)}(E;x)\right]_1 = \phi_+(E_1;x)$$
$$- 2\pi i \phi_+(E_1;x) \sum_{m=1}^{2} \int \phi_-(E_1;x')V_{1m}(x')\left[\psi_{+,1}^{(+)}(E;x')\right]_m dx', \tag{3.55a}$$

$$\left[\psi_{+,1}^{(+)}(E;x)\right]_2 = -2\pi i \phi_+(E_2;x) \sum_{m=1}^{2} \int \phi_-(E_2;x')V_{2m}(x')\left[\psi_{+,1}^{(+)}(E;x')\right]_m dx', \tag{3.55b}$$

giving the transmission amplitudes

$$t_{11} = 1 - 2\pi i \sum_{m=1}^{2} \int \phi_-(E_1;x)V_{1m}(x)\left[\psi_{+,1}^{(+)}(E;x)\right]_m dx, \tag{3.56a}$$

$$t_{21} = -2\pi i \sum_{m=1}^{2} \int \phi_-(E_2;x)V_{2m}(x)\left[\psi_{+,1}^{(+)}(E;x)\right]_m dx, \tag{3.56b}$$

with similar expressions for $t_{12}$ and $t_{22}$, corresponding to incidence on channel 2.

(ii) *One open and one closed channel*
If channel 1 is open and channel 2 is closed, i.e.,

$$K_1^2 < k^2 < K_2^2, \tag{3.57a}$$
$$k_1^2 = k^2 - K_1^2 > 0, \quad k_2^2 = k^2 - K_2^2 = -\kappa_2^2 < 0, \tag{3.57b}$$

*incidence can only be from channel 1* and we have the two coupled equations

$$\left[\psi_{+,1}^{(+)}(E;x)\right]_1 = \phi_+(E_1;x) + \sum_{m=1}^{2} \int \frac{e^{ik_1|x-x'|}}{2ik_1} U_{1m}(x')\left[\psi_{+,1}^{(+)}(E;x')\right]_m dx', \tag{3.58a}$$

$$\left[\psi_{+,1}^{(+)}(E;x)\right]_2 = \sum_{m=1}^{2} \int \frac{e^{-\kappa_2|x-x'|}}{-2\kappa_2} U_{2m}(x')\left[\psi_{+,1}^{(+)}(E;x')\right]_m dx'. \tag{3.58b}$$

We observe that eqns (3.58), and hence also their solutions, can be obtained from eqns (3.52) through the analytic continuation (3.24) for $n = 2$. Notice though that there is no $\sqrt{\kappa_2}$ occurring in eqn (3.58), so that (3.24b) is not required.

Outside the potential region, and using the definitions (3.11) and (3.21), we have, for $x < -L/2$ (and outgoing wave boundary conditions),

$$\left[\psi_{+,1}^{(+)}(E;x)\right]_1 = \phi_+(E_1;x)$$
$$- 2\pi i \phi_-(E_1;x) \sum_{m=1}^{2} \int \phi_+(E_1;x') V_{1m}(x') \left[\psi_{+,1}^{(+)}(E;x')\right]_m dx', \qquad (3.59a)$$

$$\left[\psi_{+,1}^{(+)}(E;x)\right]_2 = -2\pi i \phi_-(-\epsilon_2;x) \sum_{m=1}^{2} \int \phi_+(-\epsilon_2;x') V_{2m}(x') \left[\psi_{+,1}^{(+)}(E;x')\right]_m dx', \qquad (3.59b)$$

giving the reflection amplitudes

$$r_{11} = -2\pi i \sum_{m=1}^{2} \int \phi_+(E_1;x) V_{1m}(x) \left[\psi_{+,1}^{(+)}(E;x)\right]_m dx, \qquad (3.60a)$$

$$r_{21} = -2\pi i \sum_{m=1}^{2} \int \phi_+(-\epsilon_2;x) V_{2m}(x) \left[\psi_{+,1}^{(+)}(E;x)\right]_m dx. \qquad (3.60b)$$

Notice that $r_{11}$ is the reflection amplitude in the open channel 1, while $r_{21}$ is the reflection amplitude in the closed channel 2.

For $x > L/2$, we have

$$\left[\psi_{+,1}^{(+)}(E;x)\right]_1 = \phi_+(E_1;x)$$
$$- 2\pi i \phi_+(E_1;x) \sum_{m=1}^{2} \int \phi_-(E_1;x') V_{1m}(x') \left[\psi_{+,1}^{(+)}(E;x')\right]_m dx', \qquad (3.61a)$$

$$\left[\psi_{+,1}^{(+)}(E;x)\right]_2 = -2\pi i \phi_+(-\epsilon_2;x) \sum_{m=1}^{2} \int \phi_-(-\epsilon_2;x') V_{1m}(x') \left[\psi_{+,1}^{(+)}(E;x')\right]_m dx', \qquad (3.61b)$$

giving the transmission amplitudes

$$t_{11} = 1 - 2\pi i \sum_{m=1}^{2} \int \phi_-(E_1;x) V_{1m}(x) \left[\psi_{+,1}^{(+)}(E;x)\right]_m dx, \qquad (3.62a)$$

$$t_{21} = -2\pi i \sum_{m=1}^{2} \int \phi_-(-\epsilon_2;x) V_{2m}(x) \left[\psi_{+,1}^{(+)}(E;x)\right]_m dx. \qquad (3.62b)$$

Here, $t_{11}$ is the transmission amplitude in the open channel 1, while $t_{21}$ is the transmission amplitude in the closed channel 2.

The reflection and the transmission amplitudes $r_{11}$, $r_{21}$ and $t_{11}$, $t_{21}$ of eqns (3.60) and (3.62), respectively, are the analytic continuation of those of eqns (3.54) and (3.56), under the rule (3.24) for $n = 2$ (see also the comment immediately following eqn (3.58)). Now the rule includes (3.24b), simply because we define $t_{21}$ and $r_{21}$ as the coefficients of some basic wave function for the closed channel, and we need a phase convention to define the latter.

---

We now wish to generalize, to the multichannel case, the discussion given in the previous chapter around eqn (2.39). We shall find it simpler to carry on the analysis for a two-channel problem—consisting first of two open channels, and then one open and one closed channel, just as in the previous exercise—the extension to an arbitrary number of open and closed channels being straightforward.

*The case of two open channels*

We have found the asymptotic behavior of the *four* linearly independent solutions $\Psi_{+,1}^{(+)}(E;x)$, $\Psi_{+,2}^{(+)}(E;x)$, $\Psi_{-,1}^{(+)}(E;x)$ and $\Psi_{-,2}^{(+)}(E;x)$ (we recall that the lower index $n_0 = 1, 2$ denotes the incidence channel; the capital $\Psi$ notation indicates the column vector containing the two components $[\psi_{+,n_0}^{(+)}(E;x)]_n$, with $n = 1, 2$) to be

$$\Psi_{+,1}^{(+)}(E;x) \sim \begin{cases} \begin{bmatrix} \phi_+(E_1;x) + r_{11}\phi_-(E_1;x) \\ r_{21}\phi_-(E_2;x) \end{bmatrix}, & x \to -\infty, \\ \begin{bmatrix} t_{11}\phi_+(E_1;x) \\ t_{21}\phi_+(E_2;x) \end{bmatrix}, & x \to +\infty, \end{cases}$$

$$\Psi_{+,2}^{(+)}(E;x) \sim \begin{cases} \begin{bmatrix} r_{12}\phi_-(E_1;x) \\ \phi_+(E_2;x) + r_{22}\phi_-(E_2;x) \end{bmatrix}, & x \to -\infty, \\ \begin{bmatrix} t_{12}\phi_+(E_1;x) \\ t_{22}\phi_+(E_2;x) \end{bmatrix}, & x \to +\infty, \end{cases}$$

$$\Psi_{-,1}^{(+)}(E;x) \sim \begin{cases} \begin{bmatrix} t'_{11}\phi_-(E_1;x) \\ t'_{21}\phi_-(E_2;x) \end{bmatrix}, & x \to -\infty, \\ \begin{bmatrix} \phi_-(E_1;x) + r'_{11}\phi_+(E_1;x) \\ r'_{21}\phi_+(E_2;x) \end{bmatrix}, & x \to +\infty, \end{cases}$$

QUASI-ONE-DIMENSIONAL SCATTERING THEORY 135

$$\Psi^{(+)}_{-,2}(E;x) \sim \begin{cases} \begin{bmatrix} t'_{12}\phi_-(E_1;x) \\ t'_{22}\phi_-(E_2;x) \end{bmatrix}, & x \to -\infty, \\ \begin{bmatrix} r'_{12}\phi_+(E_1;x) \\ \phi_-(E_2;x) + r'_{22}\phi_+(E_2;x) \end{bmatrix}, & x \to +\infty. \end{cases}$$
(3.63)

The various reflection and transmission amplitudes indicated above are given in Exercise 3.1. From the linearity of the Schrödinger equation, it follows that the linear combination

$$\Psi(E;x) = \sum_{n_0=1}^{2} \left[ a^{(1)}_{n_0} \Psi^{(+)}_{+,n_0}(E;x) + a^{(2)}_{n_0} \Psi^{(+)}_{-,n_0}(E;x) \right]$$
(3.64)

of the above four solutions is a solution of the Schrödinger equation for the same energy. For arbitrary values of the coefficients $a^{(1)}_{n_0}$ and $a^{(2)}_{n_0}$, this wave function represents the *most general solution of the Schrödinger equation* at the energy $E$. Its asymptotic behavior is

$$\Psi(E;x) \sim \begin{cases} \begin{bmatrix} a^{(1)}_1 \phi_+(E_1;x) + \left(r_{11}a^{(1)}_1 + r_{12}a^{(1)}_2 + t'_{11}a^{(2)}_1 + t'_{12}a^{(2)}_2\right) \phi_-(E_1;x) \\ a^{(1)}_2 \phi_+(E_2;x) + \left(r_{21}a^{(1)}_1 + r_{22}a^{(1)}_2 + t'_{21}a^{(2)}_1 + t'_{22}a^{(2)}_2\right) \phi_-(E_2;x) \end{bmatrix} \\ = \begin{bmatrix} a^{(1)}_1 \phi_+(E_1;x) + b^{(1)}_1 \phi_-(E_1;x) \\ a^{(1)}_2 \phi_+(E_2;x) + b^{(1)}_2 \phi_-(E_2;x) \end{bmatrix}, \quad x \to -\infty, \\[2ex] \begin{bmatrix} a^{(2)}_1 \phi_-(E_1;x) + \left(t_{11}a^{(1)}_1 + t_{12}a^{(1)}_2 + r'_{11}a^{(2)}_1 + r'_{12}a^{(2)}_2\right) \phi_+(E_1;x) \\ a^{(2)}_2 \phi_-(E_2;x) + \left(t_{21}a^{(1)}_1 + t_{22}a^{(1)}_2 + r'_{21}a^{(2)}_1 + r'_{22}a^{(2)}_2\right) \phi_+(E_2;x) \end{bmatrix} \\ = \begin{bmatrix} a^{(2)}_1 \phi_-(E_1;x) + b^{(2)}_1 \phi_+(E_1;x) \\ a^{(2)}_2 \phi_-(E_2;x) + b^{(2)}_2 \phi_+(E_2;x) \end{bmatrix}, \quad x \to +\infty. \end{cases}$$
(3.65)

The coefficients $a^{(1)}_n$ and $a^{(2)}_n$ represent the amplitudes of the incoming waves in channel $n$ on either side, and $b^{(1)}_n$ and $b^{(2)}_n$ represent the amplitudes of the outgoing waves. The above equation shows that the two sets of amplitudes are *linearly related*, according to the equation

$$\begin{bmatrix} b^{(1)}_1 \\ b^{(1)}_2 \\ b^{(2)}_1 \\ b^{(2)}_2 \end{bmatrix} = \left[ \begin{array}{cc|cc} r_{11} & r_{12} & t'_{11} & t'_{12} \\ r_{21} & r_{22} & t'_{21} & t'_{22} \\ \hline t_{11} & t_{12} & r'_{11} & r'_{12} \\ t_{21} & t_{22} & r'_{21} & r'_{22} \end{array} \right] \begin{bmatrix} a^{(1)}_1 \\ a^{(1)}_2 \\ a^{(2)}_1 \\ a^{(2)}_2 \end{bmatrix}.$$
(3.66)

We are thus free to specify the *four* ($= 2N$, $N = 2$ being the number of open channels) incoming amplitudes $a^{(1)}_n$ and $a^{(2)}_n$ ($n = 1, 2$); the four outgoing amplitudes

$b_n^{(1)}$ and $b_n^{(2)}$ ($n = 1, 2$) are then dictated by the Schrödinger equation and are given by eqn (3.66). The structure of the most general solution of the Schrödinger equation outside the potential region for this two-open-channel problem is shown in Table 3.1.

*The case of one open and one closed channel*

Since we want our wave function to be normalizable, we do not allow incidence on the closed channel 2. We thus have *two* linearly independent solutions, $\Psi_{+,1}^{(+)}(E;x)$ and $\Psi_{-,1}^{(+)}(E;x)$, whose asymptotic behavior is

$$\Psi_{+,1}^{(+)}(E;x) \sim \begin{cases} \begin{bmatrix} \phi_+(E_1;x) + r_{11}\phi_-(E_1;x) \\ r_{21}\phi_-(-\epsilon_2;x) \end{bmatrix}, & x \to -\infty, \\ \begin{bmatrix} t_{11}\phi_+(E_1;x) \\ t_{21}\phi_+(-\epsilon_2;x) \end{bmatrix}, & x \to +\infty, \end{cases}$$

$$\Psi_{-,1}^{(+)}(E;x) \sim \begin{cases} \begin{bmatrix} t'_{11}\phi_-(E_1;x) \\ t'_{21}\phi_-(-\epsilon_2;x) \end{bmatrix}, & x \to -\infty, \\ \begin{bmatrix} \phi_-(E_1;x) + r'_{11}\phi_+(E_1;x) \\ r'_{21}\phi_+(-\epsilon_2;x) \end{bmatrix}, & x \to +\infty. \end{cases}$$
(3.67)

The linear combination

$$\Psi(E;x) = a_1^{(1)}\Psi_{+,1}^{(+)}(E;x) + a_1^{(2)}\Psi_{-,1}^{(+)}(E;x) \tag{3.68}$$

of the above two solutions is a solution of the Schrödinger equation for the same energy. Its asymptotic behavior is

TABLE 3.1. The most general asymptotic wave function for a quasi-one-dimensional scattering problem with two open channels.

| | $x < -L/2$ | $x > L/2$ |
|---|---|---|
| $[\psi(E;x)]_1$ | $a_1^{(1)}\phi_+(E_1;x) + b_1^{(1)}\phi_-(E_1;x)$ | $a_1^{(2)}\phi_-(E_1;x) + b_1^{(2)}\phi_+(E_1;x)$ |
| $[\psi(E;x)]_2$ | $a_2^{(1)}\phi_+(E_2;x) + b_2^{(1)}\phi_-(E_2;x)$ | $a_2^{(2)}\phi_-(E_2;x) + b_2^{(2)}\phi_+(E_2;x)$ |

$$\psi(E;x) \sim \begin{cases} \begin{bmatrix} a_1^{(1)}\phi_+(E_1;x) + \left(r_{11}a_1^{(1)} + t'_{11}a_1^{(2)}\right)\phi_-(E_1;x) \\ \left(r_{21}a_1^{(1)} + t'_{21}a_1^{(2)}\right)\phi_-(-\epsilon_2;x) \end{bmatrix} \\ \quad = \begin{bmatrix} a_1^{(1)}\phi_+(E_1;x) + b_1^{(1)}\phi_-(E_1;x) \\ b_2^{(1)}\phi_-(-\epsilon_2;x) \end{bmatrix}, \qquad x \to -\infty, \\ \\ \begin{bmatrix} a_1^{(2)}\phi_-(E_1;x) + \left(t_{11}a_1^{(1)} + r'_{11}a_1^{(2)}\right)\phi_+(E_1;x) \\ \left(t_{21}a_1^{(1)} + r'_{21}a_1^{(2)}\right)\phi_+(-\epsilon_2;x) \end{bmatrix} \\ \quad = \begin{bmatrix} a_1^{(2)}\phi_-(E_1;x) + b_1^{(2)}\phi_+(E_1;x) \\ b_2^{(2)}\phi_+(-\epsilon_2;x) \end{bmatrix}, \qquad x \to +\infty. \end{cases}$$
(3.69)

This wave function represents the *most general solution of the Schrodinger equation* at the energy $E$; $a_1^{(1)}$ and $a_1^{(2)}$ denote the amplitudes of the incoming waves in channel 1 on either side, $b_1^{(1)}$ and $b_1^{(2)}$ denote the amplitudes of the outgoing waves in channel 1, and $b_2^{(1)}$ and $b_2^{(2)}$ denote the amplitudes of the evanescent waves (channel 2) on either side. Outside the potential region this wave function is as shown in Table 3.2. The above equation shows that the two sets of amplitudes are *linearly related*, according to the equations

$$\begin{bmatrix} b_1^{(1)} \\ b_1^{(2)} \end{bmatrix} = \begin{bmatrix} r_{11} & t'_{11} \\ t_{11} & r'_{11} \end{bmatrix} \begin{bmatrix} a_1^{(1)} \\ a_1^{(2)} \end{bmatrix} \qquad (3.70)$$

and

$$\begin{bmatrix} b_2^{(1)} \\ b_2^{(2)} \end{bmatrix} = \begin{bmatrix} r_{21} & t'_{21} \\ t_{21} & r'_{21} \end{bmatrix} \begin{bmatrix} a_1^{(1)} \\ a_1^{(2)} \end{bmatrix}. \qquad (3.71)$$

We are thus free to specify the *two* ($= 2N$, $N = 1$ being the number of open channels) incoming amplitudes $a_1^{(1)}$ and $a_1^{(2)}$; the outgoing amplitudes $b_1^{(1)}$ and $b_1^{(2)}$ for the open channel and $b_2^{(1)}$ and $b_2^{(2)}$ for the evanescent mode are then dictated by the Schrödinger equation and are given by eqns (3.70) and (3.71), respectively.

TABLE 3.2. The most general asymptotic wave function for a quasi-one-dimensional scattering problem with one open and one closed channel.

| | $x < -L/2$ | $x > L/2$ |
|---|---|---|
| $[\psi(E;x)]_1$ | $a_1^{(1)}\phi_+(E_1;x) + b_1^{(1)}\phi_-(E_1;x)$ | $a_1^{(2)}\phi_-(E_1;x) + b_1^{(2)}\phi_+(E_1;x)$ |
| $[\psi(E;x)]_2$ | $b_2^{(1)}\phi_-(-\epsilon_2;x)$ | $b_2^{(2)}\phi_+(-\epsilon_2;x)$ |

*The case of an arbitrary number of channels*

If we have $N$ open channels and an arbitrary number, infinite in general, of closed channels, then we can easily extend the arguments given earlier. The present problem allows $2N$ linearly independent solutions, a linear combination of which, with arbitrary coefficients $a_n^{(1)}$ and $a_n^{(2)}$ ($n = 1, \ldots, N$), gives the most general solution at energy $E$ shown in Table 3.3. The $2N$ outgoing amplitudes $b_n^{(1)}$ and $b_n^{(2)}$ ($n = 1, \ldots, N$) can be expressed linearly in terms of the $2N$ incoming amplitudes $a_n^{(1)}$ and $a_n^{(2)}$ ($n = 1, \ldots, N$) as

$$\begin{bmatrix} \boldsymbol{b}^{(1)} \\ \boldsymbol{b}^{(2)} \end{bmatrix} = \begin{bmatrix} r & t' \\ t & r' \end{bmatrix} \begin{bmatrix} \boldsymbol{a}^{(1)} \\ \boldsymbol{a}^{(2)} \end{bmatrix}, \qquad (3.72)$$

where $\boldsymbol{b}^{(1)}$, $\boldsymbol{b}^{(2)}$, $\boldsymbol{a}^{(1)}$ and $\boldsymbol{a}^{(2)}$ (as usual, the upper indices (1) and (2) indicate the left-hand and the right-hand side of the potential, respectively) are the $N$-dimensional column vectors

$$\boldsymbol{b}^{(1)} = \begin{bmatrix} b_1^{(1)} \\ \vdots \\ b_N^{(1)} \end{bmatrix}, \quad \boldsymbol{a}^{(1)} = \begin{bmatrix} a_1^{(1)} \\ \vdots \\ a_N^{(1)} \end{bmatrix}, \qquad (3.73)$$

$$\boldsymbol{b}^{(2)} = \begin{bmatrix} b_1^{(2)} \\ \vdots \\ b_N^{(2)} \end{bmatrix}, \quad \boldsymbol{a}^{(2)} = \begin{bmatrix} a_1^{(2)} \\ \vdots \\ a_N^{(2)} \end{bmatrix}. \qquad (3.74)$$

The $r$ and $t$ in eqn (3.72) are the open-channel reflection and transmission matrices when incidence is from the left, and $r'$ and $t'$ are the corresponding matrices when incidence is from the right; they are all $N \times N$ matrices.

### 3.1.2 The S matrix

We can generalize eqns (2.87)–(2.90) which define the *on-shell scattering matrix* by

$$\left\langle \psi_{s'n'_0}^{(-)}(E') \middle| \psi_{sn_0}^{(+)}(E) \right\rangle = S_{s'n'_0,sn_0}(E)\delta(E' - E), \qquad (3.75)$$

TABLE 3.3. The most general asymptotic wave function for a multichannel quasi-one-dimensional scattering problem.

| | $x < -L/2$ | $x > L/2$ |
|---|---|---|
| $[\psi(E;x)]_n$ $(1 \leq n \leq N)$ | $a_n^{(1)}\phi_+(E_n;x) + b_n^{(1)}\phi_-(E_n;x)$ | $a_n^{(2)}\phi_-(E_n;x) + b_n^{(2)}\phi_+(E_n;x)$ |
| $[\psi(E;x)]_n$ $(n \geq N+1)$ | $b_n^{(1)}\phi_-(-\epsilon_n;x)$ | $b_n^{(2)}\phi_+(-\epsilon_n;x)$ |

where
$$S_{s'n_0',sn_0}(E) = \delta_{s's}\delta_{n_0'n_0} - 2\pi i \left\langle \phi_{s'}(E_{n_0'})\chi_{n_0'} | V | \psi_{sn_0}^{(+)}(E) \right\rangle. \tag{3.76}$$

The indices $n_0$ and $n_0'$ indicate any one of the $N$ open channels; the dimensionality of the $S$ matrix is thus $2N$.

We notice that, using the expansion of eqn (3.16), we can also write the definition (3.76) of the $S$-matrix elements as

$$S_{s'n_0',sn_0}(E) = \delta_{s's}\delta_{n_0'n_0} - 2\pi i \sum_{m=1}^{\infty} \left( \phi_{s'}(E_{n_0'};x), V_{n_0'm}(x) \left[\psi_{sn_0}^{(+)}(E;x)\right]_m \right). \tag{3.77}$$

We thus see that the outgoing wave states at energy $E$ can be expressed as linear combinations of the incoming wave states at the same energy, the expansion coefficients being the $S$-matrix elements we have just defined, i.e.,

$$\left|\psi_{sn_0}^{(+)}(E)\right\rangle = \sum_{n_0'=1}^{N(E)} \sum_{s'} \left|\psi_{s'n_0'}^{(-)}(E)\right\rangle S_{s'n_0',sn_0}(E), \tag{3.78}$$

where $N(E)$ is the number of open channels at the energy $E$. The matrix $S$ must then represent a *unitary* transformation [141], so that

$$SS^\dagger = I, \tag{3.79}$$

with $I$ denoting the $2N$-dimensional unit matrix. We recall that for finite-dimensional matrices the left and the right inverses coincide.

Despite the fact that incidence in one (open) channel feeds, in general, all channels (open and closed), expressions (3.49) and (3.50) show that the $S$-matrix elements defined in (3.76) can in fact be identified with *the reflection and the transmission amplitudes* to *open* channels only, inasmuch as they refer to the asymptotic behavior of the scattering wave function, i.e.,

$$\begin{aligned} r_{n_0'n_0} &= S_{-n_0',+n_0}, & t'_{n_0'n_0} &= S_{-n_0',-n_0}, \\ t_{n_0'n_0} &= S_{+n_0',+n_0}, & r'_{n_0'n_0} &= S_{+n_0',-n_0}. \end{aligned} \tag{3.80}$$

As a result, the matrix that appears in eqn (3.72) and relates the incoming to the outgoing amplitudes is precisely the $S$ matrix we have just defined, i.e.,

$$\boldsymbol{b} = S\boldsymbol{a}, \tag{3.81}$$

where
$$S = \begin{bmatrix} r & t' \\ t & r' \end{bmatrix} \tag{3.82}$$

and
$$\boldsymbol{b} = \begin{bmatrix} \boldsymbol{b}^{(1)} \\ \boldsymbol{b}^{(2)} \end{bmatrix}, \qquad \boldsymbol{a} = \begin{bmatrix} \boldsymbol{a}^{(1)} \\ \boldsymbol{a}^{(2)} \end{bmatrix}. \tag{3.83}$$

As usual, we notice that the unitarity of $S$ implies *flux conservation*, as expected.

Later on, in Section 3.1.7, we shall need the wave function in the potential-free region *between* two scatterers, inside an array of scatterers. The most general wave function in this region contains, in its open-channel part, waves travelling in either direction; in its closed-channel part, in contrast to the second line of Table 3.3, it contains *increasing as well as decreasing* real exponentials with, in general, complex coefficients. As we shall see, it will be of interest to define an extended (infinite-dimensional) $S$ matrix $\widetilde{S}$, which relates the vector of all 'incoming' amplitudes to that of all 'outgoing' amplitudes, including those for open as well as closed channels. In this section, though, we restrict our definitions to the ones given above.

The potential appearing in the previous equations was assumed to be real, so that the Hamiltonian is *time-reversal invariant*. In the absence of spin, the $S$ matrix, besides being unitary, is *symmetric* [67, 112, 123, 141, 147], i.e.,

$$S = S^\top, \tag{3.84}$$

as can be seen from an argument entirely analogous to that leading to eqn (2.132). Note, however, that for an arbitrary choice of phases of the transverse wave functions of eqn (3.6), the above symmetry of the $S$ matrix would not hold true. Indeed, one would obtain, in general,

$$S = D S^\top D^*, \tag{3.85}$$

where $D$ is a diagonal phase matrix, corresponding to the choice of phases just mentioned. This is the *orthogonal* case, also denoted by $\beta = 1$ in connection with random-matrix theory [124, 152]. When time-reversal symmetry is broken, as in the presence of a magnetic field, and in the absence of other symmetries, the only requirement on the $S$ matrix is unitarity. This is the so-called *unitary* case, also denoted in the literature on random-matrix theory by $\beta = 2$. The *symplectic* case ($\beta = 4$) arises in the presence of half-integral spin, time-reversal invariance and in the absence of other symmetries, but will not be considered here.

*Parametrization of the S matrix*

The matrix elements of $S$ are not independent of one another because of the requirement of unitarity and, if applicable, symmetry. It is convenient to express $S$ in terms of independent parameters. A very convenient parametrization is given by the so-called 'polar representation' [14, 87, 132, 134]

$$S = \begin{bmatrix} v^{(1)} & 0 \\ 0 & v^{(2)} \end{bmatrix} \begin{bmatrix} -\sqrt{1-\tau} & \sqrt{\tau} \\ \sqrt{\tau} & \sqrt{1-\tau} \end{bmatrix} \begin{bmatrix} v^{(3)} & 0 \\ 0 & v^{(4)} \end{bmatrix} \equiv VRW, \tag{3.86}$$

where $\tau$ stands for the $N \times N$ diagonal matrix constructed from the eigenvalues $\tau_a$ of the Hermitian matrix $tt^\dagger$ and the $v^{(i)}$ are arbitrary unitary matrices for $\beta = 2$, with the restriction that $v^{(3)} = (v^{(1)})^\top$ and $v^{(4)} = (v^{(2)})^\top$ for $\beta = 1$. It is readily verified that any matrix of the form (3.86) satisfies the appropriate requirements of an $S$ matrix for $\beta = 1, 2$. The converse statement can also be proved.

### 3.1.3 The transfer matrix

We generalize here the presentation given in Section 2.1.5 for the one-dimensional case. We discuss the one-open-one-closed-channel case, its extension to any number of open and closed channels being straightforward.

Consider the two independent combinations of $\Psi^{(+)}_{+,1}(E;x)$ and $\Psi^{(+)}_{-,1}(E;x)$ given by

$$\Psi_{L,+,1}(E;x) = \Psi^{(+)}_{+,1}(E;x) - \frac{r_{11}}{t'_{11}}\Psi^{(+)}_{-,1}(E;x), \qquad (3.87a)$$

$$\Psi_{L,-,1}(E;x) = \frac{1}{t'_{11}}\Psi^{(+)}_{-,1}(E;x). \qquad (3.87b)$$

From the asymptotic behavior of $\Psi^{(+)}_{+,1}(E;x)$ and $\Psi^{(+)}_{-,1}(E;x)$ given in eqn (3.67), we find that of $\Psi_{L,+,1}(E;x)$ and $\Psi_{L,-,1}(E;x)$ to be

$$\Psi_{L,+,1}(E;x) \sim \begin{cases} \begin{bmatrix} \phi_+(E_1;x) \\ (r_{21} - r_{11}t'_{21}/t'_{11})\phi_-(-\epsilon_2;x) \end{bmatrix}, & x \to -\infty, \\ \begin{bmatrix} (t_{11} - r_{11}r'_{11}/t'_{11})\phi_+(E_1;x) - (r_{11}/t'_{11})\phi_-(E_1;x) \\ (t_{21} - r_{11}r'_{21}/t'_{11})\phi_+(-\epsilon_2;x) \end{bmatrix}, & x \to +\infty, \end{cases}$$

$$\Psi_{L,-,1}(E;x) \sim \begin{cases} \begin{bmatrix} \phi_-(E_1;x) \\ (t'_{21}/t'_{11})\phi_-(-\epsilon_2;x) \end{bmatrix}, & x \to -\infty, \\ \begin{bmatrix} (r'_{11}/t'_{11})\phi_+(E_1;x) + (1/t'_{11})\phi_-(E_1;x) \\ (r'_{21}/t'_{11})\phi_+(-\epsilon_2;x) \end{bmatrix}, & x \to +\infty. \end{cases}$$
$$(3.88)$$

We observe that on the left of the potential, i.e., in region I, the *open-channel* (channel 1) component of $\Psi_{L,+,1}(E;x)$ (i.e., the first component of the 'spinor') *only contains a wave traveling to the right* (as indicated by the lower index '+'), whereas that of $\Psi_{L,-,1}(E;x)$ *only contains a wave traveling to the left* (as indicated by the lower index '−').

The linear combination

$$\Psi(E;x) = a_1^{(1)}\Psi_{L,+,1}(E;x) + b_1^{(1)}\Psi_{L,-,1}(E;x) \qquad (3.89)$$

of the above two solutions is a solution of the Schrödinger equation for the same energy. For arbitrary values of $a_1^{(1)}$ and $b_1^{(1)}$ (which represent, on the left of the potential, the amplitudes of the waves in the open channel 1 traveling in the positive and negative directions, respectively), the wave function (3.89) is the *most general solution of the Schrödinger equation* at the energy $E$. Its asymptotic behavior is

$$\psi(E;x) \sim \begin{cases} \begin{bmatrix} a_1^{(1)}\phi_+(E_1;x) + b_1^{(1)}\phi_-(E_1;x) \\ \left[(r_{21} - r_{11}t'_{21}/t'_{11})\,a_1^{(1)} + (t'_{21}/t'_{11})b_1^{(1)}\right]\phi_-(-\epsilon_2;x) \end{bmatrix} \\ \qquad = \begin{bmatrix} a_1^{(1)}\phi_+(E_1;x) + b_1^{(1)}\phi_-(E_1;x) \\ b_2^{(1)}\phi_-(-\epsilon_2;x) \end{bmatrix}, \qquad x \to -\infty, \\[2ex] \begin{bmatrix} \left[(t_{11} - r_{11}r'_{11}/t'_{11})\,a_1^{(1)} + (r'_{11}/t'_{11})b_1^{(1)}\right]\phi_+(E_1;x) \\ + \left[-(r_{11}/t'_{11})a_1^{(1)} + (1/t'_{11})b_1^{(1)}\right]\phi_-(E_1;x) \\ \left[(t_{21} - r_{11}r'_{21}/t'_{11})\,a_1^{(1)} + (r'_{21}/t'_{11})b_1^{(1)}\right]\phi_+(-\epsilon_2;x) \end{bmatrix} \\ \qquad = \begin{bmatrix} b_1^{(2)}\phi_+(E_1;x) + a_1^{(2)}\phi_-(E_1;x) + \\ b_2^{(2)}\phi_+(-\epsilon_2;x) \end{bmatrix}, \qquad x \to +\infty. \end{cases}$$

(3.90)

We have thus shown that the *open-channel amplitudes* on the right of the potential, $b_1^{(2)}$ and $a_1^{(2)}$, are linearly related to the *open-channel amplitudes* on the left, $a_1^{(1)}$ and $b_1^{(1)}$, according to the equation

$$\begin{bmatrix} b_1^{(2)} \\ a_1^{(2)} \end{bmatrix} = \begin{bmatrix} \alpha & \beta \\ \gamma & \delta \end{bmatrix} \begin{bmatrix} a_1^{(1)} \\ b_1^{(1)} \end{bmatrix}, \qquad (3.91)$$

where

$$\alpha = t_{11} - \frac{r_{11}r'_{11}}{t'_{11}} = \frac{1}{t^*_{11}}, \qquad \beta = \frac{r'_{11}}{t'_{11}}, \qquad (3.92a)$$

$$\gamma = -\frac{r_{11}}{t'_{11}}, \qquad \delta = \frac{1}{t'_{11}}. \qquad (3.92b)$$

We can thus fix arbitrarily the *two* ($= 2N$, $N = 1$ being the number of open channels) amplitudes $a_1^{(1)}$ and $b_1^{(1)}$ in region I of Fig. 2.2; the two amplitudes $a_1^{(2)}$ and $b_1^{(2)}$ in region II are then uniquely determined by the Schrödinger equation, according to the linear relation (3.91).

The above argument can clearly be generalized to the case of $N$ open channels and an arbitrary number of closed channels. We can thus write

$$\boldsymbol{c}^{(2)} = M\boldsymbol{c}^{(1)}, \qquad (3.93)$$

where the $2N$-dimensional vectors $\boldsymbol{c}^{(1)}$ and $\boldsymbol{c}^{(2)}$ are defined by

$$c^{(2)} = \begin{bmatrix} b^{(2)} \\ a^{(2)} \end{bmatrix}, \quad c^{(1)} = \begin{bmatrix} a^{(1)} \\ b^{(1)} \end{bmatrix}. \tag{3.94}$$

The matrix $M$ has the form

$$M = \begin{bmatrix} \alpha & \beta \\ \gamma & \delta \end{bmatrix}, \tag{3.95}$$

a generalization of eqn (2.170), with $\alpha$, etc. being $N \times N$ matrices.

By an argument entirely analogous to that leading to eqn (2.173), one can prove that *current conservation* leads to the requirement that

$$M^\dagger \Sigma_z M = \Sigma_z, \tag{3.96}$$

where now

$$\Sigma_z = \begin{bmatrix} I_N & 0_N \\ 0_N & -I_N \end{bmatrix}, \tag{3.97}$$

with $I_N$ and $0_N$ being the unit and zero $N$-dimensional matrices, respectively. The $M$ matrices satisfying eqn (3.96) correspond to the *unitary* case, or $\beta = 2$, in the notation introduced above. They form a group which is called, in the literature, the *pseudounitary group* $U(N, N)$ [18, 178].

The relation between the $N$-dimensional blocks of the scattering and the transfer matrices is

$$r = -\delta^{-1}\gamma, \quad t' = \delta^{-1}, \tag{3.98a}$$
$$t = \left(\alpha^\dagger\right)^{-1}, \quad r' = \beta\delta^{-1}. \tag{3.98b}$$

These relations can be inverted to give

$$\alpha = \left(t^\dagger\right)^{-1}, \quad \beta = r'\left(t'\right)^{-1}, \tag{3.99a}$$
$$\gamma = -\left(t'\right)^{-1} r, \quad \delta = \left(t'\right)^{-1}. \tag{3.99b}$$

If the system is *time-reversal invariant (TRI)*, then we can generalize to $N$ channels the argument leading to eqn (2.179), to give the property

$$M^* = \Sigma_x M \Sigma_x, \tag{3.100}$$

with

$$\Sigma_x = \begin{bmatrix} 0_N & I_N \\ I_N & 0_N \end{bmatrix}. \tag{3.101}$$

From this condition, the transfer matrix (3.95) takes the form

$$M = \begin{bmatrix} \alpha & \beta \\ \beta^* & \alpha^* \end{bmatrix}. \tag{3.102}$$

Matrices $M$ satisfying eqns (3.96) and (3.100) correspond to the orthogonal, or $\beta = 1$, case and also form a group, which we call $\mathcal{G}$. From these two equations we find

$$|\det M| = 1, \tag{3.103a}$$
$$\det M = \text{real}, \tag{3.103b}$$

so that $\det M = \pm 1$. Since every $M$ should be related continuously to the unit matrix, we have

$$\det M = 1. \tag{3.103c}$$

We thus find that $\mathcal{G} \subset SU(N, N)$. One can show [132] that every $2N \times 2N$ matrix $M$ satisfying eqns (3.96) and (3.100) can be placed in one-to-one correspondence with the $2N \times 2N$ matrices of the real symplectic group $Sp(2N, \mathcal{R})$. When $N = 1$, we have $Sp(2, \mathcal{R}) \overset{1:1}{\sim} SU(1, 1)$, as indicated in Section 2.1.5; for $N > 1$, we have $Sp(2N, \mathcal{R}) \subset SU(N, N)$ (see [18,178]). When $TRI$ applies, relations (3.98) reduce to

$$r = -(\alpha^*)^{-1} \beta^*, \qquad t' = (\alpha^*)^{-1}, \tag{3.104a}$$
$$t = (\alpha^\dagger)^{-1}, \qquad r' = \beta(\alpha^*)^{-1}, \tag{3.104b}$$

and the inverse relations (3.99) reduce to

$$\alpha = (t^\dagger)^{-1}, \qquad \beta = -(t^\dagger)^{-1} r^*, \tag{3.105a}$$
$$\gamma = -(t^\top)^{-1} r, \qquad \delta = (t^\top)^{-1}. \tag{3.105b}$$

For $\beta = 2$, any transfer matrix can be parametrized in the *polar representation* by

$$M = \begin{bmatrix} u^{(1)} & 0 \\ 0 & u^{(2)} \end{bmatrix} \begin{bmatrix} \sqrt{1+\lambda} & \sqrt{\lambda} \\ \sqrt{\lambda} & \sqrt{1+\lambda} \end{bmatrix} \begin{bmatrix} v^{(1)} & 0 \\ 0 & v^{(2)} \end{bmatrix} \tag{3.106a}$$
$$= U \Lambda V, \tag{3.106b}$$

the $u^{(i)}$ and $v^{(i)}$ being arbitrary $N$-dimensional unitary matrices; $\lambda$ is a real, non-negative and diagonal $N$-dimensional matrix, related to $\tau$ of eqn (3.86) by

$$\tau = \frac{1}{1+\lambda}. \tag{3.107}$$

For $\beta = 1$ we have the constraints

$$u^{(2)} = [u^{(1)}]^*, \qquad v^{(2)} = [v^{(1)}]^*. \tag{3.108}$$

One can express the various quantities of interest in terms of the parameters of eqn (3.106). For instance, using eqns (3.98) we see that the $N \times N$ reflection and transmission matrices $r$ and $t$ are given by

$$r = -[v^{(2)}]^\dagger \sqrt{\frac{\lambda}{1+\lambda}} \, v^{(1)}, \tag{3.109a}$$

$$t = u^{(1)} \frac{1}{\sqrt{1+\lambda}} v^{(1)}. \qquad (3.109b)$$

From these equations we can see the meaning of the parameters introduced in eqns (3.106). Suppose that a wave is incident in just one channel from the left. The first factor in eqns (3.109), i.e., the unitary matrix $v^{(1)}$, acts as a 'mixer'; it generates a normalized linear combination of the $N$ components, with arbitrary (complex) amplitudes, so that there will be a nonzero amplitude in every channel. The middle factor in eqns (3.109) does not mix channels (the matrix $\lambda$ is diagonal), but describes a reflection by the fraction $[\lambda_a/(1+\lambda_a)]^{1/2}$ in channel $a$, and a transmission by the fraction $(1+\lambda_a)^{-1/2}$. Finally, the unitary matrices $[v^{(2)}]^\dagger$ in eqn (3.109a) and $u^{(1)}$ in eqn (3.109b) mix channels amongst themselves once again.

In Section 2.1.5 we saw that, in the one-channel case, two-dimensional $M$ matrices associated with non-overlapping potentials can be multiplied to give the total $M$ matrix. Here, the existence of evanescent modes complicates the situation; we shall see in Section 3.1.7 that, in general, the transfer matrices to be multiplied are the objects that we shall call extended $\widetilde{M}$ matrices, which include the closed channels in the description. In the presence of closed channels, the wave function in the region between two successive scatterers contains, in addition to running waves, real exponentials of both signs with, in general, complex coefficients. Nonetheless, we shall find in Section 3.1.7 that we can dispose of the closed channels and multiply the standard, open-channel $M$ matrices, provided that $\kappa_n d \gg 1$, with $\kappa_n x$ being the exponent of the real exponentials defined in eqn (3.21) and $d$ being the distance between successive scatterers. This result is plausible since, under these circumstances, an exponential that decreases away from a given scatterer is hardly 'seen' by the next one. Then eqn (2.188) can be taken over to the present situation, i.e.,

$$M = M_2 M_1, \qquad (3.110)$$

with $M_1$, $M_2$ and $M$ now being $2N$-dimensional matrices.

### 3.1.4 Combining the S matrices for two scatterers in series

Exactly the same comments made in the previous paragraph apply to the combination of $S$ matrices for two scatterers in series. This issue will be discussed in detail in Section 3.1.7. Assuming that the condition mentioned in the previous paragraph is fulfilled, we can repeat the derivation given in Section 2.1.6 for $N=1$, bearing in mind that $r_1, r_2$, etc. are now $N \times N$ matrices which do not commute, so the order must be kept. We arrive at eqn (2.193), which now reads

$$S = \begin{bmatrix} r_1 & 0 \\ 0 & r_2' \end{bmatrix} + \begin{bmatrix} t_1' & 0 \\ 0 & t_2 \end{bmatrix} \frac{1}{\Sigma_x - \begin{bmatrix} r_1' & 0 \\ 0 & r_2 \end{bmatrix}} \begin{bmatrix} t_1 & 0 \\ 0 & t_2' \end{bmatrix}, \qquad (3.111)$$

where $\Sigma_x$ is the $2N$-dimensional Pauli matrix given in eqn (3.101). We can easily show that the inverse occurring in eqn (3.111) is given by

so that

$$\begin{bmatrix} -r'_1 & I_N \\ I_N & -r_2 \end{bmatrix}^{-1} = \begin{bmatrix} r_2/(I_N - r'_1 r_2) & 1/(I_N - r_2 r'_1) \\ 1/(I_N - r'_1 r_2) & r'_1/(I_N - r_2 r'_1) \end{bmatrix},$$

$$S = \begin{bmatrix} r & t' \\ t & r' \end{bmatrix} = \begin{bmatrix} r_1 + t'_1 r_2 \left(1/(I_N - r'_1 r_2)\right) t_1 & t'_1 \left(1/(I_N - r_2 r'_1)\right) t'_2 \\ t_2 \left(1/(I_N - r'_1 r_2)\right) t_1 & r'_2 + t_2 r'_1 \left(1/(I_N - r_2 r'_1)\right) t'_2 \end{bmatrix}. \tag{3.112}$$

This result can again be very easily understood in terms of the multiple scattering matrix series generated from the construction of Fig. 2.5, *keeping the order* of the various factors that appear.

The alternative derivation given in Section 2.1.6 in terms of the projection operators $P$ and $Q$ can also be carried through for the present $N$-channel case. We arrive at the same result (2.200), with the Pauli matrix $\sigma_x$ replaced by $\Sigma_x$ of eqn (3.101), i.e.,

$$S = S_{12}^{PP} + S_{12}^{PQ} \frac{1}{\Sigma_x - S_{12}^{QQ}} S_{12}^{QP}. \tag{3.113}$$

### 3.1.5 *Transformation of the scattering and transfer matrices under a translation*

If the potential is translated a distance $d$, then the new $S$ and $M$ matrices can be related to the original matrices $\mathring{S}$ and $\mathring{M}$, through a multichannel extension of the arguments presented in Section 2.1.7. For $N$ open channels only, the result takes the form of eqns (2.206)–(2.213), where the matrix $D$ is now $2N$-dimensional, i.e.,

$$D(\boldsymbol{k}d) = \begin{bmatrix} T(\boldsymbol{k}d) & 0 \\ 0 & T^{-1}(\boldsymbol{k}d) \end{bmatrix}, \tag{3.114}$$

with $T(\boldsymbol{k}d)$ being the $N$-dimensional diagonal matrix

$$T(\boldsymbol{k}d) = \begin{bmatrix} e^{ik_1 d} & & 0 \\ & \ddots & \\ 0 & & e^{ik_N d} \end{bmatrix}. \tag{3.115}$$

In particular, eqns (2.209) and (2.214) become

$$S = \begin{bmatrix} r & t' \\ t & r' \end{bmatrix} = \begin{bmatrix} T(\boldsymbol{k}d)\mathring{r}T(\boldsymbol{k}d) & T(\boldsymbol{k}d)\mathring{t}'T^{-1}(\boldsymbol{k}d) \\ T^{-1}(\boldsymbol{k}d)\mathring{t}T(\boldsymbol{k}d) & T^{-1}(\boldsymbol{k}d)\mathring{r}'T^{-1}(\boldsymbol{k}d) \end{bmatrix} \tag{3.116}$$

and

$$M = \begin{bmatrix} \alpha & \beta \\ \gamma & \delta \end{bmatrix} = \begin{bmatrix} T^{-1}(\boldsymbol{k}d)\mathring{\alpha}T(\boldsymbol{k}d) & T^{-1}(\boldsymbol{k}d)\mathring{\beta}T^{-1}(\boldsymbol{k}d) \\ T(\boldsymbol{k}d)\mathring{\gamma}T(\boldsymbol{k}d) & T(\boldsymbol{k}d)\mathring{\delta}T^{-1}(\boldsymbol{k}d) \end{bmatrix}, \tag{3.117}$$

respectively. In the polar representation we have

$$M = \begin{bmatrix} u^{(1)}\sqrt{1+\lambda}\, v^{(1)} & u^{(1)}\sqrt{\lambda}\, v^{(2)} \\ u^{(2)}\sqrt{\lambda}\, v^{(1)} & u^{(2)}\sqrt{1+\lambda}\, v^{(2)} \end{bmatrix}$$

$$= \begin{bmatrix} T^{-1}\mathring{u}^{(1)}\sqrt{1+\mathring{\lambda}}\, \mathring{v}^{(1)}T & T^{-1}\mathring{u}^{(1)}\sqrt{\mathring{\lambda}}\, \mathring{v}^{(2)}T^{-1} \\ T\mathring{u}^{(2)}\sqrt{\mathring{\lambda}}\, \mathring{v}^{(1)}T & T\mathring{u}^{(2)}\sqrt{1+\mathring{\lambda}}\, \mathring{v}^{(2)}T^{-1} \end{bmatrix}, \quad (3.118)$$

where $T = T(\mathbf{k}d)$ of eqn (3.115). Thus

$$\lambda = \mathring{\lambda}, \qquad (3.119\text{a})$$
$$u^{(1)} = T^{-1}\mathring{u}^{(1)}, \qquad v^{(1)} = \mathring{v}^{(1)}T, \qquad (3.119\text{b})$$
$$u^{(2)} = T\mathring{u}^{(2)}, \qquad v^{(2)} = \mathring{v}^{(2)}T^{-1}. \qquad (3.119\text{c})$$

### 3.1.6 Exactly soluble example for the two-channel problem
*Two open channels*

For two open channels, eqns (3.25) become the pair of coupled equations

$$[\psi(x)]_1'' + k_1^2 [\psi(x)]_1 = U_{11}(x) [\psi(x)]_1 + U_{12}(x) [\psi(x)]_2, \qquad (3.120\text{a})$$
$$[\psi(x)]_2'' + k_2^2 [\psi(x)]_2 = U_{21}(x) [\psi(x)]_1 + U_{22}(x) [\psi(x)]_2, \qquad (3.120\text{b})$$

which can be written in matrix form as

$$\begin{bmatrix} \partial_x^2 + k_1^2 - U_{11}(x) & -U_{12}(x) \\ -U_{21}(x) & \partial_x^2 + k_2^2 - U_{22}(x) \end{bmatrix} \begin{bmatrix} [\psi(x)]_1 \\ [\psi(x)]_2 \end{bmatrix} = 0. \qquad (3.121)$$

As an example, we choose the delta potential

$$U_{ab}(x) = u_{ab}\delta(x), \qquad (3.122\text{a})$$
$$u_{11} = u_{11}^*, \quad u_{22} = u_{22}^*, \qquad (3.122\text{b})$$
$$u_{21} = u_{12}^*. \qquad (3.122\text{c})$$

Then, for $x \neq 0$ we have

$$\begin{bmatrix} \partial_x^2 + k_1^2 & 0 \\ 0 & \partial_x^2 + k_2^2 \end{bmatrix} \begin{bmatrix} [\psi(x)]_1 \\ [\psi(x)]_2 \end{bmatrix} = 0. \qquad (3.123)$$

At $x = 0$, continuity of the wave function gives the following boundary conditions for the two components:

$$\lim_{\epsilon \to 0} [[\psi(x)]_1]_{-\epsilon}^{+\epsilon} = 0, \qquad (3.124\text{a})$$
$$\lim_{\epsilon \to 0} [[\psi(x)]_2]_{-\epsilon}^{+\epsilon} = 0. \qquad (3.124\text{b})$$

The delta potential causes a discontinuity in the slope of the two components at $x = 0$ as follows:

$$\lim_{\epsilon \to 0} [[\psi'(x)]_1]_{-\epsilon}^{+\epsilon} = u_{11} [\psi(0)]_1 + u_{12} [\psi(0)]_2, \qquad (3.125a)$$

$$\lim_{\epsilon \to 0} [[\psi'(x)]_2]_{-\epsilon}^{+\epsilon} = u_{21} [\psi(0)]_1 + u_{22} [\psi(0)]_2. \qquad (3.125b)$$

The most general wave function has the asymptotic structure given in Table 3.1. Applying to it the boundary conditions (3.124) and (3.125), we obtain the four equations

$$a_1^{(1)} + b_1^{(1)} = a_1^{(2)} + b_1^{(2)}, \qquad (3.126a)$$

$$a_2^{(1)} + b_2^{(1)} = a_2^{(2)} + b_2^{(2)}, \qquad (3.126b)$$

$$\mathrm{i}\sqrt{k_1}(b_1^{(2)} - a_1^{(2)}) - \mathrm{i}\sqrt{k_1}(a_1^{(1)} - b_1^{(1)}) = u_{11} \frac{a_1^{(1)} + b_1^{(1)}}{\sqrt{k_1}} + u_{12} \frac{a_2^{(1)} + b_2^{(1)}}{\sqrt{k_2}}, \qquad (3.126c)$$

$$\mathrm{i}\sqrt{k_2}(b_2^{(2)} - a_2^{(2)}) - \mathrm{i}\sqrt{k_2}(a_2^{(1)} - b_2^{(1)}) = u_{21} \frac{a_1^{(1)} + b_1^{(1)}}{\sqrt{k_1}} + u_{22} \frac{a_2^{(1)} + b_2^{(1)}}{\sqrt{k_2}}. \qquad (3.126d)$$

Assuming that the *four* incoming amplitudes are given, eqns (3.126) represent four equations for the four unknowns $b_1^{(1)}$, $b_2^{(1)}$, $b_1^{(2)}$ and $b_2^{(2)}$. Solving eqns (3.126) for the $b$s in terms of the $a$s, we find the $4 \times 4$ $S$ matrix of eqns (3.66) and (3.81), whose $2 \times 2$ reflection and transmission matrices are

$$r = r' = \frac{1}{\Delta} \begin{bmatrix} (u_{11}/2\mathrm{i}k_1)(1 - u_{22}/2\mathrm{i}k_2) - |u_{12}|^2/4k_1k_2 & -\mathrm{i}u_{12}/2\sqrt{k_1k_2} \\ -\mathrm{i}u_{21}/2\sqrt{k_1k_2} & (u_{22}/2\mathrm{i}k_2)(1 - u_{11}/2\mathrm{i}k_1) - |u_{12}|^2/4k_1k_2 \end{bmatrix}, \qquad (3.127a)$$

$$t = t' = \frac{1}{\Delta} \begin{bmatrix} 1 - u_{22}/2\mathrm{i}k_2 & -\mathrm{i}u_{12}/2\sqrt{k_1k_2} \\ -\mathrm{i}u_{21}/2\sqrt{k_1k_2} & 1 - u_{11}/2\mathrm{i}k_1 \end{bmatrix}, \qquad (3.127b)$$

where

$$\Delta = \left(1 - \frac{u_{11}}{2\mathrm{i}k_1}\right)\left(1 - \frac{u_{22}}{2\mathrm{i}k_2}\right) + \frac{u_{12}u_{21}}{4k_1k_2}. \qquad (3.127c)$$

A number of comments on this result are in order.

(i) The $S$ matrix formed with the reflection and transmission blocks of eqns (3.127) can be verified to be unitary.

(ii) That $r = r'$ and $t = t'$ is due to the left–right symmetry of the potential. In fact, the result $r = r'$ and $t = t'$ can be proved in general for a Hermitian potential matrix $\|u_{ab}(x)\|$ which is invariant with respect to the operation $x \to -x$, which gives rise to a Schrödinger equation with left–right symmetry.

(iii) For TRI, $u_{12}$ is real and eqns (3.127) give $r = r^\top$ and $t = t^\top$, as expected.
(iv) If we decouple the two channels, setting $u_{12} = 0$, then we obtain $r_{12} = t_{12} = 0$, while $r_{aa}$ and $t_{aa}$ reduce to the values obtained for the one-dimensional delta potential of eqn (2.218), with $u_0$ replaced by $u_{aa}$.
(v) The threshold behavior is analyzed in the following exercise.

**Exercise 3.2** Study how the various scattering quantities behave as the total energy $E$ approaches the energy $\mathcal{E}_2$ of the second mode from above, and thus the longitudinal energy in channel 2, $E_2 \to 0^+$.

In terms of wave numbers, $k$ approaches $K_2$ and the longitudinal momentum in channel 2, $k_2 \to 0^+$. We find, from eqns (3.127),

$$r = r' = \begin{bmatrix} r_{11} = -(|u_{12}|^2 - u_{11}u_{22})/[(|u_{12}|^2 - u_{11}u_{22}) + 2ik_1 u_{22}] & r_{12} = 0 \\ r_{21} = 0 & r_{22} = -1 \end{bmatrix}, \tag{3.128a}$$

$$t = t' = \begin{bmatrix} t_{11} = 2ik_1 u_{22}/[(|u_{12}|^2 - u_{11}u_{22}) + 2ik_1 u_{22}] & t_{12} = 0 \\ t_{21} = 0 & t_{22} = 0 \end{bmatrix}. \tag{3.128b}$$

At the threshold for channel 2 we thus have the following results.
(a) In channel 2 there is *complete reflection* back to the same channel.
(b) From unitarity, the remaining elements of the second row and column of the $S$ matrix must vanish. Therefore, when entrance is from channel 2, there is no reflection to channel 1, nor transmission to any of the two channels; when incidence is from channel 1 there is also no response in channel 2, i.e., there is *no inelastic scattering*.
(c) In channel 1 there is *elastic scattering* described by the $2 \times 2$ unitary $S$ matrix formed with $r_{11} = r'_{11}$ and $t_{11} = t'_{11}$ given in eqns (3.128) above.

Solving eqns (3.126) for the left coefficients in terms of the right coefficients we find, for the $4 \times 4$ $M$ matrix,

$$M = \left[ \begin{array}{cc|cc} 1 + u_{11}/2ik_1 & (u_{12}/2ik_1)\sqrt{k_1/k_2} & u_{11}/2ik_1 & (u_{12}/2ik_1)\sqrt{k_1/k_2} \\ (u_{21}/2ik_1)\sqrt{k_1/k_2} & 1 + u_{22}/2ik_2 & (u_{21}/2ik_1)\sqrt{k_1/k_2} & u_{22}/2ik_2 \\ \hline -u_{11}/2ik_1 & -(u_{12}/2ik_1)\sqrt{k_1/k_2} & 1 - u_{11}/2ik_1 & -(u_{12}/2ik_1)\sqrt{k_1/k_2} \\ -(u_{21}/2ik_1)\sqrt{k_1/k_2} & -u_{22}/2ik_2 & -(u_{21}/2ik_1)\sqrt{k_1/k_2} & 1 - u_{22}/2ik_2 \end{array} \right]. \tag{3.129}$$

If we decouple the two channels, setting $u_{12} = 0$, then we obtain $\alpha_{12} = \beta_{12} = \gamma_{12} = \delta_{12} = 0$, while $\alpha_{aa}$, etc. reduce to the values obtained for the one-dimensional delta potential of eqn (2.220), with $u_0$ replaced by $u_{aa}$.

*One open and one closed channel*

We consider the same potential as above, given in eqn (3.122). Equations (3.120)–(3.123) can be taken over to the present case of one open and one closed channel, using the analytic continuation of eqn (3.24) for $n = 2$. The most general wave function has the structure given in Table 3.2. Applying the boundary conditions (3.124) and (3.125) to this wave function we obtain the four equations

$$a_1^{(1)} + b_1^{(1)} = a_1^{(2)} + b_1^{(2)}, \qquad (3.130\text{a})$$

$$b_2^{(1)} = b_2^{(2)}, \qquad (3.130\text{b})$$

$$\mathrm{i}\sqrt{k_1}(b_1^{(2)} - a_1^{(2)}) - \mathrm{i}\sqrt{k_1}(a_1^{(1)} - b_1^{(1)}) = u_{11}\frac{a_1^{(1)} + b_1^{(1)}}{\sqrt{k_1}} + u_{12}\frac{b_2^{(1)}}{\mathrm{e}^{\mathrm{i}\pi/4}\sqrt{\kappa_2}}, \qquad (3.130\text{c})$$

$$(-\mathrm{e}^{-\mathrm{i}\pi/4}\sqrt{\kappa_2}b_2^{(2)}) - (\mathrm{e}^{-\mathrm{i}\pi/4}\sqrt{\kappa_2}b_2^{(1)}) = u_{21}\frac{a_1^{(1)} + b_1^{(1)}}{\sqrt{k_1}} + u_{22}\frac{b_2^{(1)}}{\mathrm{e}^{\mathrm{i}\pi/4}\sqrt{\kappa_2}}. \qquad (3.130\text{d})$$

Assuming that the *two* incoming amplitudes $a_1^{(1)}$ and $a_1^{(2)}$ are given, eqns (3.130) constitute four equations for the four unknowns $b_1^{(1)}$, $b_2^{(1)}$, $b_1^{(2)}$ and $b_2^{(2)}$. Notice that the two amplitudes $b_2^{(1)}$ and $b_2^{(2)}$ associated with the closed channel 2 can be eliminated from eqns (3.130), leaving us with two equations for the two remaining unknowns $b_1^{(1)}$ and $b_1^{(2)}$ associated with the open channel 1. We can thus express the outgoing amplitudes $b_1^{(1)}$ and $b_1^{(2)}$ for the open channel 1 in terms of the incoming amplitudes $a_1^{(1)}$ and $a_1^{(2)}$ for that same channel. Had we attempted a similar elimination of $b_2^{(1)}$ and $b_2^{(2)}$ from eqns (3.126) for the two-open-channel case, we would have arrived at two equations, expressing $b_1^{(1)}$ and $b_1^{(2)}$ in terms of $a_1^{(1)}$, $a_1^{(2)}$, $a_2^{(1)}$ and $a_2^{(2)}$. The latter four amplitudes being arbitrary, it is clear that it would not have been possible to express $b_1^{(1)}$ and $b_1^{(2)}$ in terms of $a_1^{(1)}$ and $a_1^{(2)}$ only; hence for two open channels there is no $2 \times 2$ $S$ matrix relating incoming and outgoing amplitudes pertaining to channel 1 only. In contrast, this *is* possible in the present case of one open and one closed channel, since the incoming amplitudes for channel 2 are given and are equal to zero.

We thus eliminate the $b_2^{(1)}$ and $b_2^{(2)}$ amplitudes from eqns (3.130) to obtain

$$a_1^{(1)} + b_1^{(1)} = a_1^{(2)} + b_1^{(2)},$$

$$\left[1 + \frac{1}{\mathrm{i}k_1}\left(u_{11} - \frac{u_{12}u_{21}}{u_{22} + 2\kappa_2}\right)\right] a_1^{(1)} \qquad (3.131)$$

$$+ \left[-1 + \frac{1}{\mathrm{i}k_1}\left(u_{11} - \frac{u_{12}u_{21}}{u_{22} + 2\kappa_2}\right)\right] b_1^{(1)} + a_1^{(2)} - b_1^{(2)} = 0.$$

These two equations can be written in matrix form as

$$\begin{bmatrix} 1 & -1 \\ -ik_1 + u_{11} - |u_{12}|^2/(u_{22}+2\kappa_2) & -ik_1 \end{bmatrix} \begin{bmatrix} b_1^{(1)} \\ b_1^{(2)} \end{bmatrix}$$
$$= \begin{bmatrix} -1 & 1 \\ -ik_1 - u_{11} + |u_{12}|^2/(u_{22}+2\kappa_2) & -ik_1 \end{bmatrix} \begin{bmatrix} a_1^{(1)} \\ a_1^{(2)} \end{bmatrix}. \tag{3.132}$$

We pass to the right-hand side the inverse of the matrix occurring on the left-hand side and identify the $S$ matrix as

$$S = \begin{bmatrix} r_{11} & t'_{11} \\ t_{11} & r'_{11} \end{bmatrix}$$
$$= \frac{1}{1 - (1/2ik_1)(u_{11} - |u_{12}|^2/(u_{22}+2\kappa_2))}$$
$$\times \begin{bmatrix} (1/2ik_1)(u_{11} - |u_{12}|^2/(u_{22}+2\kappa_2)) & 1 \\ 1 & (1/2ik_1)(u_{11} - |u_{12}|^2/(u_{22}+2\kappa_2)) \end{bmatrix}. \tag{3.133}$$

In order to find the transfer matrix $M$, we return to eqns (3.131) and write them as

$$\begin{bmatrix} 1 & 1 \\ ik_1 & -ik_1 \end{bmatrix} \begin{bmatrix} b_1^{(2)} \\ a_1^{(2)} \end{bmatrix}$$
$$= \begin{bmatrix} 1 & 1 \\ ik_1 + u_{11} - u_{12}u_{21}/(u_{22}+2\kappa) & -ik_1 + u_{11} - u_{12}u_{21}/(u_{22}+2\kappa) \end{bmatrix} \begin{bmatrix} a_1^{(1)} \\ b_1^{(1)} \end{bmatrix}.$$

Taking to the right-hand side the inverse of the matrix on the left-hand side, we find

$$M = \begin{bmatrix} 1 + (1/2ik_1)(u_{11} - |u_{12}|^2/(u_{22}+2\kappa_2)) & (1/2ik_1)(u_{11} - |u_{12}|^2/(u_{22}+2\kappa_2)) \\ -(1/2ik_1)(u_{11} - |u_{12}|^2/(u_{22}+2\kappa_2)) & 1 - (1/2ik_1)(u_{11} - |u_{12}|^2/(u_{22}+2\kappa_2)) \end{bmatrix}. \tag{3.134}$$

It is easily verified that the $M$ matrix of eqn (3.134) satisfies the pseudounitarity requirement of eqn (2.173). In the absence of channel coupling it reduces to the one-dimensional result, eqn (2.220).

A number of comments are appropriate at this point.

(i) The $2 \times 2$ $S$ matrix of eqn (3.133) is clearly unitary.
(ii) That $r = r'$ and $t = t'$ follows from the left–right symmetry of the interaction.
(iii) In the absence of channel coupling, i.e., when $u_{12} = 0$, the matrix $S$ of eqn (3.133) reduces to the matrix found in eqn (2.218), with $u_0$ replaced by $u_{11}$.

(iv) From the comment made immediately after eqns (3.58) and the last paragraph of Exercise 3.1, the $r_{11}$ and $t_{11}$ of eqn (3.133) must be the analytic continuation of the $r_{11}$ and $t_{11}$ of eqns (3.127) under the rule (3.24) for $k_2$. That this property is fulfilled can be verified directly.

(v) Of particular interest is the effect of the so-called *bound states in the continuum* (BSC). This is studied in the following exercise.

**Exercise 3.3** Assume that the delta potential in channel 2 is attractive, i.e.,

$$u_{22} = -w_0 < 0. \tag{3.135}$$

Thus channel 2, if it were uncoupled to channel 1, would admit one bound state. Show that by choosing the parameters properly, such a bound state would occur at positive energy, i.e., it would lie in the continuum of channel 1 (BSC). Analyze the scattering properties of such a system at energies such that channel 2 is closed, when the coupling between the two channels is nonzero.

For simplicity, assume that $K_1 = 0$ and write $K_2 = K$. Then $k_1 = k$ and $\kappa_2^2 = K^2 - k^2$. The Schrödinger equation for channel 2, uncoupled to channel 1, would read

$$\left[-\frac{\partial^2}{\partial x^2} - w_0 \delta(x) + K^2\right] \psi(x) = k^2 \psi(x). \tag{3.136}$$

Should $K = 0$, eqn (3.136) would admit a bound state at the energy

$$E_{\text{BS}}^0 = -\frac{\hbar^2}{2m}\left(\frac{w_0}{2}\right)^2. \tag{3.137}$$

For the actual case $K \neq 0$, the bound state is at the energy

$$E_{\text{BS}} = \frac{\hbar^2 k_{\text{BS}}^2}{2m}, \tag{3.138a}$$

$$k_{\text{BS}}^2 = K^2 - \left(\frac{w_0}{2}\right)^2. \tag{3.138b}$$

If $K^2 > (w_0/2)^2$, then $E_{\text{BS}} > 0$ and we speak of a bound state in the continuum (BSC), as shown in Fig. 3.3. Of course, the $S$ matrix can have no poles on the real axis, so that a zero of $S$ at the same energy is required to make $S$ regular there. In the situation we have just discussed, channels 1 and 2 are uncoupled. The moment we couple them, the BSC that would have occurred in channel 2 would show up as a *resonance* in channel 1. This we now study.

Consider the reflection amplitude $r_{11}$ of eqn (3.133). We rewrite it as

$$r_{11} = \frac{(2\kappa_2 - w_0) u_{11} - |u_{12}|^2}{(2ik - u_{11})(2\kappa_2 - w_0) + |u_{12}|^2}. \tag{3.139}$$

To simplify the discussion, let us assume that $u_{11} = 0$; should there be no coupling to the closed channel 2, we would have full transmission and no reflection. We write the above reflection amplitude as

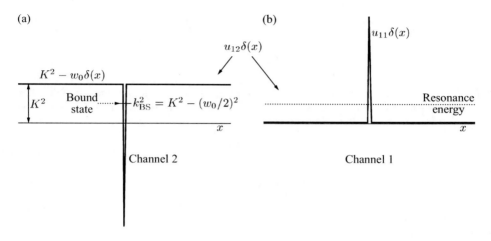

FIG. 3.3. (a) Schematic illustration of the bound state admitted by an attractive delta potential shifted upwards by the amount $K^2$, as it would occur in channel 2 if that channel were uncoupled from channel 1; $K^2$ is sufficiently large for the bound state to lie in the continuum of channel 1. We speak of a 'bound state in the continuum' (BSC). (b) Schematic representation of the potential $u_{11}\delta(x)$ seen in channel 1. This potential has been turned off in the example solved in Exercise 3.3. Also shown schematically is the coupling between the two channels, provided by the potential $u_{12}\delta(x)$. When incidence is in channel 1, we see a resonance at the energy of the 'would be BSC'.

$$r = -\frac{|u_{12}|^2/2ik}{2\kappa_2 - w_0 + |u_{12}|^2/2ik}. \tag{3.140}$$

We see that the effect of the coupling is to have a nonzero reflection, with a maximum at the *resonance* momentum $k = k_{\text{BS}}$, which is reached when $\kappa_2 = w_0/2$; also, as we go away from resonance, the reflection goes to zero again. To find the actual shape of the resonance we write, in the vicinity of the resonance,

$$k = k_{\text{BS}} + \eta, \tag{3.141a}$$

so that

$$\begin{aligned}\kappa_2^2 &= K^2 - k^2 \\ &= \left(\frac{w_0}{2}\right)^2 \left[1 - 2\frac{k_{\text{BS}}}{(w_0/2)^2}\eta - \frac{\eta^2}{(w_0/2)^2}\right]\end{aligned} \tag{3.141b}$$

and

$$\kappa_2 = \frac{w_0}{2}\left[1 - \frac{k_{\text{BS}}}{(w_0/2)^2}\eta + O(\eta^2)\right],$$

$$2\kappa_2 - w_0 = -4\frac{k_{\text{BS}}}{w_0}\eta + O(\eta^2), \tag{3.141c}$$

for $\eta \ll k_{\text{BS}}$. Substituting this last result into eqn (3.140), we finally find the following expressions for the reflection amplitude $r_{11}$ and the reflection coefficient $R_{11} = |r_{11}|^2$:

$$r_{11} = -\frac{i\Gamma^{(k)}/2}{\eta + i\Gamma^{(k)}/2}, \tag{3.142a}$$

$$R_{11} = \frac{\left(\Gamma^{(k)}/2\right)^2}{\eta^2 + (\Gamma^{(k)}/2)^2}, \tag{3.142b}$$

a Breit–Wigner form with a width

$$\Gamma^{(k)} = \frac{w_0}{4k_{\text{BS}}^2}|u_{12}|^2. \tag{3.142c}$$

Similarly, the transmission amplitude $t_{11}$ and the reflection coefficient $T_{11} = |t_{11}|^2$ are given by

$$t_{11} = \frac{\eta}{\eta + i\Gamma^{(k)}/2}, \tag{3.143a}$$

$$T_{11} = \frac{\eta^2}{\eta^2 + (\Gamma^{(k)}/2)^2}. \tag{3.143b}$$

This is commonly called a *Fano resonance* in atomic spectroscopy.

We compare the above result for the width with the standard one obtained very generally (see, for instance, [72]) for a BSC, i.e.,

$$\Gamma_s = 2\pi \left|\langle\phi_s(E)|V_{12}|\psi_{\text{BS}}\rangle\right|^2. \tag{3.144}$$

This expression gives the partial width (in energy) for the bound state $\psi_{\text{BS}}$ in channel 2 to decay to the state $\phi_s(E)$ in channel 1, via the interaction $V_{12}$. In our case we have

$$\psi_{\text{BS}}(x) = \sqrt{\frac{w_0}{2}}\exp\left(-\frac{w_0}{2}|x|\right), \tag{3.145a}$$

$$\phi_s(E; x) = \frac{e^{iskx}}{\sqrt{2\pi\hbar^2 k/m}}, \tag{3.145b}$$

$$V_{12}(x) = \frac{\hbar^2}{2m}u_{12}\delta(x), \tag{3.145c}$$

so that the partial width $\Gamma_s$ and the total width $\Gamma$ are given by

$$\Gamma_s = \frac{\hbar^2 w_0}{8mk_{\text{BS}}}|u_{12}|^2, \tag{3.146}$$

$$\Gamma = \sum_s \Gamma_s = \frac{\hbar^2 w_0}{4mk_{\text{BS}}}|u_{12}|^2. \tag{3.147}$$

The width in wave number space can be found to be

$$\Gamma^{(k)} \approx \frac{\Gamma}{(\hbar^2/2m)2k_{\rm BS}} \qquad (3.148{\rm a})$$

$$= \frac{w_0}{4k_{\rm BS}^2} |u_{12}|^2, \qquad (3.148{\rm b})$$

which coincides with the result obtained above (eqn (3.142c)).

### 3.1.7 Extension of the S and M matrices to include open and closed channels

So far we have studied the $S$ and $M$ matrices which are relevant to open channels only, even in cases where the system admits closed channels; this is all we need for the description of the asymptotic wave function in a scattering problem. When we combine potentials in series using the individual $S$ and $M$ matrices we need, in general, the extension of these matrices to include closed channels as well. The reason is that in the region between two potentials the wave function which is a solution of the Schrödinger equation consists, in general, of a linear combination of incoming and outgoing waves for the open channels, and increasing and decreasing real exponentials for the closed channels. The extension of the $S$ and $M$ matrices to include both open and closed channels, their relation with the open-channel matrices, and the actual combination of $M$ and $S$ matrices are discussed below.

One open and one closed channel are good enough to illustrate the ideas. Following the notation of Table 3.2, we can write the most general solution of the Schrödinger equation in a potential-free region as

$$\Psi(x) = \begin{bmatrix} a\phi_+(E_1;x) + b\phi_-(E_1;x) \\ c\phi_+(-\epsilon_2;x) + d\phi_-(-\epsilon_2;x) \end{bmatrix}$$

$$= \begin{bmatrix} Ae^{ik_1 x} + Be^{-ik_1 x} \\ Ce^{-\kappa_2 x} + De^{\kappa_2 x} \end{bmatrix}. \qquad (3.149)$$

Firstly, we calculate the particle current associated with this wave function. In the integration over the transverse direction $y$, we make use of the orthogonality of the transverse wave functions associated with two different channels and obtain, after an elementary calculation,

$$I = \frac{\hbar}{m}\left[ k_1\left(|A|^2 - |B|^2\right) + i\kappa_2\left(CD^* - C^*D\right)\right]$$

$$= \frac{1}{h}\left[\left(|a|^2 - |b|^2\right) + i\left(cd^* - c^*d\right)\right]. \qquad (3.150)$$

If either $c$ or $d$ vanishes—as it occurs fully outside the array of scattering potentials—or if the product $cd^*$ is real, then the closed channel does not contribute to the current. In the potential-free region between two scatterers *the open-channel part of the wave function does not suffice, in general, for the calculation of the current.*

For this reason we now select one of the potentials to be combined and, in the presence of open *and* closed channels, formally introduce the most general solution (to be called the *extended* solution) of the Schrödinger equation for that potential at a given energy, independently of whether that solution, on either side of the potential (regions I and II), is finite or divergent as $x \to \pm\infty$. All we require is the connection between the sets of coefficients of the exponential wave functions (complex *and* real) in the two regions. The final solution for the full array of scatterers will certainly be normalizable in the standard way. In preparation for this study, we first point out that, if we do not pay attention to the normalization of the wave function, in a one-open-one-closed-channel case we have *four* linearly independent solutions, just as in the case of two-open channels, eqn (3.63). In regions I and II on either side of the potential these *extended* solutions behave as follows:

$$\Psi^{(+)}_{+,1}(E;x) \sim \begin{cases} \begin{bmatrix} \phi_+(E_1;x) + \widetilde{r}_{11}\phi_-(E_1;x) \\ \widetilde{r}_{21}\phi_-(-\epsilon_2;x) \end{bmatrix}, & x \in \text{I}, \\ \begin{bmatrix} \widetilde{t}_{11}\phi_+(E_1;x) \\ \widetilde{t}_{21}\phi_+(-\epsilon_2;x) \end{bmatrix}, & x \in \text{II}, \end{cases}$$

$$\Psi^{(+)}_{+,2}(E;x) \sim \begin{cases} \begin{bmatrix} \widetilde{r}_{12}\phi_-(E_1;x) \\ \phi_+(-\epsilon_2;x) + \widetilde{r}_{22}\phi_-(-\epsilon_2;x) \end{bmatrix}, & x \in \text{I}, \\ \begin{bmatrix} \widetilde{t}_{12}\phi_+(E_1;x) \\ \widetilde{t}_{22}\phi_+(-\epsilon_2;x) \end{bmatrix}, & x \in \text{II}, \end{cases}$$

$$\Psi^{(+)}_{-,1}(E;x) \sim \begin{cases} \begin{bmatrix} \widetilde{t}'_{11}\phi_-(E_1;x) \\ \widetilde{t}'_{21}\phi_-(-\epsilon_2;x) \end{bmatrix}, & x \in \text{I}, \\ \begin{bmatrix} \phi_-(E_1;x) + \widetilde{r}'_{11}\phi_+(E_1;x) \\ \widetilde{r}'_{21}\phi_+(-\epsilon_2;x) \end{bmatrix}, & x \in \text{II}, \end{cases}$$

$$\Psi^{(+)}_{-,2}(E;x) \sim \begin{cases} \begin{bmatrix} \widetilde{t}'_{12}\phi_-(E_1;x) \\ \widetilde{t}'_{22}\phi_-(-\epsilon_2;x) \end{bmatrix}, & x \in \text{I}, \\ \begin{bmatrix} \widetilde{r}'_{12}\phi_+(E_1;x) \\ \phi_-(-\epsilon_2;x) + \widetilde{r}'_{22}\phi_+(-\epsilon_2;x) \end{bmatrix}, & x \in \text{II}. \end{cases}$$

(3.151)

We have indicated with a tilde the reflection and transmission amplitudes associated with the extended solutions under discussion. It is clear that we can carry on the same analysis as for the two-open-channel case (see eqns (3.63) and the following equations), just taking the analytic continuation defined by the rule

(3.24). Thus, the most general solution on either side of the potential can be obtained from the analytic continuation of the one shown in Table 3.1, i.e.,

$$\Psi^{(1)}(E;x) = \begin{bmatrix} a_1^{(1)}\phi_+(E_1;x) + b_1^{(1)}\phi_-(E_1;x) \\ a_2^{(1)}\phi_+(-\epsilon_2;x) + b_2^{(1)}\phi_-(-\epsilon_2;x) \end{bmatrix}, \quad (3.152)$$

$$\Psi^{(2)}(E;x) = \begin{bmatrix} a_1^{(2)}\phi_-(E_1;x) + b_1^{(2)}\phi_+(E_1;x) \\ a_2^{(2)}\phi_-(-\epsilon_2;x) + b_2^{(2)}\phi_+(-\epsilon_2;x) \end{bmatrix}. \quad (3.153)$$

(Notice that, for the closed channel, the 'outgoing' wave function $\phi_s(-\epsilon_2;x)$ of eqn (3.21), to which we have assigned a $b$ coefficient, decays to zero as $x \to \pm\infty$, while the 'incoming' wave function diverges as $x \to \pm\infty$. The incoming part of the wave function associated with the closed channel was absent in Table 3.2.) Just as in the two-open-channel case, we reach the conclusion that the four outgoing amplitudes $b_1^{(1)}$, $b_2^{(1)}$, $b_1^{(2)}$ and $b_2^{(2)}$ are linearly related to the four incoming amplitudes $a_1^{(1)}$, $a_2^{(1)}$, $a_1^{(2)}$ and $a_2^{(2)}$ through a matrix that we call the *extended S matrix* $\widetilde{S}$ (which is now *non-unitary*), i.e.,

$$\begin{bmatrix} b_1^{(1)} \\ b_2^{(1)} \\ b_1^{(2)} \\ b_2^{(2)} \end{bmatrix} = \widetilde{S} \begin{bmatrix} a_1^{(1)} \\ a_2^{(1)} \\ a_1^{(2)} \\ a_2^{(2)} \end{bmatrix}. \quad (3.154)$$

The matrix $\widetilde{S}$ is $4 \times 4$ and can be written in the standard fashion, so as to define $2 \times 2$ extended reflection and transmission matrices as follows:

$$\widetilde{S} = \begin{bmatrix} \widetilde{r} & \widetilde{t}' \\ \widetilde{t} & \widetilde{r}' \end{bmatrix} = \left[\begin{array}{cc|cc} \widetilde{r}_{11} & \widetilde{r}_{12} & \widetilde{t}'_{11} & \widetilde{t}'_{12} \\ \widetilde{r}_{21} & \widetilde{r}_{22} & \widetilde{t}'_{21} & \widetilde{t}'_{22} \\ \hline \widetilde{t}_{11} & \widetilde{t}_{12} & \widetilde{r}'_{11} & \widetilde{r}'_{12} \\ \widetilde{t}_{21} & \widetilde{t}_{22} & \widetilde{r}'_{21} & \widetilde{r}'_{22} \end{array}\right]. \quad (3.155)$$

According to eqn (3.154), if we are given the four coefficients $a_1^{(1)}$, $a_2^{(1)}$, $a_1^{(2)}$ and $a_2^{(2)}$, then the $4 \times 4$ extended matrix $\widetilde{S}$ allows us to find the four coefficients $b_1^{(1)}$, $b_2^{(1)}$, $b_1^{(2)}$ and $b_2^{(2)}$. Should we be dealing with the full system, i.e., the full array of scatterers, we could think that we are given the amplitudes of the incoming waves $a_1^{(1)}$ and $a_1^{(2)}$ in the open channel as arbitrary complex numbers, and that the amplitudes of the 'incoming' waves for the closed channel are set equal to zero, i.e., $a_2^{(1)} = a_2^{(2)} = 0$; we would thus have the full information, from which the four $b$s could be found. In contrast, if we fix our attention to *two consecutive potential-free regions* I and II in an array of potentials, then it is not enough to know the open-channel incoming coefficients $a_1^{(1)}$ and $a_1^{(2)}$ in order to find the two open-channel outgoing coefficients $b_1^{(1)}$ and $b_1^{(2)}$, the reason being that the closed-channel 'incoming' amplitudes $a_2^{(1)}$ and $a_2^{(2)}$ are, in general, nonzero and they will contribute to the outgoing amplitudes. Thus, in such a case *there is no 2 × 2 matrix relating only open-channel coefficients.*

Similarly, we define the *extended M matrix*, namely $\widetilde{M}$, in terms of the left and the right coefficients as

$$\begin{bmatrix} b_1^{(2)} \\ b_2^{(2)} \\ a_1^{(2)} \\ a_2^{(2)} \end{bmatrix} = \widetilde{M} \begin{bmatrix} a_1^{(1)} \\ a_2^{(1)} \\ b_1^{(1)} \\ b_2^{(1)} \end{bmatrix}. \tag{3.156}$$

Here $M$ is $4 \times 4$ and can be written in terms of $2 \times 2$ blocks as follows:

$$\widetilde{M} = \begin{bmatrix} \widetilde{\alpha} & \widetilde{\beta} \\ \widetilde{\gamma} & \widetilde{\delta} \end{bmatrix} = \begin{bmatrix} \widetilde{\alpha}_{11} & \widetilde{\alpha}_{12} & \widetilde{\beta}_{11} & \widetilde{\beta}_{12} \\ \widetilde{\alpha}_{21} & \widetilde{\alpha}_{22} & \widetilde{\beta}_{21} & \widetilde{\beta}_{22} \\ \widetilde{\gamma}_{11} & \widetilde{\gamma}_{12} & \widetilde{\delta}_{11} & \widetilde{\delta}_{12} \\ \widetilde{\gamma}_{21} & \widetilde{\gamma}_{22} & \widetilde{\delta}_{21} & \widetilde{\delta}_{22} \end{bmatrix}. \tag{3.157}$$

*The relation between the open-channel S matrix and the extended $\widetilde{S}$ matrix*

We expand eqn (3.154) using eqn (3.155), to obtain

$$b_1^{(1)} = \widetilde{r}_{11} a_1^{(1)} + \widetilde{r}_{12} a_2^{(1)} + \widetilde{t}'_{11} a_1^{(2)} + \widetilde{t}'_{12} a_2^{(2)}, \tag{3.158a}$$

$$b_2^{(1)} = \widetilde{r}_{21} a_1^{(1)} + \widetilde{r}_{22} a_2^{(1)} + \widetilde{t}'_{21} a_1^{(2)} + \widetilde{t}'_{22} a_2^{(2)}, \tag{3.158b}$$

$$b_1^{(2)} = \widetilde{t}_{11} a_1^{(1)} + \widetilde{t}_{12} a_2^{(1)} + \widetilde{r}'_{11} a_1^{(2)} + \widetilde{r}'_{12} a_2^{(2)}, \tag{3.158c}$$

$$b_2^{(2)} = \widetilde{t}_{21} a_1^{(1)} + \widetilde{t}_{22} a_2^{(1)} + \widetilde{r}'_{21} a_1^{(2)} + \widetilde{r}'_{22} a_2^{(2)}. \tag{3.158d}$$

We now assume that the upper indices (1) and (2) refer to the asymptotic regions *fully outside the array of scatterers*; given the extended $\widetilde{S}$ matrix, we wish to find the standard $2 \times 2$ open-channel $S$ matrix. This means that we want to retain the open-channel variables

$$b_1^{(1)}, \quad a_1^{(1)},$$
$$b_1^{(2)}, \quad a_1^{(2)}$$

and eliminate the closed-channel variables

$$b_2^{(1)}, \quad a_2^{(1)},$$
$$b_2^{(2)}, \quad a_2^{(2)}.$$

Recall that now $a_2^{(1)} = a_2^{(2)} = 0$, by definition of the open-channel $S$ matrix, just as was explained immediately after eqn (3.130) in relation to the soluble example. Thus we only have two variables to eliminate, namely $b_2^{(1)}$ and $b_2^{(2)}$, after which we are left with two equations in the two unknowns $b_1^{(1)}$ and $b_1^{(2)}$, assuming that $a_1^{(1)}$ and $a_1^{(2)}$ are given. This elimination can be done trivially, by discarding the only equations where they appear, eqns (3.158b) and (3.158d). Equations (3.158a) and (3.158c) then give

$$b_1^{(1)} = \widetilde{r}_{11} a_1^{(1)} + \widetilde{t}'_{11} a_1^{(2)},$$
$$b_1^{(2)} = \widetilde{t}_{11} a_1^{(1)} + \widetilde{r}'_{11} a_1^{(2)}.$$

The standard, or *open-channel, S matrix*, defined through

$$\begin{bmatrix} b_1^{(1)} \\ b_1^{(2)} \end{bmatrix} = S \begin{bmatrix} a_1^{(1)} \\ a_1^{(2)} \end{bmatrix}, \qquad (3.159)$$

is thus

$$S \equiv \begin{bmatrix} r & t' \\ t & r' \end{bmatrix} = \begin{bmatrix} \widetilde{r}_{11} & \widetilde{t}'_{11} \\ \widetilde{t}_{11} & \widetilde{r}'_{11} \end{bmatrix}. \qquad (3.160)$$

We thus reach the result that *S is just the $2 \times 2$ open-channel submatrix of the extended $4 \times 4$ $\widetilde{S}$ matrix* of eqn (3.155).

*The relation between the open-channel M matrix and the extended $\widetilde{M}$ matrix*

We expand eqn (3.156) using eqn (3.157), to obtain

$$b_1^{(2)} = \widetilde{\alpha}_{11} a_1^{(1)} + \widetilde{\alpha}_{12} a_2^{(1)} + \widetilde{\beta}_{11} b_1^{(1)} + \widetilde{\beta}_{12} b_2^{(1)}, \qquad (3.161a)$$
$$b_2^{(2)} = \widetilde{\alpha}_{21} a_1^{(1)} + \widetilde{\alpha}_{22} a_2^{(1)} + \widetilde{\beta}_{21} b_1^{(1)} + \widetilde{\beta}_{22} b_2^{(1)}, \qquad (3.161b)$$
$$a_1^{(2)} = \widetilde{\gamma}_{11} a_1^{(1)} + \widetilde{\gamma}_{12} a_2^{(1)} + \widetilde{\delta}_{11} b_1^{(1)} + \widetilde{\delta}_{12} b_2^{(1)}, \qquad (3.161c)$$
$$a_2^{(2)} = \widetilde{\gamma}_{21} a_1^{(1)} + \widetilde{\gamma}_{22} a_2^{(1)} + \widetilde{\delta}_{21} b_1^{(1)} + \widetilde{\delta}_{22} b_2^{(1)}. \qquad (3.161d)$$

We again assume that the upper indices (1) and (2) refer to the asymptotic regions *fully outside the array of scatterers*; given the extended $\widetilde{M}$ matrix, we wish to find the standard $2 \times 2$ open-channel $M$ matrix. This means that, just as before, we want to retain the open-channel variables

$$b_1^{(2)}, \quad a_1^{(1)},$$
$$a_1^{(2)}, \quad b_1^{(1)}$$

and eliminate the closed-channel variables

$$b_2^{(2)}, \quad a_2^{(1)},$$
$$a_2^{(2)}, \quad b_2^{(1)}.$$

Again, $a_2^{(1)} = a_2^{(2)} = 0$, by definition of the open-channel $M$ matrix; $b_2^{(2)}$ appears only in eqn (3.161b), which we just discard. There only remains $b_2^{(1)}$, which will be solved for from eqn (3.161d) and substituted into eqns (3.161a) and (3.161c), to obtain

$$b_1^{(2)} = \left[\widetilde{\alpha}_{11} - \frac{\widetilde{\beta}_{12}\widetilde{\gamma}_{21}}{\widetilde{\delta}_{22}}\right] a_1^{(1)} + \left[\widetilde{\beta}_{11} - \frac{\widetilde{\beta}_{12}\widetilde{\delta}_{21}}{\widetilde{\delta}_{22}}\right] b_1^{(1)}, \qquad (3.162)$$

$$a_1^{(2)} = \left[\widetilde{\gamma}_{11} - \frac{\widetilde{\delta}_{12}\widetilde{\gamma}_{21}}{\widetilde{\delta}_{22}}\right] a_1^{(1)} + \left[\widetilde{\delta}_{11} - \frac{\widetilde{\delta}_{12}\widetilde{\delta}_{21}}{\widetilde{\delta}_{22}}\right] b_1^{(1)}. \qquad (3.163)$$

We again end up with two equations in the two unknowns $b_1^{(2)}$ and $a_1^{(2)}$, assuming that $a_1^{(1)}$ and $b_1^{(1)}$ are given.

The standard, *open-channel* $2 \times 2$ $M$ *matrix* is defined as

$$\begin{bmatrix} b_1^{(2)} \\ a_1^{(2)} \end{bmatrix} = M \begin{bmatrix} a_1^{(1)} \\ b_1^{(1)} \end{bmatrix}, \qquad (3.164)$$

and is then found to be

$$M \equiv \begin{bmatrix} \alpha & \beta \\ \gamma & \delta \end{bmatrix} = \begin{bmatrix} \widetilde{\alpha}_{11} - \widetilde{\beta}_{12}\widetilde{\gamma}_{21}/\widetilde{\delta}_{22} & \widetilde{\beta}_{11} - \widetilde{\beta}_{12}\widetilde{\delta}_{21}/\widetilde{\delta}_{22} \\ \widetilde{\gamma}_{11} - \widetilde{\delta}_{12}\widetilde{\gamma}_{21}/\widetilde{\delta}_{22} & \widetilde{\delta}_{11} - \widetilde{\delta}_{12}\widetilde{\delta}_{21}/\widetilde{\delta}_{22} \end{bmatrix}. \qquad (3.165)$$

We observe that $M$ *is not just the open-channel part of the extended $\widetilde{M}$ matrix* of eqn (3.157). This statement should be contrasted with the one made in relation to the $S$ matrix immediately below eqn (3.160).

Let us check that the matrix $M$ of eqn (3.165) is consistent with the matrix $S$ of eqn (3.160). For instance, from eqn (2.171), or eqn (3.98) with $N = 1$, we should have the relation

$$t' = \frac{1}{\delta} \qquad (3.166)$$

between $t'$ found in eqn (3.160) and $\delta$ found in eqn (3.165). To verify this relation, notice that between the block $\widetilde{t}'$ of the extended $\widetilde{S}$ matrix and the block $\widetilde{\delta}$ of the extended $\widetilde{M}$ matrix there is the identity

$$\widetilde{t}' = \frac{1}{\widetilde{\delta}},$$

or

$$\begin{bmatrix} \widetilde{t}'_{11} & \widetilde{t}'_{12} \\ \widetilde{t}'_{21} & \widetilde{t}'_{22} \end{bmatrix} = \begin{bmatrix} \widetilde{\delta}_{11} & \widetilde{\delta}_{12} \\ \widetilde{\delta}_{21} & \widetilde{\delta}_{22} \end{bmatrix}^{-1}$$

$$= \frac{1}{\widetilde{\delta}_{11}\widetilde{\delta}_{22} - \widetilde{\delta}_{12}\widetilde{\delta}_{21}} \begin{bmatrix} \widetilde{\delta}_{22} & -\widetilde{\delta}_{12} \\ -\widetilde{\delta}_{21} & \widetilde{\delta}_{11} \end{bmatrix}.$$

In particular,

$$\widetilde{t}'_{11} = \frac{\widetilde{\delta}_{22}}{\widetilde{\delta}_{11}\widetilde{\delta}_{22} - \widetilde{\delta}_{12}\widetilde{\delta}_{21}},$$

which, from eqns (3.160) and (3.165), leads to (3.166).

*An exactly soluble example*

From the above discussion it is clear that, within the soluble coupled-channel example of Section 3.1.6, the extended $\widetilde{S}$ and $\widetilde{M}$ matrices for the case of one open and one closed channel can be simply obtained from the $S$ and $M$ matrices for the two-open-channel problem by making use of the analytic continuation (3.24).

From eqn (3.127) we thus obtain, for the blocks of the extended $\widetilde{S}$ matrix of eqn (3.155),

$$\widetilde{r} = \widetilde{r}' = \frac{1}{\Delta} \begin{bmatrix} (u_{11}/2ik_1)(1+u_{22}/2\kappa_2) - u_{12}u_{21}/4ik_1\kappa_2 & -iu_{12}/2e^{i\pi/4}\sqrt{k_1\kappa_2} \\ -iu_{21}/2e^{i\pi/4}\sqrt{k_1\kappa_2} & -(u_{22}/2\kappa_2)(1-u_{11}/2ik_1) - u_{12}u_{21}/4ik_1\kappa_2 \end{bmatrix}, \quad (3.167)$$

$$\widetilde{t} = \widetilde{t}' = \frac{1}{\Delta} \begin{bmatrix} 1 + u_{22}/2\kappa_2 & -iu_{12}/2e^{i\pi/4}\sqrt{k_1\kappa_2} \\ -iu_{21}/2e^{i\pi/4}\sqrt{k_1\kappa_2} & 1 - u_{11}/2ik_1 \end{bmatrix}, \quad (3.168)$$

where

$$\Delta = \left(1 - \frac{u_{11}}{2ik_1}\right)\left(1 + \frac{u_{22}}{2\kappa_2}\right) + \frac{u_{12}u_{21}}{4ik_1\kappa_2}. \quad (3.169)$$

We explicitly verify, in this example, the statement made immediately after eqn (3.160), i.e., the $2 \times 2$ standard, open-channel, $S$ matrix of eqn (3.133) is simply the open-channel submatrix (i.e., the elements $S_{11}$, $S_{13}$, $S_{31}$ and $S_{33}$) of the extended $S$ matrix of eqns (3.167) and (3.168), and *it is that submatrix that is unitary*.

Similarly, from the $M$ matrix for two open channels, eqn (3.129), we obtain the extended $M$ matrix of eqn (3.157) as follows:

$$M = \left[\begin{array}{cc|cc} 1+u_{11}/2ik_1 & (u_{12}/2ik_1)\sqrt{k_1/i\kappa} & u_{11}/2ik_1 & (u_{12}/2ik_1)\sqrt{k_1/i\kappa} \\ (u_{21}/2ik_1)\sqrt{k_1/i\kappa} & 1-u_{22}/2\kappa & (u_{21}/2ik_1)\sqrt{k_1/i\kappa} & -u_{22}/2\kappa \\ \hline -u_{11}/2ik_1 & -(u_{12}/2ik_1)\sqrt{k_1/i\kappa} & 1-u_{11}/2ik_1 & -(u_{12}/2ik_1)\sqrt{k_1/i\kappa} \\ -(u_{21}/2ik_1)\sqrt{k_1/i\kappa} & u_{22}/2\kappa & -(u_{21}/2ik_1)\sqrt{k_1/i\kappa} & 1+u_{22}/2\kappa \end{array}\right]. \quad (3.170)$$

We observe that the standard, open-channel, $M$ matrix of eqn (3.134) is not simply the open-channel submatrix (i.e., the elements $M_{11}$, $M_{13}$, $M_{31}$ and $M_{33}$) of the extended $M$ matrix of eqn (3.170) above. We can verify directly that the relation (3.165) is satisfied.

*Combining transfer and scattering matrices in the presence of open and closed channels*

As indicated at the beginning of this section, when combining two potentials in a problem that admits the presence of closed channels, the transfer and the scattering matrices to be used are the extended ones studied above. In the following two exercises we illustrate such a combination explicitly, again for the case of one open and one closed channel, and present an important simplification that occurs when the two scatterers are much further apart than the characteristic length of the decaying or growing exponentials associated with the closed channel.

**Exercise 3.4** Consider two scatterers a distance $d$ apart, each described by one open and one closed channel. Combine the extended transfer matrices for the two scatterers, assuming that $\kappa_2 d \gg 1$, where $\kappa_2$ is the usual quantity defined for a closed channel in eqns (3.24).

Consider two scatterers, as illustrated in Fig. 2.4 for the one-dimensional case, described by the extended transfer matrices $\widetilde{M}_1$ and $\widetilde{M}_2$. The first scatterer is in the vicinity of the origin, say; its $\widetilde{M}_1$ can be written as in eqn (3.157), i.e.,

$$\widetilde{M}_1 = \begin{bmatrix} \widetilde{\alpha}_1 & \widetilde{\beta}_1 \\ \widetilde{\gamma}_1 & \widetilde{\delta}_1 \end{bmatrix}, \tag{3.171}$$

where the various blocks $\widetilde{\alpha}_1$, etc. are now two-dimensional. The second scatterer is displaced a distance $d$ from the first; before displacement it is described by the transfer matrix $\overset{\circ}{M}_2$ (see Fig. 2.6). The relation between $\widetilde{M}_2$ and $\overset{\circ}{M}_2$ can be found from Section 3.1.5. Equation (3.115) is intended to apply to the open-channel case. As usual, for closed channels we replace $k_n$, say, by $i\kappa_n$. The two matrices

$$\widetilde{M}_2 = \begin{bmatrix} \widetilde{\alpha}_2 & \widetilde{\beta}_2 \\ \widetilde{\gamma}_2 & \widetilde{\delta}_2 \end{bmatrix}, \qquad \overset{\circ}{M}_2 = \begin{bmatrix} \overset{\circ}{\alpha}_2 & \overset{\circ}{\beta}_2 \\ \overset{\circ}{\gamma}_2 & \overset{\circ}{\delta}_2 \end{bmatrix}$$

are then related by

$$\widetilde{M}_2 = \begin{bmatrix} \widetilde{T}(kd) & 0 \\ 0 & \widetilde{T}^{-1}(kd) \end{bmatrix}^{-1} \overset{\circ}{M}_2 \begin{bmatrix} \widetilde{T}(kd) & 0 \\ 0 & \widetilde{T}^{-1}(kd) \end{bmatrix}, \tag{3.172}$$

where

$$\widetilde{T}(kd) = \begin{bmatrix} e^{ik_1 d} & 0 \\ 0 & e^{-\kappa_2 d} \end{bmatrix}. \tag{3.173}$$

We thus write

$$\begin{bmatrix} \widetilde{\alpha}_2 & \widetilde{\beta}_2 \\ \widetilde{\gamma}_2 & \widetilde{\delta}_2 \end{bmatrix} = \begin{bmatrix} \widetilde{T}^{-1}(kd) & 0 \\ 0 & \widetilde{T}(kd) \end{bmatrix} \begin{bmatrix} \overset{\circ}{\alpha}_2 & \overset{\circ}{\beta}_2 \\ \overset{\circ}{\gamma}_2 & \overset{\circ}{\delta}_2 \end{bmatrix} \begin{bmatrix} \widetilde{T}(kd) & 0 \\ 0 & \widetilde{T}^{-1}(kd) \end{bmatrix}$$

$$= \begin{bmatrix} \widetilde{T}^{-1}(kd)\overset{\circ}{\widetilde{\alpha}}_2\widetilde{T}(kd) & \widetilde{T}^{-1}(kd)\overset{\circ}{\widetilde{\beta}}_2\widetilde{T}^{-1}(kd) \\ \widetilde{T}(kd)\overset{\circ}{\widetilde{\gamma}}_2\widetilde{T}(kd) & \widetilde{T}(kd)\overset{\circ}{\widetilde{\delta}}_2\widetilde{T}^{-1}(kd) \end{bmatrix}. \qquad (3.174)$$

The total $\widetilde{M}$ is thus given by

$$\widetilde{M} = \widetilde{M}_2\widetilde{M}_1, \qquad (3.175)$$

or

$$\begin{bmatrix} \widetilde{\alpha} & \widetilde{\beta} \\ \widetilde{\gamma} & \widetilde{\delta} \end{bmatrix} = \begin{bmatrix} \widetilde{\alpha}_2 & \widetilde{\beta}_2 \\ \widetilde{\gamma}_2 & \widetilde{\delta}_2 \end{bmatrix} \begin{bmatrix} \widetilde{\alpha}_1 & \widetilde{\beta}_1 \\ \widetilde{\gamma}_1 & \widetilde{\delta}_1 \end{bmatrix}$$

$$= \left[ \begin{array}{c|c} \widetilde{T}^{-1}(kd)\overset{\circ}{\widetilde{\alpha}}_2\widetilde{T}(kd)\widetilde{\alpha}_1 & \widetilde{T}^{-1}(kd)\overset{\circ}{\widetilde{\alpha}}_2\widetilde{T}(kd)\widetilde{\beta}_1 \\ +\widetilde{T}^{-1}(kd)\overset{\circ}{\widetilde{\beta}}_2\widetilde{T}^{-1}(kd)\widetilde{\gamma}_1 & +\widetilde{T}^{-1}(kd)\overset{\circ}{\widetilde{\beta}}_2\widetilde{T}^{-1}(kd)\widetilde{\delta}_1 \\ \hline \widetilde{T}(kd)\overset{\circ}{\widetilde{\gamma}}_2\widetilde{T}(kd)\widetilde{\alpha}_1 & \widetilde{T}(kd)\overset{\circ}{\widetilde{\gamma}}_2\widetilde{T}(kd)\widetilde{\beta}_1 \\ +\widetilde{T}(kd)\overset{\circ}{\widetilde{\delta}}_2\widetilde{T}^{-1}(kd)\widetilde{\gamma}_1 & +\widetilde{T}(kd)\overset{\circ}{\widetilde{\delta}}_2\widetilde{T}^{-1}(kd)\widetilde{\delta}_1 \end{array} \right].$$

$$(3.176)$$

In particular, we concentrate on the block $\widetilde{\delta}$. A similar analysis can be performed for the other blocks. We have

$$\widetilde{\delta} = \widetilde{T}(kd)\overset{\circ}{\widetilde{\gamma}}_2\widetilde{T}(kd)\widetilde{\beta}_1 + \widetilde{T}(kd)\overset{\circ}{\widetilde{\delta}}_2\widetilde{T}^{-1}(kd)\widetilde{\delta}_1, \qquad (3.177)$$

or, in terms of its matrix elements,

$$\begin{bmatrix} \widetilde{\delta}_{11} & \widetilde{\delta}_{12} \\ \widetilde{\delta}_{21} & \widetilde{\delta}_{22} \end{bmatrix}$$

$$= \begin{bmatrix} e^{ik_1d} & 0 \\ 0 & e^{-\kappa_2 d} \end{bmatrix} \begin{bmatrix} (\overset{\circ}{\widetilde{\gamma}}_2)_{11} & (\overset{\circ}{\widetilde{\gamma}}_2)_{12} \\ (\overset{\circ}{\widetilde{\gamma}}_2)_{21} & (\overset{\circ}{\widetilde{\gamma}}_2)_{22} \end{bmatrix} \begin{bmatrix} e^{ik_1d} & 0 \\ 0 & e^{-\kappa_2 d} \end{bmatrix} \begin{bmatrix} (\widetilde{\beta}_1)_{11} & (\widetilde{\beta}_1)_{12} \\ (\widetilde{\beta}_1)_{21} & (\widetilde{\beta}_1)_{22} \end{bmatrix}$$

$$+ \begin{bmatrix} e^{ik_1d} & 0 \\ 0 & e^{-\kappa_2 d} \end{bmatrix} \begin{bmatrix} (\overset{\circ}{\widetilde{\delta}}_2)_{11} & (\overset{\circ}{\widetilde{\delta}}_2)_{12} \\ (\overset{\circ}{\widetilde{\delta}}_2)_{21} & (\overset{\circ}{\widetilde{\delta}}_2)_{22} \end{bmatrix} \begin{bmatrix} e^{-ik_1d} & 0 \\ 0 & e^{\kappa_2 d} \end{bmatrix} \begin{bmatrix} (\widetilde{\delta}_1)_{11} & (\widetilde{\delta}_1)_{12} \\ (\widetilde{\delta}_1)_{21} & (\widetilde{\delta}_1)_{22} \end{bmatrix}$$

$$= \begin{bmatrix} e^{2ik_1 d}\left(\overset{\circ}{\tilde{\gamma}}_2\right)_{11}\left(\tilde{\beta}_1\right)_{11} & e^{2ik_1 d}\left(\overset{\circ}{\tilde{\gamma}}_2\right)_{11}\left(\tilde{\beta}_1\right)_{12} \\ + e^{(ik_1 - \kappa_2)d}\left(\overset{\circ}{\tilde{\gamma}}_2\right)_{12}\left(\tilde{\beta}_1\right)_{21} & + e^{(ik_1 - \kappa_2)d}\left(\overset{\circ}{\tilde{\gamma}}_2\right)_{12}\left(\tilde{\beta}_1\right)_{22} \\ + \left(\overset{\circ}{\tilde{\delta}}_2\right)_{11}\left(\tilde{\delta}_1\right)_{11} & + \left(\overset{\circ}{\tilde{\delta}}_2\right)_{11}\left(\tilde{\delta}_1\right)_{12} \\ + e^{(ik_1 + \kappa_2)d}\left(\overset{\circ}{\tilde{\delta}}_2\right)_{12}\left(\tilde{\delta}_1\right)_{21} & + e^{(ik_1 + \kappa_2)d}\left(\overset{\circ}{\tilde{\delta}}_2\right)_{12}\left(\tilde{\delta}_1\right)_{22} \\ \hline e^{(ik_1 - \kappa_2)d}\left(\overset{\circ}{\tilde{\gamma}}_2\right)_{21}\left(\tilde{\beta}_1\right)_{11} & e^{(ik_1 - \kappa_2)d}\left(\overset{\circ}{\tilde{\gamma}}_2\right)_{21}\left(\tilde{\beta}_1\right)_{12} \\ + e^{-2\kappa_2 d}\left(\overset{\circ}{\tilde{\gamma}}_2\right)_{22}\left(\tilde{\beta}_1\right)_{21} & + e^{-2\kappa_2 d}\left(\overset{\circ}{\tilde{\gamma}}_2\right)_{22}\left(\tilde{\beta}_1\right)_{22} \\ + e^{-(ik_1 + \kappa_2)d}\left(\overset{\circ}{\tilde{\delta}}_2\right)_{21}\left(\tilde{\delta}_1\right)_{11} & + e^{-(ik_1 + \kappa_2)d}\left(\overset{\circ}{\tilde{\delta}}_2\right)_{21}\left(\tilde{\delta}_1\right)_{12} \\ + \left(\overset{\circ}{\tilde{\delta}}_2\right)_{22}\left(\tilde{\delta}_1\right)_{21} & + \left(\overset{\circ}{\tilde{\delta}}_2\right)_{22}\left(\tilde{\delta}_1\right)_{22} \end{bmatrix}. \tag{3.178}$$

We calculate, for instance, $\delta$ of eqn (3.165). We find

$$\begin{aligned}\delta &= \tilde{\delta}_{11} - \frac{\tilde{\delta}_{21}\tilde{\delta}_{12}}{\tilde{\delta}_{22}} \\ &= \left[ e^{2ik_1 d}\left(\overset{\circ}{\tilde{\gamma}}_2\right)_{11}\left(\tilde{\beta}_1\right)_{11} + e^{(ik_1 - \kappa_2)d}\left(\overset{\circ}{\tilde{\gamma}}_2\right)_{12}\left(\tilde{\beta}_1\right)_{21} + \left(\overset{\circ}{\tilde{\delta}}_2\right)_{11}\left(\tilde{\delta}_1\right)_{11} \right. \\ &\qquad\left. + e^{(ik_1 + \kappa_2)d}\left(\overset{\circ}{\tilde{\delta}}_2\right)_{12}\left(\tilde{\delta}_1\right)_{21} \right] \\ &\quad - \left[ e^{(ik_1 - \kappa_2)d}\left(\overset{\circ}{\tilde{\gamma}}_2\right)_{21}\left(\tilde{\beta}_1\right)_{12} + e^{-2\kappa_2 d}\left(\overset{\circ}{\tilde{\gamma}}_2\right)_{22}\left(\tilde{\beta}_1\right)_{22} + e^{-(ik_1 + \kappa_2)d}\left(\overset{\circ}{\tilde{\delta}}_2\right)_{21}\left(\tilde{\delta}_1\right)_{12} \right.\\ &\qquad\left. + \left(\overset{\circ}{\tilde{\delta}}_2\right)_{22}\left(\tilde{\delta}_1\right)_{22} \right]^{-1} \\ &\quad \times \left[ e^{(ik_1 - \kappa_2)d}\left(\overset{\circ}{\tilde{\gamma}}_2\right)_{21}\left(\tilde{\beta}_1\right)_{11} + e^{-2\kappa_2 d}\left(\overset{\circ}{\tilde{\gamma}}_2\right)_{22}\left(\tilde{\beta}_1\right)_{21} + e^{-(ik_1 + \kappa_2)d}\left(\overset{\circ}{\tilde{\delta}}_2\right)_{21}\left(\tilde{\delta}_1\right)_{11} \right.\\ &\qquad\left. + \left(\overset{\circ}{\tilde{\delta}}_2\right)_{22}\left(\tilde{\delta}_1\right)_{21} \right] \\ &\quad \times \left[ e^{2ik_1 d}\left(\overset{\circ}{\tilde{\gamma}}_2\right)_{11}\left(\tilde{\beta}_1\right)_{12} + e^{(ik_1 - \kappa_2)d}\left(\overset{\circ}{\tilde{\gamma}}_2\right)_{12}\left(\tilde{\beta}_1\right)_{22} + \left(\overset{\circ}{\tilde{\delta}}_2\right)_{11}\left(\tilde{\delta}_1\right)_{12} \right.\\ &\qquad\left. + e^{(ik_1 + \kappa_2)d}\left(\overset{\circ}{\tilde{\delta}}_2\right)_{12}\left(\tilde{\delta}_1\right)_{22} \right]. \end{aligned} \tag{3.179}$$

When $\kappa_2 d \gg 1$ we can neglect the exponentially decreasing terms that behave as $e^{-\kappa_2 d}$ and $e^{-2\kappa_2 d}$. One also finds that *the exponentially increasing terms that behave as* $e^{\kappa_2 d}$ *cancel out*, so that, as $\kappa_2 d \to \infty$,

$$\delta \to e^{2ik_1d}\left[\left(\overset{\circ}{\tilde{\gamma}}_2\right)_{11}\left(\tilde{\beta}_1\right)_{11} - \frac{\left(\overset{\circ}{\tilde{\delta}}_2\right)_{12}\left(\overset{\circ}{\tilde{\gamma}}_2\right)_{21}\left(\tilde{\beta}_1\right)_{11}}{\left(\overset{\circ}{\tilde{\delta}}_2\right)_{22}} - \frac{\left(\overset{\circ}{\tilde{\gamma}}_2\right)_{11}\left(\tilde{\beta}_1\right)_{12}\left(\tilde{\delta}_1\right)_{21}}{\left(\tilde{\delta}_1\right)_{22}}\right.$$

$$\left.+ \frac{\left(\overset{\circ}{\tilde{\delta}}_2\right)_{12}\left(\overset{\circ}{\tilde{\gamma}}_2\right)_{21}\left(\tilde{\beta}_1\right)_{12}\left(\tilde{\delta}_1\right)_{21}}{\left(\overset{\circ}{\tilde{\delta}}_2\right)_{22}\left(\tilde{\delta}_1\right)_{22}}\right]$$

$$+ \left(\overset{\circ}{\tilde{\delta}}_2\right)_{11}\left(\tilde{\delta}_1\right)_{11} - \frac{\left(\overset{\circ}{\tilde{\delta}}_2\right)_{12}\left(\overset{\circ}{\tilde{\delta}}_2\right)_{21}\left(\tilde{\delta}_1\right)_{11}}{\left(\overset{\circ}{\tilde{\delta}}_2\right)_{22}} - \frac{\left(\tilde{\delta}_1\right)_{21}\left(\overset{\circ}{\tilde{\delta}}_2\right)_{11}\left(\tilde{\delta}_1\right)_{12}}{\left(\tilde{\delta}_1\right)_{22}}$$

$$+ \frac{\left(\overset{\circ}{\tilde{\delta}}_2\right)_{12}\left(\tilde{\delta}_1\right)_{21}\left(\overset{\circ}{\tilde{\delta}}_2\right)_{21}\left(\tilde{\delta}_1\right)_{12}}{\left(\overset{\circ}{\tilde{\delta}}_2\right)_{22}\left(\tilde{\delta}_1\right)_{22}}.$$

(3.180)

This result can be factorized as

$$\delta = e^{2ik_1d}\left[\left(\overset{\circ}{\tilde{\gamma}}_2\right)_{11} - \frac{\left(\overset{\circ}{\tilde{\delta}}_2\right)_{12}\left(\overset{\circ}{\tilde{\gamma}}_2\right)_{21}}{\left(\overset{\circ}{\tilde{\delta}}_2\right)_{22}}\right]\left[\left(\tilde{\beta}_1\right)_{11} - \frac{\left(\tilde{\beta}_1\right)_{12}\left(\tilde{\delta}_1\right)_{21}}{\left(\tilde{\delta}_1\right)_{22}}\right]$$

$$+ \left[\left(\overset{\circ}{\tilde{\delta}}_2\right)_{11} - \frac{\left(\overset{\circ}{\tilde{\delta}}_2\right)_{12}\left(\overset{\circ}{\tilde{\delta}}_2\right)_{21}}{\left(\overset{\circ}{\tilde{\delta}}_2\right)_{22}}\right]\left[\left(\tilde{\delta}_1\right)_{11} - \frac{\left(\tilde{\delta}_1\right)_{12}\left(\tilde{\delta}_1\right)_{21}}{\left(\tilde{\delta}_1\right)_{22}}\right].$$

(3.181)

Recalling the expression (3.165) for the elements of the $M$ matrix in terms of those of $\widetilde{M}$, we have

$$\delta \approx e^{2ik_1d}\overset{\circ}{\gamma}_2\beta_1 + \overset{\circ}{\delta}_2\delta_1.$$

Using eqn (2.213), which relates a $2 \times 2$ open-channel $M$ matrix before and after a translation of the potential, we have

$$\delta = \gamma_2\beta_1 + \delta_2\delta_1$$
$$= \left\{\begin{bmatrix} \alpha_2 & \beta_2 \\ \gamma_2 & \delta_2 \end{bmatrix}\begin{bmatrix} \alpha_1 & \beta_1 \\ \gamma_1 & \delta_1 \end{bmatrix}\right\}_{22}. \tag{3.182}$$

A similar calculation can be performed for $\alpha$, $\beta$ and $\gamma$.

We thus reach the following conclusion. *If the two scatterers are so far apart that $\kappa_2 d \gg 1$, then the total standard (open-channel) transfer matrix can be found by multiplying the standard (open-channel) transfer matrices of the individual scatterers*, i.e., there is no need to use the extended transfer matrices. Thus

$$M \approx M_2 M_1, \quad \kappa_2 d \gg 1. \tag{3.183}$$

**Exercise 3.5** Consider two scatterers a distance $d$ apart, each described by one open and one closed channel. Combine the extended scattering matrices for the two scatterers, assuming that $d \gg \kappa_2$, where $\kappa_2$ is the usual quantity defined for a closed channel in eqns (3.24).

Here we analyze, in terms of the $\widetilde{S}$ matrix, the same problem that was treated in the previous exercise in terms of the $\widetilde{M}$ matrix. The two scatterers are described in terms of the extended scattering matrices $\widetilde{S}_1$ and $\widetilde{S}_2$. We write $\widetilde{S}_1$ as

$$\widetilde{S}_1 = \begin{bmatrix} \widetilde{r}_1 & \widetilde{t}'_1 \\ \widetilde{t}_1 & \widetilde{r}'_1 \end{bmatrix}, \qquad (3.184)$$

where the various blocks $\widetilde{r}_1$, etc. are two-dimensional. Just as above, the second scatterer is displaced a distance $d$ from the first. Before displacement it is described by the matrix $\overset{\circ}{\widetilde{S}}_2$ (see Fig. 2.6). The relation between $\widetilde{S}_2$ and $\overset{\circ}{\widetilde{S}}_2$ can be found from Section 3.1.5, replacing again, for the closed channel, $k_2$ by $i\kappa_2$ in eqn (3.115). The two matrices

$$\widetilde{S}_2 = \begin{bmatrix} \widetilde{r}_2 & \widetilde{t}'_2 \\ \widetilde{t}_2 & \widetilde{r}'_2 \end{bmatrix}, \qquad \overset{\circ}{\widetilde{S}}_2 = \begin{bmatrix} \overset{\circ}{\widetilde{r}}_2 & \overset{\circ}{\widetilde{t}}'_2 \\ \overset{\circ}{\widetilde{t}}_2 & \overset{\circ}{\widetilde{r}}'_2 \end{bmatrix}$$

are then related by

$$\widetilde{S}_2 = \begin{bmatrix} \widetilde{T}(kd) & 0 \\ 0 & \widetilde{T}^{-1}(kd) \end{bmatrix} \overset{\circ}{\widetilde{S}}_2 \begin{bmatrix} \widetilde{T}(kd) & 0 \\ 0 & \widetilde{T}^{-1}(kd) \end{bmatrix}, \qquad (3.185)$$

where $\widetilde{T}(kd)$ is given in eqn (3.173). We thus write

$$\begin{bmatrix} \widetilde{r}_2 & \widetilde{t}'_2 \\ \widetilde{t}_2 & \widetilde{r}'_2 \end{bmatrix} = \begin{bmatrix} \widetilde{T}(kd) & 0 \\ 0 & \widetilde{T}^{-1}(kd) \end{bmatrix} \begin{bmatrix} \overset{\circ}{\widetilde{r}}_2 & \overset{\circ}{\widetilde{t}}'_2 \\ \overset{\circ}{\widetilde{t}}_2 & \overset{\circ}{\widetilde{r}}'_2 \end{bmatrix} \begin{bmatrix} \widetilde{T}(kd) & 0 \\ 0 & \widetilde{T}^{-1}(kd) \end{bmatrix}$$

$$= \begin{bmatrix} \widetilde{T}(kd)\overset{\circ}{\widetilde{r}}_2\widetilde{T}(kd) & \widetilde{T}(kd)\overset{\circ}{\widetilde{t}}'_2\widetilde{T}^{-1}(kd) \\ \widetilde{T}^{-1}(kd)\overset{\circ}{\widetilde{t}}_2\widetilde{T}(kd) & \widetilde{T}^{-1}(kd)\overset{\circ}{\widetilde{r}}'_2\widetilde{T}^{-1}(kd) \end{bmatrix}. \qquad (3.186)$$

The total $\widetilde{S}$ matrix can be found from eqn (3.111). In particular, we concentrate on the block $\widetilde{t}$. A similar analysis can be done for the other blocks. We have

$$\widetilde{t} = \widetilde{t}_2 \frac{1}{I_2 - \widetilde{r}'_1 \widetilde{r}_2} \widetilde{t}_1,$$

or, in terms of its matrix elements,

# QUASI-ONE-DIMENSIONAL SCATTERING THEORY

$$\begin{bmatrix} \widetilde{t}_{11} & \widetilde{t}_{12} \\ \widetilde{t}_{21} & \widetilde{t}_{22} \end{bmatrix}$$

$$= \begin{bmatrix} \left(\overset{\circ}{t}_2\right)_{11} & e^{-(ik_1+\kappa_2)d}\left(\overset{\circ}{t}_2\right)_{12} \\ e^{(ik_1+\kappa_2)d}\left(\overset{\circ}{t}_2\right)_{21} & \left(\overset{\circ}{t}_2\right)_{22} \end{bmatrix}$$

$$\times \left\{ \begin{bmatrix} 1 & 0 \\ 0 & 1 \end{bmatrix} - \begin{bmatrix} (\widetilde{r}'_1)_{11} & (\widetilde{r}'_1)_{12} \\ (\widetilde{r}'_1)_{21} & (\widetilde{r}'_1)_{22} \end{bmatrix} \begin{bmatrix} e^{2ik_1 d}\left(\overset{\circ}{r}_2\right)_{11} & e^{(ik_1-\kappa_2)d}\left(\overset{\circ}{r}_2\right)_{12} \\ e^{(ik_1-\kappa_2)d}\left(\overset{\circ}{r}_2\right)_{21} & e^{-2\kappa_2 d}\left(\overset{\circ}{r}_2\right)_{22} \end{bmatrix} \right\}^{-1}$$

$$\times \begin{bmatrix} (\widetilde{t}_1)_{11} & (\widetilde{t}_1)_{12} \\ (\widetilde{t}_1)_{21} & (\widetilde{t}_1)_{22} \end{bmatrix}. \tag{3.187}$$

The inverse indicated in the last equation is given by

$$\frac{1}{\Delta}\begin{bmatrix} \begin{array}{c} 1 - e^{(ik_1-\kappa_2)d}(\widetilde{r}'_1)_{21}\left(\overset{\circ}{r}_2\right)_{12} \\ -e^{-2\kappa_2 d}(\widetilde{r}'_1)_{22}\left(\overset{\circ}{r}_2\right)_{22} \end{array} & \begin{array}{c} e^{(ik_1-\kappa_2)d}(\widetilde{r}'_1)_{11}\left(\overset{\circ}{r}_2\right)_{12} \\ + e^{-2\kappa_2 d}(\widetilde{r}'_1)_{12}\left(\overset{\circ}{r}_2\right)_{22} \end{array} \\ \begin{array}{c} e^{2ik_1 d}(\widetilde{r}'_1)_{21}\left(\overset{\circ}{r}_2\right)_{11} \\ + e^{(ik_1-\kappa_2)d}(\widetilde{r}'_1)_{22}\left(\overset{\circ}{r}_2\right)_{21} \end{array} & \begin{array}{c} 1 - e^{2ik_1 d}(\widetilde{r}'_1)_{11}\left(\overset{\circ}{r}_2\right)_{11} \\ - e^{(ik_1-\kappa_2)d}(\widetilde{r}'_1)_{12}\left(\overset{\circ}{r}_2\right)_{21} \end{array} \end{bmatrix},$$

$\Delta$ being the determinant of the matrix to be inverted, which, as $\kappa_2 d \to \infty$, becomes

$$\Delta \to 1 - e^{2ik_1 d}(\widetilde{r}'_1)_{11}\left(\overset{\circ}{r}_2\right)_{11}.$$

Now, the standard, open-channel, $S$ matrix is the open-channel submatrix of the extended $\widetilde{S}$, i.e.,

$$S = \begin{bmatrix} r & t' \\ t & r' \end{bmatrix} = \begin{bmatrix} \widetilde{r}_{11} & \widetilde{t}'_{11} \\ \widetilde{t}_{11} & \widetilde{r}'_{11} \end{bmatrix}.$$

In particular, from eqn (3.187) we find, for $\kappa_2 d \gg 1$,

$$t = \widetilde{t}_{11} \approx \frac{\left(\overset{\circ}{t}_2\right)_{11}(\widetilde{t}_1)_{11}}{1 - e^{2ik_1 d}(\widetilde{r}'_1)_{11}\left(\overset{\circ}{r}_2\right)_{11}} = \frac{\overset{\circ}{t}_2 t_1}{1 - e^{2ik_1 d} r'_1 \overset{\circ}{r}_2},$$

and from eqn (2.209) we have $e^{2ik_1 d}\overset{\circ}{r}_2 = r_2$, so that

$$t = \frac{t_2 t_1}{1 - r'_1 r_2}. \tag{3.188}$$

From eqn (2.194), this is precisely the transmission amplitude obtained by combining the two open-channel $S$ matrices $S_1$ and $S_2$, which, in turn, are just the open-channel submatrices of the extended $\widetilde{S}_1$ and $\widetilde{S}_2$. This result can be extended to the other elements of the $S$ matrix.

We thus reach the following conclusion. *If the two scatterers are so far apart that $\kappa_2 d \gg 1$, then the total standard, open-channel, $S$ matrix can be found combining the standard, open-channel, $S$ matrices of the individual scatterers.*

We would like to end this section with some general remarks on the occurrence of closed channels in quasi-one-dimensional systems. Clearly, this has to do with the transverse bandwidth of the system *vis-à-vis* the energy of the wave, e.g., the Fermi energy at zero temperature. For a quasi-one-dimensional system with an infinite transverse bandwidth, e.g., a waveguide with uniform cross-section, closed channels must be considered, as explained above. For a quasi-one-dimensional scattering system, bandwidth-limited in the transverse direction, however, the closed channels are disallowed for wave energies exceeding the transverse bandwidth. An example of such a transverse-bandwidth-limited quasi-one-dimensional system would be a bundle of strictly one-dimensional parallel conductors (physical channels) coupled weakly by the inter-channel tunneling. This is readily modeled by the tight-binding Hamiltonian written in the channel ($i$) basis, namely

$$H_{ij}(x) = \left(-\frac{\hbar^2}{2m}\frac{\partial^2}{\partial x^2} + V_i(x)\right)\delta_{ij} - t(\delta_{i,j+1} + \delta_{i,j-1}), \qquad (3.189)$$

with $4|t|$ as the transverse bandwidth. Such highly-anisotropic low-dimensional systems constitute an important class of quasi-one-dimensional conductors.

## 3.2 Scattering by a cavity with an arbitrary number of waveguides

### 3.2.1 Statement of the problem

We are interested in studying the scattering of a particle with energy $E = \hbar^2 k^2/2m$ in the interior of the structure shown schematically in Fig. 3.4. The structure consists of a cavity, connected to the outside by $L$ waveguides, or leads, ideally of infinite length. Although not indicated in the figure, we may allow for the possibility of scattering by impurities located inside the cavity, but not in the leads. Only the case with TRI will be considered explicitly in what follows. Generalization to the non-TRI case will be commented upon in the last paragraph of this chapter (see also the comments following eqn (3.85)).

The $l$th lead ($l = 1, \ldots, L$) has width $W_l$. We are interested in the (scattering) solutions of the Schrödinger equation, subject to the ideal boundary condition that the walls of the cavity and the leads are completely impenetrable; hence the wave function should vanish there. In the presence of impurities, one could describe the scattering by the present system in terms of Lippmann–Schwinger theory, $H_0$ being the kinetic energy operator inside the structure, i.e., $H$ in the absence of impurities, and $V$ being the potential produced by the latter. The peculiar situation now is that $H_0$ itself produces nonzero scattering (see the

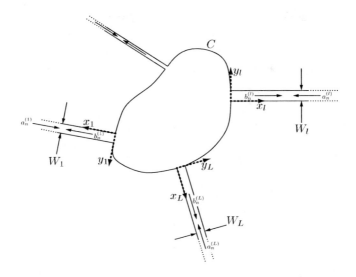

FIG. 3.4. The two-dimensional cavity studied in the text. The cavity is connected to the outside via $L$ waveguides, or leads. The arrows inside the waveguides indicate incoming or outgoing waves of the type (3.191). In waveguide $l$ there can be $N_l$ such incoming or outgoing waves; this is indicated in the figure by the amplitudes $a_n^{(l)}$ or $b_n^{(l)}$, respectively, where $n = 1, \ldots, N_l$.

discussion of a simpler, but similar problem in Section 2.1.9). In the absence of impurities, $H = H_0$ and we are left with the scattering produced by the impenetrable walls. We cannot describe this problem using Lippmann–Schwinger theory, at least as it was presented in this chapter, because we cannot represent the impenetrable walls by a potential with finite range and strength. In Section 3.2.2 we shall thus study the problem directly, starting from the Schrödinger equation.

In waveguide $l$ we now introduce a system of coordinates $x_l$, $y_l$, as indicated in Fig. 3.4. The $x_l$-axis runs along the waveguide and points *away* from the cavity. The $y_l$-axis runs in the transverse direction, and is tangential to the cavity wall and its continuation across the lead. In the $y_l$ direction the system could be, in principle, $(d-1)$-dimensional; to be specific, though, we consider the $d = 2$ case in what follows, the extension to higher dimensionalities being straightforward. Thus $y_l$ takes on the values 0 and $W_l$ on the two walls of the $l$th waveguide.

We now generalize to an arbitrary number of leads some of the concepts introduced in the previous section in connection with quasi-one-dimensional systems, such as the basis of functions in the transverse direction and the concept of open and closed channels. In lead $l$ and for $x_l > 0$, we have the elementary solutions of the Schrödinger equation given by

$$\phi_s\left(E_n^{(l)}; x_l\right) \chi_n^{(l)}(y_l), \tag{3.190}$$

where the longitudinal part $\phi_s(E_n^{(l)}; x_l)$ of the wave function is given, as in eqn (3.11), by

$$\phi_s\left(E_n^{(l)}; x_l\right) = \frac{e^{isk_n^{(l)} x_l}}{\sqrt{2\pi\hbar^2 k_n^{(l)}/m}}, \tag{3.191}$$

and the transverse functions $\chi_n^{(l)}(y_l)$ are given, as in eqn (3.6), by

$$\chi_n^{(l)}(y_l) = \sqrt{\frac{2}{W_l}} \sin K_n^{(l)} y_l, \quad K_n^{(l)} = \frac{n\pi}{W_l}, \tag{3.192}$$

with energy

$$\mathcal{E}_n^{(l)} = \frac{\hbar^2}{2m} \left[K_n^{(l)}\right]^2. \tag{3.193}$$

Again, $n$ is referred to as a mode or channel in waveguide $l$. The functions $\chi_n^{(l)}(y_l)$ vanish on the two walls of waveguide $l$ and form a complete orthonormal set of functions for the variable $y_l$, i.e.,

$$\left\langle \chi_n^{(l)} | \chi_n^{(l)} \right\rangle = \delta_{nm}. \tag{3.194}$$

Between 'longitudinal' and 'transverse' wave numbers and energies we have the following relations, which are similar to those of eqns (3.15):

$$\left[k_n^{(l)}\right]^2 + \left[K_n^{(l)}\right]^2 = k^2, \tag{3.195a}$$

$$E_n^{(l)} + \mathcal{E}_n^{(l)} = E, \tag{3.195b}$$

where $E_n^{(l)}$ and $k_n^{(l)}$ are related by

$$E_n^{(l)} = \frac{\hbar^2}{2m} \left[k_n^{(l)}\right]^2. \tag{3.196}$$

Again, provided that $[K_n^{(l)}]^2 < k^2$ we have $[k_n^{(l)}]^2 > 0$, $k_n^{(l)}$ is real and the $e^{\pm i k_n^{(l)} x_l}$ occurring in eqn (3.191) represent running waves along waveguide $l$; we speak of running modes or open channels. When $[K_n^{(l)}]^2 > k^2$ we have $[k_n^{(l)}]^2 < 0$ and $k_n^{(l)}$ is purely imaginary ($k_n^{(l)} = i\kappa_n^{(l)}$), giving rise to exponentially decaying waves along the waveguide; we thus have evanescent modes or closed channels. If

$$N_l < \frac{kW_l}{\pi} < N_l + 1, \tag{3.197}$$

then there are $N_l$ open channels in waveguide $l$. At a large distance along the waveguide, i.e., as $x_l \to \infty$, only the open channels contribute to the wave function.

At this point we find it advantageous to use the *lead index* notation introduced in Section 2.1.9 in relation to the one-dimensional step-potential problem. Thus

$$\left|\psi_{ln_0}^{(+)}(E)\right\rangle \tag{3.198a}$$

will denote the wave function that has an *incoming wave in lead l and channel $n_0$ only*, with outgoing waves in all the other channels of the same lead and of the other leads, while

$$\left|\psi_{ln_0}^{(-)}(E)\right\rangle \tag{3.198b}$$

has an *outgoing wave in lead l and channel $n_0$ only*, with incoming waves in all the other channels of all the leads. This situation is indicated schematically in Fig. 3.5.

### 3.2.2 The S matrix; the reflection and transmission amplitudes

Just as in the previous sections, we can expand the states (3.198a), defined with outgoing wave boundary conditions, as linear combinations of the incoming wave states for the same energy $E$, eqn (3.198b); the expansion coefficients are the elements of the on-the-energy-shell $S$ matrix. Thus, eqn (3.75) translates to the present case as

$$\left\langle \psi_{l'_0 n'_0}^{(-)}(E') \middle| \psi_{l_0 n_0}^{(+)}(E) \right\rangle = S_{n'_0, n_0}^{l'_0, l_0}(E)\delta(E' - E). \tag{3.199}$$

---

**Exercise 3.6** Derive the relation between the scattering amplitudes and the scalar product of a $\psi^{(-)}$ and a $\psi^{(+)}$ state.

For reference, see Exercise 2.6. We consider, for simplicity, a two-waveguide problem. The wave functions $\psi_{l_0 n_0}^{(+)}(E; \boldsymbol{r})$ and $[\psi_{l'_0 n'_0}^{(-)}(E'; \boldsymbol{r})]^*$ satisfy the equations

$$\left(\nabla^2 + k^2\right)\psi_{l_0 n_0}^{(+)}(E; \boldsymbol{r}) = 0, \tag{3.200a}$$

$$\left(\nabla^2 + k'^2\right)\left[\psi_{l'_0 n'_0}^{(-)}(E'; \boldsymbol{r})\right]^* = 0, \tag{3.200b}$$

with the boundary condition

$$\psi_{l_0 n_0}^{(\pm)}(E; \boldsymbol{r}) = 0, \quad \boldsymbol{r} \in \mathcal{C}, \tag{3.201}$$

$\mathcal{C}$ being the contour of the {cavity + waveguides} system shown in Fig. 3.4.

We multiply eqn (3.200a) by $[\psi_{l'_0 n'_0}^{(-)}(E'; \boldsymbol{r})]^*$, eqn (3.200b) by $\psi_{l_0 n_0}^{(+)}(E; \boldsymbol{r})$ and subtract the two equations to obtain

$$\left(k'^2 - k^2\right)\left[\psi_{l'_0 n'_0}^{(-)}(E'; \boldsymbol{r})\right]^* \psi_{l_0 n_0}^{(+)}(E; \boldsymbol{r})$$
$$= \nabla \cdot \left\{\left[\psi_{l'_0 n'_0}^{(-)}(E'; \boldsymbol{r})\right]^* \nabla \psi_{l_0 n_0}^{(+)}(E; \boldsymbol{r}) - \psi_{l_0 n_0}^{(+)}(E; \boldsymbol{r})\nabla\left[\psi_{l'_0 n'_0}^{(-)}(E'; \boldsymbol{r})\right]^*\right\}. \tag{3.202}$$

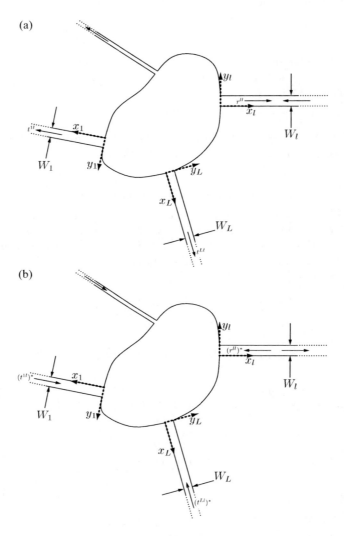

FIG. 3.5. (a) Schematic representation of the wave function $|\psi_{ln_0}^{(+)}(E)\rangle$ of eqn (3.198a), consisting of an incoming wave in lead $l$ and channel $n_0$, and outgoing waves in all the other channels of the same lead and of the other leads. (b) Schematic representation of the wave function $|\psi_{ln_0}^{(-)}(E)\rangle$ of eqn (3.198b), consisting of an outgoing wave in lead $l$ and channel $n_0$, and incoming waves in all the other channels of the same lead and of the other leads.

We first multiply the two sides of the above equation by the 'convergence factor' $f_\epsilon(r)$, defined as

$$f_\epsilon(r) = \begin{cases} 1, & r \in \mathcal{S}_0, \\ e^{-\epsilon x_l}, & r \in \mathcal{S}_l, \end{cases} \quad (3.203)$$

where $\mathcal{S}_0$ is the surface inside the cavity proper and $\mathcal{S}_l$ is the surface inside the $l$th waveguide (or 'lead'), where the coordinate $x_l$ is defined as in Fig. 3.4, increasing away from the cavity. We integrate the resulting equation over the whole surface $\mathcal{S}$ contained inside $\mathcal{C}$, i.e., inside the {cavity + waveguides} system, from infinitely far away in one lead to infinitely far away in the other. Due to the convergence factor the integrals are well defined; at the end of the calculation we take the limit as $\epsilon \to 0^+$. We have

$$(k'^2 - k^2) \lim_{\epsilon \to 0} \int_\mathcal{S} f_\epsilon(r) \left[\psi_{l'_0 n'_0}^{(-)}(E'; r)\right]^* \psi_{l_0 n_0}^{(+)}(E; r) \, d^2 r$$
$$= \lim_{\epsilon \to 0} \int_\mathcal{S} f_\epsilon(r) \nabla \cdot \{\cdots\} \, d^2 r \equiv R, \quad (3.204)$$

where $\{\cdots\}$ indicates, for brevity, the curly bracket on the right-hand side of eqn (3.202). Also, the right-hand side of eqn (3.204) is denoted by $R$.

Integration of $R$ by parts gives

$$R = \lim_{\epsilon \to 0} \left\{ \int_\mathcal{C} f_\epsilon(r) \{\cdots\} \cdot dt - \int_\mathcal{S} \nabla f_\epsilon(r) \cdot \{\cdots\} \, d^2 r \right\}, \quad (3.205)$$

where, for a point on the contour $\mathcal{C}$, the differential $dt$ is perpendicular to $\mathcal{C}$. Due to the boundary condition (3.201), the curly bracket $\{\cdots\}$ vanishes on $\mathcal{C}$ and so does the first integral in (3.205). For the second integral in (3.205) we need the gradient of the convergence factor, which is

$$\nabla f_\epsilon(r) = \begin{cases} 0, & r \in \mathcal{S}_0, \\ -\epsilon e^{-\epsilon x_l} i_l, & r \in \mathcal{S}_l, \end{cases} \quad (3.206)$$

$i_l$ being a unit vector along the $x_l$-axis. Thus, in the second term in eqn (3.205) the cavity proper does not contribute and we are left with

$$R = R_1 + R_2, \quad (3.207)$$

$R_l$ being the contribution of the $l$th waveguide given by

$$R_l = \lim_{\epsilon \to 0} \epsilon \int_0^\infty dx \int_0^{W_l} dy \, e^{-\epsilon x} \left\{ \left[\psi_{l'_0 n'_0}^{(-)}(E'; r)\right]^* \partial_x \psi_{l_0 n_0}^{(+)}(E; r) \right.$$
$$\left. - \psi_{l_0 n_0}^{(+)}(E; r) \partial_x \left[\psi_{l'_0 n'_0}^{(-)}(E'; r)\right]^* \right\}_{r \in \mathcal{S}_l}. \quad (3.208)$$

In this last result the integrals involved can be carried out, since the structure of the wave functions in $\mathcal{S}_l$ is known.

From now on we treat the particular case of $l_0 = 1$, $l'_0 = 2$. The wave functions we need have the form

$$\psi_{1n_0}^{(+)}(E;\boldsymbol{r}) = \begin{cases} \phi_-(E_{n_0}^{(1)};x)\chi_{n_0}^{(1)}(y) + \sum_{n=1}^{N_1} r_{nn_0}^{11}(E)\phi_+(E_n^{(1)};x)\chi_n^{(1)}(y), & \boldsymbol{r} \in \mathcal{S}_1, \\ \sum_{n=1}^{N_2} t_{nn_0}^{21}(E)\phi_+(E_n^{(2)};x)\chi_n^{(2)}(y), & \boldsymbol{r} \in \mathcal{S}_2, \end{cases}$$
(3.209)

$$\left[\psi_{2n_0'}^{(-)}(E';\boldsymbol{r})\right]^* = \psi_{2n_0'}^{(+)}(E';\boldsymbol{r})$$
$$= \begin{cases} \sum_{n=1}^{N_1'} t_{nn_0'}^{12}(E')\phi_+(E_n^{(1)'};x)\chi_n^{(1)}(y), & \boldsymbol{r} \in \mathcal{S}_1, \\ \phi_-(E_{n_0'}^{(2)'};x)\chi_{n_0'}^{(2)}(y) + \sum_{n=1}^{N_2'} r_{nn_0'}^{22}(E')\phi_+(E_n^{(2)'};x)\chi_n^{(2)}(y), & \boldsymbol{r} \in \mathcal{S}_2. \end{cases}$$
(3.210)

Here we have neglected the contribution of the evanescent modes in the two waveguides. We have used the notation $N_l = N_l(E)$ and $N_l' = N_l(E')$. The incident channels $n_0$ and $n_0'$ must satisfy

$$1 \leqslant n_0 \leqslant N_1,$$
$$1 \leqslant n_0' \leqslant N_2'.$$
(3.211)

Substituting these expressions for the wave functions into eqn (3.208), we find

$$R_1 = \lim_{\epsilon \to 0} \epsilon \int_0^\infty dx\, e^{-\epsilon x} \int_0^{W_1} dy\, \Big\{ \left[\psi_{2n_0'}^{(-)}(E';\boldsymbol{r})\right]^* \partial_x \psi_{1n_0}^{(+)}(E;\boldsymbol{r})$$
$$- \psi_{1n_0}^{(+)}(E;\boldsymbol{r})\partial_x \left[\psi_{2n_0'}^{(-)}(E';\boldsymbol{r})\right]^* \Big\}_{\mathcal{S}_1}$$
$$= \lim_{\epsilon \to 0} \epsilon \int_0^\infty dx\, e^{-\epsilon x} \int_0^{W_1} dy\, \Bigg\{ \left[\sum_{n'=1}^{N_1'} t_{n'n_0'}^{12}(E')\phi_+(E_{n'}^{(1)'};x)\chi_{n'}^{(1)}(y)\right]$$
$$\times \left[\partial_x \phi_-(E_{n_0}^{(1)};x)\chi_{n_0}^{(1)}(y) + \sum_{n=1}^{N_1} r_{nn_0}^{11}(E)\partial_x \phi_+(E_n^{(1)};x)\chi_n^{(1)}(y)\right]$$
$$- \left[\phi_-(E_{n_0}^{(1)};x)\chi_{n_0}^{(1)}(y) + \sum_{n=1}^{N_1} r_{nn_0}^{11}(E)\phi_+(E_n^{(1)};x)\chi_n^{(1)}(y)\right]$$
$$\times \left[\sum_{n'=1}^{N_1'} t_{n'n_0'}^{12}(E')\partial_x \phi_+(E_{n'}^{(1)'};x)\chi_{n'}^{(1)}(y)\right] \Bigg\}.$$
(3.212)

The $\chi$ functions appear in expressions of the following type (with $1 \leqslant n_0 \leqslant N_1$ and $1 \leqslant n' \leqslant N_1'$):

$$\sum_{n'=1}^{N_1'} F(n')\left(\chi_{n'}^{(1)}(y), \chi_{n_0}^{(1)}(y)\right) = \sum_{n'=1}^{N_1'} F(n')\delta_{n'n_0}$$
$$= F(n_0)\theta\left(N_1' - n_0\right),$$
(3.213)

where

$$\theta(x) = \begin{cases} 1, & x \geqslant 0, \\ 0, & \text{otherwise} \end{cases}$$

is the step function. The $\chi$ functions also appear in expressions of the type

$$\sum_{n'=1}^{N_1'} \sum_{n=1}^{N_1} F(n,n') \left(\chi_{n'}^{(1)}(y), \chi_n^{(1)}(y)\right) = \sum_{n'=1}^{N_1'} \sum_{n=1}^{N_1} F(n,n') \delta_{nn'}$$

$$= \sum_{n=1}^{\min(N_1, N_1')} F(n,n).$$

We thus have

$$R_1 = \lim_{\epsilon \to 0} \epsilon \int_0^\infty dx\, e^{-\epsilon x} \Bigg\{ \left[t_{n_0 n_0'}^{12}(E') \phi_+(E_{n_0}^{(1)'}; x)\right] \left[-ik_{n_0}^{(1)} \phi_-(E_{n_0}^{(1)}; x)\right] \theta\left(N_1' - n_0\right)$$

$$+ \sum_{n=1}^{\min(N_1, N_1')} \left[t_{n n_0'}^{12}(E') \phi_+(E_n^{(1)'}; x)\right] \left[r_{n n_0}^{11}(E) ik_n^{(1)} \phi_+(E_n^{(1)}; x)\right]$$

$$- \left[\phi_-(E_{n_0}^{(1)}; x)\right] \left[t_{n_0 n_0'}^{12}(E') ik_{n_0}^{(1)'} \phi_+(E_{n_0}^{(1)'}; x)\right] \theta\left(N_1' - n_0\right)$$

$$- \sum_{n=1}^{\min(N_1, N_1')} \left[r_{n n_0}^{11}(E) \phi_+(E_n^{(1)}; x)\right] \left[t_{n n_0'}^{12}(E') ik_n^{(1)'} \phi_+(E_n^{(1)'}; x)\right] \Bigg\}. \tag{3.214}$$

Collecting terms, we find

$$R_1 = \lim_{\epsilon \to 0} \epsilon \Bigg\{ -i t_{n_0 n_0'}^{12}(E') \left(k_{n_0}^{(1)} + k_{n_0}^{(1)'}\right) \theta\left(N_1' - n_0\right) \int_0^\infty dx\, e^{-\epsilon x} \phi_+(E_{n_0}^{(1)'}; x) \phi_-(E_{n_0}^{(1)}; x)$$

$$+ i \sum_{n=1}^{\min(N_1, N_1')} t_{n_0 n_0'}^{12}(E') \left(k_n^{(1)} - k_n^{(1)'}\right) r_{n n_0}^{11}(E) \int_0^\infty dx\, e^{-\epsilon x} \phi_+(E_n^{(1)'}; x) \phi_+(E_n^{(1)}; x) \Bigg\}. \tag{3.215}$$

For the contribution $R_2$ from lead 2 we have

$$R_2 = \lim_{\epsilon \to 0} \epsilon \int_0^\infty dx\, e^{-\epsilon x} \int_0^{W_2} dy \Bigg\{ \left[\psi_{2n_0'}^{(-)}(E'; \boldsymbol{r})\right]^* \partial_x \psi_{1n_0}^{(+)}(E; \boldsymbol{r})$$

$$- \psi_{1n_0}^{(+)}(E; \boldsymbol{r}) \partial_x \left[\psi_{2n_0'}^{(-)}(E'; \boldsymbol{r})\right]^* \Bigg\}_{S_2}$$

$$= \lim_{\epsilon \to 0} \epsilon \int_0^\infty dx\, e^{-\epsilon x} \int_0^{W_2} dy \left\{ \left[ \phi_-(E_{n_0'}^{(2)'}; x) \chi_{n_0'}^{(2)}(y) \right.\right.$$

$$\left. + \sum_{n'=1}^{N_2'} r_{n'n_0'}^{22}(E') \phi_+(E_{n'}^{(2)'}; x) \chi_{n'}^{(2)}(y) \right]$$

$$\times \left[ \sum_{n=1}^{N_2} t_{nn_0}^{21}(E) \partial_x \phi_+(E_n^{(2)}; x) \chi_n^{(2)}(y) \right]$$

$$- \left[ \sum_{n=1}^{N_2} t_{nn_0}^{21}(E) \phi_+(E_n^{(2)}; x) \chi_n^{(2)}(y) \right] \partial_x \phi_-(E_{n_0'}^{(2)'}; x) \chi_{n_0'}^{(2)}(y)$$

$$\left. + \sum_{n'=1}^{N_2'} r_{n'n_0'}^{22}(E') \partial_x \phi_+(E_{n'}^{(2)'}; x) \chi_{n'}^{(2)}(y) \right\}. \tag{3.216}$$

The orthogonality of the $\chi$ functions gives

$$R_2 = \lim_{\epsilon \to 0} \epsilon \int_0^\infty dx\, e^{-\epsilon x} \left\{ \left[ \phi_-(E_{n_0'}^{(2)'}; x) \right] \left[ t_{n_0'n_0}^{21}(E) i k_{n_0'}^{(2)} \phi_+(E_{n_0'}^{(2)}; x) \right] \theta(N_2 - n_0') \right.$$

$$+ \sum_{n=1}^{\min(N_2, N_2')} \left[ r_{nn_0'}^{22}(E') \phi_+(E_n^{(2)'}; x) \right] \left[ t_{nn_0}^{21}(E) i k_n^{(2)} \phi_+(E_n^{(2)}; x) \right]$$

$$- \left[ t_{n_0'n_0}^{21}(E) \phi_+(E_{n_0'}^{(2)}; x) \right] \left[ (-i k_{n_0'}^{(2)'}) \phi_-(E_{n_0'}^{(2)'}; x) \right] \theta(N_2 - n_0')$$

$$\left. - \sum_{n=1}^{\min(N_2, N_2')} \left[ t_{nn_0}^{21}(E) \phi_+(E_n^{(2)}; x) \right] \left[ r_{nn_0'}^{22}(E') i k_n^{(2)'} \phi_+(E_n^{(2)'}; x) \right] \right\}. \tag{3.217}$$

Collecting terms, we find

$$R_2 = \lim_{\epsilon \to 0} \epsilon \left\{ i t_{n_0'n_0}^{21}(E) \left( k_{n_0'}^{(2)} + k_{n_0'}^{(2)'} \right) \int_0^\infty dx\, e^{-\epsilon x} \phi_-(E_{n_0'}^{(2)'}; x) \phi_+(E_{n_0'}^{(2)}; x) \theta(N_2 - n_0') \right.$$

$$\left. + i \sum_{n=1}^{\min(N_2, N_2')} t_{nn_0}^{21}(E) \left( k_n^{(2)} - k_n^{(2)'} \right) r_{nn_0'}^{22}(E') \int_0^\infty dx\, e^{-\epsilon x} \phi_+(E_n^{(2)'}; x) \phi_+(E_n^{(2)}; x) \right\}. \tag{3.218}$$

We now need the following integrals:

$$\int_0^\infty dx\, e^{-\epsilon x} \phi_+(E_{n_0}^{(1)'}; x) \phi_-(E_{n_0}^{(1)}; x) = \frac{im}{2\pi \hbar^2 \sqrt{k_{n_0}^{(1)'} k_{n_0}^{(1)}}} \frac{1}{k_{n_0}^{(1)'} - k_{n_0}^{(1)} + i\epsilon},$$

$$\int_0^\infty dx\, e^{-\epsilon x} \phi_+(E_n^{(1)'}; x) \phi_+(E_n^{(1)}; x) = \frac{im}{2\pi \hbar^2 \sqrt{k_n^{(1)'} k_n^{(1)}}} \frac{1}{k_n^{(1)'} + k_n^{(1)} + i\epsilon},$$

$$\int_0^\infty \mathrm{d}x\, e^{-\epsilon x} \phi_-(E_{n_0'}^{(2)'}; x) \phi_+(E_{n_0}^{(2)}; x) = \frac{im}{2\pi\hbar^2 \sqrt{k_{n_0'}^{(2)'} k_{n_0}^{(2)}}} \frac{1}{-k_{n_0'}^{(2)'} + k_{n_0}^{(2)} + i\epsilon},$$

$$\int_0^\infty \mathrm{d}x\, e^{-\epsilon x} \phi_+(E_n^{(2)'}; x) \phi_+(E_n^{(2)}; x) = \frac{im}{2\pi\hbar^2 \sqrt{k_n^{(2)'} k_n^{(2)}}} \frac{1}{k_{n_0'}^{(2)'} + k_{n_0}^{(2)} + i\epsilon},$$

which have to be introduced into eqns (3.215) and (3.218) for $R_1$ and $R_2$, respectively, to give $R = R_1 + R_2$. From eqn (3.204) we then find

$$\lim_{\epsilon \to 0} \int_S f_\epsilon(\mathbf{r}) \left[\psi_{l_0' n_0'}^{(-)}(E'; \mathbf{r})\right]^* \psi_{l_0 n_0}^{(+)}(E; \mathbf{r})\, \mathrm{d}^2 \mathbf{r}$$

$$= \frac{m}{2\pi\hbar^2} \lim_{\epsilon \to 0} \epsilon \left\{ \frac{t_{n_0 n_0'}^{12}(E')}{\sqrt{k_{n_0}^{(1)} k_{n_0'}^{(1)'}} \left(k_{n_0}^{(1)'} - k_{n_0}^{(1)}\right)\left(k_{n_0}^{(1)'} - k_{n_0}^{(1)} + i\epsilon\right)} \theta(N_1' - n_0) \right.$$

$$+ \sum_{n=1}^{\min(N_1, N_1')} \frac{t_{nn_0'}^{12}(E') r_{nn_0}^{11}(E)}{\sqrt{k_n^{(1)} k_n^{(1)'}} \left(k_n^{(1)'} + k_n^{(1)}\right)\left(k_n^{(1)'} + k_n^{(1)} + i\epsilon\right)}$$

$$- \frac{t_{n_0' n_0}^{21}(E)}{\sqrt{k_{n_0'}^{(2)} k_{n_0'}^{(2)'}} \left(k_{n_0'}^{(2)'} - k_{n_0'}^{(2)}\right)\left(-k_{n_0'}^{(2)'} + k_{n_0'}^{(2)} + i\epsilon\right)} \theta(N_2 - n_0')$$

$$\left. - \sum_{n=1}^{\min(N_2, N_2')} \frac{t_{nn_0}^{21}(E) r_{nn_0'}^{22}(E')}{\sqrt{k_n^{(2)'} k_n^{(2)}} \left(k_n^{(2)'} + k_n^{(2)}\right)\left(k_n^{(2)'} + k_n^{(2)} + i\epsilon\right)} \right\}. \quad (3.219)$$

We have used the relations

$$k'^2 - k^2 = \left\{\left[k_{n_0}^{(1)'}\right]^2 + \left[K_{n_0}^{(1)}\right]^2\right\} - \left\{\left[k_{n_0}^{(1)}\right]^2 + \left[K_{n_0}^{(1)}\right]^2\right\}$$

$$= \left[k_{n_0}^{(1)'}\right]^2 - \left[k_{n_0}^{(1)}\right]^2 = \left[k_n^{(1)'}\right]^2 - \left[k_n^{(1)}\right]^2, \quad (3.220)$$

and similarly

$$k'^2 - k^2 = \left[k_{n_0'}^{(2)'}\right]^2 - \left[k_{n_0'}^{(2)}\right]^2 = \left[k_n^{(2)'}\right]^2 - \left[k_n^{(2)}\right]^2. \quad (3.221)$$

The second and fourth lines inside the curly bracket in eqn (3.219) are finite, since

$$k_n^{(1)'}, k_n^{(1)}, k_n^{(2)'}, k_n^{(2)} > 0,$$

and thus do not contribute to the limit as $\epsilon \to 0$. Thus

$$\lim_{\epsilon \to 0} \int_S f_\epsilon(\mathbf{r}) \left[\psi_{2,n_0'}^{(-)}(E'; \mathbf{r})\right]^* \psi_{1,n_0}^{(+)}(E; \mathbf{r})\, \mathrm{d}^2\mathbf{r} = \frac{m}{2\pi\hbar^2} \lim_{\epsilon \to 0} F_\epsilon(k, k'), \quad (3.222)$$

where we have defined the following function (we apply TRI, which implies that $t_{n_0 n_0'}^{12}(E') = t_{n_0' n_0}^{21}(E')$):

$$F_\epsilon(k,k') = \frac{\epsilon\, t^{21}_{n'_0 n_0}(E')}{\sqrt{k^{(1)}_{n_0} k^{(1)'}_{n_0}}\left(k^{(1)}_{n_0} - k^{(1)'}_{n_0}\right)\left(k^{(1)}_{n_0} - k^{(1)'}_{n_0} - i\epsilon\right)} \theta\left(N'_1 - n_0\right)$$

$$+ \frac{\epsilon\, t^{21}_{n'_0 n_0}(E)}{\sqrt{k^{(2)}_{n'_0} k^{(2)'}_{n'_0}}\left(k^{(2)}_{n'_0} - k^{(2)'}_{n'_0}\right)\left(k^{(2)}_{n'_0} - k^{(2)'}_{n'_0} + i\epsilon\right)} \theta\left(N_2 - n'_0\right)$$

$$= \frac{\epsilon\, t^{21}_{n'_0 n_0}(E')}{\sqrt{k^{(1)}_{n_0} k^{(1)'}_{n_0}}\left(k^{(1)}_{n_0} - k^{(1)'}_{n_0}\right)} \frac{k^{(1)}_{n_0} - k^{(1)'}_{n_0} + i\epsilon}{\left(k^{(1)}_{n_0} - k^{(1)'}_{n_0}\right)^2 + \epsilon^2} \theta\left(N'_1 - n_0\right)$$

$$+ \frac{\epsilon\, t^{21}_{n'_0 n_0}(E)}{\sqrt{k^{(2)}_{n'_0} k^{(2)'}_{n'_0}}\left(k^{(2)}_{n'_0} - k^{(2)'}_{n'_0}\right)} \frac{k^{(2)}_{n'_0} - k^{(2)'}_{n'_0} - i\epsilon}{\left(k^{(2)}_{n'_0} - k^{(2)'}_{n'_0}\right)^2 + \epsilon^2} \theta\left(N_2 - n'_0\right)$$

$$= \frac{t^{21}_{n'_0 n_0}(E')}{\sqrt{k^{(1)}_{n_0} k^{(1)'}_{n_0}}} \left[\frac{\epsilon}{\left(k^{(1)}_{n_0} - k^{(1)'}_{n_0}\right)^2 + \epsilon^2}\right.$$

$$\left. + \frac{i}{k^{(1)}_{n_0} - k^{(1)'}_{n_0}} \frac{\epsilon^2}{\left(k^{(1)}_{n_0} - k^{(1)'}_{n_0}\right)^2 + \epsilon^2}\right] \theta\left(N'_1 - n_0\right)$$

$$+ \frac{t^{21}_{n'_0 n_0}(E)}{\sqrt{k^{(2)}_{n'_0} k^{(2)'}_{n'_0}}} \left[\frac{\epsilon}{\left(k^{(2)}_{n'_0} - k^{(2)'}_{n'_0}\right)^2 + \epsilon^2}\right.$$

$$\left. - \frac{i}{k^{(2)}_{n'_0} - k^{(2)'}_{n'_0}} \frac{\epsilon^2}{\left(k^{(2)}_{n'_0} - k^{(2)'}_{n'_0}\right)^2 + \epsilon^2}\right] \theta\left(N_2 - n'_0\right)$$

$$\equiv F^{(1)}_\epsilon(k,k') + F^{(2)}_\epsilon(k,k'). \tag{3.223}$$

We have denoted by $F^{(1)}_\epsilon(k,k')$ the sum of the first term on the last line and the first term on the last but one line in eqn (3.223), and by $F^{(2)}_\epsilon(k,k')$ the sum of the second terms on each of those two lines.

We first concentrate on $F^{(1)}_\epsilon(k,k')$. We find

$$\lim_{\epsilon \to 0^+} F^{(1)}_\epsilon(k,k') = \pi \frac{t^{21}_{n'_0 n_0}(E')}{\sqrt{k^{(1)}_{n_0} k^{(1)'}_{n_0}}} \delta(k^{(1)}_{n_0} - k^{(1)'}_{n_0}) \theta\left(N'_1 - n_0\right)$$

$$+ \pi \frac{t^{21}_{n'_0 n_0}(E)}{\sqrt{k^{(2)}_{n'_0} k^{(2)'}_{n'_0}}} \delta(k^{(2)}_{n'_0} - k^{(2)'}_{n'_0}) \theta\left(N_2 - n'_0\right)$$

$$= 2\pi t^{21}_{n'_0 n_0}(E) \delta\left(\frac{1}{2}(k^2 - k'^2)\right)$$

$$= 2\pi \frac{\hbar^2}{m} t^{21}_{n'_0 n_0}(E) \delta(E' - E). \tag{3.224}$$

We have used the identities (3.220) and (3.221). Notice that because of the delta functions we can replace $N'_1 \equiv N_1(E')$ by $N_1 \equiv N_1(E)$ and $N_2 \equiv N_2(E)$ by $N'_2 \equiv N_2(E')$;

the step functions in the first two lines of eqn (3.224) are then equal to 1 because of the inequalities (3.211).

We now turn to $F_\epsilon^{(2)}(k,k')$. Writing, for brevity, $k_{n_0}^{(1)} = k_1$, $k_{n_0}^{(1)'} = k_1'$, $k_{n_0'}^{(2)} = k_2$, $k_{n_0'}^{(2)'} = k_2'$, $t_{n_0'n_0}^{21}(E') = t(E')$ and $t_{n_0'n_0}^{21}(E) = t(E)$, we have

$$iF_\epsilon^{(2)}(k,k') = -\frac{t(E')\theta(N_1(E') - n_0)}{\sqrt{k_1 k_1'}(k_1 - k_1')} \frac{\epsilon^2}{(k_1 - k_1')^2 + \epsilon^2}$$
$$+ \frac{t(E)\theta(N_2(E) - n_0')}{\sqrt{k_2 k_2'}(k_2 - k_2')} \frac{\epsilon^2}{(k_2 - k_2')^2 + \epsilon^2}. \qquad (3.225)$$

We have the identities

$$k^2 = k_1^2 + K_1^2 = k_2^2 + K_2^2, \qquad (3.226)$$
$$k'^2 = k_1'^2 + K_1^2 = k_2'^2 + K_2^2, \qquad (3.227)$$

where we have written, for brevity, $K_{n_0}^{(1)} = K_1$ and $K_{n_0'}^{(2)} = K_2$. We can thus write eqn (3.225) entirely in terms of $k$ and $k'$ as

$$iF_\epsilon^{(2)}(k,k') = \frac{2}{(k'-k)(k'+k)}$$
$$\times \left[ t(E') \frac{\sqrt{k^2 - K_1^2} + \sqrt{k'^2 - K_1^2}}{2(k^2 - K_1^2)^{1/4}(k'^2 - K_1^2)^{1/4}} \theta(N_1(E') - n_0) \right.$$
$$\times \frac{\epsilon^2}{\left(\sqrt{k^2 - K_1^2} - \sqrt{k'^2 - K_1^2}\right)^2 + \epsilon^2}$$
$$- t(E) \frac{\sqrt{k^2 - K_2^2} + \sqrt{k'^2 - K_2^2}}{2(k^2 - K_2^2)^{1/4}(k'^2 - K_2^2)^{1/4}} \theta(N_2(E) - n_0')$$
$$\left. \times \frac{\epsilon^2}{\left(\sqrt{k^2 - K_2^2} - \sqrt{k'^2 - K_2^2}\right)^2 + \epsilon^2} \right]. \qquad (3.228)$$

We now proceed in a way analogous to that followed at the end of Exercise 2.6. We consider the above expression as a function of $k'$ for small but finite $\epsilon$. This function is finite for all $k' \neq k$; when $k' \to k$, it tends to the *finite* limit $k^{-1} \partial t(k)/\partial k$. Thus, if we construct the integral

$$I = \int_a^b F_\epsilon^{(2)}(k,k') \phi(k') \, dk', \qquad (3.229)$$

where $k$ is contained inside the interval $(a,b)$ and $\phi(k')$ is an arbitrary but continuous function of $k'$, then we find that $I$ exists and can be written as the principal value integral

$$\mathcal{P} \int_a^b F_\epsilon^{(2)}(k,k') \phi(k') \, dk' = \mathcal{P} \int_a^b \left[ F_\epsilon^{(2)}(k,k') \right]_1 \phi(k') \, dk'$$
$$+ \mathcal{P} \int_a^b \left[ F_\epsilon^{(2)}(k,k') \right]_2 \phi(k') \, dk', \qquad (3.230)$$

where the two terms arise from having split the integrand into the two terms contained in the square bracket of eqn (3.228). Notice that the principal value, which could have been omitted on the left-hand side of eqn (3.230), is essential for each one of the two summands on its right-hand side (where the integrand diverges at $k' = k$, as is clear from the above eqn (3.228)) to exist. Each of the two integrals in eqn (3.230) has the structure

$$\mathcal{P} \int_a^b \left[ F_\epsilon^{(2)}(k, k') \right]_{1,2} \phi(k')\, \mathrm{d}k' \equiv \mathcal{P} \int_a^b \frac{g_\epsilon(k, k')}{k' - k}\, \mathrm{d}k'. \qquad (3.231)$$

The principal value integral in the last equation exists for any finite $\epsilon$; in this case the function $g_\epsilon(k, k')$ is non-negligible and almost constant in a region of $O(\epsilon)$; thus this principal value integral vanishes as we take the limit as $\epsilon \to 0$ at the end. It is in this sense that we say that

$$\lim_{\epsilon \to 0} F_\epsilon^{(2)}(k, k') = 0. \qquad (3.232)$$

We are thus left with $F_\epsilon^{(1)}(k, k')$ only, and eqns (3.224) and (3.222) show that

$$\left( \psi_{2,n_0'}^{(-)}(E'; \boldsymbol{r}), \psi_{1,n_0}^{(+)}(E; \boldsymbol{r}) \right) = t_{n_0' n_0}^{21}(E) \delta \left( E - E' \right). \qquad (3.233)$$

A similar procedure could be carried out for the other values of $s$ and $s'$. We encourage the intrepid reader to try this out!

We can thus generalize Table 3.3 to the present situation. The problem now admits

$$M = \sum_{l=1}^{L} N_l \qquad (3.234)$$

linearly independent solutions, a linear combination of which generates the most general solution (at the given energy $E$), which, in lead $l$ and for the open channel $n$, has the form (see Fig. 3.4)

$$a_n^{(l)} \phi_- \left( E_n^{(l)}; x_l \right) + b_n^{(l)} \phi_+ \left( E_n^{(l)}; x_l \right), \quad n = 1, \ldots, N_l. \qquad (3.235)$$

The coefficients $a_n^{(l)}$ and $b_n^{(l)}$ represent, respectively, the amplitudes of the incoming and the outgoing waves for the open channel $n$ in lead $l$. We define the $N_l$-dimensional vector

$$\boldsymbol{a}^{(l)} = \begin{bmatrix} a_1^{(l)} \\ \vdots \\ a_{N_l}^{(l)} \end{bmatrix}, \qquad (3.236)$$

which contains all the $N_l$ incoming amplitudes in lead $l$ ($l = 1, \ldots, N_l$). Putting all of the $\boldsymbol{a}^{(l)}$ ($l = 1, \ldots, L$) together, we form the $M$-dimensional vector

$$\boldsymbol{a} = \begin{bmatrix} \boldsymbol{a}^{(1)} \\ \vdots \\ \boldsymbol{a}^{(L)} \end{bmatrix}. \qquad (3.237)$$

We can make similar definitions for the outgoing wave amplitudes. The incoming and outgoing amplitudes are linearly related through the *scattering matrix S* by

$$\boldsymbol{b} = S\boldsymbol{a}. \tag{3.238}$$

In terms of individual waveguides, we can write

$$S = \begin{bmatrix} r^{11} & t^{12} & \cdots & t^{1L} \\ t^{21} & r^{22} & \cdots & t^{2L} \\ \vdots & \vdots & \ddots & \vdots \\ t^{L1} & t^{L2} & \cdots & r^{LL} \end{bmatrix}. \tag{3.239}$$

Here, $r^{ll}$ is an $N_l \times N_l$ matrix, containing the reflection amplitudes from the $N_l$ channels of waveguide $l$ back to the same waveguide; $t^{lm}$ is an $N_l \times N_m$ matrix, containing the transmission amplitudes from the $N_m$ channels of waveguide $m$ to the $N_l$ channels of waveguide $l$. The $S$ matrix is thus a square matrix with dimensionality $M$. It has the property of unitarity, i.e.,

$$SS^\dagger = I, \tag{3.240}$$

thus ensuring, as usual, flux conservation. Again, this is the only requirement in the absence of other symmetries (the *unitary* case). In the presence of *time-reversal invariance* (as is the case in the absence of a magnetic field) and no spin, the matrix $S$, besides being unitary, is *symmetric* (the *orthogonal* case), i.e.,

$$S = S^\top. \tag{3.241}$$

For a two-waveguide problem the $S$ matrix has the structure

$$S = \begin{bmatrix} r^{11} & t^{12} \\ t^{21} & r^{22} \end{bmatrix} \equiv \begin{bmatrix} r & t' \\ t & r' \end{bmatrix}. \tag{3.242}$$

Here, leads 1 and 2 support channels $N_1$ and $N_2$, respectively; $r$ and $t$ are the reflection and transmission matrices when incidence is from lead 1, while $r'$ and $t'$ correspond to incidence from lead 2. In the particular case of $N_1 = N_2 = N$, i.e., when the two leads have the same number of channels $N$ as in the quasi-one-dimensional problem of Section 3.1, the four blocks $r$, $t$, $r'$ and $t'$ are $N \times N$ and the $S$ matrix is $2N \times 2N$.

## 3.3 The *R*-matrix theory of two-dimensional scattering

In this section we develop Wigner's *R*-matrix theory of scattering for the two-dimensional cavity structure studied in the previous section and illustrated in Fig. 3.4. This will be the two-dimensional extension of the theory presented in Section 2.2.7 for semi-infinite one-dimensional scattering.

For simplicity, we consider only one lead ($L = 1$), which can support $N$ open channels. The result can be easily reinterpreted to cover the many-lead case. We restrict the analysis to the spinless case with TRI, so that the resulting $N$-dimensional $S$ matrix is, from eqn (3.241), symmetric.

As explained in the previous section, we pose the scattering problem through the Schrödinger equation

$$\left(\nabla^2 + k^2\right) \psi(\boldsymbol{r}) = U(\boldsymbol{r})\psi(\boldsymbol{r}) \qquad (3.243\text{a})$$

defined inside the structure, with the Dirichlet condition

$$\psi(\boldsymbol{r}) = 0, \quad \boldsymbol{r} \in C \qquad (3.243\text{b})$$

on the boundary $C$. The most general $n$-channel component of the wave function in the waveguide has the form (see Table 3.3 and eqn (3.235))

$$[\psi(E; x > 0)]_n = a_n \phi_-(E_n; x) + b_n \phi_+(E_n; x), \quad 1 \leqslant n \leqslant N, \qquad (3.244\text{a})$$
$$[\psi(E; x > 0)]_n = b_n \phi_+(-\epsilon_n; x), \quad n > N, \qquad (3.244\text{b})$$

the two sets of open-channel incoming and outgoing coefficients being related, as usual, by the $N$-dimensional $S$ matrix, i.e.,

$$\boldsymbol{b} = S\boldsymbol{a}. \qquad (3.245)$$

As explained in Section 2.2.7, the main idea of $R$-matrix theory is to close the cavity first and set up, inside the cavity, the Schrödinger equation with appropriate boundary conditions, that give rise to solutions with a discrete spectrum. The scattering solution of eqns (3.243) inside the cavity is expanded in terms of them and then joined to the external wave function (3.244) to find the $S$ matrix.

If we close up the cavity, its full boundary $C_{\text{cav}}$ can be considered as the union of the wall $C_w$ (the subscript '$w$' meaning wall) and the part $C_o$ (the subscript '$o$' meaning opening) drawn through the opening. We define the internal problem through the Schrödinger equation

$$\left(\nabla^2 + k_\lambda^2\right) \psi_\lambda(\boldsymbol{r}) = U(\boldsymbol{r})\psi_\lambda(\boldsymbol{r}), \qquad (3.246\text{a})$$

with the *mixed* boundary conditions

$$\psi_\lambda(\boldsymbol{r}) = 0, \quad \boldsymbol{r} \in C_w, \qquad (3.246\text{b})$$
$$\frac{\partial \psi_\lambda}{\partial n} = 0, \quad \boldsymbol{r} \in C_o, \qquad (3.246\text{c})$$

which extend to the two-dimensional case the boundary conditions (2.397b) and (2.397c), respectively, for the semi-infinite one-dimensional case.

To prove the orthogonality of the $\psi_\lambda(r)$s, we write the complex conjugate of eqn (3.246) for $\psi_{\lambda'}(r)$, say

$$\left(\nabla^2 + k_{\lambda'}^2\right)\psi_{\lambda'}^*(r) = U(r)\psi_{\lambda'}^*(r), \qquad (3.247a)$$

with the boundary conditions

$$\psi_{\lambda'}(r) = 0, \quad r \in C_w, \qquad (3.247b)$$

$$\frac{\partial \psi_{\lambda'}}{\partial n} = 0, \quad r \in C_o. \qquad (3.247c)$$

We have written explicitly the complex conjugate of the wave function, although we know that, for TRI, $U(r)$ is real and the wave functions $\psi_\lambda(r)$ can also be chosen to be real. We multiply eqn (3.246a) by $\psi_{\lambda'}^*(r)$, eqn (3.247a) by $\psi_\lambda(r)$ and integrate over the internal region of the cavity. We subtract the two equations to obtain ($d\sigma$ indicates a surface element inside the cavity)

$$\int_{\text{cav}} \left[\psi_{\lambda'}^* \nabla^2 \psi_\lambda - \psi_\lambda \nabla^2 \psi_{\lambda'}^*\right] d\sigma = \left(k_{\lambda'}^2 - k_\lambda^2\right)(\psi_{\lambda'}, \psi_\lambda), \qquad (3.248)$$

and using the divergence theorem ($ds$ indicates a line element along $C_{\text{cav}}$) we have

$$\oint_{C_{\text{cav}}} \left[\psi_{\lambda'}^* \frac{\partial \psi_\lambda}{\partial n} - \psi_\lambda \frac{\partial \psi_{\lambda'}^*}{\partial n}\right] ds = \left(k_{\lambda'}^2 - k_\lambda^2\right)(\psi_{\lambda'}, \psi_\lambda). \qquad (3.249)$$

The boundary conditions (3.246b) and (3.247b) on $C_w$, and (3.246c) and (3.247c) on $C_o$ imply the vanishing of the line integral along $C_{\text{cav}}$, so that

$$\left(k_{\lambda'}^2 - k_\lambda^2\right)(\psi_{\lambda'}, \psi_\lambda) = 0$$

and two states with $k_{\lambda'}^2 \neq k_\lambda^2$, and hence $E_\lambda \neq E_{\lambda'}$, are automatically orthogonal, i.e., $(\psi_{\lambda'}, \psi_\lambda) = 0$. If we have degenerate states, they can always be orthonormalized by taking an appropriate linear combination of them, so that finally

$$(\psi_{\lambda'}, \psi_\lambda) = \delta_{\lambda'\lambda}. \qquad (3.250)$$

The states $\psi_\lambda(r)$ thus form a complete set of orthonormal states in the internal region of the cavity.

The *actual* scattering wave function, the solution of eqn (3.243), can thus be expanded, inside the cavity, in terms of the $\psi_\lambda(r)$s, i.e.,

$$\psi(r) = \sum_\lambda A_\lambda \psi_\lambda(r), \quad r \in \text{cavity}. \qquad (3.251)$$

We multiply eqn (3.243a) by $\psi_\lambda^*(r)$, the complex conjugate of eqn (3.246a) by $\psi(r)$, subtract the two equations and integrate over the surface defining the cavity to obtain

$$\int_{\text{cav}} \left[ \psi_\lambda^* \nabla^2 \psi - \psi \nabla^2 \psi_\lambda^* \right] d\sigma = \left( k_\lambda^2 - k^2 \right) (\psi_\lambda, \psi). \tag{3.252}$$

Then, using the divergence theorem,

$$\oint_{C_{\text{cav}}} \left[ \psi_\lambda^* \frac{\partial \psi}{\partial n} - \psi \frac{\partial \psi_\lambda^*}{\partial n} \right] ds = \left( k_\lambda^2 - k^2 \right) A_\lambda. \tag{3.253}$$

The line integral on the left-hand side can be split into an integral over $C_w$ and one over $C_o$, i.e.,

$$\oint_{C_{\text{cav}}} \left[ \psi_\lambda^* \frac{\partial \psi}{\partial n} - \psi \frac{\partial \psi_\lambda^*}{\partial n} \right] ds$$
$$= \int_{C_w} \left[ \psi_\lambda^* \frac{\partial \psi}{\partial n} - \psi \frac{\partial \psi_\lambda^*}{\partial n} \right] ds + \int_{C_o} \left[ \psi_\lambda^* \frac{\partial \psi}{\partial n} - \psi \frac{\partial \psi_\lambda^*}{\partial n} \right] ds$$
$$= \int_{C_o} \psi_\lambda^* \frac{\partial \psi}{\partial n} ds,$$

where we have used the boundary conditions (3.243b), (3.246b) and (3.246c). Substituting into eqn (3.253), we find the coefficients $A_\lambda$ to be

$$A_\lambda = \frac{1}{E_\lambda - E} \left[ \frac{\hbar^2}{2m} \int_{C_o} \psi_\lambda^* \frac{\partial \psi}{\partial n} ds \right]. \tag{3.254}$$

On the opening $C_o$, i.e., for $x = 0$, we can expand the scattering wave function $\psi$ and its normal derivative $\partial \psi / \partial n$ in terms of the functions $\chi_n(y)$ defined in eqn (3.192) as follows:

$$\psi(\mathbf{r} \in C_o) = \psi(0, y) = \sum_{n=1}^{\infty} V_n \chi_n(y), \tag{3.255}$$

$$\left( \frac{\partial \psi(\mathbf{r})}{\partial n} \right)_{C_o} = \sum_{n=1}^{\infty} D_n \chi_n(y), \tag{3.256}$$

where we have defined the $n$th component *value* $V_n$ and *derivative* $D_n$ of the scattering wave function to be

$$V_n = \int_{C_o} \chi_n^*(y) \psi(0, y) \, dy, \tag{3.257}$$

$$D_n = \int_{C_o} \chi_n^*(y) \left( \frac{\partial \psi(x, y)}{\partial x} \right)_{x=0} dy. \tag{3.258}$$

Notice that the above expansions involve *all* channels, open and closed.

We write the square bracket in eqn (3.254) as

$$\sqrt{\frac{\hbar^2}{2m}} \sum_n D_n \gamma^*_{\lambda n},$$

where we have defined the *reduced amplitude*

$$\gamma_{\lambda n} = \sqrt{\frac{\hbar^2}{2m}} \int_{C_o} \chi^*_n(y) \psi_\lambda(\boldsymbol{r}) \, \mathrm{d}y \qquad (3.259)$$

for level $\lambda$ into channel $n$. The coefficients $A_\lambda$ of eqn (3.254) thus become

$$A_\lambda = \frac{1}{E_\lambda - E} \sqrt{\frac{\hbar^2}{2m}} \sum_n D_n \gamma^*_{\lambda n}. \qquad (3.260)$$

Inside the cavity, the expansion (3.251) of the scattering wave function $\psi(\boldsymbol{r})$ thus becomes

$$\psi(\boldsymbol{r}) = \sqrt{\frac{\hbar^2}{2m}} \sum_{n=1}^{\infty} \left[ \sum_\lambda \frac{\psi_\lambda(\boldsymbol{r}) \gamma^*_{\lambda n}}{E_\lambda - E} \right] D_n, \quad \boldsymbol{r} \in \text{cavity}. \qquad (3.261)$$

We multiply both sides of this last equation by $\chi^*_{n'}(y)$ and integrate over $y$ across the opening; using eqns (3.257) and (3.259) we obtain

$$V_{n'} = \sum_{n=1}^{\infty} \left[ \sum_\lambda \frac{\gamma_{\lambda n'} \gamma^*_{\lambda n}}{E_\lambda - E} \right] D_n, \quad 1 \leqslant n' < \infty. \qquad (3.262)$$

From now on we use explicitly TRI, implying reality of $\psi_\lambda(\boldsymbol{r})$, $\chi_n(y)$ and $\gamma_{\lambda n}$. The $R$ matrix is defined to be

$$R_{n'n}(E) = \sum_\lambda \frac{\gamma_{\lambda n'} \gamma_{\lambda n}}{E_\lambda - E}, \quad 1 \leqslant n', n < \infty. \qquad (3.263)$$

In terms of this $R$ matrix, the relation (3.262) becomes

$$V_{n'} = \sum_{n=1}^{\infty} R_{n'n} D_n, \quad 1 \leqslant n' < \infty. \qquad (3.264)$$

From eqns (3.244a,b), (3.257) and (3.258) we find

$$V_n = \frac{a_n + b_n}{\sqrt{2\pi \hbar^2 k_n/m}}, \qquad D_n = \mathrm{i} k_n \frac{-a_n + b_n}{\sqrt{2\pi \hbar^2 k_n/m}}, \qquad 1 \leqslant n \leqslant N, \qquad (3.265\mathrm{a})$$

$$V_n = \frac{b_n}{\mathrm{e}^{\mathrm{i}\pi/4} \sqrt{2\pi \hbar^2 \mathrm{i} \kappa_n/m}}, \qquad D_n = -\kappa_n \frac{b_n}{\mathrm{e}^{\mathrm{i}\pi/4} \sqrt{2\pi \hbar^2 \kappa_n/m}}, \quad n > N. \qquad (3.265\mathrm{b})$$

It will be convenient to condense these relations as follows:

$$V_n = \frac{a_n + b_n}{\sqrt{2\pi\hbar^2 k_n/m}}, \quad D_n = \mathrm{i}k_n \frac{-a_n + b_n}{\sqrt{2\pi\hbar^2 k_n/m}}, \quad 1 \leqslant n < \infty, \tag{3.266}$$

with the understanding that, for $n > N$,

$$a_n = 0, \quad k_n = \mathrm{i}\kappa_n, \quad \sqrt{k_n} = \mathrm{e}^{\mathrm{i}\pi/4}\sqrt{\kappa_n}, \quad n > N. \tag{3.267}$$

Then, relation (3.264) between the $V_n$ and the $D_n$ reads

$$a_{n'} + b_{n'} = \mathrm{i}\sum_{n=1}^{\infty} \sqrt{k_{n'}} R_{n'n} \sqrt{k_n} \left(-a_n + b_n\right).$$

In matrix notation,

$$\boldsymbol{a}^\infty + \boldsymbol{b}^\infty = -\mathrm{i}\sqrt{k}R\sqrt{k}(\boldsymbol{a}^\infty - \boldsymbol{b}^\infty), \tag{3.268}$$

where

$$\boldsymbol{a}^\infty = (a_1, \ldots, a_N, a_{N+1}, \ldots)^\mathsf{T},$$
$$\boldsymbol{b}^\infty = (b_1, \ldots, b_N, b_{N+1}, \ldots)^\mathsf{T},$$
$$k_{n'n} = k_n \delta_{n'n}, \quad 1 \leqslant n', n < \infty.$$

We solve eqn (3.268) for $\boldsymbol{b}^\infty$ to give

$$\boldsymbol{b}^\infty = -\frac{1}{1 - \mathrm{i}\sqrt{k}R\sqrt{k}} \left(1 + \mathrm{i}\sqrt{k}R\sqrt{k}\right) \boldsymbol{a}^\infty. \tag{3.269}$$

The standard, open-channel, $S$ matrix is thus the $N \times N$ submatrix of

$$S = -\frac{1 + \mathrm{i}\sqrt{k}R\sqrt{k}}{1 - \mathrm{i}\sqrt{k}R\sqrt{k}}. \tag{3.270}$$

The expression (3.270) clearly reduces, for one channel, to eqn (2.415) of the previous chapter.

A comment is in order here, in relation to a non-TRI situation, as it occurs in the presence of a magnetic field, which we shall consider to be nonzero inside the cavity and zero in the waveguides. Through the gauge transformation $A \Longrightarrow A' = A + \nabla\Gamma$ and a proper choice of the gauge function $\Gamma$, it is always possible to ensure that the problem in the *internal* region is self-adjoint, by imposing $A'_\perp = 0$ at the opening of the waveguides. Moreover, it is also possible to ensure that $A'$ itself vanishes in the interior of the waveguides, so that we have to deal with plane waves in the latter region. The $R$-matrix theory can then be set up in the usual fashion.

# 4

# LINEAR RESPONSE THEORY OF QUANTUM ELECTRONIC TRANSPORT

The problem of electronic transport is a *non-equilibrium* one. The traditional method used for its study, known as *linear response theory* (LRT), starts from the equilibrium ensemble and then treats the field that drives the system out of equilibrium as a first-order perturbation. This method was devised many years ago by R. Kubo [63,107,108]. It has a remarkable feature, namely that it expresses the linear response to an external perturbation in terms of certain *equilibrium correlations* evaluated in the absence of that perturbation. Of course, a linear response by definition implies that the response be a property of the system in equilibrium, i.e., in the absence of the perturbation. The point of LRT is that it shows that such a response really exists, and gives it in terms of correlations of the equilibrium fluctuations. LRT has been very successful in the study of *intensive* electronic transport properties (such as the *conductivity*) of materials in the *thermodynamic limit* (infinite size).

In mesoscopic physics the interest shifts to the study of *extensive*, or *global*, quantities (such as the *electrical conductance*) of microstructures—*finite-size* samples whose shape and geometry are under control. Here one finds *sample-specific* results which are no longer probe independent, but rather *probe dependent*, so much so that the sample and the probes used to perform the measurement ought to be considered together as one system. The experimental set-up used to measure electronic transport in microstructures can often be represented as the sample under study connected to macroscopic probes, between which a potential difference is established. Within such a conception, Landauer [111] proposed in 1970 a novel approach to the study of the conductance of a microstructure. His idea was aimed at understanding the electrical conductance of a sample in terms of its *scattering properties*. The problem of electronic transport is thus converted into that of solving a quantum mechanical scattering problem, with the whole sample acting as a giant scatterer. This *scattering approach to quantum electronic transport* has been studied and extended by a number of authors. The literature on this topic is, by now, very large; the reader is referred to a number of important representative articles, such as [49, 50, 52], and to the references cited therein, as well as to a comprehensive discussion in review articles and books, such as [24, 61, 93]. The method used in these approaches is not the Kubo LRT, but rather a clever way of counting the charges transferred from one probe to the other, the latter being treated as charge reservoirs.

The problem of quantum transport through a microstructure can also be treated using the Kubo LRT and one expects, of course, to obtain the same answer. The relation between the two approaches has been, in the early stages of mesoscopic physics, a subject of controversy. In fact, the notions—such as those of leads, contacts and reservoirs—that are employed in the two methods that one is trying to compare have a deep physical significance and ought to be treated with great care. The reader may find an account of this relation, for instance, in [17, 68, 100, 114, 118, 119, 165, 169]. Extensions to include a.c. conductances and capacitances can be found in [51].

Since the scattering approach to the problem has already been discussed in a large variety of publications, we have decided to present here the Kubo LRT point of view. Our discussion will follow very closely the physical scheme presented by Levinson [118] and Levinson and Shapiro [119]. This chapter is organized as follows. Section 4.1 describes the geometry of the system in equilibrium, while Section 4.2 introduces the external field that drives the system out of equilibrium. The linear response to this field is treated in Sections 4.3 and 4.4 in different gauges. In order to make the discussion self-contained, a brief account of the time-dependent perturbation theory in quantum mechanics is presented in Exercise 4.1. The actual evaluation of the conductance is carried out in Section 4.5. Appendices A, B and C give the details of some of the calculations.

## 4.1 The system in equilibrium

We consider an electron gas in a three-dimensional geometry such as the one shown schematically in Fig. 4.1. The treatment, however, readily generalizes to arbitrary dimensionality. The system to be studied consists of a constriction between two 'horns', which expand up to a very large distance $R_h$. As a simple illustrative example, the figure shows two large spheres connected at the ends, which are *kept at fixed voltages* $V_1$ and $V_2$ by a van de Graaf machine. Of course, a more practical alternative is to connect the system to a battery. A constant current then flows along the system.

We first suppose that no external voltage is applied, so that the whole system—the sample plus the contacts—is in equilibrium. We only consider the case in which the conduction electrons can be described through a *time-reversal invariant (TRI) Hamiltonian*. In particular, *no static magnetic field is assumed to be present*. The inclusion of a static magnetic field in the analysis can be found, for instance, in [17]. The Hamiltonian of the system is taken to be of the form

$$H_0 = \sum_i \left[ \frac{p_i^2}{2m} - e \sum_I \frac{e_I}{|\boldsymbol{r}_i - \boldsymbol{R}_I|} - e \sum_j \frac{e_j}{|\boldsymbol{r}_i - \boldsymbol{R}_j|} \right] + \sum_{i<j} \frac{e^2}{|\boldsymbol{r}_i - \boldsymbol{r}_j|} \quad (4.1\text{a})$$

$$= T + V_{eI} + V_{e,\text{imp}} + V_{e-e}. \quad (4.1\text{b})$$

Here, $-e$ is the electronic charge, $\boldsymbol{r}_i$ is the position variable of the $i$th *conduction* electron, $\boldsymbol{R}_I$ is the (static) position of the $I$th ion with positive charge $e_I$

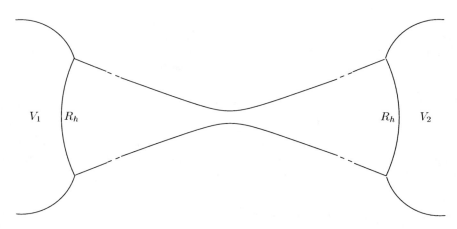

FIG. 4.1. The geometry for the electronic conduction problem studied in the text. The system consists of a constriction between two expanding 'horns', which extend up to a very large distance $R_h$. As an illustration, the system is connected to two large spheres (indicated schematically) at its ends, *kept at fixed voltages* $V_1$ and $V_2$ by a van de Graaf machine. Alternatively, the system can be considered to be connected to the two terminals of a battery. A d.c. current flows along the system.

(screened by the bound electrons), and $e_j$ and $\boldsymbol{R}_j$ are, respectively, the charge and position of the $j$th impurity, considered to be static. Thus $H_0$ contains the kinetic energy, the interaction of the electrons with the ions and the impurities, and the electron–electron interaction represented by the last term in eqn (4.1), for which no approximation is assumed for the time being. No degree of freedom is assigned to the bound electrons. The $I$th positive ion with the corresponding bound electrons is treated as a point charge $e_I$ at the position $\boldsymbol{R}_I$; the possibility of polarization of the bound electronic cloud is thus disregarded in this treatment, although it could easily be incorporated.

The expectation value of an observable $\hat{O}$ over an *equilibrium* statistical mechanical ensemble specified by the density matrix $\hat{\rho}_0(\beta,\mu)$ arising from the Hamiltonian $H_0$ will be denoted by $\langle \hat{O}\rangle_0$, i.e.,

$$\left\langle \hat{O} \right\rangle_0 = \mathrm{Tr}\left[\hat{\rho}_0(\beta,\mu)\hat{O}\right]. \tag{4.2}$$

The density matrix will be taken to be the one describing a grand canonical ensemble, i.e.,

$$\hat{\rho}_0(\beta,\mu) = \frac{\mathrm{e}^{-\beta(\hat{H}_0-\mu\hat{N})}}{\mathrm{Tr}\,\mathrm{e}^{-\beta(\hat{H}_0-\mu\hat{N})}}. \tag{4.3}$$

Here, $\hat{N}$ is the electron number operator and the various traces are thus over the entire Fock space. In the above, $\beta = 1/k_B T$ and $\mu$ is the chemical potential. We now consider a number of physically important observables.

The electronic charge density is described by the operator

$$\hat{\rho}(\boldsymbol{r}) \equiv -e\hat{n}(\boldsymbol{r})$$
$$= -e \sum_i \delta(\boldsymbol{r} - \boldsymbol{r}_i), \qquad (4.4)$$

where $\hat{n}(\boldsymbol{r})$ is the electron number density operator. The expression in the last line of eqn (4.4) is written using the coordinate representation. We denote by $\rho_0(\boldsymbol{r})$ its equilibrium expectation value, namely

$$\rho_0(\boldsymbol{r}) \equiv \langle \hat{\rho}(\boldsymbol{r}) \rangle_0 = \left\langle -e \sum_i \delta(\boldsymbol{r} - \boldsymbol{r}_i) \right\rangle_0. \qquad (4.5)$$

The electronic contribution to the scalar potential at the point $\boldsymbol{r}$ is described by the operator (in the coordinate representation)

$$\hat{\phi}(\boldsymbol{r}) = \sum_i \frac{-e}{|\boldsymbol{r} - \boldsymbol{r}_i|}. \qquad (4.6)$$

The pair of operators $\hat{\rho}(\boldsymbol{r})$ and $\hat{\phi}(\boldsymbol{r})$ satisfy Poisson's equation

$$\nabla^2 \hat{\phi}(\boldsymbol{r}) = -4\pi \hat{\rho}(\boldsymbol{r}). \qquad (4.7)$$

We denote by $\phi_0(\boldsymbol{r})$ the equilibrium expectation value of $\hat{\phi}(\boldsymbol{r})$, i.e.,

$$\phi_0(\boldsymbol{r}) \equiv \langle \hat{\phi}(\boldsymbol{r}) \rangle_0 = \left\langle \sum_i \frac{-e}{|\boldsymbol{r} - \boldsymbol{r}_i|} \right\rangle_0. \qquad (4.8)$$

From the relation (4.7), we see that the potential of eqn (4.8) and the charge density of eqn (4.5) satisfy Poisson's equation, i.e.,

$$\nabla^2 \phi_0(\boldsymbol{r}) = -4\pi \rho_0(\boldsymbol{r}). \qquad (4.9)$$

The charge density $\rho_0(\boldsymbol{r})$ and, therefore, the potential $\phi_0(\boldsymbol{r})$, are static, but may depend on position, due to the effect of the ionic and impurity potentials.

If we were to include the possibility of polarization of the bound electrons around each ion, then the right-hand side of Poisson's equation would be modified by the factor $1/\epsilon$, where $\epsilon$ is the static dielectric constant *not* related to the conduction electrons.

The electronic current density operator at the point $\boldsymbol{r}$, appropriate to the Hamiltonian $H_0$ of eqn (4.1), is given by

$$\hat{\boldsymbol{j}}(\boldsymbol{r}) = -\frac{e}{2m} \sum_i [\delta(\boldsymbol{r} - \boldsymbol{r}_i) \hat{\boldsymbol{p}}_i + \hat{\boldsymbol{p}}_i \delta(\boldsymbol{r} - \boldsymbol{r}_i)], \qquad (4.10)$$

the summation being, as before, over the conduction electrons. Its expectation value in the present equilibrium situation, and in the absence of a magnetic field, vanishes everywhere, i.e.,

$$J_0(r) \equiv \left\langle \hat{j}(r) \right\rangle_0 \equiv 0. \tag{4.11}$$

## 4.2 Application of an external electromagnetic field

The application of an external electromagnetic field with (circular) frequency $\omega$ causes a current with that frequency to flow in the system. It is our aim to calculate this within the framework of LRT, which will be summarized later.

We describe the external electromagnetic field, $E_{\text{ext}}$ and $B_{\text{ext}}$, through the scalar and the vector potentials $\phi_{\text{ext}}(r,t)$ and $A_{\text{ext}}(r,t)$, i.e.,

$$E_{\text{ext}} = -\nabla \phi_{\text{ext}} - \frac{1}{c}\frac{\partial A_{\text{ext}}}{\partial t}, \tag{4.12}$$

$$B_{\text{ext}} = \nabla \times A_{\text{ext}}. \tag{4.13}$$

Starting from the Hamiltonian $H_0$ of eqn (4.1), we write the resulting Hamiltonian in the presence of the external potentials as

$$H = \sum_i \left\{ \frac{1}{2m}\left[p_i + \frac{e}{c}A_{\text{ext}}(r_i,t)\right]^2 - e\sum_I \frac{e_I}{|r_i - R_I|} \right.$$
$$\left. - e\sum_j \frac{e_j}{|r_i - R_j|} - e\phi_{\text{ext}}(r_i,t) \right\} + \sum_{i<j} \frac{e^2}{|r_i - r_j|}. \tag{4.14}$$

Notice that the external electromagnetic potentials in eqn (4.14) and in what follows are given classical fields; they are not dynamical variables and will thus be treated as c-numbers.

It will be convenient to write these potentials as $\mathring{\phi}_{\text{ext}}(r,t)$ and $\mathring{A}_{\text{ext}}(r,t)$ and recall that they can be subjected to the usual gauge transformation that leaves the fields $E_{\text{ext}}$ and $B_{\text{ext}}$ invariant [94, 122], i.e.,

$$\mathring{\phi}_{\text{ext}} \Longrightarrow \mathring{\phi}_{\text{ext}} - \frac{1}{c}\frac{\partial \Gamma}{\partial t}, \tag{4.15a}$$

$$\mathring{A}_{\text{ext}} \Longrightarrow \mathring{A}_{\text{ext}} + \nabla \Gamma, \tag{4.15b}$$

$\Gamma$ being a scalar gauge function.

Notice first that the vector potential $\mathring{A}_{\text{ext}}$ can always be represented as the sum of a longitudinal and a transverse part, i.e.,

$$\mathring{A}_{\text{ext}} = \mathring{A}^l_{\text{ext}} + \mathring{A}^t_{\text{ext}}, \tag{4.16}$$

where $\mathring{A}^l_{\text{ext}} \equiv \nabla V_{\text{ext}}$ and $\mathring{A}^t_{\text{ext}} \equiv \nabla \times C_{\text{ext}}$. The magnetic field is then given by

$$\boldsymbol{B}_{\text{ext}} = \nabla \times \mathring{\boldsymbol{A}}^{t}_{\text{ext}}. \tag{4.17}$$

As a first choice of gauge, we see from eqn (4.15b) that the choice

$$\Gamma = -V_{\text{ext}} \tag{4.18}$$

cancels exactly the longitudinal part $\mathring{\boldsymbol{A}}^{l}_{\text{ext}}$, leaving only the transverse part $\mathring{\boldsymbol{A}}^{t}_{\text{ext}}$. We can then write the potentials as

$$\phi_{\text{ext}} = \mathring{\phi}_{\text{ext}} + \frac{1}{c}\frac{\partial V_{\text{ext}}}{\partial t}, \tag{4.19a}$$

$$\boldsymbol{A}_{\text{ext}} = \boldsymbol{A}^{t}_{\text{ext}} = \mathring{\boldsymbol{A}}^{t}_{\text{ext}}, \tag{4.19b}$$

and the electric field as

$$\boldsymbol{E}_{\text{ext}}(\boldsymbol{r},t) = -\nabla \phi_{\text{ext}}(\boldsymbol{r},t) - \frac{1}{c}\frac{\partial \boldsymbol{A}^{t}_{\text{ext}}}{\partial t}. \tag{4.20}$$

This is normally referred to as the Coulomb, or transverse, gauge [94, 122].

As a second choice of gauge, we notice that, by an appropriate choice of $\Gamma$, we can have a description in which the scalar potential vanishes identically. This is achieved by choosing the gauge function to be

$$\Gamma'(\boldsymbol{r},t) = c\int^{t} \mathring{\phi}_{\text{ext}}(\boldsymbol{r},t')\,dt', \tag{4.21}$$

so that, from eqns (4.15), the new scalar and vector potentials are

$$\phi'_{\text{ext}} \equiv 0, \tag{4.22a}$$

$$\boldsymbol{A}'_{\text{ext}}(\boldsymbol{r},t) = \mathring{\boldsymbol{A}}_{\text{ext}} + c\nabla \int^{t} \mathring{\phi}_{\text{ext}}(\boldsymbol{r},t')\,dt', \tag{4.22b}$$

respectively. Notice that, in general, the vector potential will possess a longitudinal part in this gauge. The electric field is then given by

$$\boldsymbol{E}_{\text{ext}}(\boldsymbol{r},t) = -\frac{1}{c}\frac{\partial \boldsymbol{A}'_{\text{ext}}(\boldsymbol{r},t)}{\partial t} \tag{4.23a}$$

$$= -\frac{1}{c}\frac{\partial \boldsymbol{A}'^{l}_{\text{ext}}}{\partial t} - \frac{1}{c}\frac{\partial \boldsymbol{A}'^{t}_{\text{ext}}}{\partial t}, \tag{4.23b}$$

where we have split $\boldsymbol{A}'_{\text{ext}}$ into its longitudinal and transverse parts.

Comparing the electric field in the two gauges, eqns (4.20) and (4.23), we see that the transverse part of the vector potential coincides in the two gauges, i.e., $\boldsymbol{A}^{t}_{\text{ext}} = \boldsymbol{A}'^{t}_{\text{ext}}$, and so $\boldsymbol{B}_{\text{ext}} = \nabla \times \boldsymbol{A}^{t}_{\text{ext}}$ (see eqn (4.17)).

In the so-called radiation gauge one has (for $\nabla \cdot \boldsymbol{E} \equiv 0$) $\phi_{\text{ext}} \equiv 0$ and $\boldsymbol{A}'^{l}_{\text{ext}} \equiv 0$, so that both $\boldsymbol{E}_{\text{ext}}$ and $\boldsymbol{B}_{\text{ext}}$ are purely transverse.

Here, on the other hand, we shall assume, from now on, that the frequency of the applied field is small enough to neglect any radiative effect. In this *quasi-static*, or *long-wavelength* approximation, we shall be concerned with a nonzero electric field, with a corresponding negligible magnetic field. We then neglect the transverse component of the vector potential $\boldsymbol{A}_{\text{ext}}^t$ in the above equations, thus giving, in the first gauge,

$$\phi_{\text{ext}} \neq 0, \quad \boldsymbol{A}_{\text{ext}} \approx 0, \qquad (4.24)$$

with

$$\boldsymbol{E}_{\text{ext}}(\boldsymbol{r}, t) \approx -\nabla \phi_{\text{ext}}(\boldsymbol{r}, t), \qquad (4.25)$$
$$\boldsymbol{B}_{\text{ext}} \approx 0. \qquad (4.26)$$

Since, under these conditions, only the scalar potential survives, we shall refer to this gauge as the *scalar potential gauge*. In the second gauge,

$$\phi'_{\text{ext}} \equiv 0, \quad \boldsymbol{A}'_{\text{ext}} \approx \boldsymbol{A}''^{l}_{\text{ext}} \neq 0, \qquad (4.27)$$

with

$$\boldsymbol{E}_{\text{ext}}(\boldsymbol{r}, t) = -\frac{1}{c}\frac{\partial \boldsymbol{A}'_{\text{ext}}(\boldsymbol{r}, t)}{\partial t}, \qquad (4.28)$$
$$\boldsymbol{B}_{\text{ext}} \approx 0. \qquad (4.29)$$

Since in this gauge the scalar potential has been cancelled exactly, we shall refer to it as the *vector potential gauge*.

For a harmonic time dependence $e^{-i\omega t}$ with frequency $\omega$, we have

$$\boldsymbol{E}_{\text{ext}}^{\omega}(\boldsymbol{r}) = -\nabla \phi_{\text{ext}}^{\omega}(\boldsymbol{r}) = \frac{i\omega}{c} \boldsymbol{A}'^{\omega}_{\text{ext}}(\boldsymbol{r}), \qquad (4.30)$$

giving the external electric field $\boldsymbol{E}_{\text{ext}}^{\omega}(\boldsymbol{r})$ in the two gauges introduced above.

Of course, general gauge invariance ensures that the physical results should be independent of the gauge. In the following sections we will evaluate the physical quantities of interest in the two gauges described above, but only in the quasi-static approximation, and show that we still recover the same physics.

## 4.3 The external field in the scalar potential gauge

From eqns (4.24) and (4.14) we write the perturbed time-dependent Hamiltonian $H(t)$ as

$$H(t) = H_0 - e \sum_i \phi_{\text{ext}}(\boldsymbol{r}_i, t). \qquad (4.31)$$

We assume that we *control* the externally applied voltages. Very far out in the horns, $\phi_{\text{ext}}(\boldsymbol{r}, t)$ becomes equal to the *spatially constant* values $V_1(t)$ and $V_2(t)$, in the left and right horns, respectively, giving rise to the potential difference

$$V(t) = V_1(t) - V_2(t), \qquad (4.32)$$

which we control by the application of an emf (electromotive force). Inside the system the external potential $\phi_{\text{ext}}(\mathbf{r}, t)$ satisfies Laplace's equation

$$\nabla^2 \phi_{\text{ext}}(\mathbf{r}, t) \equiv 0. \qquad (4.33)$$

Our aim now is to find, in the new situation, the charge density $\rho(\mathbf{r}, t)$, the potential profile $\phi(\mathbf{r}, t)$ and the current density $\mathbf{J}(\mathbf{r}, t)$, within the framework of LRT.

### 4.3.1 *The charge density and the potential profile*

In first-order time-dependent perturbation theory the external field can be taken into account as explained in the following exercise.

**Exercise 4.1** Consider a quantum mechanical system whose Hamiltonian $H$ can be split into the sum of an unperturbed part $H_0$ and a time-dependent perturbation $H'(t)$, i.e.,

$$H = H_0 + H'(t). \qquad (4.34)$$

The state $|\psi(t)\rangle$ evolves according to the time-dependent Schrödinger equation

$$(i\hbar \partial_t - H_0)|\psi(t)\rangle = H'(t)|\psi(t)\rangle, \qquad (4.35)$$

supplemented by the initial condition at $t = t_0$ given by

$$|\psi(t = t_0)\rangle = |0\rangle. \qquad (4.36)$$

Use time-dependent perturbation theory to find, to first order in the perturbation $H'(t)$, the time evolution of the state vector $|\psi(t)\rangle$ and of the expectation value of an operator $\hat{\mathcal{O}}$.

We write the retarded solution of the Schrödinger equation (4.35) for $t > t_0$ in terms of the retarded unperturbed Green function $G_0^{(+)}(t - t')$ as

$$|\psi(t)\rangle = |\psi_0(t)\rangle + \int_{t_0}^{\infty} dt'\, G_0^{(+)}(t - t') H'(t') |\psi(t')\rangle, \qquad (4.37)$$

where the evolution of $|\psi_0(t)\rangle$ is governed by the unperturbed part $H_0$ of the Hamiltonian, i.e.,

$$|\psi_0(t)\rangle = e^{-(i/\hbar)H_0(t - t_0)} |0\rangle. \qquad (4.38)$$

Since the retarded Green function $G_0^{(+)}$ is given by

$$G_0^{(+)}(t - t') = -\frac{i}{\hbar} e^{-(i/\hbar)H_0(t - t')} \theta(t - t'), \qquad (4.39)$$

with $\theta(\tau)$ denoting the Heaviside step function, we write the solution (4.37) as

$$|\psi(t)\rangle = |\psi_0(t)\rangle - \frac{i}{\hbar} \int_{t_0}^{\infty} dt'\, \theta(t-t') e^{-(i/\hbar)H_0(t-t')} H'(t') |\psi(t')\rangle. \tag{4.40}$$

This is an integral equation for $|\psi(t)\rangle$ that can be iterated to obtain a perturbation expansion. To first order in the perturbation $H'$, we have

$$|\psi(t)\rangle = |\psi_0(t)\rangle - \frac{i}{\hbar} \int_{t_0}^{\infty} dt'\, \theta(t-t') e^{-(i/\hbar)H_0(t-t')} H'(t') |\psi_0(t')\rangle + \cdots. \tag{4.41}$$

The expectation value $\langle \hat{O} \rangle_t$ of an observable $\hat{O}$ with respect to the solution $|\psi(t)\rangle$ can thus be written, *to first order in the perturbation*, as

$$\begin{aligned}
\langle \hat{O} \rangle_t &\equiv \langle \psi(t) | \hat{O} | \psi(t) \rangle \\
&= \langle \hat{O} \rangle_{0t} + \frac{i}{\hbar} \int_{t_0}^{\infty} dt'\, \theta(t-t') \langle \psi_0(t') | H'(t') e^{(i/\hbar)H_0(t-t')} \hat{O} | \psi_0(t) \rangle \\
&\quad - \frac{i}{\hbar} \int_{t_0}^{\infty} dt'\, \theta(t-t') \langle \psi_0(t) | \hat{O} e^{-(i/\hbar)H_0(t-t')} H'(t') | \psi_0(t') \rangle + \cdots,
\end{aligned} \tag{4.42}$$

where

$$\langle \hat{O} \rangle_{0t} = \langle \psi_0(t) | \hat{O} | \psi_0(t) \rangle \tag{4.43}$$

represents the—in general time-dependent—expectation value of the operator $\hat{O}$, should the wave function evolve according to the *unperturbed* Hamiltonian $H_0$, i.e., as in eqn (4.38).

Consider the last matrix element, which we may call $I$, occurring in eqn (4.42). We write explicitly the evolution of $|\psi_0(t)\rangle$ occurring in this matrix element, to obtain

$$I = \langle 0 | e^{(i/\hbar)H_0(t-t_0)} \hat{O} e^{-(i/\hbar)H_0(t-t')} H'(t') e^{-(i/\hbar)H_0(t'-t_0)} | 0 \rangle. \tag{4.44}$$

We now assume that *the initial state $|0\rangle$ that was introduced in eqn (4.36) is an eigenstate of $H_0$*. The phase arising from $t_0$ in the previous equation then drops out and we can write

$$I = \langle 0 | \widetilde{O}(t) \widetilde{H}'(t') | 0 \rangle, \tag{4.45}$$

where we have defined the operators in the *interaction representation* as

$$\widetilde{O}(t) \equiv e^{(i/\hbar)H_0 t} \hat{O} e^{-(i/\hbar)H_0 t}. \tag{4.46}$$

With a similar consideration for the matrix element occurring in the second row of eqn (4.42), we finally write

$$\langle \hat{O} \rangle_t = \langle \hat{O} \rangle_{0t} - \frac{i}{\hbar} \int_{t_0}^{\infty} dt'\, \theta(t-t') \langle 0 | \left[ \widetilde{O}(t), \widetilde{H}'(t') \right] | 0 \rangle + \cdots. \tag{4.47}$$

The above treatment which uses the eigenstate $|0\rangle$ of $H_0$ as the initial state can be readily generalized to a thermal ensemble average replacing the single-initial-state average, so that

$$\langle \hat{O} \rangle_t = \langle \hat{O} \rangle_{0t} - \frac{i}{\hbar} \int_{t_0}^{\infty} dt'\, \theta(t-t') \left\langle \left[ \tilde{O}(t), \widetilde{H}'(t') \right] \right\rangle_0 + \cdots, \qquad (4.48)$$

where the notation $\langle \cdots \rangle_0$ indicates an average over an *equilibrium* statistical mechanical ensemble defined by the Hamiltonian $H_0$.

A simple application of the foregoing formalism for a linear chain of harmonic oscillators driven by an external interaction is given in the following exercise.

**Exercise 4.2** A linear chain of harmonic oscillators is described by the Hamiltonian

$$H_0 = \sum_{n=1}^{N} \frac{p_n^2}{2\mu} + \frac{1}{2}\mu\omega_0^2 \sum_{n,m=1}^{N} C_{nm} x_n x_m, \qquad (4.49)$$

where $C_{nm}$ are the coupling coefficients. For instance, for nearest-neighbor coupling of identical masses by identical springs, the matrix $C$ is tridiagonal, with 2 along the diagonal and $-1$ along the two adjacent secondary diagonals. The chain is driven by the external interaction

$$H' = -\mu \sum_n x_n f_n(t). \qquad (4.50)$$

The system is prepared, at time $t_0$, in the ground state $|0\rangle$ of $H_0$, and allowed to evolve thereafter according to the full Hamiltonian $H = H_0 + H'$. Using first-order time-dependent perturbation theory, find the expectation value $\langle \hat{v}_n \rangle_t$ of the velocity of the $n$th oscillator for $t > t_0$.

From eqn (4.47) of Exercise 4.1, we find

$$\langle \hat{v}_n \rangle_t = \frac{i}{\hbar} \mu \sum_m \int_{t_0}^{t} dt'\, \langle 0 | [\tilde{v}_n(t), \tilde{x}_m(t')] | 0 \rangle f_m(t') + \cdots. \qquad (4.51)$$

Using the definition (4.46) of an operator in the interaction representation, and denoting by $|\nu\rangle$ the eigenstates of $H_0$ and by $E_\nu$ the corresponding energies, i.e.,

$$H_0 |\nu\rangle = E_\nu |\nu\rangle, \qquad (4.52)$$

we write the above equation as

$$\langle \hat{v}_n \rangle_t = \frac{i}{\hbar} \mu \sum_m \int_{t_0}^{t} dt' \sum_\nu \left[ e^{(i/\hbar)(E_\nu - E_0)(t'-t)} \langle 0|\hat{v}_n|\nu\rangle \langle \nu|\hat{x}_m|0\rangle \right.$$
$$\left. - e^{(i/\hbar)(E_\nu - E_0)(t-t')} \langle 0|\hat{x}_m|\nu\rangle \langle \nu|\hat{v}_n|0\rangle \right] f_m(t') + \cdots. \qquad (4.53)$$

The Hamiltonian $H_0$ can be brought into the normal-mode form

$$H_0 = \sum_{i=1}^{N} \frac{\pi_i^2}{2\mu} + \frac{1}{2} \sum_{i=1}^{N} \mu \omega_i^2 \xi_i^2 \qquad (4.54)$$

## THE EXTERNAL FIELD IN THE SCALAR POTENTIAL GAUGE

by an orthogonal transformation $O$ of coordinates and momenta (the orthogonal transformation that diagonalizes the real symmetric matrix $C$ of eqn (4.49)), i.e.,

$$\hat{x}_m = \sum_{i=1}^{N} O_{mi}\xi_i \equiv \sum_{i=1}^{N} O_{mi}\sqrt{\frac{\hbar}{\mu\omega_i}}\frac{\alpha_i + \alpha_i^\dagger}{\sqrt{2}}, \tag{4.55a}$$

$$\hat{p}_m = \sum_{i=1}^{N} O_{mi}\pi_i \equiv \sum_{i=1}^{N} O_{mi}\sqrt{\hbar\mu\omega_i}\frac{\alpha_i - \alpha_i^\dagger}{\sqrt{2\mathrm{i}}}, \tag{4.55b}$$

where we have also introduced the creation and annihilation operators $\alpha_i^\dagger$ and $\alpha_i$ for the normal mode $i$. The eigenstate $|\nu\rangle$ of $H_0$ can be written as

$$|\nu\rangle = |\nu_1\nu_2\cdots\nu_N\rangle, \tag{4.56}$$

$\nu_i$ being the number of quanta in the normal mode $i$. The matrix elements needed in eqn (4.53) are then given by

$$\langle 0|\hat{v}_n|\nu\rangle = \frac{1}{\sqrt{2\mathrm{i}}}\sum_i O_{ni}\sqrt{\frac{\hbar\omega_i}{\mu}}\delta_{\nu_1 0}\cdots\delta_{\nu_{i-1}0}\delta_{\nu_i 1}\delta_{\nu_{i+1}0}\cdots\delta_{\nu_N 0}, \tag{4.57a}$$

$$\langle \nu|\hat{x}_m|0\rangle = \frac{1}{\sqrt{2}}\sum_i O_{mi}\sqrt{\frac{\hbar}{\mu\omega_i}}\delta_{\nu_1 0}\cdots\delta_{\nu_{i-1}0}\delta_{\nu_i 1}\delta_{\nu_{i+1}0}\cdots\delta_{\nu_N 0}. \tag{4.57b}$$

Thus

$$\langle 0|\hat{v}_n|\nu\rangle\langle\nu|\hat{x}_m|0\rangle = \frac{\hbar}{2\mathrm{i}\mu}\sum_i O_{ni}O_{mi}\delta_{\nu_1 0}\cdots\delta_{\nu_{i-1}0}\delta_{\nu_i 1}\delta_{\nu_{i+1}0}\cdots\delta_{\nu_N 0}$$

$$= \frac{\hbar}{2\mathrm{i}\mu}\left[O_{n1}O_{m1}\delta_{\nu_1 1}\delta_{\nu_2 0}\cdots\delta_{\nu_{N-1}0}\delta_{\nu_N 0}\right.$$

$$\left.+\cdots+O_{nN}O_{mN}\delta_{\nu_1 0}\delta_{\nu_2 0}\cdots\delta_{\nu_{N-1}0}\delta_{\nu_N 1}\right] \tag{4.58}$$

and

$$\sum_{\nu}\left[\mathrm{e}^{-(\mathrm{i}/\hbar)(E_\nu - E_0)(t-t')}\langle 0|\hat{v}_n|\nu\rangle\langle\nu|\hat{x}_m|0\rangle - \mathrm{c.c.}\right]$$

$$= \frac{\hbar}{2\mathrm{i}\mu}\sum_{\nu_1,\ldots,\nu_N}\mathrm{e}^{-(\mathrm{i}/\hbar)(E_{\nu_1}+\cdots+E_{\nu_i}+\cdots+E_{\nu_N}-E_0)(t-t')}$$

$$\times\left[O_{n1}O_{m1}\delta_{\nu_1 1}\cdots\delta_{\nu_N 0}+\cdots+O_{nN}O_{mN}\delta_{\nu_1 0}\cdots\delta_{\nu_N 1}\right] - \mathrm{c.c.}$$

$$= \frac{\hbar}{2\mathrm{i}\mu}\left[O_{n1}O_{m1}\mathrm{e}^{-(\mathrm{i}/\hbar)(E_{\nu_1=1}-E_{\nu_1=0})(t-t')}\right.$$

$$\left.+\cdots+O_{nN}O_{mN}\mathrm{e}^{-(\mathrm{i}/\hbar)(E_{\nu_N=1}-E_{\nu_N=0})(t-t')}\right] - \mathrm{c.c.}$$

$$= \frac{\hbar}{\mathrm{i}\mu}\sum_{i=1}^{N}O_{ni}O_{mi}\cos\omega_i(t-t'). \tag{4.59}$$

Here, the abbreviation c.c. means 'complex conjugate'. The final expression for the $\langle \hat{v}_n \rangle_t$ of eqn (4.53) is then

$$\langle \hat{v}_n \rangle_t = \sum_m \int_{t_0}^{\infty} \sigma_{nm}(t-t') f_m(t') \, \mathrm{d}t', \tag{4.60a}$$

where we have introduced the *response function* $\sigma_{nm}(t-t')$, defined by

$$\sigma_{nm}(t-t') = \theta(t-t') \sum_{i=1}^{N} O_{ni} O_{mi} \cos \omega_i (t-t'). \tag{4.60b}$$

We notice, incidentally, that the result of eqns (4.60), although deduced in first-order perturbation theory, is exact for the present case of the linear chain of harmonic oscillators.

In our applications of the effect of an external field on a mesoscopic system we first choose, as the operator $\hat{O}$ in eqn (4.48), the electronic charge density operator $\hat{\rho}(\boldsymbol{r})$ of eqn (4.4). According to eqn (4.31), the perturbation, i.e., the operator $H'$ in eqn (4.48), is given by

$$H'(t) = -e \sum_i \phi_{\text{ext}}(\boldsymbol{r}_i, t)$$

$$= \int \hat{\rho}(\boldsymbol{r}') \phi_{\text{ext}}(\boldsymbol{r}', t) \, \mathrm{d}^3 \boldsymbol{r}' \tag{4.61}$$

in the Schrödinger picture, and

$$\widetilde{H'}(t) = \int \widetilde{\rho}(\boldsymbol{r}', t) \phi_{\text{ext}}(\boldsymbol{r}', t) \, \mathrm{d}^3 \boldsymbol{r}' \tag{4.62}$$

in the interaction picture. From eqn (4.48) we find the perturbed electronic charge density to be

$$\rho(\boldsymbol{r}, t) = \rho_0(\boldsymbol{r}) - \frac{\mathrm{i}}{\hbar} \lim_{\eta \to 0^+} \int_{-\infty}^{+\infty} \mathrm{d}t' \, e^{-\eta(t-t')} \theta(t-t')$$
$$\times \int \mathrm{d}^3 \boldsymbol{r}' \, \langle [\widetilde{\rho}(\boldsymbol{r}, t), \widetilde{\rho}(\boldsymbol{r}', t')] \rangle_0 \, \phi_{\text{ext}}(\boldsymbol{r}', t'). \tag{4.63}$$

Here we have defined

$$\rho(\boldsymbol{r}, t) = \langle \hat{\rho}(\boldsymbol{r}) \rangle_t, \tag{4.64}$$

where the right-hand side is as in eqn (4.48). The first term $\rho_0(\boldsymbol{r})$ on the right-hand side of eqn (4.63) is the equilibrium density of eqn (4.5). Again, the brackets $\langle \cdots \rangle_0$ indicate an average over the *equilibrium* statistical mechanical ensemble defined by the Hamiltonian $H_0$ of eqn (4.1) in the absence of the external potential, as in eqn (4.2).

# THE EXTERNAL FIELD IN THE SCALAR POTENTIAL GAUGE 199

The convergence factor $e^{-\eta(t-t')}$ in eqn (4.63) and the associated limit $\eta \to 0^+$ assume that the system is *weakly coupled to a very large heat bath*, to which the energy transferred from the external field to the system is finally disposed of *irreversibly*. Without this feature, our treatment of the problem as given in Exercise 4.1 would be purely quantum mechanical and would thus show no irreversibility.

In order to fully appreciate this point, recall that a finite $d$-dimensional system of volume $\Omega$ typically has nonzero energy-level spacings $\Delta \sim 1/\Omega$. As discussed in Chapter 1, this granularity defines the Heisenberg time-scale $\tau_H \sim \hbar/\Delta$, beyond which the transport under strictly elastic scatterings effectively ceases to be diffusive (or irreversible)—recurrence sets in. Irreversibility then requires inelastic processes to cause real transitions involving energy exchanges greater than $\Delta$. Such non-adiabaticity is physically equivalent to a lifetime broadening of the energy levels, and can be formally included as a small damping factor $e^{-\eta t}$, as in the expressions above, with $\hbar\eta > \Delta$. (For a macroscopic (bulk) system, typically $\Delta \sim 10^{-22}\,\text{eV}$ ($\sim 10^{-18}\,\text{K}$ in temperature units), and thus $\eta \geqslant 10^{-7}\,\text{s}^{-1}$ would suffice.) Such a tiny damping is almost always present because the inelastic scattering rate $1/\tau_{\text{in}}$ is seldom less than $10^6\,\text{s}^{-1}$, even at millikelvin temperatures. For a small metallic system, however, we can easily have $\hbar\eta \lesssim \Delta$, implying adiabaticity and hence no irreversibility [93, pp. 89–92]. (We note in passing that the introduction of the condition $\hbar\eta > \Delta$ here to ensure non-adiabaticity is just the opposite of the mathematical artifice employed to ensure adiabatic switching of the interaction in many-body physics.) Formally, as we shall see, it is advantageous to take the limit $\eta \to 0$, as indicated in the above equations, but only after the *thermodynamic limit* $\Omega \to \infty$ has been taken, in order to ensure that $\hbar\eta > \Delta$ at all times. This point is further illustrated in the following exercise. In the present problem depicted in Fig. 4.1, the thermodynamic limit will be attained by taking $R_h \to \infty$, as illustrated in Fig. 4.2 on p. 217.

**Exercise 4.3** Consider the driven linear chain of oscillators described in Exercise 4.2, fixing attention on the central oscillator, to be labeled as $n = 0$, and assuming, for simplicity, that this is the only one to be excited. The excitation will be taken to be

$$f_0(t') = \begin{cases} f_0, & 0 \leqslant t' \leqslant T, \\ 0, & t' < 0,\ t' > T. \end{cases} \tag{4.65}$$

Find the work per unit time $P(t)$, as well as the total work $W$, performed on the chain by the external agent when the chain is, firstly, isolated and, secondly, weakly coupled to an external bath, as described in the above paragraphs; verify the setting in of irreversibility in the second case.

For the external field of eqn (4.65), we find, from eqn (4.60), the response

$$\langle \hat{v}_0 \rangle_t = f_0 \sum_{i=1}^{N} O_{0i}^2 \frac{\sin \omega_i t}{\omega_i}, \quad 0 < t < T. \tag{4.66}$$

The work performed per unit time $P(t)$ on the oscillator is

$$P(t) = \mu f(t) \langle \hat{v}_0 \rangle_t$$
$$= \mu f_0^2 \sum_{i=1}^{N} O_{0i}^2 \frac{\sin \omega_i t}{\omega_i}, \quad 0 < t < T, \tag{4.67}$$

and the total work is

$$W = \int_0^T P(t)\, dt$$
$$= 2\mu f_0^2 \sum_{i=1}^{N} O_{0i}^2 \frac{\sin^2 \omega_i T/2}{\omega_i^2}. \tag{4.68}$$

We see that, for every mode $i$, energy is periodically absorbed by the system and given back to the external agent, a purely reactive behavior.

Coupling to an external bath is achieved as explained in the previous paragraphs, in connection with eqn (4.63). Let us calculate explicitly its effect in the present context. From

$$\langle \hat{v}_0 \rangle_t = f_0 \sum_{i=1}^{N} O_{0i}^2 \int_0^t e^{-\eta(t-t')} \cos \omega_i (t-t')\, dt', \tag{4.69}$$

we find

$$P(t) = \mu f_0^2 \sum_{i=1}^{N} O_{0i}^2 \left\{ \frac{1}{2}\left[\frac{e^{(i\omega_i - \eta)t}}{i\omega_i - \eta} - \frac{e^{-(i\omega_i + \eta)t}}{i\omega_i + \eta}\right] + \frac{\eta}{\omega_i^2 + \eta^2} \right\} \tag{4.70}$$

and

$$W = -\frac{1}{2}\mu f_0^2 \sum_{i=1}^{N} O_{0i}^2 \left[\frac{e^{(i\omega_i - \eta)T} - 1}{(\omega_i + i\eta)^2} + \frac{e^{-(i\omega_i + \eta)T} - 1}{(\omega_i - i\eta)^2}\right] + \pi\mu f_0^2 T \sum_{i=1}^{N} O_{0i}^2 \frac{\eta/\pi}{\omega_i^2 + \eta^2}$$
$$= W_R + W_D. \tag{4.71}$$

When $\eta = 0$, $W_R$ again gives the previous result, while $W_D$ vanishes. If, however, *before* letting $\eta \to 0$ in order to describe a very weak coupling to an external bath, we take the limit of an *infinite chain*, which represents the *thermodynamic limit* in this case, then we can approximate as well as we want the sums by integrals and thus obtain, for $W_D$, the *nonzero* result

$$W_D \approx \pi\mu f_0^2 T \lim_{\eta \to 0} \int_0^\infty \rho_0(\omega) \frac{\eta/\pi}{\omega^2 + \eta^2}\, d\omega$$
$$= \left[\pi\mu f_0^2 \rho_0(0)\right] T. \tag{4.72}$$

This is the *dissipative*, or *resistive, part*, that grows linearly with the time $T$, while $W_R$ becomes the *reactive part*. In the above equation, $\rho_0(\omega)$ is the local density of modes, i.e., the density $\rho(\omega)$ modified by the presence of the overlap weight factor $O_{0i}^2$. For a linear chain with nearest neighbor couplings, we have

$$\rho(\omega) = \frac{2N/\pi}{\sqrt{(2\omega_0)^2 - \omega^2}},$$

while $O_{0i}^2 = 1/N$, so that $\rho_0(0) = 1/\pi\omega_0$.

It is important to clarify here once again the role of $\eta$ as a purely mathematical device to suppress a recurrence which is quite unphysical for any real system having a quasi-continuous excitation spectrum. The point is that any such system will evolve, for all practical purposes, irreversibly, due to the appearance of a large number of frequency components, incommensurate in general, as can be seen explicitly in the expression (4.66), for a physical quantity of interest. However, this irreversibility will be cut off on a time-scale set by the smallest energy-level spacing $\sim \hbar/\Delta$, which, as we have already noted, is very long. Beyond this cut-off there will be persistent recurrences. Now, almost any inelastic coupling to the environment (bath), however weak, will smear out this cut-off, eliminating the unphysical recurrences mentioned above. It is this latter physical process which is implemented by the mathematical device $\eta$. We emphasize here that $\eta$ itself never enters into any physically relevant relaxation time-scale, which is entirely determined by the many-body Hamiltonian of the system. Finally, one can also relate this $\eta$ to the $\eta$ appearing in the scattering theory of Chapters 2 and 3, where, e.g., $\eta^+$ determines the boundary condition giving an 'irreversible' evolution of an incoming wave packet as the time tends to infinity.

We now return to eqn (4.63) and write it in the more compact form

$$\rho_{\text{ind}}(\mathbf{r}, t) = \lim_{\eta \to 0^+} \int_{-\infty}^{+\infty} dt' \int d^3 r' \, \Pi_0(\mathbf{r}, \mathbf{r}'; t, t') \phi_{\text{ext}}(\mathbf{r}', t'), \tag{4.73a}$$

in terms of the kernel

$$\Pi_0(\mathbf{r}, \mathbf{r}'; t, t') = -\frac{i}{\hbar} e^{-\eta(t-t')} \theta(t - t') \langle [\widetilde{\rho}(\mathbf{r}, t), \widetilde{\rho}(\mathbf{r}', t')] \rangle_0 , \tag{4.73b}$$

where the subscript 0 reminds us again that the expectation value on the right-hand side is to be taken with respect to an equilibrium ensemble associated with the Hamiltonian $H_0$. We have also introduced the *induced* electronic charge density

$$\rho_{\text{ind}}(\mathbf{r}, t) = \rho(\mathbf{r}, t) - \rho_0(\mathbf{r}). \tag{4.74}$$

In what follows we shall be interested in the response of our system to an excitation with frequency $\omega$; it is thus useful, at this point, to construct the Fourier transform in time of the various quantities that appear in the above equations. For the potential $\phi_{\text{ext}}(\mathbf{r}, t)$, we define

$$\phi_{\text{ext}}^\omega(\mathbf{r}) = \int_{-\infty}^{+\infty} dt \, e^{i\omega t} \phi_{\text{ext}}(\mathbf{r}, t), \tag{4.75}$$

and, for the induced charge density $\rho_{\text{ind}}(\mathbf{r}, t)$, we define

$$\rho_{\text{ind}}^\omega(\mathbf{r}) = \int_{-\infty}^{+\infty} dt \, e^{i\omega t} \rho_{\text{ind}}(\mathbf{r}, t). \tag{4.76}$$

In equilibrium, the bracket appearing in eqn (4.73b), and thus the whole kernel $\Pi_0(\boldsymbol{r}, \boldsymbol{r}'; t - t')$, depend only on the time difference $\tau = t - t'$. Applying the convolution theorem, we can thus write

$$\rho_{\text{ind}}^\omega(\boldsymbol{r}) = \int \mathrm{d}^3 r' \, \Pi_0^\omega(\boldsymbol{r}, \boldsymbol{r}') \phi_{\text{ext}}^\omega(\boldsymbol{r}'), \tag{4.77a}$$

where we have defined the kernel

$$\Pi_0^\omega(\boldsymbol{r}, \boldsymbol{r}') = -\frac{\mathrm{i}}{\hbar} \lim_{\eta \to 0^+} \int_0^\infty \mathrm{d}\tau \, \mathrm{e}^{\mathrm{i}\omega\tau - \eta\tau} \langle [\widetilde{\rho}(\boldsymbol{r}, \tau), \widetilde{\rho}(\boldsymbol{r}', 0)] \rangle_0. \tag{4.77b}$$

It is assumed here that $\hbar\omega$ lies within the bandwidth of the system in question. Should this not be the case, then $\eta$ will serve no purpose. In practice, $\hbar\omega$ almost always lies within the band.

So far we have concentrated on the electronic charge density. The potential profile $\phi(\boldsymbol{r}, t)$ that is attained in the presence of the *applied field* $\phi_{\text{ext}}(\boldsymbol{r}, t)$ is also of great interest. Clearly, the *total* correction to the equilibrium value of the field consists not only of the externally applied field, but also contains an additional *induced field*, $\phi_{\text{ind}}(\boldsymbol{r}, t)$, unknown a priori. Before attempting its calculation, we make a number of qualitative considerations.

First notice that the application to a system in equilibrium (with charge density $\rho_0(\boldsymbol{r})$ and potential $\phi_0(\boldsymbol{r})$) of an external, global potential $V$, constant everywhere in space and in time, would have no physical effect on the problem; the system would still be in the same state of equilibrium and the resulting charge density, potential and current density would be

$$\rho(\boldsymbol{r}) = \rho_0(\boldsymbol{r}), \tag{4.78a}$$

$$\phi(\boldsymbol{r}) = \phi_0(\boldsymbol{r}) + V, \tag{4.78b}$$

$$\boldsymbol{J}(\boldsymbol{r}) \equiv 0. \tag{4.78c}$$

In the problem of interest here the applied external potential is *not* constant in space. However, far out in the horns it is constant over a very large region. The expanding horns may be considered to be in *local equilibrium*, even when a finite total current flows in response to the application of the potential difference (4.32); in fact, the cross-sectional area of the horns increases indefinitely as we go further away from the constriction and thus the local current density $\boldsymbol{J}(\boldsymbol{r})$ eventually tends to zero. Thus, sufficiently far inside the horns the system remains practically undisturbed by the external potential; in those asymptotic regions we expect, to a good approximation (and for very low frequencies), that

$$\rho(\boldsymbol{r}) \approx \rho_0(\boldsymbol{r}), \tag{4.79a}$$

$$\phi(\boldsymbol{r}) \approx \phi_0(\boldsymbol{r}) + \begin{cases} V_1(t) \\ V_2(t) \end{cases}, \tag{4.79b}$$

$$\boldsymbol{J}(\boldsymbol{r}) \approx 0, \tag{4.79c}$$

where $V_1(t)$ and $V_2(t)$ denote the asymptotic values of the external potential very far away on the left and on the right, respectively. Of course, these last equations do not apply in the region of the constriction, where a finite $J(r,t)$ exists and induced potentials and a redistribution of charges are expected to occur. They would not apply anywhere either if our system had a finite constant cross-section instead of going asymptotically to widening horns, because then we would have a nonzero $J(r,t)$ everywhere.

We finally note that, inasmuch as the current density is essentially nonzero only in the region of the constriction, i.e., the sample, the total potential drop occurs mainly across the sample. This potential difference is approximately the external emf if voltage drops in other parts of the system are disregarded.

The geometry of a constriction terminating in the gradually expanding horns at the two ends, as depicted in Fig. 4.2 on p. 217, is an idealization, quite appropriate to the experimental situation for a two-probe conductance measurement on a mesoscopic sample connected to massive terminals through the probe leads. The horns act asymptotically as ideal *charge reservoirs*, specified only and completely by the two parameters $\beta$ and $\mu$, in that (i) they stay in equilibrium despite a nonzero current passing through them, inasmuch as the current density itself becomes vanishingly small, and (ii) they do not reflect an incoming electron wave as the horns expand adiabatically on the length-scale set by the de Broglie wavelength. Moreover, as the potential drops almost entirely across the constriction (the sample), one measures essentially the two-probe sample-specific conductance, which is calculable—the horns finally drop out as mere scaffolding. It is apt to point out, as mentioned above, that replacing the horns by some waveguide leads of uniform cross-section and infinite length (or of a finite length but terminating abruptly in a massive contact) will not do—it will suffer terminally from the leads not being in thermodynamic equilibrium, because of the nonzero current density (or from the reflection at the lead–terminal interface). This will make the results of calculations and measurements dependent on the details of the probe arrangement.

A comparison of the above *mesoscopic* system analysis with that for a *macroscopic* system seems in order here. One may ask, for instance, where does the dissipation really occur? Now, in a mesoscopic system, the transport through the sample, the constriction, is supposed to be *coherent*, with elastic scattering only—it can even be ballistic—and the dissipation takes place entirely in the charge reservoirs at the two far ends, where the energy, in the amount of the chemical potential difference per electron, is finally dumped. In contrast to this, in the usual quantum transport through a macroscopic sample there is dissipation throughout the bulk, involving inelastic interaction with phonons, say, that act as the thermal bath. In this case, the role of the terminal leads and scattering at the contacts is relatively negligible, *vis-à-vis* that of the scattering in the bulk. And so it happens that for a mesoscopic system one does measure and talk meaningfully of a sample-specific conductance rather than any material-specific conductivity, while for the usual macroscopic systems one usually measures a material-specific conductivity.

We now return to eqns (4.63) and (4.73) which give the charge density $\rho(r,t)$

in terms of the *external* potential $\phi_{\text{ext}}(\boldsymbol{r},t)$. We have already remarked that the kernel appearing in eqn (4.73) is calculated with a thermal ensemble obtained from the unperturbed Hamiltonian $H_0$ of eqn (4.1). However, $H_0$ contains the electron–electron interaction $V_{\text{e-e}}$, whose two-body nature makes an exact solution unattainable; $V_{\text{e-e}}$ can only be treated within an approximation scheme, which we now develop.

Once the external field is turned on, the system is governed by the total Hamiltonian $H$ of eqn (4.31), i.e.,

$$H = T + V_{eI} + V_{e,\text{imp}} + V_{\text{e-e}} - e\sum_i \phi_{\text{ext}}(\boldsymbol{r}_i, t). \tag{4.80}$$

Due to the electron–electron interaction

$$V_{\text{e-e}} = \sum_{i<j} \frac{e^2}{|\boldsymbol{r}_i - \boldsymbol{r}_j|}, \tag{4.81}$$

the $i$th electron interacts with all the other electrons in the system. We simplify the description and adopt the point of view that what the $i$th electron 'sees' is, approximately, an 'electron cloud' whose density is $\rho(\boldsymbol{r},t)$ (one of our unknown quantities); within this approximate conception each electron interacts with the average electron density of the rest of the electrons rather than with each electron individually. This corresponds to a mean-field (or Hartree) approximation and is known in the literature as the *time-dependent Hartree approximation* or, for historical reasons, the *random-phase approximation* (RPA) (see, e.g., [63, Section 6.5] and [106,119,122]), and has a well-known diagrammatic formulation in many-body theory.

With this idea in mind, the *two-body* operator $V_{\text{e-e}}$ is replaced, approximately, by the *one-body* operator

$$V_{\text{e-e}}^{(1)} = -e\sum_i \int \frac{\rho(\boldsymbol{r}',t)}{|\boldsymbol{r}_i - \boldsymbol{r}'|} \, d^3\boldsymbol{r}'. \tag{4.82}$$

We now separate $\rho(\boldsymbol{r},t)$ into two contributions as follows:

$$\rho(\boldsymbol{r},t) = \rho_{\text{sc}}(\boldsymbol{r}) + \delta\rho(\boldsymbol{r},t), \tag{4.83}$$

the first term representing the electronic charge density (in a Hartree, self-consistent (sc) approximation, whose calculation is indicated below) in the *absence* of the external field, and the second term representing the time-dependent electronic charge density induced by the external field. Accordingly, $V_{\text{e-e}}^{(1)}$ can be written as

$$\begin{aligned}V_{\text{e-e}}^{(1)} &= -e\sum_i \int \frac{\rho_{\text{sc}}(\boldsymbol{r}')}{|\boldsymbol{r}_i - \boldsymbol{r}'|} \, d^3\boldsymbol{r}' - e\sum_i \int \frac{\delta\rho(\boldsymbol{r}',t)}{|\boldsymbol{r}_i - \boldsymbol{r}'|} \, d^3\boldsymbol{r}' \\ &= -e\sum_i \phi_{\text{sc}}(\boldsymbol{r}_i) - e\sum_i \phi_{\text{ind}}(\boldsymbol{r}_i, t).\end{aligned} \tag{4.84}$$

Here, $\phi_{\rm sc}(\boldsymbol{r})$ is the potential produced by the electronic charge density $\rho_{\rm sc}(\boldsymbol{r})$, while $\phi_{\rm ind}(\boldsymbol{r},t)$ represents the potential generated by the induced charges $\delta\rho(\boldsymbol{r},t)$. We thus write, approximately, the total Hamiltonian $H$ of eqn (4.80) as the one-body operator [19]

$$H^{(1)} = \left[T + V_{eI} + V_{e,\rm imp} - e\sum_i \phi_{\rm sc}(\boldsymbol{r}_i)\right] - e\left[\sum_i \phi_{\rm ext}(\boldsymbol{r}_i,t) + \sum_i \phi_{\rm ind}(\boldsymbol{r}_i,t)\right]$$

$$= \left[T - e\sum_i \phi_{\rm sc}^{\rm tot}(\boldsymbol{r}_i)\right] - e\sum_i \delta\phi(\boldsymbol{r}_i,t)$$

$$= H_{\rm sc} - e\sum_i \delta\phi(\boldsymbol{r}_i,t), \tag{4.85}$$

where

$$\phi_{\rm sc}^{\rm tot}(\boldsymbol{r}) = \phi_{eI}(\boldsymbol{r}) + \phi_{e,\rm imp}(\boldsymbol{r}) + \phi_{\rm sc}(\boldsymbol{r}) \tag{4.86}$$

is the *total self-consistent field* felt by an electron (in the absence of the external field), consisting of the electron ions potential and the electron impurities potential (the $\sum_I$ and $\sum_j$ inside the square bracket in eqn (4.1)), plus the self-consistent contribution arising from the electron–electron interaction. On the other hand,

$$\delta\phi(\boldsymbol{r},t) = \phi_{\rm ext}(\boldsymbol{r},t) + \phi_{\rm ind}(\boldsymbol{r},t) \tag{4.87}$$

represents the total correction arising from the applied field, which consists of the applied field plus the field generated by the induced charges.

The idea now is to commence with the model Hamiltonian $H_{\rm sc}$ of eqn (4.85) and perform perturbation theory on the second term of that equation [19]. Notice that $H_{\rm sc}$ is not known and must be calculated self-consistently, using the fact that $\phi_{\rm sc}(\boldsymbol{r})$ is the expectation value of (4.6) with respect to a thermodynamic ensemble defined with $H_{\rm sc}$ itself, i.e.,

$$\phi_{\rm sc}(\boldsymbol{r}) = \left\langle -\sum_i \frac{e}{|\boldsymbol{r}-\boldsymbol{r}_i|}\right\rangle_{\rm sc}, \tag{4.88}$$

where

$$\left\langle \hat{O}\right\rangle_{\rm sc} = \frac{1}{\mathcal{Z}_{\rm sc}} {\rm Tr}\left[\hat{O}e^{-\beta(\hat{H}_{\rm sc}-\mu\hat{N})}\right] \tag{4.89}$$

and the grand partition function $\mathcal{Z}_{\rm sc}$ is given by

$$\mathcal{Z}_{\rm sc} = {\rm Tr}\left[e^{-\beta(\hat{H}_{\rm sc}-\mu\hat{N})}\right]. \tag{4.90}$$

Treating the second term in the last line of eqn (4.85) in LRT (which is permitted, as $\phi_{\rm ind}(\boldsymbol{r},t)$ is linear in $\phi_{\rm ext}(\boldsymbol{r},t)$) we obtain, for $\delta\rho(\boldsymbol{r},t)$ of eqn (4.83), the expression

$$\delta\rho(\bm{r},t) = \lim_{\eta\to 0^+} \int_{-\infty}^{+\infty} \mathrm{d}t' \int \mathrm{d}^3 r'\, \Pi_{\mathrm{sc}}(\bm{r},\bm{r}';t,t')\delta\phi(\bm{r}',t'), \tag{4.91a}$$

in terms of the kernel

$$\Pi_{\mathrm{sc}}(\bm{r},\bm{r}';t,t') = -\frac{\mathrm{i}}{\hbar}\mathrm{e}^{-\eta(t-t')}\theta(t-t')\,\langle[\widetilde{\rho}(\bm{r},t),\widetilde{\rho}(\bm{r}',t')]\rangle_{\mathrm{sc}}. \tag{4.91b}$$

In the following, except where there is a possibility of confusion, we shall drop the subscript 'sc' in this kernel, with the understanding that the expectation value occurring in it is to be taken with respect to an equilibrium ensemble associated with the Hamiltonian $H_{\mathrm{sc}}$. Also, the operators in the interaction representation are defined as in eqn (4.46), with $H_0$ replaced by $H_{\mathrm{sc}}$, i.e.,

$$\widetilde{\mathcal{O}}(t) = \mathrm{e}^{(\mathrm{i}/\hbar)H_{\mathrm{sc}}t}\hat{\mathcal{O}}\mathrm{e}^{-(\mathrm{i}/\hbar)H_{\mathrm{sc}}t}. \tag{4.92}$$

Equation (4.91a) contains the two unknowns $\delta\rho(\bm{r},t)$ and $\delta\phi(\bm{r},t)$. It must be complemented with Poisson's equation (according to eqn (4.87), $\delta\phi(\bm{r},t)$ contains the external field in addition to the induced one; we assume $\nabla^2\phi_{\mathrm{ext}}(\bm{r},t) = 0$, eqn (4.33))

$$\nabla^2\delta\phi(\bm{r},t) = -4\pi\delta\rho(\bm{r},t), \tag{4.93a}$$

subject to the boundary conditions

$$\delta\phi(\mp\infty,t) = \begin{cases} V_1(t) \\ V_2(t) \end{cases}. \tag{4.93b}$$

(See Fig. 4.2 on p. 217.)

We take a time Fourier transform of both sides of eqn (4.91) (with definitions similar to those of eqns (4.75) and (4.76)) and use the convolution theorem to find

$$\delta\rho^\omega(\bm{r}) = \int \mathrm{d}^3 r'\, \Pi^\omega(\bm{r},\bm{r}')\delta\phi^\omega(\bm{r}'), \tag{4.94a}$$

where we have defined the kernel

$$\Pi^\omega(\bm{r},\bm{r}') = -\frac{\mathrm{i}}{\hbar}\lim_{\eta\to 0^+}\int_0^\infty \mathrm{d}\tau\, \mathrm{e}^{\mathrm{i}\omega\tau-\eta\tau}\langle[\widetilde{\rho}(\bm{r},\tau),\widetilde{\rho}(\bm{r}',0)]\rangle_{\mathrm{sc}}, \tag{4.94b}$$

which depends entirely on the properties of the system itself and not on the applied field.

Similarly, the Fourier transform of Poisson's equation (4.93a) gives

$$\nabla^2\delta\phi^\omega(\bm{r}) = -4\pi\delta\rho^\omega(\bm{r}), \tag{4.95a}$$

with the boundary conditions

$$\delta\phi^\omega(\mp\infty) = \begin{cases} V_1^\omega(t) \\ V_2^\omega(t) \end{cases}. \tag{4.95b}$$

Combining eqns (4.94a) and (4.95a), we obtain the integro-differential equation

$$\nabla^2 \delta\phi^\omega(\boldsymbol{r}) = -4\pi \int \mathrm{d}^3 r'\, \Pi^\omega(\boldsymbol{r}, \boldsymbol{r}') \delta\phi^\omega(\boldsymbol{r}'), \tag{4.96}$$

which has to be solved with the boundary conditions (4.95b). Lateral boundary conditions do not have to be imposed additionally in a Hamiltonian formulation with electrons and ions, where confining potentials arise automatically, and $J_n = 0$ would be satisfied automatically.

An application of the above ideas to a simple example is given in the exercise below [160].

**Exercise 4.4** An external static charge $Q$ is introduced in a metallic medium of infinite extent in all directions. Find, in LRT and in the RPA, the induced electronic charge density and the potential of the charge screened by the induced charges.

If the external charge is introduced at the origin of a system of coordinates, the time-independent external potential $\phi_{\text{ext}}(\boldsymbol{r})$ is

$$\phi_{\text{ext}}(\boldsymbol{r}) = \frac{Q}{r}. \tag{4.97}$$

The induced charge density is $\rho_{\text{ind}}(\boldsymbol{r}) = \rho(\boldsymbol{r}) - \rho_0(\boldsymbol{r})$, where $\rho_0(\boldsymbol{r})$ and $\rho(\boldsymbol{r})$ are, respectively, the electronic charge density before and after the introduction of the external charge $Q$. From eqns (4.73) we have

$$\rho_{\text{ind}}(\boldsymbol{r}) = \int \Pi_0^0(\boldsymbol{r}, \boldsymbol{r}') \phi_{\text{ext}}(\boldsymbol{r}')\, \mathrm{d}^3 r', \tag{4.98a}$$

where the kernel $\Pi_0^0(\boldsymbol{r}, \boldsymbol{r}')$ is given by eqn (4.77b) for $\omega = 0$ (indicated by the superscript '0'), i.e.,

$$\Pi_0^0(\boldsymbol{r}, \boldsymbol{r}') = -\frac{\mathrm{i}}{\hbar} \lim_{\eta \to 0^+} \int_0^\infty \mathrm{d}\tau\, \mathrm{e}^{-\eta\tau} \langle [\widetilde{\rho}(\boldsymbol{r}, \tau), \widetilde{\rho}(\boldsymbol{r}', 0)] \rangle_0; \tag{4.98b}$$

the subscript '0' indicates, as usual, that expectation values and the interaction representation are defined with respect to the Hamiltonian $H_0$. In RPA we write

$$\delta\rho(\boldsymbol{r}) = \int \mathrm{d}^3 r'\, \Pi_{\text{sc}}^0(\boldsymbol{r}, \boldsymbol{r}') \delta\phi(\boldsymbol{r}'), \tag{4.99a}$$

where

$$\Pi_{\text{sc}}^0(\boldsymbol{r}, \boldsymbol{r}') = -\frac{\mathrm{i}}{\hbar} \lim_{\eta \to 0^+} \int_0^\infty \mathrm{d}\tau\, \mathrm{e}^{-\eta\tau} \langle [\widetilde{\rho}(\boldsymbol{r}, \tau), \widetilde{\rho}(\boldsymbol{r}', 0)] \rangle_{\text{sc}}, \tag{4.99b}$$

$\delta\rho(\boldsymbol{r})$ and $\delta\phi(\boldsymbol{r})$ are defined as in eqns (4.83) and (4.87), respectively, without the time dependence, and $\Pi_{\text{sc}}^0(\boldsymbol{r}, \boldsymbol{r}')$ is given in eqn (4.94b) with $\omega = 0$; i.e., $\delta\phi(\boldsymbol{r})$ now contains the induced field in addition to the external one. The expectation value in eqn (4.99b) is evaluated, as explained above, using $H_{\text{sc}}$ of eqns (4.85) and (4.86). The above equation has to be supplemented with Poisson's equation

$$\nabla^2 \delta\phi(\boldsymbol{r}) = -4\pi \delta\rho(\boldsymbol{r}) - 4\pi Q \delta(\boldsymbol{r}), \tag{4.99c}$$

the boundary condition now being $\delta\phi(\boldsymbol{r}) = 0$ at infinity.

For a homogeneous, i.e., translationally invariant, medium, the kernel in eqn (4.99) depends only on the difference of position vectors, i.e., $\Pi_{\text{sc}}^0(\bm{r},\bm{r}') = \Pi_{\text{sc}}^0(\bm{r}-\bm{r}')$. In this case, taking the spatial Fourier transform, with $\bm{q}$ being the conjugate variable, and using the convolution theorem, the two equations above become

$$\delta\rho(\bm{q}) = \Pi_{\text{sc}}^0(\bm{q})\delta\phi(\bm{q}), \tag{4.100a}$$

$$q^2 \delta\phi(\bm{q}) = 4\pi\delta\rho(\bm{q}) + 4\pi Q. \tag{4.100b}$$

We can solve these two equations for $\delta\phi(\bm{q})$ and $\delta\rho(\bm{q})$, with the result

$$\delta\phi(\bm{q}) = \frac{4\pi Q}{q^2 - 4\pi\Pi_{\text{sc}}^0(\bm{q})}, \tag{4.101a}$$

$$\delta\rho(\bm{q}) = \frac{\Pi_{\text{sc}}^0(\bm{q})}{q^2 - 4\pi\Pi_{\text{sc}}^0(\bm{q})} 4\pi Q. \tag{4.101b}$$

The result for $\delta\phi(\bm{q})$ is found to coincide with the 'effective potential' obtained in RPA in [63, pp. 148–9], if the quantity $\mathcal{P}_0^R(\bm{q},\omega=0)$ defined there is identified with our $\Pi_{\text{sc}}^0(\bm{q})$. In particular, in the long-wavelength limit $q \to 0$, one finds the result in the Thomas–Fermi approximation, i.e.,

$$\delta\phi(\bm{q}) \approx \frac{4\pi Q}{q^2 + k_s^2}, \tag{4.102}$$

whose inverse Fourier transform is $\delta\phi(\bm{r}) = \mathrm{e}^{-k_s r}/r$, with $1/k_s$ being the screening length, where $k_s = \sqrt{4me^2 k_F/\pi\hbar^2}$.

### 4.3.2 The current density

The current density operator associated with the total Hamiltonian $H$ of eqn (4.31) is the operator $\hat{\bm{j}}(\bm{r})$ of eqn (4.10), which we choose as the operator $\hat{O}$ in eqn (4.48). The perturbation is again given by eqns (4.61) and (4.62). The current density in equilibrium, eqn (4.11), vanishes and eqn (4.48) gives

$$J_\alpha(\bm{r},t) = \left\langle \hat{j}_\alpha(\bm{r}) \right\rangle_t$$

$$= \lim_{\eta \to 0^+} \int_{-\infty}^{+\infty} \mathrm{d}t' \int \mathrm{d}^3 r' \, \Gamma_{0,\alpha}(\bm{r},\bm{r}'; t-t')\phi_{\text{ext}}(\bm{r}',t'), \tag{4.103a}$$

where the kernel $\Gamma_{0,\alpha}(\bm{r},\bm{r}';t-t')$ is given by

$$\Gamma_{0,\alpha}(\bm{r},\bm{r}';t-t') = -\frac{\mathrm{i}}{\hbar}\mathrm{e}^{-\eta(t-t')}\theta(t-t')\left\langle \left[\tilde{j}_\alpha(\bm{r},t),\tilde{\rho}(\bm{r}',t')\right]\right\rangle_0. \tag{4.103b}$$

Just as in the calculation of the charge density, the expectation value appearing in eqn (4.103b) has to be calculated with a thermal ensemble arising from the Hamiltonian $H_0$, eqn (4.1).

We follow the same philosophy as above and restrict the analysis to the RPA. We find

$$J_\alpha(\boldsymbol{r},t) = \lim_{\eta\to 0^+} \int_{-\infty}^{+\infty} dt' \int d^3r'\, \Gamma_\alpha(\boldsymbol{r},\boldsymbol{r}';t-t')\delta\phi(\boldsymbol{r}',t'), \qquad (4.104a)$$

where the kernel

$$\Gamma_\alpha(\boldsymbol{r},\boldsymbol{r}';t-t') = -\frac{i}{\hbar} e^{-\eta(t-t')} \theta(t-t') \left\langle \left[\widetilde{j}_\alpha(\boldsymbol{r},t), \widetilde{\rho}(\boldsymbol{r}',t')\right]\right\rangle_{\text{sc}} \qquad (4.104b)$$

is to be calculated in terms of the self-consistent Hamiltonian $H_{\text{sc}}$ of eqn (4.85). For the $\omega$ component of $J_\alpha(\boldsymbol{r},t)$, i.e.,

$$J_\alpha^\omega(\boldsymbol{r}) = \int_{-\infty}^{+\infty} dt\, e^{i\omega t} J_\alpha(\boldsymbol{r},t), \qquad (4.105)$$

we find

$$J_\alpha^\omega(\boldsymbol{r}) = \int d^3r'\, \Gamma_\alpha^\omega(\boldsymbol{r},\boldsymbol{r}')\, \delta\phi^\omega(\boldsymbol{r}'), \qquad (4.106a)$$

where the kernel $\Gamma_\alpha^\omega(\boldsymbol{r},\boldsymbol{r}')$ is given by

$$\Gamma_\alpha^\omega(\boldsymbol{r},\boldsymbol{r}') = -\frac{i}{\hbar} \lim_{\eta\to 0^+} \int_0^\infty d\tau\, e^{i\omega\tau - \eta\tau} \left\langle \left[\widetilde{j}_\alpha(\boldsymbol{r},\tau), \widetilde{\rho}(\boldsymbol{r}',0)\right]\right\rangle_{\text{sc}}. \qquad (4.106b)$$

Thus, once eqns (4.94) and (4.95) have been solved for $\delta\phi^\omega(\boldsymbol{r})$ and $\delta\rho^\omega(\boldsymbol{r})$, $\delta\phi^\omega(\boldsymbol{r})$ can be inserted into eqn (4.106a) to find the current density $J_\alpha^\omega(\boldsymbol{r})$.

## 4.4 The external field in the vector potential gauge

In this section we repeat the above calculation for the current, choosing the vector potential gauge described in Section 4.2. The final result should certainly be the same as the one found in the previous section where we used the scalar potential gauge.

The total Hamiltonian, eqn (4.14), is now

$$H = \sum_i \left\{ \frac{1}{2m}\left[\boldsymbol{p}_i + \frac{e}{c}\boldsymbol{A}'_{\text{ext}}(\boldsymbol{r}_i,t)\right]^2 - e\sum_I \frac{e_I}{|\boldsymbol{r}_i - \boldsymbol{R}_I|} - e\sum_j \frac{e_j}{|\boldsymbol{r}_i - \boldsymbol{R}_j|} \right\} + \sum_{i<j} \frac{e^2}{|\boldsymbol{r}_i - \boldsymbol{r}_j|}, \qquad (4.107)$$

where $\boldsymbol{A}'_{\text{ext}}(\boldsymbol{r}_i,t)$ is given in eqn (4.22).

The current density operator associated with the Hamiltonian (4.107) is

$$\hat{J}(r,t) = -\frac{e}{2m}\sum_i \left\{ \delta(r-r_i)\left[p_i + \frac{e}{c}A'_{\text{ext}}(r_i,t)\right] \right.$$
$$\left. + \left[p_i + \frac{e}{c}A'_{\text{ext}}(r_i,t)\right]\delta(r-r_i) \right\}$$
$$= \hat{j}(r) - \frac{e^2}{mc}\hat{n}(r)A'_{\text{ext}}(r,t), \tag{4.108}$$

where $\hat{j}(r)$ is defined in eqn (4.10).

From eqn (4.107) we see that, to first order in $A'(r_i,t)$, the perturbation is now

$$H'(t) = \sum_i \frac{e}{2mc}\left[p_i \cdot A'_{\text{ext}}(r_i,t) + A'_{\text{ext}}(r_i,t)\cdot p_i\right], \tag{4.109}$$

which can be rewritten in terms of $\hat{j}(r)$ as

$$H'(t) = -\frac{1}{c}\int \hat{j}(r)\cdot A'_{\text{ext}}(r,t)\,\mathrm{d}^3 r \tag{4.110}$$

in the Schrödinger picture, and as

$$\widetilde{H}'(t) = -\frac{1}{c}\int \widetilde{j}(r,t)\cdot A'_{\text{ext}}(r,t)\,\mathrm{d}^3 r \tag{4.111}$$

in the interaction picture.

Expressions (4.108) and (4.111) must now be substituted into eqn (4.48) of Exercise 4.1 as $\hat{O}$ and $\widetilde{H}'(t)$, respectively, in order to find the current in LRT, with the result that (recall that the expectation value of the operator $\hat{j}_\alpha(r)$ vanishes in equilibrium)

$$J_\alpha(r,t) = \left\langle \hat{J}_\alpha(r,t) \right\rangle_t$$
$$= -\frac{e^2}{mc}\langle \hat{n}(r)\rangle_0 A'_{\text{ext},\alpha}(r,t) + \frac{i}{\hbar}\lim_{\eta\to 0^+}\int_{-\infty}^{+\infty}\mathrm{d}t'\, e^{-\eta(t-t')}\theta(t-t')$$
$$\times \int \mathrm{d}^3 r' \left\langle \left[\widetilde{J}_\alpha(r,t),\widetilde{j}_\beta(r',t')\right]\right\rangle_0 A'_{\text{ext},\beta}(r',t'). \tag{4.112}$$

We have used and shall use hereinafter the Einstein convention of summation over repeated indices. The expectation value on the last line of eqn (4.112) can be written as

$$\left\langle \left[\widetilde{J}_\alpha(r,t),\widetilde{j}_\beta(r',t')\right]\right\rangle_0 = \left\langle \left[\widetilde{j}_\alpha(r,t) - \frac{e^2}{mc}\widetilde{n}(r,t)A'_{\text{ext}}(r,t),\widetilde{j}_\beta(r',t')\right]\right\rangle_0$$
$$\approx \left\langle \left[\widetilde{j}_\alpha(r,t),\widetilde{j}_\beta(r',t')\right]\right\rangle_0, \tag{4.113}$$

to zeroth order in $A'_{\text{ext}}(r,t)$, so as to give a linear response in eqn (4.112). We can thus write the current $J_\alpha(r,t)$, eqn (4.112), as

$$J_\alpha(r,t) = \lim_{\eta\to 0^+} \int_{-\infty}^{+\infty} dt' \int d^3r' \, R_{0,\alpha\beta}(r,r';t-t') A'_{\text{ext},\beta}(r',t'), \quad (4.114a)$$

where we have introduced the kernel

$$R_{0,\alpha\beta}(r,r';t-t') = \frac{i}{\hbar c} e^{-\eta(t-t')} \theta(t-t') \left\langle \left[ \tilde{j}_\alpha(r,t), \tilde{j}_\beta(r',t') \right] \right\rangle_0$$
$$- \frac{e^2}{mc} n_0(r) \delta_{\alpha\beta} \delta(r-r') \delta(t-t')$$
$$\equiv \mathcal{R}_{0,\alpha\beta}(r,r';t-t') - \frac{e^2}{mc} n_0(r) \delta_{\alpha\beta} \delta(r-r') \delta(t-t').$$
$$(4.114b)$$

Here

$$n_0(r) = \langle \hat{n}(r) \rangle_0 \quad (4.115)$$

is the equilibrium electron density, $\hat{n}(r)$ having been defined in eqn (4.4).

Proceeding as in the last section, we invoke the RPA. The starting point is the one-body Hamiltonian of eqn (4.85), where $H_{\text{sc}}$ is taken to be the unperturbed Hamiltonian and $-e\sum_i \delta\phi(r_i,t)$ is the perturbation. In this description we thus have the scalar potential $\delta\phi(r,t)$ and a vanishing vector potential $\delta A(r,t)$. If we subject $\delta\phi(r,t)$ and $\delta A(r,t)$ to the gauge transformation (4.15), choosing $\Gamma(r,t)$ as

$$\Gamma(r,t) = c \int^t \delta\phi(r,t') \, dt', \quad (4.116)$$

then

$$\delta\phi(r,t) \Longrightarrow \delta\phi'(r,t) = 0, \quad (4.117a)$$

$$\delta A(r,t) \Longrightarrow \delta A'(r,t) = c\nabla \int^t \delta\phi(r,t) \, dt. \quad (4.117b)$$

In this gauge, $H^{(1)}$ of eqn (4.85) takes the form

$$H^{(1)} = \frac{1}{2m} \sum_i \left[ p_i + \frac{e}{c} \delta A'(r_i,t) \right]^2 + V_{eI} + V_{e,\text{imp}} - e\sum_i \phi_{\text{sc}}(r_i), \quad (4.118a)$$

which we split as

$$H^{(1)} = H_{\text{sc}} + H', \quad (4.118b)$$

where

$$H_{\text{sc}} = \sum_i \frac{p_i^2}{2m} + V_{eI} + V_{e,\text{imp}} - e\sum_i \phi_{\text{sc}}(r_i), \quad (4.118c)$$

just as in eqn (4.85), and

$$H' = \frac{e}{2mc} \sum_i [\boldsymbol{p}_i \cdot \delta \boldsymbol{A}'(\boldsymbol{r}_i,t) + \delta \boldsymbol{A}'(\boldsymbol{r}_i,t) \cdot \boldsymbol{p}_i], \quad (4.118d)$$

to first order in $\delta \boldsymbol{A}'$. The idea now is to do LRT on $H'$. We thus find, for the current density,

$$J_\alpha(\boldsymbol{r},t) = \lim_{\eta \to 0^+} \int_{-\infty}^{+\infty} dt' \int d^3r' \, R_{\alpha\beta}(\boldsymbol{r},\boldsymbol{r}';t-t') \delta A'_\beta(\boldsymbol{r}',t'), \quad (4.119a)$$

where we have defined the *retarded response function*

$$R_{\alpha\beta}(\boldsymbol{r},\boldsymbol{r}';t-t') = \frac{i}{\hbar c} e^{-\eta(t-t')} \theta(t-t') \left\langle \left[ \tilde{j}_\alpha(\boldsymbol{r},t), \tilde{j}_\beta(\boldsymbol{r}',t') \right] \right\rangle_{\text{sc}}$$

$$- \frac{e^2}{mc} n_{\text{sc}}(\boldsymbol{r}) \delta_{\alpha\beta} \delta(\boldsymbol{r}-\boldsymbol{r}') \delta(t-t')$$

$$\equiv \mathcal{R}_{\alpha\beta}(\boldsymbol{r},\boldsymbol{r}';t-t') - \frac{e^2}{mc} n_{\text{sc}}(\boldsymbol{r}) \delta_{\alpha\beta} \delta(\boldsymbol{r}-\boldsymbol{r}') \delta(t-t'),$$
$$(4.119b)$$

and $n_{\text{sc}}$ is the self-consistent electron density in equilibrium.

We take the Fourier transform of both sides of eqn (4.119), and use eqns (4.105) and (see eqns (4.117b) and (4.30))

$$\delta \boldsymbol{A}'^\omega(\boldsymbol{r}) = -\frac{c}{i\omega} \nabla \delta \phi^\omega(\boldsymbol{r}) \quad (4.120)$$

to obtain

$$J_\alpha^\omega(\boldsymbol{r}) = -\frac{c}{i\omega} \int d^3r' \, R_{\alpha\beta}^\omega(\boldsymbol{r},\boldsymbol{r}') \partial'_\beta \delta\phi^\omega(\boldsymbol{r}'), \quad (4.121a)$$

where we have defined the kernel

$$R_{\alpha\beta}^\omega(\boldsymbol{r},\boldsymbol{r}') = \frac{i}{\hbar c} \lim_{\eta \to 0^+} \int_0^\infty d\tau \, e^{i\omega\tau - \eta\tau} \left\langle \left[ \tilde{j}_\alpha(\boldsymbol{r},\tau), \tilde{j}_\beta(\boldsymbol{r}',0) \right] \right\rangle_{\text{sc}}$$

$$- \frac{e^2 n_{\text{sc}}(\boldsymbol{r})}{mc} \delta(\boldsymbol{r}-\boldsymbol{r}') \delta_{\alpha\beta} \quad (4.121b)$$

$$\equiv \mathcal{R}_{\alpha\beta}^\omega(\boldsymbol{r},\boldsymbol{r}') - \frac{e^2 n_{\text{sc}}(\boldsymbol{r})}{mc} \delta(\boldsymbol{r}-\boldsymbol{r}') \delta_{\alpha\beta}. \quad (4.121c)$$

An important symmetry property of $\mathcal{R}_{\alpha\beta}^\omega(\boldsymbol{r},\boldsymbol{r}')$ is now proved. We use the identity

$$\{i \langle [A,B] \rangle\}^* = i \langle [A,B] \rangle, \quad (4.122)$$

which is valid for two Hermitian operators $A$ and $B$. Then the above definition of $\mathcal{R}_{\alpha\beta}^\omega(\boldsymbol{r},\boldsymbol{r}')$ can easily be used to show that

$$\left[ \mathcal{R}_{\alpha\beta}^\omega(\boldsymbol{r},\boldsymbol{r}') \right]^* = \mathcal{R}_{\alpha\beta}^{-\omega}(\boldsymbol{r},\boldsymbol{r}'), \quad (4.123a)$$

or, separating the real and imaginary parts,

$$\operatorname{Re} \mathcal{R}^{\omega}_{\alpha\beta}(\boldsymbol{r},\boldsymbol{r}') = \operatorname{Re} \mathcal{R}^{-\omega}_{\alpha\beta}(\boldsymbol{r},\boldsymbol{r}'), \qquad (4.123\text{b})$$

$$\operatorname{Im} \mathcal{R}^{\omega}_{\alpha\beta}(\boldsymbol{r},\boldsymbol{r}') = -\operatorname{Im} \mathcal{R}^{-\omega}_{\alpha\beta}(\boldsymbol{r},\boldsymbol{r}'). \qquad (4.123\text{c})$$

From eqns (4.121) we identify the *conductivity tensor* $\sigma^{\omega}_{\alpha\beta}(\boldsymbol{r},\boldsymbol{r}')$, which we write as

$$\begin{aligned}\sigma^{\omega}_{\alpha\beta}(\boldsymbol{r},\boldsymbol{r}') &= \frac{c}{\mathrm{i}\omega} R^{\omega}_{\alpha\beta}(\boldsymbol{r},\boldsymbol{r}') \\ &= \frac{1}{\hbar\omega} \lim_{\eta \to 0^{+}} \int_{0}^{\infty} \mathrm{d}\tau\, \mathrm{e}^{\mathrm{i}\omega\tau - \eta\tau} \left\langle \left[ \tilde{j}_{\alpha}(\boldsymbol{r},\tau), \tilde{j}_{\beta}(\boldsymbol{r}',0) \right] \right\rangle_{\mathrm{sc}} \\ &\quad + \mathrm{i} \frac{e^{2} n_{\mathrm{sc}}(\boldsymbol{r})}{m\omega} \delta(\boldsymbol{r}-\boldsymbol{r}')\delta_{\alpha\beta},\end{aligned} \qquad (4.124\text{a})$$

in terms of which the current density takes the ohmic form

$$J^{\omega}_{\alpha}(\boldsymbol{r}) = -\int \mathrm{d}^3 r'\, \sigma^{\omega}_{\alpha\beta}(\boldsymbol{r},\boldsymbol{r}')\, \partial'_{\beta} \delta\phi^{\omega}(\boldsymbol{r}'). \qquad (4.124\text{b})$$

It is to be noted here that the 'internal' conductivity tensor $\sigma^{\omega}_{\alpha\beta}(\boldsymbol{r},\boldsymbol{r}')$ defined through eqn (4.124) in terms of the *total* (as distinct from the external) field is a mesoscopic sample-specific property not amenable to direct measurement.

The solution $\delta\phi^{\omega}(\boldsymbol{r}')$ of eqns (4.94) and (4.95) can be inserted into eqn (4.124b) to find the current density $J^{\omega}_{\alpha}(\boldsymbol{r})$.

Expression (4.114) must be consistent with eqn (4.103) of the previous section on general grounds of gauge invariance. We now show that this consistency also holds within RPA, i.e., eqns (4.124) are consistent with eqns (4.106) of the previous section. Integrating eqn (4.124b) by parts, we find

$$J^{\omega}_{\alpha}(\boldsymbol{r}) = \int \mathrm{d}^3 r'\, \left[\partial'_{\beta}\sigma^{\omega}_{\alpha\beta}(\boldsymbol{r},\boldsymbol{r}')\right]\delta\phi^{\omega}(\boldsymbol{r}') - \oint_{S'} \mathrm{d}S'_{\beta}\, \sigma^{\omega}_{\alpha\beta}(\boldsymbol{r},\boldsymbol{r}')\delta\phi^{\omega}(\boldsymbol{r}'). \qquad (4.125)$$

In the following two exercises we prove the identity

$$\partial'_{\beta}\sigma^{\omega}_{\alpha\beta}(\boldsymbol{r},\boldsymbol{r}') = \Gamma^{\omega}_{\alpha}(\boldsymbol{r},\boldsymbol{r}'), \qquad (4.126)$$

which relates the divergence of the conductivity tensor to the kernel defined in eqn (4.106b).

**Exercise 4.5** As a first step to arrive at the identity (4.126), prove the continuity equation

$$\nabla \cdot \tilde{\boldsymbol{j}}(\boldsymbol{r},t) + \frac{\partial \tilde{\rho}(\boldsymbol{r},t)}{\partial t} = 0. \qquad (4.127)$$

The charge density and the current operators ($i$ labels the conduction electrons), i.e.,

$$\tilde{\rho}(\boldsymbol{r},t) = -\mathrm{e}^{(\mathrm{i}/\hbar)H_0 t} \sum_{i} e\delta(\boldsymbol{r}-\boldsymbol{r}_i)\mathrm{e}^{-(\mathrm{i}/\hbar)H_0 t} \qquad (4.128)$$

and

$$\tilde{\boldsymbol{j}}(\boldsymbol{r},t) = -\mathrm{e}^{(\mathrm{i}/\hbar)H_0 t}\frac{e}{2m}\sum_i \Big[\delta(\boldsymbol{r}-\boldsymbol{r}_i)\boldsymbol{p}_i + \boldsymbol{p}_i\delta(\boldsymbol{r}-\boldsymbol{r}_i)\Big]\mathrm{e}^{-(\mathrm{i}/\hbar)H_0 t}, \tag{4.129}$$

respectively, are said to be in the Heisenberg representation with regards to the Hamiltonian $H_0$, or in the interaction representation with regards to the full Hamiltonian $H = H_0 + H'$. Obviously, the proof carries on if $H_0$ is replaced by $H_{\mathrm{sc}}$ in the RPA approximation.

For the divergence of the current, we find

$$\nabla \cdot \tilde{\boldsymbol{j}}(\boldsymbol{r},t) = -\mathrm{e}^{(\mathrm{i}/\hbar)H_0 t}\frac{e}{2m}\sum_i \Big[\nabla\delta(\boldsymbol{r}-\boldsymbol{r}_i)\cdot\boldsymbol{p}_i + \boldsymbol{p}_i\cdot\nabla\delta(\boldsymbol{r}-\boldsymbol{r}_i)\Big]\mathrm{e}^{-(\mathrm{i}/\hbar)H_0 t}. \tag{4.130}$$

For the time derivative of the charge density we have

$$\partial_t \tilde{\rho}(\boldsymbol{r},t) = -\frac{\mathrm{i}}{\hbar}\mathrm{e}^{(\mathrm{i}/\hbar)H_0 t}e\sum_i \Big[H_0\delta(\boldsymbol{r}-\boldsymbol{r}_i) - \delta(\boldsymbol{r}-\boldsymbol{r}_i)H_0\Big]\mathrm{e}^{-(\mathrm{i}/\hbar)H_0 t}$$

$$= -\frac{\mathrm{i}}{\hbar}\mathrm{e}^{(\mathrm{i}/\hbar)H_0 t}e\sum_i \Big[H_0,\delta(\boldsymbol{r}-\boldsymbol{r}_i)\Big]\mathrm{e}^{-(\mathrm{i}/\hbar)H_0 t}. \tag{4.131}$$

We calculate the commutator in (4.131) to be

$$[H_0, f(\boldsymbol{r}_i)] = \frac{1}{2m}\sum_j [\boldsymbol{p}_j^2, f(\boldsymbol{r}_i)]$$

$$= \frac{1}{2m}\Big\{p_{i\alpha}\big[p_{i\alpha}, f(\boldsymbol{r}_i)\big] + \big[p_{i\alpha}, f(\boldsymbol{r}_i)\big]p_{i\alpha}\Big\}$$

$$= -\frac{\mathrm{i}\hbar}{2m}\Big\{p_{i\alpha}\Big[\frac{\partial}{\partial x_{i\alpha}}, f(\boldsymbol{r}_i)\Big] + \Big[\frac{\partial}{\partial x_{i\alpha}}, f(\boldsymbol{r}_i)\Big]p_{i\alpha}\Big\}$$

$$= -\frac{\mathrm{i}\hbar}{2m}\big[\boldsymbol{p}_i\cdot\nabla_i f(\boldsymbol{r}_i) + \nabla_i f(\boldsymbol{r}_i)\cdot\boldsymbol{p}_i\big], \tag{4.132}$$

which, when substituted into eqn (4.131), gives

$$\partial_t \tilde{\rho}(\boldsymbol{r},t) = -\frac{e}{2m}\mathrm{e}^{(\mathrm{i}/\hbar)H_0 t}\sum_i \Big[\boldsymbol{p}_i\cdot\nabla_i\delta(\boldsymbol{r}-\boldsymbol{r}_i) + \nabla_i\delta(\boldsymbol{r}-\boldsymbol{r}_i)\cdot\boldsymbol{p}_i\Big]\mathrm{e}^{-(\mathrm{i}/\hbar)H_0 t}$$

$$= \frac{e}{2m}\mathrm{e}^{(\mathrm{i}/\hbar)H_0 t}\sum_i \Big[\boldsymbol{p}_i\cdot\nabla\delta(\boldsymbol{r}-\boldsymbol{r}_i) + \nabla\delta(\boldsymbol{r}-\boldsymbol{r}_i)\cdot\boldsymbol{p}_i\Big]\mathrm{e}^{-(\mathrm{i}/\hbar)H_0 t}$$

$$= -\nabla\cdot\tilde{\boldsymbol{j}}(\boldsymbol{r},t), \tag{4.133}$$

the last equality being obtained by comparison with eqn (4.130). We have thus proved the continuity equation (4.127).

**Exercise 4.6** As an application of the continuity equation (4.127), prove the identity (4.126).

We first calculate the divergence of the kernel $R^\omega_{\alpha\beta}(\boldsymbol{r},\boldsymbol{r}')$ of eqn (4.121). We find

$$\partial'_\beta R^\omega_{\alpha\beta}(\boldsymbol{r},\boldsymbol{r}') = \frac{\mathrm{i}}{\hbar c}\int_0^\infty \mathrm{d}\tau\, \mathrm{e}^{\mathrm{i}\omega\tau - \eta\tau}\left\langle\left[\tilde{j}_\alpha(\boldsymbol{r},\tau), \partial'_\beta \tilde{j}_\beta(\boldsymbol{r}',0)\right]\right\rangle - \frac{e^2 n(\boldsymbol{r})}{mc}\partial'_\beta \delta(\boldsymbol{r}-\boldsymbol{r}')\delta_{\alpha\beta}, \tag{4.134}$$

the limit $\eta \to 0^+$ being understood. In the definition of the operators in the interaction representation and of the expectation values, one uses $H_{\mathrm{sc}}$ in the RPA, as in eqn (4.121), and $H_0$ in the exact LRT formulation. For this reason the subscript 'sc', or '0', is omitted from the angled brackets of the previous equation and in the electronic density $n(\boldsymbol{r})$.

Using eqn (4.127), we can write the expectation value under the integral sign as (with $\tau = t - t'$)

$$\left\langle\left[\tilde{j}_\alpha(\boldsymbol{r},\tau), \partial'_\beta \tilde{j}_\beta(\boldsymbol{r}',0)\right]\right\rangle = \left\langle\left[\tilde{j}_\alpha(\boldsymbol{r},t), \partial'_\beta \tilde{j}_\beta(\boldsymbol{r}',t')\right]\right\rangle$$
$$= -\frac{\partial}{\partial t'}\left\langle\left[\tilde{j}_\alpha(\boldsymbol{r},t), \tilde{\rho}(\boldsymbol{r}',t')\right]\right\rangle = \frac{\partial}{\partial \tau}\left\langle\left[\tilde{j}_\alpha(\boldsymbol{r},\tau), \tilde{\rho}(\boldsymbol{r}',0)\right]\right\rangle, \tag{4.135}$$

so that

$$\partial'_\beta R^\omega_{\alpha\beta}(\boldsymbol{r},\boldsymbol{r}') = \frac{\mathrm{i}}{\hbar c}\int_0^\infty \mathrm{d}\tau\, \mathrm{e}^{\mathrm{i}\omega\tau - \eta\tau}\frac{\partial}{\partial \tau}\left\langle\left[\tilde{j}_\alpha(\boldsymbol{r},\tau), \tilde{\rho}(\boldsymbol{r}',0)\right]\right\rangle + \frac{e^2 n_0(\boldsymbol{r})}{mc}\partial_\beta \delta(\boldsymbol{r}-\boldsymbol{r}')\delta_{\alpha\beta}. \tag{4.136}$$

Upon integrating by parts, we have

$$\partial'_\beta R^\omega_{\alpha\beta}(\boldsymbol{r},\boldsymbol{r}') = \frac{\mathrm{i}}{\hbar c}\left[\mathrm{e}^{\mathrm{i}\omega\tau - \eta\tau}\left\langle\left[\tilde{j}_\alpha(\boldsymbol{r},\tau), \tilde{\rho}(\boldsymbol{r}',0)\right]\right\rangle\right]_0^\infty$$
$$- \frac{\mathrm{i}}{\hbar c}\int_0^\infty \mathrm{d}\tau\,(\mathrm{i}\omega - \eta)\mathrm{e}^{\mathrm{i}\omega\tau - \eta\tau}\left\langle\left[\tilde{j}_\alpha(\boldsymbol{r},\tau), \tilde{\rho}(\boldsymbol{r}',0)\right]\right\rangle$$
$$+ \frac{e^2 n_0(\boldsymbol{r})}{mc}\partial_\alpha \delta(\boldsymbol{r}-\boldsymbol{r}') \tag{4.137}$$
$$= -\frac{\mathrm{i}}{\hbar c}\left\langle\left[\tilde{j}_\alpha(\boldsymbol{r},0), \tilde{\rho}(\boldsymbol{r}',0)\right]\right\rangle + \frac{\mathrm{i}\omega - \eta}{c}\Gamma^\omega_\alpha(\boldsymbol{r},\boldsymbol{r}') + \frac{e^2 n(\boldsymbol{r})}{mc}\partial_\alpha \delta(\boldsymbol{r}-\boldsymbol{r}'). \tag{4.138}$$

The commutator on the last line can be shown to give

$$\left[\tilde{j}_\alpha(\boldsymbol{r},0), \tilde{\rho}(\boldsymbol{r}',0)\right] = -\mathrm{i}\hbar\frac{e^2 \hat{n}(\boldsymbol{r})}{m}\partial_\alpha \delta(\boldsymbol{r}-\boldsymbol{r}'). \tag{4.139}$$

In fact,

$$\left[\tilde{j}_\alpha(\boldsymbol{r},0), \tilde{\rho}(\boldsymbol{r}',0)\right] = \left[\hat{j}_\alpha(\boldsymbol{r}), \hat{\rho}(\boldsymbol{r}')\right]$$
$$= \frac{e^2}{2m}\sum_{ij}\left[\delta(\boldsymbol{r}-\boldsymbol{r}_i)p_{i\alpha} + p_{i\alpha}\delta(\boldsymbol{r}-\boldsymbol{r}_i), \delta(\boldsymbol{r}'-\boldsymbol{r}_j)\right]$$
$$= \frac{e^2}{2m}\sum_{i}\left[\delta(\boldsymbol{r}-\boldsymbol{r}_i)p_{i\alpha} + p_{i\alpha}\delta(\boldsymbol{r}-\boldsymbol{r}_i), \delta(\boldsymbol{r}'-\boldsymbol{r}_i)\right]$$

$$= \frac{e^2}{2m}\sum_i\left\{\delta(\bm{r}-\bm{r}_i)\left[p_{i\alpha},\delta(\bm{r}'-\bm{r}_i)\right]+\left[p_{i\alpha},\delta(\bm{r}'-\bm{r}_i)\right]\delta(\bm{r}-\bm{r}_i)\right\}$$

$$= -i\hbar\frac{e^2}{m}\sum_i\delta(\bm{r}-\bm{r}_i)\frac{\partial\delta(\bm{r}'-\bm{r}_i)}{\partial x_{i\alpha}}$$

$$= i\hbar\frac{e^2}{m}\sum_i\delta(\bm{r}-\bm{r}_i)\frac{\partial\delta(\bm{r}'-\bm{r}_i)}{\partial x'_\alpha}$$

$$= i\hbar\frac{e^2}{m}\sum_i\delta(\bm{r}-\bm{r}_i)\frac{\partial\delta(\bm{r}'-\bm{r})}{\partial x'_\alpha}$$

$$= -i\hbar\frac{e^2}{m}\sum_i\delta(\bm{r}-\bm{r}_i)\frac{\partial\delta(\bm{r}'-\bm{r})}{\partial x_\alpha}$$

$$= -i\hbar\frac{e^2\hat{n}(\bm{r})}{m}\partial_\alpha\delta(\bm{r}-\bm{r}'). \tag{4.140}$$

Substituting this result into eqn (4.138), we find

$$\partial'_\beta R^\omega_{\alpha\beta}(\bm{r},\bm{r}') = \frac{i\omega-\eta}{c}\Gamma^\omega_\alpha(\bm{r},\bm{r}') - \frac{e^2}{mc}n(\bm{r})\partial_\alpha\delta(\bm{r}-\bm{r}') + \frac{e^2}{mc}n(\bm{r})\partial_\alpha\delta(\bm{r}-\bm{r}'). \tag{4.141}$$

The last two terms cancel and, in the limit as $\eta \to 0^+$, we find

$$\partial'_\beta R^\omega_{\alpha\beta}(\bm{r},\bm{r}') = \frac{i\omega}{c}\Gamma^\omega_\alpha(\bm{r},\bm{r}'). \tag{4.142}$$

Recalling the expression (4.124a) for the conductivity in terms of the kernel $R^\omega_{\alpha\beta}$, we thus find, as a result, the identity (4.126) which we wanted to prove.

Substituting eqn (4.126) into (4.125), we have

$$J^\omega_\alpha(\bm{r}) = \int d^3r'\,\Gamma^\omega_\alpha(\bm{r},\bm{r}')\delta\phi^\omega(\bm{r}') - \oint_{S'} dS'_\beta\,\sigma^\omega_{\alpha\beta}(\bm{r},\bm{r}')\,\delta\phi^\omega(\bm{r}'). \tag{4.143}$$

The conductivity tensor is evaluated in RPA in eqn (B.37) of Appendix B. It is expressed in terms of the single-particle wave functions of $H_{\text{sc}}$ of eqn (4.85), which, due to the confining potential, decay to zero as we move laterally outside the structure shown in Fig. 4.2. Thus the volume integral in the above eqn (4.143) is effectively restricted to the region indicated in the figure, while the lateral contribution to the surface integral decays to zero as we move the surface of integration away from that structure. There remains the contribution from the two surfaces $S'(R')$ of radius $R'$ located inside the metal, far away on the left and on the right of the structure, in the limit as $R' \to \infty$. Notice that far away inside the horns $\delta\phi^\omega(\bm{r}')$ tends to constant values; the resulting integral of $\sigma^\omega_{\alpha\beta}(\bm{r},\bm{r}')$ then vanishes according to the statement made in [119] that, for fixed $\bm{r}$ and $\bm{r}' \to \infty$, $\sigma^\omega_{\alpha\beta}(\bm{r},\bm{r}')$ tends to zero faster than $(r')^{-2}$ for three dimensions. Thus the surface integral in eqn (4.143) vanishes and the expression (4.143) reduces to (4.106a).

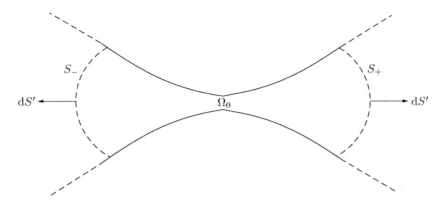

FIG. 4.2. The regions of integration in eqn (4.145). Beyond the surfaces $S_+$ and $S_-$ the potential $\delta\phi^\omega$ takes on constant values. Notice that the surface element $\mathrm{d}S'$ points outwards from the constriction, while $\mathrm{d}S''$ appearing in eqn (4.147) points inwards.

To summarize, once eqns (4.94) and (4.95) have been solved for the potential profile $\delta\phi^\omega(\mathbf{r})$ and the charge density $\delta\rho^\omega(\mathbf{r})$, then $\delta\phi^\omega(\mathbf{r})$ can be inserted into eqn (4.106a), or eqn (4.124b), which we have just shown to be equivalent, in order to find the current density $J_\alpha^\omega(\mathbf{r})$. One could then compute the total current $I^\omega$, and thus the conductance $G^\omega = I^\omega/V^\omega$. This is a formidable problem. A great simplification, however, occurs in the d.c. limit, and we discuss this in what follows.

In fact, important progress can be made due to a result obtained by Kane, Serota and Lee [100], to be referred to as the *KSL theorem*, which states that *in the d.c. limit the conductivity tensor of a TRI system is divergenceless*, i.e.,

$$\lim_{\omega \to 0} \partial'_\beta \sigma^\omega_{\alpha\beta}(\mathbf{r}, \mathbf{r}') = \lim_{\omega \to 0} \Gamma^\omega_\alpha(\mathbf{r}, \mathbf{r}') = 0. \qquad (4.144)$$

The KSL theorem is proved in Appendix A for $\mathrm{Re}\,\sigma^\omega_{\alpha\beta}(\mathbf{r}, \mathbf{r}')$, the imaginary part vanishing in the d.c. limit anyway (see eqn (4.149)). Notice, however, that this result, introduced in eqn (4.106a), would lead to a vanishing d.c. current, which is clearly wrong! The reason for the (apparent) contradiction is that the operation just outlined involves taking both the $\omega \to 0$ limit and the limit of an *infinite domain of integration* (for fixed $\mathbf{r}$). These two limits do not commute [119], so that the transition to zero frequency should be exercised with care. As a mathematical example with similar features, consider the function $f^\omega(x) = \mathrm{e}^{-(\omega/v)x}$, for $0 \leqslant x < \infty$. Its derivative is $g^\omega(x) = \partial_x f^\omega(x) = -(\omega/v)\mathrm{e}^{-(\omega/v)x}$. Suppose that we are interested in the following integral of the function $g^\omega(x)$:

$$J^\omega(x) = \lim_{X \to \infty} \int_0^X g^\omega(x)\,\mathrm{d}x = -1;$$

this result is independent of $\omega$ and is thus unchanged if we take the further limit $\omega \to 0$. On the other hand, we see that $\lim_{\omega \to 0} g^\omega(x) = 0$; had we taken the limit $\omega \to 0$ inside the integrand we would have obtained a vanishing result. It is thus clear, in this example, that the two limits, i.e., $X \to \infty$ and $\omega \to 0$, do not commute.

With these ideas in mind, we return to eqn (4.124b). We stated earlier that, sufficiently far away from the constriction, i.e., at the surfaces $S_-$ and $S_+$ and beyond (see Fig. 4.2), $\delta\phi^\omega(\mathbf{r}')$ tends to the spatially constant values $V_1^\omega$ and $V_2^\omega$, respectively. The integral in eqn (4.124b) can thus be stopped at the surfaces $S_+$ and $S_-$, so that in eqn (4.125) the region of integration for the volume integral can be restricted to that inside the horns bounded by $S_+$ and $S_-$, called $\Omega_0$ in Fig. 4.2, while the surface integral is over the surface consisting of $S_+$, $S_-$ and the lateral boundaries. With the latter integral vanishing, we have

$$J_\alpha^\omega(\mathbf{r}) = \int_{\Omega_0} d^3r'\, \Gamma_\alpha^\omega(\mathbf{r},\mathbf{r}')\delta\phi^\omega(\mathbf{r}') - V_2^\omega \int_{S_+} dS'_\beta\, \sigma_{\alpha\beta}^\omega(\mathbf{r},\mathbf{r}') \\ - V_1^\omega \int_{S_-} dS'_\beta\, \sigma_{\alpha\beta}^\omega(\mathbf{r},\mathbf{r}'). \tag{4.145}$$

Here, the surface element $dS'$ points *outwards* from the constriction. Notice that $\Omega_0$ is a large, but *finite* volume.

**Exercise 4.7** Show that eqn (4.145) can be derived from eqn (4.106a).

The integral in eqn (4.106a) can be split into an integral over $\Omega_0$, that coincides with the one in (4.145), plus an integral over the rest of the space. In the latter integral we use the identity (4.126) and notice that $\delta\phi^\omega(\mathbf{r}')$ can be taken out of the integral sign as $V_2^\omega$ and $V_1^\omega$, in the right and left portions, respectively, i.e.,

$$J_\alpha^\omega(\mathbf{r}) = \int_{\Omega_0} d^3r'\, \Gamma_\alpha^\omega(\mathbf{r},\mathbf{r}')\, \delta\phi^\omega(\mathbf{r}') + V_2^\omega \int_{\Omega_+} d^3r'\, [\partial'_\beta \sigma_{\alpha\beta}^\omega(\mathbf{r},\mathbf{r}')] \\ + V_1^\omega \int_{\Omega_-} d^3r'\, [\partial'_\beta \sigma_{\alpha\beta}^\omega(\mathbf{r},\mathbf{r}')]. \tag{4.146}$$

In this last equation, $\Omega_+$ denotes the volume on the right of $S_+$ and $\Omega_-$ denotes the volume on the left of $S_-$. The last two integrals can now be written as surface integrals, i.e., over $S_+$ and the surface at infinity on the right and over $S_-$ and the surface at infinity on the left, respectively, and over the corresponding lateral surfaces. The integrals over the lateral surfaces and the surfaces at infinity vanish, and we are left with

$$J_\alpha^\omega(\mathbf{r}) = \int_{\Omega_0} d^3r'\, \Gamma_\alpha^\omega(\mathbf{r},\mathbf{r}')\, \delta\phi^\omega(\mathbf{r}') + V_2^\omega \int_{S_+} dS''_\beta\, \sigma_{\alpha\beta}^\omega(\mathbf{r},\mathbf{r}') \\ + V_1^\omega \int_{S_-} dS''_\beta\, \sigma_{\alpha\beta}^\omega(\mathbf{r},\mathbf{r}'). \tag{4.147}$$

Here, the surface element $dS''$ points *inwards* from the constriction. This result thus coincides with (4.145), as stated.

We now consider eqn (4.145) and expand $\Gamma_\alpha^\omega(\mathbf{r},\mathbf{r}')$, *inside* the integral, as a Taylor series in powers of the frequency $\omega$, which is allowed because, as we have already remarked, the volume integral is now realized over the large, but *finite* volume $\Omega_0$. We first expand the conductivity tensor in powers of $\omega$, through an expansion of the quantity $\mathcal{R}_{\alpha\beta}^\omega(\mathbf{r},\mathbf{r}')$ (related to $\sigma_{\alpha\beta}^\omega(\mathbf{r},\mathbf{r}')$ by eqns (4.124a) and (4.121c)), as follows:

$$\sigma_{\alpha\beta}^\omega(\mathbf{r},\mathbf{r}') = \frac{c}{i\omega}\left[\mathcal{R}_{\alpha\beta}^\omega(\mathbf{r},\mathbf{r}') - \frac{e^2 n_{\text{sc}}(\mathbf{r})}{mc}\delta(\mathbf{r}-\mathbf{r}')\delta_{\alpha\beta}\right]$$

$$= \frac{c}{i\omega}\left\{\mathcal{R}_{\alpha\beta}^{(0)}(\mathbf{r},\mathbf{r}') + \mathcal{R}_{\alpha\beta}^{(2)}(\mathbf{r},\mathbf{r}')\frac{\omega^2}{2!} + \cdots\right.$$

$$\left. + i\left[\mathcal{R}_{\alpha\beta}^{(1)}(\mathbf{r},\mathbf{r}')\omega + \mathcal{R}_{\alpha\beta}^{(3)}(\mathbf{r},\mathbf{r}')\frac{\omega^3}{3!} + \cdots\right]\right.$$

$$\left. - \frac{e^2 n_{\text{sc}}(\mathbf{r})}{mc}\delta(\mathbf{r}-\mathbf{r}')\delta_{\alpha\beta}\right\}.$$
(4.148)

We have used the symmetry relation, eqn (4.123), from which we see that the various functions $\mathcal{R}_{\alpha\beta}^{(k)}(\mathbf{r},\mathbf{r}')$ are real. The first and the last terms (independent of $\omega$) must cancel, in order to give a finite d.c. conductivity. We thus find

$$\sigma_{\alpha\beta}^\omega(\mathbf{r},\mathbf{r}') = c\mathcal{R}_{\alpha\beta}^{(1)}(\mathbf{r},\mathbf{r}') - i\frac{\omega}{2!}c\mathcal{R}_{\alpha\beta}^{(2)}(\mathbf{r},\mathbf{r}') + \cdots, \qquad (4.149)$$

and

$$\Gamma_\alpha^\omega(\mathbf{r},\mathbf{r}') = \partial_\beta'\sigma_{\alpha\beta}^\omega(\mathbf{r},\mathbf{r}')$$

$$= c\partial_\beta'\mathcal{R}_{\alpha\beta}^{(1)}(\mathbf{r},\mathbf{r}') - i\frac{\omega}{2!}c\partial_\beta'\mathcal{R}_{\alpha\beta}^{(2)}(\mathbf{r},\mathbf{r}') + \cdots$$

$$= -i\frac{\omega}{2!}c\partial_\beta'\mathcal{R}_{\alpha\beta}^{(2)}(\mathbf{r},\mathbf{r}') + \cdots. \qquad (4.150)$$

In obtaining the last line we have used the KSL theorem, eqn (4.144). Substituting the expansions (4.150) and (4.149) into eqn (4.145) for the current density, we have

$$J_\alpha^\omega(\mathbf{r}) = \int_{\Omega_0} d^3\mathbf{r}'\left[-i\frac{\omega}{2!}c\partial_\beta'\mathcal{R}_{\alpha\beta}^{(2)}(\mathbf{r},\mathbf{r}') + O(\omega^2)\right]\delta\phi^\omega(\mathbf{r}')$$

$$- V_2^\omega \int_{S_+} dS_\beta'\left[c\mathcal{R}_{\alpha\beta}^{(1)}(\mathbf{r},\mathbf{r}') - i\frac{\omega}{2!}c\mathcal{R}_{\alpha\beta}^{(2)}(\mathbf{r},\mathbf{r}') + O(\omega^2)\right]$$

$$- V_1^\omega \int_{S_-} dS_\beta'\left[c\mathcal{R}_{\alpha\beta}^{(1)}(\mathbf{r},\mathbf{r}') - i\frac{\omega}{2!}c\mathcal{R}_{\alpha\beta}^{(2)}(\mathbf{r},\mathbf{r}') + O(\omega^2)\right].$$
(4.151)

Thus, as $\omega \to 0$, we have

$$\lim_{\omega \to 0} J_\alpha^\omega(\mathbf{r}) = J_\alpha^{\omega=0}(\mathbf{r})$$
$$= -V_1^{\omega=0} \int_{S_-} \mathrm{d}S'_\beta\, \sigma_{\alpha\beta}^{(0)}(\mathbf{r}, \mathbf{r}') - V_2^{\omega=0} \int_{S_+} \mathrm{d}S'_\beta\, \sigma_{\alpha\beta}^{(0)}(\mathbf{r}, \mathbf{r}'), \quad (4.152)$$

where $\sigma_{\alpha\beta}^{(0)}(\mathbf{r}, \mathbf{r}') = c\mathcal{R}_{\alpha\beta}^{(1)}(\mathbf{r}, \mathbf{r}')$.

In the RPA that we are employing here, the electron–electron interaction enters through the self-consistent field of eqn (4.86), which in turn affects (i) the conductivity tensor $\sigma_{\alpha\beta}^\omega(\mathbf{r}, \mathbf{r}')$ (as seen from its definition (4.124a)), and (ii) the potential profile $\delta\phi^\omega(\mathbf{r})$ (the solution of eqns (4.94) and (4.95)), which, from its definition (4.87), contains, in addition to $\phi_{\mathrm{ext}}^\omega(\mathbf{r})$, the field $\phi_{\mathrm{ind}}^\omega(\mathbf{r})$ generated by the induced charges. Both (i) and (ii) enter eqn (4.124b) for the current density. In the d.c. limit, though, the result (4.152) demonstrates that the electron–electron interaction affects the current density only through (i); (ii) is not needed, only the asymptotic values of the potential $V_1^{\omega=0}$ and $V_2^{\omega=0}$ (eqn (4.95b)) being relevant.

Restating the above remarks, we point out that, given the self-consistent field, the conductivity tensor $\sigma_{\alpha\beta}^\omega(\mathbf{r}, \mathbf{r}')$ is fully determined by eqn (4.124a); in fact, the same will be calculated in Appendix B. Note, however, that an essential simplification occurs in the calculation of the d.c. current density $J_\alpha^{\omega=0}(\mathbf{r})$, inasmuch as $\delta\phi^{\omega=0}(\mathbf{r}')$ is not needed, because the first term in eqn (4.151) containing $\delta\phi^\omega(\mathbf{r})$ drops out as $\omega \to 0$. On the other hand, for $\omega \neq 0$, $\delta\phi^\omega(\mathbf{r}')$ would have to be determined from the integro-differential equation (4.96).

The tensor $\sigma_{\alpha\beta}^{(0)}(\mathbf{r}, \mathbf{r}')$ being divergenceless, as in eqn (4.144), we have

$$\int_{S_+} \mathrm{d}S'_\beta\, \sigma_{\alpha\beta}^{(0)}(\mathbf{r}, \mathbf{r}') = -\int_{S_-} \mathrm{d}S'_\beta\, \sigma_{\alpha\beta}^{(0)}(\mathbf{r}, \mathbf{r}'), \quad (4.153)$$

and so

$$J_\alpha^{\omega=0}(\mathbf{r}) = -(V_1^{\omega=0} - V_2^{\omega=0}) \int_{S_-} \mathrm{d}S'_\beta\, \sigma_{\alpha\beta}^{(0)}(\mathbf{r}, \mathbf{r}'), \quad (4.154)$$

and the result depends only upon the potential difference $V = V_1^{\omega=0} - V_2^{\omega=0}$.

The total d.c. current crossing any surface $S$ can thus be written as

$$I^0 = -V \int_S \mathrm{d}S_\alpha \int_{S_-} \mathrm{d}S'_\beta\, \sigma_{\alpha\beta}^0(\mathbf{r}, \mathbf{r}'). \quad (4.155)$$

Here, the surface element $\mathrm{d}S_\alpha$ is related to the variable $\mathbf{r}$, while $\mathrm{d}S'_\beta$ is related to the variable $\mathbf{r}'$. Again, due to the divergenceless condition (4.144), the integral over $S$ can be replaced by an integral over the surface $S_+$ of Fig. 4.2, i.e.,

$$I^0 = -V \int_{S_+} \mathrm{d}S_\alpha \int_{S_-} \mathrm{d}S'_\beta\, \sigma_{\alpha\beta}^0(\mathbf{r}, \mathbf{r}'). \quad (4.156)$$

The d.c. conductance $G = I_0/V$ is thus

$$G = -\int_{S_+} dS_\alpha \int_{S_-} dS'_\beta\, \sigma^0_{\alpha\beta}(\boldsymbol{r},\boldsymbol{r}'). \qquad (4.157)$$

We recall again that the surface elements $dS_\alpha$ and $dS'_\beta$ point *outwards* from the constriction, as in eqn (4.145).

Equation (4.157) is remarkable for its formal simplicity and is valid within linear response theory in the random-phase approximation.

## 4.5 Evaluation of the conductance

The conductivity tensor $\sigma^\omega_{\alpha\beta}(\boldsymbol{r},\boldsymbol{r}')$ of eqn (4.124a) is evaluated from first principles, i.e., without recourse to the many-body formalism [63, 122], in Appendix B for a TRI system in the d.c. limit, $\omega \to 0$, in terms of single-particle Green functions defined for $H_{\rm sc}$, with the result (see eqn (B.40))

$$\lim_{\omega \to 0} \sigma^\omega_{\alpha\beta}(\boldsymbol{r},\boldsymbol{r}')$$
$$= \pi \frac{e^2 \hbar^3}{2m^2} \int d\epsilon \left(-\frac{\partial f(\epsilon)}{\partial \epsilon}\right) \left\{\mathcal{N}(\epsilon;\boldsymbol{r},\boldsymbol{r}')\left[\partial_\alpha \partial'_\beta \mathcal{N}(\epsilon;\boldsymbol{r},\boldsymbol{r}')\right]\right.$$
$$\left. - \left[\partial_\alpha \mathcal{N}(\epsilon;\boldsymbol{r},\boldsymbol{r}')\right]\left[\partial'_\beta \mathcal{N}(\epsilon;\boldsymbol{r},\boldsymbol{r}')\right]\right\}. \qquad (4.158)$$

Here, $f(\epsilon)$ is the Fermi function of eqn (B.21) and $\mathcal{N}(\epsilon;\boldsymbol{r},\boldsymbol{r}')$, defined in eqn (B.28), is related to the single-particle Green functions as follows:

$$\mathcal{N}(\epsilon;\boldsymbol{r},\boldsymbol{r}') = -\frac{1}{2\pi i}\left[G^{(+)}(\epsilon;\boldsymbol{r},\boldsymbol{r}') - G^{(-)}(\epsilon;\boldsymbol{r},\boldsymbol{r}')\right]. \qquad (4.159)$$

Notice that $\mathcal{N}(\epsilon;\boldsymbol{r},\boldsymbol{r}')$ is the 'off-diagonal local density' of single-particle states at the energy $\epsilon$. The quantity $\mathcal{N}(\epsilon;\boldsymbol{r},\boldsymbol{r}')$ can also be considered to be the matrix element

$$\mathcal{N}(\epsilon;\boldsymbol{r},\boldsymbol{r}') = \langle \boldsymbol{r}|\hat{\mathcal{N}}(\epsilon)|\boldsymbol{r}'\rangle \qquad (4.160)$$

of the operator $\hat{\mathcal{N}}(\epsilon)$ defined in eqn (B.31).

We are dealing with the case in which the system is invariant under the operation of time reversal (TRI); it is seen in Appendix B, eqns (B.28) and (B.32), that under these circumstances $\mathcal{N}(\epsilon;\boldsymbol{r},\boldsymbol{r}')$ is real and symmetric. This can also be seen as follows. From eqn (2.66) we have the relation

$$G^{(-)}(\epsilon;\boldsymbol{r},\boldsymbol{r}') = \left[G^{(+)}(\epsilon;\boldsymbol{r}',\boldsymbol{r})\right]^*, \qquad (4.161)$$

which is valid even in the absence of TRI. With TRI we have the additional symmetry relation (2.58), namely

$$G^{(+)}(\epsilon;\boldsymbol{r},\boldsymbol{r}') = G^{(+)}(\epsilon;\boldsymbol{r}',\boldsymbol{r}), \qquad (4.162)$$

so that
$$\mathcal{N}(\epsilon;\boldsymbol{r},\boldsymbol{r}') = -\frac{1}{2\pi\mathrm{i}}\left\{G^{(+)}(\epsilon;\boldsymbol{r},\boldsymbol{r}') - \left[G^{(+)}(\epsilon;\boldsymbol{r},\boldsymbol{r}')\right]^*\right\}, \tag{4.163}$$
and our statement follows.

**Exercise 4.8** Verify that in the d.c. limit the conductivity tensor given in eqn (4.158) satisfies the KSL theorem, eqn (4.144).

Denote by $\tau_{\alpha\beta}(\boldsymbol{r},\boldsymbol{r}')$ the quantity in curly brackets in eqn (4.158). Its divergence is given by
$$\partial'_\beta \tau_{\alpha\beta}(\boldsymbol{r},\boldsymbol{r}') = \mathcal{N}(\epsilon;\boldsymbol{r},\boldsymbol{r}')\left[\partial_\alpha \partial'_\beta \partial'_\beta \mathcal{N}(\epsilon;\boldsymbol{r},\boldsymbol{r}')\right] - \left[\partial_\alpha \mathcal{N}(\epsilon;\boldsymbol{r},\boldsymbol{r}')\right]\left[\partial'_\beta \partial'_\beta \mathcal{N}(\epsilon;\boldsymbol{r},\boldsymbol{r}')\right]. \tag{4.164}$$
The single-particle wave functions $\psi_\rho(\boldsymbol{r})$ that enter the expression for $\mathcal{N}(\epsilon;\boldsymbol{r},\boldsymbol{r}')$, eqn (B.28), satisfy the Schrödinger equation
$$\left(\nabla^2 + k_\rho^2\right)\psi_\rho(\boldsymbol{r}) = -eU_{\mathrm{sc}}(\boldsymbol{r})\psi_\rho(\boldsymbol{r}), \tag{4.165}$$
where the total self-consistent potential $U_{\mathrm{sc}} = 2mV_{\mathrm{sc}}/\hbar^2$ was defined in eqn (4.86). Thus
$$\nabla'^2 \mathcal{N}(\epsilon;\boldsymbol{r},\boldsymbol{r}') = \left[-k^2 - eU_{\mathrm{sc}}(\boldsymbol{r}')\right]\mathcal{N}(\epsilon;\boldsymbol{r},\boldsymbol{r}'). \tag{4.166}$$
Substituting into eqn (4.164), we thus find
$$\partial'_\beta \tau_{\alpha\beta}(\boldsymbol{r},\boldsymbol{r}') = \mathcal{N}(\epsilon;\boldsymbol{r},\boldsymbol{r}')\left[-k^2 - eU_{\mathrm{sc}}(\boldsymbol{r}')\right]\left[\partial_\alpha \mathcal{N}(\epsilon;\boldsymbol{r},\boldsymbol{r}')\right]$$
$$- \left[\partial_\alpha \mathcal{N}(\epsilon;\boldsymbol{r},\boldsymbol{r}')\right]\left[-k^2 - eU_{\mathrm{sc}}(\boldsymbol{r}')\right]\mathcal{N}(\epsilon;\boldsymbol{r},\boldsymbol{r}')$$
$$= 0. \tag{4.167}$$
This proves the statement.

Our aim is to calculate the conductance given by eqn (4.157). To simplify the calculation, assume that the horns in Fig. 4.2 are replaced, from a little closer than $S_\pm$ all the way to infinity, by infinitely long pipes whose width $W'$ will eventually be taken to be infinite. Then the surfaces $S_\pm$ are flat and perpendicular to the $x$ direction, with
$$\mathrm{d}S'_x = -\mathrm{d}S', \quad \mathrm{d}S_x = \mathrm{d}S, \tag{4.168}$$
so that
$$G = -\int_{S_+} \mathrm{d}S_x \int_{S_-} \mathrm{d}S'_x\, \sigma^0_{xx}(\boldsymbol{r},\boldsymbol{r}')$$
$$= \int_{S_+} \mathrm{d}S \int_{S_-} \mathrm{d}S'\, \sigma^0_{xx}(\boldsymbol{r},\boldsymbol{r}'), \tag{4.169}$$

where

$$\sigma_{xx}^0(\mathbf{r},\mathbf{r}') = \pi \frac{e^2 \hbar^3}{2m^2} \int d\epsilon \left(-\frac{\partial f(\epsilon)}{\partial \epsilon}\right) \left\{ \mathcal{N}(\epsilon;\mathbf{r},\mathbf{r}')\left[\partial_x \partial'_x \mathcal{N}(\epsilon;\mathbf{r},\mathbf{r}')\right] \right.$$
$$\left. - \left[\partial_x \mathcal{N}(\epsilon;\mathbf{r},\mathbf{r}')\right]\left[\partial'_x \mathcal{N}(\epsilon;\mathbf{r},\mathbf{r}')\right] \right\}. \quad (4.170)$$

Let us write $\mathcal{N}(\epsilon;\mathbf{r},\mathbf{r}')$, with $\mathbf{r} \in S_-$ and $\mathbf{r}' \in S_+$ ($S$ stands for $(y,z)$ and $S'$ for $(y',z')$), as follows:

$$\mathcal{N}(\epsilon;\mathbf{r},\mathbf{r}') = \langle x, S | \hat{\mathcal{N}}(\epsilon) | x', S' \rangle$$
$$= \sum_{ab} \langle S | \chi_a \rangle \left\langle \chi_a \left| \hat{\mathcal{N}}(\epsilon; x, x') \right| \chi_b \right\rangle \langle \chi_b | S' \rangle$$
$$= \sum_{ab} \chi_a(S) \mathcal{N}_{ab}(\epsilon; x, x') \chi_b(S'), \quad (4.171)$$

where the functions $\chi_a(S) \equiv \chi_a(y,z)$ form a complete set of orthonormal functions (which are real, because of TRI) on the surface $S$, just as, for example, they are defined in eqn (3.6) for $d=2$. The second line of eqn (4.171) defines the operator $\hat{\mathcal{N}}(\epsilon; x, x')$. The quantity $\mathcal{N}_{ab}(\epsilon; x, x')$ is given by

$$\mathcal{N}_{ab}(\epsilon; x, x') = \left\langle \chi_a \left| \hat{\mathcal{N}}(\epsilon; x, x') \right| \chi_b \right\rangle$$
$$= \iint \chi_a(S) \mathcal{N}(\epsilon;\mathbf{r},\mathbf{r}') \chi_b(S') \, dS \, dS'. \quad (4.172)$$

Substituting eqn (4.171) into (4.170) and the result into (4.169), we have

$$G = \frac{\pi e^2 \hbar^3}{2m^2} \int d\epsilon \left(-\frac{\partial f(\epsilon)}{\partial \epsilon}\right) \int_{S_+} dS \int_{S_-} dS' \sum_{abcd} \chi_b(S') \chi_a(S) \chi_c(S) \chi_d(S')$$
$$\times \left\{ \mathcal{N}_{ab}(\epsilon; x, x')\left[\partial_x \partial_{x'} \mathcal{N}_{cd}(\epsilon; x, x')\right] - \left[\partial_x \mathcal{N}_{ab}(\epsilon; x, x')\right]\left[\partial_{x'} \mathcal{N}_{cd}(\epsilon; x, x')\right] \right\}. \quad (4.173)$$

Integrating first over $dS$ and $dS'$ we obtain $\delta_{ac}\delta_{bd}$, so that

$$G = \frac{\pi e^2 \hbar^3}{2m^2} \int d\epsilon \left(-\frac{\partial f(\epsilon)}{\partial \epsilon}\right) \sum_{ab} \left\{ \mathcal{N}_{ab}(\epsilon; x, x')\left[\partial_x \partial_{x'} \mathcal{N}_{ab}(\epsilon; x, x')\right] \right.$$
$$\left. - \left[\partial_x \mathcal{N}_{ab}(\epsilon; x, x')\right]\left[\partial_{x'} \mathcal{N}_{ab}(\epsilon; x, x')\right] \right\}. \quad (4.174)$$

The function $\mathcal{N}(\epsilon;\mathbf{r},\mathbf{r}')$ is defined in eqn (4.159); similarly, for the function $\mathcal{N}_{ab}(\epsilon; x, x')$ we have

$$\mathcal{N}_{ab}(\epsilon; x, x') = -\frac{1}{2\pi i}\left[G_{ab}^{(+)}(\epsilon; x, x') - G_{ba}^{(-)}(\epsilon; x, x')\right]. \quad (4.175)$$

Here, $G_{ab}^{(+)}(\epsilon; x, x')$ is defined in terms of $G_{ab}^{(+)}(\epsilon;\mathbf{r},\mathbf{r}')$ in a manner similar to $\mathcal{N}_{ab}(\epsilon; x, x')$ of eqn (4.172).

The summations over $a$ and $b$ in eqn (4.174) are from 1 to $\infty$, since the functions $\chi_a(S)$ form a complete set on $S$ only when the index $a$ runs over its full range. However, the relationship of the functions $\mathcal{N}_{ab}(\epsilon; x, x')$ to the Green function given in eqn (4.175) makes it clear that the summations in question in eqn (4.174) run, effectively, over the open channels only, as we are in the asymptotic region where the Green function for the closed channels is exponentially small; this is illustrated, for instance, for the free Green function in eqn (3.43).

The Green function can be written in terms of elements of the $S$ matrix, as shown in the one-dimensional case in eqn (2.162). For the multichannel case see [165]. When $x$ and $x'$ are far away from the sample, on the right and the left of it, respectively, eqn (77) of [165], translated to the present notation, gives

$$G_{ab}^{(+)}(\epsilon; x, x') = -\frac{i}{\hbar v_b} t_{ab}^{(\text{tot})}(\epsilon) \sqrt{\frac{k_b}{k_a}} e^{i(k_a x - k_b x')}, \quad \begin{cases} x \to +\infty \\ x' \to -\infty \end{cases}, \tag{4.176}$$

where $t_{ab}^{(\text{tot})}(\epsilon)$ denotes the transmission amplitude from the channel $b$ in the 'pipe' associated with the *far left horn* to the channel $a$ in the *far right horn*. Equation (4.176) reduces (interchanging $x$ and $x'$) to eqn (2.162) in the one-channel case. Thus, from eqn (4.175) we find

$$\mathcal{N}_{ab}(\epsilon; x, x') = \frac{m}{2\pi\hbar^2 \sqrt{k_a k_b}} \left[ t_{ab}^{(\text{tot})}(\epsilon) e^{i(k_a x - k_b x')} + t_{ab}^{(\text{tot})*}(\epsilon) e^{i(k_b x' - k_a x)} \right],$$

$$\begin{cases} x \to +\infty \\ x' \to -\infty \end{cases}. \tag{4.177}$$

We thus find, for the expression in the curly brackets in eqn (4.174),

$$\mathcal{N}_{ab}(\epsilon; x, x') \left[ \partial_x \partial_{x'} \mathcal{N}_{ab}(\epsilon; x, x') \right] - \left[ \partial_x \mathcal{N}_{ab}(\epsilon; x, x') \right] \left[ \partial_{x'} \mathcal{N}_{ab}(\epsilon; x, x') \right]$$
$$= \frac{m^2}{\pi^2 \hbar^4} \left| t_{ab}^{(\text{tot})}(\epsilon) \right|^2. \tag{4.178}$$

Substituting eqn (4.178) into (4.174), we finally obtain

$$G = \frac{e^2}{h} g, \tag{4.179}$$

where the *dimensionless conductance* $g$ is given by

$$g = 2 \int d\epsilon \left( -\frac{\partial f(\epsilon)}{\partial \epsilon} \right) T^{(\text{tot})}(\epsilon), \tag{4.180}$$

and

$$T^{(\text{tot})}(\epsilon) = \sum_{ab} \left| t_{ab}^{(\text{tot})}(\epsilon) \right|^2 \tag{4.181}$$

is the transmission coefficient $|t_{ab}^{(\text{tot})}|^2$ summed over the initial and final channels. In eqn (4.180) we have included a factor 2 to take into account the two spin directions. We again remark that the summation in the above equation is over open channels only.

This is the equation obtained, for instance, in [50]. We emphasize that $t_{ab}^{(\text{tot})}(\epsilon)$ is the transmission matrix from deep inside the left horn to deep inside the right one. The conductance can be expressed entirely in terms of the *transmission matrix $t(\epsilon)$ associated with the sample itself* when the reflection coefficient back to the sample of a wave travelling from the sample to the horns is negligible. We show in Appendix C that, in this case,

$$g = 2 \int d\epsilon \left( -\frac{\partial f(\epsilon)}{\partial \epsilon} \right) T(\epsilon), \qquad (4.182)$$

where

$$T(\epsilon) = \sum_{ab} |t_{ab}(\epsilon)|^2. \qquad (4.183)$$

The number of open channels entering this equation is the one associated with the sample itself. In terms of the parametrization (3.86) of the $S$ matrix, we can write the quantity $T$ in the above equation as

$$T = \sum_{a} \tau_a. \qquad (4.184)$$

The function $-\partial f(\epsilon)/\partial \epsilon$ is peaked around the Fermi energy $\epsilon_F$. In the zero-temperature limit we have

$$-\frac{\partial f(\epsilon)}{\partial \epsilon} \to \delta(\epsilon - \epsilon_F), \qquad (4.185)$$

so that the formula for the conductance, eqn (4.182), reduces finally to

$$g = 2T(\epsilon_F). \qquad (4.186)$$

Relations (4.180), (4.182) and (4.186) are various versions of the so-called Landauer's formula.

# 5

# THE MAXIMUM-ENTROPY APPROACH: AN INFORMATION-THEORETIC VIEWPOINT

In the following chapters of this book we shall be dealing with a variety of quantum mechanical problems in which the underlying complexity washes out most of the microscopic details and, as a result, several physical properties exhibit features that are common within large families of systems. The underlying complexity alluded to above is produced, in the cases to be examined later, either by, e.g., the system being spatially random, causing extensive multiple scattering, as in a disordered conductor, or by the underlying classical dynamics of the system being completely chaotic, a property that refers to the *long-time* behavior of the system. The physical properties exhibiting common features are all those that actually probe the long-time dynamical behavior, the reason being that the system can then explore (in a classical sense) most of the phase space that is available to it. Such properties are robust against the sample-specific peculiarities.

We shall see that the behavior of those common physical properties is, to a large extent, governed by the *symmetries* of the systems in question, and by a rather limited number of (system-dependent) *relevant parameters*—generally of a macroscopic nature and having a clear physical significance—being otherwise *insensitive to other details* of the problem. In the examples treated in the following chapters, the relevant parameters are represented, e.g., by the average, or optical, scattering matrix in the case of ballistic chaotic cavities (the average being taken over an ensemble of cavities at a fixed energy $E$, or, ergodically equivalently, over the energy axis for a given cavity), or by the elastic mean free path in the case of disordered conductors. It is indeed remarkable that a few such relevant average parameters determine the full statistics of the sample-to-sample fluctuations, that turn out to be universal within a class.

When there are no such system-dependent parameters (such as the mean free path) relevant to the properties in question, or they amount to a trivial rescaling, we speak of *universal* properties; otherwise, we refer to *quasi-universal* properties. An example of a universal property would be one depending on the dimensionality only.

In a number of cases one has discovered a generalized *central-limit theorem* (CLT)—acting at a more microscopic level—responsible for certain universal, or quasi-universal, properties. In these situations, a more microscopic calculation might end up as being mere 'scaffolding', due to the final insensitivity to most of the details. It is then conceptually appealing to construct directly the statistical distribution of the quantities of interest by imposing the constraints of symmetry and the values of the relevant parameters alluded to above. This procedure may determine the statistical distribution uniquely; and when it does not, we have found very often that a selection made on the basis of a *maximum-entropy* criterion *captures the universal, or quasi-universal, features* in question. We could describe the *maximum-entropy approach* (MEA) as providing an 'ansatz' that picks 'the most probable distribution' among those that satisfy the given constraints; such a distribution is as close to an 'equal-a-priori distribution' (to be discussed below) as is allowed by these constraints.

There are features of the data arising from actual physical systems that deviate from the predictions of a maximum-entropy model. In several circumstances one has been able to identify the cause for that discrepancy; data that probe the system on a short time-scale fall within this category. For other features it may be that one has failed to recognize and incorporate a number of physically relevant constraints when maximizing the entropy; these cases are still open.

The notion of information entropy for a statistical distribution has been discussed at length in various publications on the subject [97, 102, 104, 117, 158, 159]. Here we merely give a simple account of it, with the idea of making the book reasonably self-contained. The power of the method will become more apparent in the chapters that follow.

In the next section we introduce the notion of information entropy for a statistical distribution. We discuss, through simple examples, the role played by the physical parameters in determining the constraints to be imposed. As in any statistical calculation, there is a point in the analysis where one makes, one way or another, an assumption of *equal-a-priori probabilities*. Technically, this corresponds to the choice of what we shall call the *prior*. The role of symmetries in motivating a natural probability measure, or prior, is discussed in Section 5.2. Applications to classical and quantum statistical mechanics are presented in Section 5.3. Finally, Section 5.4 briefly discusses the maximum-entropy method in the usual context of statistical data analysis—statistics *sans* underlying mechanics.

## 5.1 Probability and information entropy: the role of the relevant physical parameters as constraints

Suppose that we wish to accommodate $N$ distinct objects in $n$ boxes. If $N_i$ denotes the number of particles in box $i$, then the weight $W$ of the configuration $N_1, N_2, \ldots, N_n$, with $\sum_{i=1}^{n} N_i = N$, is [88]

$$W = \frac{N!}{N_1! N_2! \cdots N_n!}. \tag{5.1}$$

In the absence of other physically relevant restrictions, the most probable configuration is the one of maximum weight: $N_1 = N_2 = \cdots = N_n = N/n$. In classical statistical mechanics one finds the Maxwell–Boltzmann distribution using a maximum-weight argument. There, the objects are particles and the boxes represent equal volume elements in $\mu$-space, with the energy $\epsilon_i$ assigned to the

$i$th box. However, one has further constraints; in the microcanonical ensemble the total number of particles $N$ and the total energy $E$ are fixed, i.e.,

$$\sum_i N_i = N, \qquad (5.2a)$$

$$\sum_i \epsilon_i N_i = E. \qquad (5.2b)$$

For $N \gg 1$, these (plus the total volume) are, for a thermodynamic description, *the only relevant physical quantities*, other dynamical details being irrelevant. For given $N$ and $E$, the configuration with maximum weight $W$ is given by the well-known Maxwell–Boltzmann expression $N_i = e^{-\alpha' - \beta \epsilon_i}$, with $\alpha'$ and $\beta$ being Lagrange multipliers, determined from the above constraints (5.2). This statement can be rephrased in terms of the usual *entropy* per particle $s = N^{-1} k_B \ln W$. Using Stirling's approximation, $s$ can be written as the Boltzmann expression[1] ($k_B$ being Boltzmann's constant)

$$s = -k_B \sum_{i=1}^n p_i \ln p_i, \qquad (5.3)$$

where $p_i = N_i/N$ is the fraction of particles (or the statistical probability to find a particle) in box $i$. In terms of the set $\{p_i\}$, we now have the conditions

$$\sum_i p_i = 1, \qquad (5.4a)$$

$$\sum_i \epsilon_i p_i = \bar{\epsilon}, \qquad (5.4b)$$

and

$$p_i = e^{-\alpha - \beta \epsilon_i} \qquad (5.5)$$

maximizes the entropy $s$.

The thermodynamic entropy in the statistical mechanical problem can be generalized outside the statistical mechanical context and assigned to an arbitrary probability distribution $p_i$, $i = 1, \ldots, n$, in which case the term entropy is sometimes qualified as the *Shannon entropy* or the *information-theoretic entropy* [97, 102, 104, 117, 158, 159]. We denote it by $\mathcal{S}$ and define it by

$$\mathcal{S} = -\sum_{i=1}^n p_i \ln p_i. \qquad (5.6)$$

(The information is defined to be $I = -\mathcal{S}$.) The distribution (5.5) is the one with *maximum entropy* or *minimum information*, consistent with the constraints

---

[1] Notice that we are not concerned here with the indistinguishability of the particles at the classical level and the related issue of avoiding the Gibbs paradox [88].

(5.4), which in turn represent the only *physically relevant quantities* in this problem.

In Fig. 5.1 we contrast the distribution of maximum entropy (or minimum information), consistent with the constraint (5.4a) alone, with a distribution of minimum entropy (or maximum information); in the former case all the $p_i$s, $i = 1, \ldots, n$, are equal to $1/n$ and $\mathcal{S} = \ln n$; in the latter case, the probability is concentrated at $i = 2$, say, and thus $\mathcal{S} = 0$. This is consistent with the usual notion of entropy as a measure of disorder.

### 5.1.1 *Properties of the entropy*

Important properties of the entropy follow from the *convexity* nature of the function $f(x) = x \ln x$ entering its definition, eqn (5.6). A real function $f(x)$ is said to be *convex downwards* in the interval $(a, b)$ if, for any set of points $x_1, \ldots, x_n$ contained in $(a, b)$,

$$f\left(\frac{1}{n}\sum_{i=1}^{n} x_i\right) \leqslant \frac{1}{n}\sum_{i=1}^{n} f(x_i). \tag{5.7}$$

It can be shown that the property (5.7) holds if and only if $f''(x) \geqslant 0$ in the interval $(a, b)$.

If the $x_i$s appearing in the inequality (5.7) are identified with the probabilities $p_i$ of the above discussion, then we find the relation

$$f\left(\frac{1}{n}\right) \leqslant \frac{1}{n}\sum_{i} f(p_i), \tag{5.8}$$

where we have used the normalization condition (5.4a).

The real function $f(x) = x \ln x$ for $x > 0$ is convex downwards. Substituted into eqn (5.8), it gives

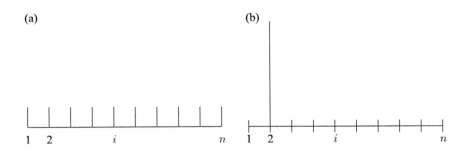

FIG. 5.1. A simple example of (a) maximum-entropy, and (b) minimum-entropy distributions.

$$\frac{1}{n}\ln\frac{1}{n} \leqslant \frac{1}{n}\sum_i p_i \ln p_i,$$

$$-\sum_i p_i \ln p_i \leqslant \ln n, \qquad (5.9)$$

or

$$\mathcal{S}(p_1,\ldots,p_n) \leqslant \mathcal{S}\left(\frac{1}{n},\ldots,\frac{1}{n}\right), \qquad (5.10)$$

showing that the entropy $\mathcal{S}(p_1,\ldots,p_n)$ takes its largest possible value when all the $p_i$s are equal, i.e., when the distribution is broadest.

Now consider two statistically independent sets of events, with probabilities

$$p_1,\ldots,p_n \qquad (5.11a)$$

and

$$q_1,\ldots,q_m. \qquad (5.11b)$$

The entropy associated with the joint probability distribution $P_{ij} = p_i q_j$ is then

$$\begin{aligned}
\mathcal{S}([P]) &= -\sum_{ij} P_{ij} \ln P_{ij} \\
&= -\sum_{ij} p_i q_j \ln p_i q_j \\
&= -\sum_i p_i \ln p_i - \sum_j q_j \ln q_j \\
&= \mathcal{S}([p]) + \mathcal{S}([q]),
\end{aligned} \qquad (5.12)$$

i.e., the sum of the two individual entropies; entropies add when probabilities multiply, for statistically independent events.

If the two sets of events are not independent, then the joint probability can be written as $P_{ij} = p_i q_{j|i}$, and its associated entropy is

$$\begin{aligned}
\mathcal{S}([P]) &= -\sum_{ij} p_i q_{j|i} \ln p_i q_{j|i} \\
&= -\sum_i p_i \ln p_i + \sum_i p_i \left(-\sum_j q_{j|i} \ln q_{j|i}\right) \\
&= \mathcal{S}([p]) + \sum_i \mathcal{S}_i([q]),
\end{aligned} \qquad (5.13)$$

where we have defined

$$\mathcal{S}_i([q]) = -\sum_j q_{j|i} \ln q_{j|i} \qquad (5.14)$$

as the *conditional entropy* of $\{q_j\}$, evaluated on the assumption that event $i$ has occurred.

It is interesting that the form (5.6) for the entropy can also be found from a *uniqueness theorem* [104, p. 9]. Suppose that we require the entropy $\mathcal{S}$ to have the following entirely reasonable properties.

(i) For given $n$ and for $\sum_{i=1}^{n} p_i = 1$, the function $\mathcal{S}(p_1, \ldots, p_n)$ takes its largest possible value for $p_i = 1/n$.
(ii) For two sets of events, not necessarily statistically independent, the combined entropy is given by eqn (5.13).
(iii) Including an impossible event does not change the entropy, i.e.,

$$\mathcal{S}(p_1, \ldots, p_n, 0) = \mathcal{S}(p_1, \ldots, p_n).$$

Then, *if $\mathcal{S}(p_1, \ldots, p_n)$ is continuous with respect to all its arguments and has the above three properties, it must have the form (5.6), up to a multiplicative constant.* This uniqueness strongly motivates, almost validates, our introduction of the Shannon entropy, eqn (5.6). Other definitions of entropy, however, are possible, where some of the above properties are relaxed; see, for example, the one proposed by Tsallis [170, 171] which has recently received much attention.

As a next step in our analysis, and to prepare the way for further applications, consider the $r$ functions

$$f_i^{(\alpha)}, \quad \alpha = 1, \ldots, r, \tag{5.15}$$

of the event variable $i$, and suppose that we are requested to find the probability distribution $p_i$ that *maximizes* the entropy (5.6), under the *constraint* that the $r$ averages

$$\left\langle f^{(\alpha)} \right\rangle_p = \sum_i f_i^{(\alpha)} p_i = C_\alpha, \quad \alpha = 1, \ldots, r, \tag{5.16}$$

be fixed numbers $C_\alpha$. The method of Lagrange multipliers tells us that the resulting $p_i$ must be of the form

$$p_i = \exp\left(-\sum_{\alpha=1}^{r} \beta_\alpha f_i^{(\alpha)}\right), \tag{5.17}$$

where the Lagrange multipliers $\beta_\alpha$ must be found so that the $r$ conditions (5.16) are fulfilled. The Maxwell–Boltzmann distribution of eqn (5.5) is clearly a particular case of eqn (5.17).

For further insight into the result (5.17), we consider another important property following from convexity, namely the so-called Gibbs' inequality

$$\mathcal{S}[q] = -\sum_i q_i \ln q_i \leqslant -\sum_i q_i \ln p_i, \tag{5.18}$$

fulfilled by two normalized probability distributions $\{p_i\}$ and $\{q_i\}$, i.e.,

$$\sum_i q_i = \sum_i p_i = 1. \tag{5.19}$$

The equality sign in (5.18) holds if and only if $q_i = p_i$, for all $i$. Notice that the left-hand side of eqn (5.18) is the entropy for $\{q_i\}$.

We can now interpret the result of eqn (5.17) from the standpoint of Gibbs' inequality. Assume that the $r$ functions (5.15) have the same average under $\{p_i\}$ as under $\{q_i\}$, i.e.,

$$\left\langle f^{(\alpha)} \right\rangle_p = \left\langle f^{(\alpha)} \right\rangle_q = C_\alpha, \quad \alpha = 1, \ldots, r. \tag{5.20}$$

Also assume $\{p_i\}$ to be of the form of eqn (5.17). Substitution of eqn (5.17) into (5.18) gives

$$\mathcal{S}[q] \leqslant -\sum_i q_i \left( -\sum_{\alpha=1}^r \beta_\alpha f_i^{(\alpha)} \right)$$

$$= \sum_{\alpha=1}^r \beta_\alpha \left\langle f^{(\alpha)} \right\rangle_q$$

$$= \sum_{\alpha=1}^r \beta_\alpha \left\langle f^{(\alpha)} \right\rangle_p$$

$$= -\sum_i p_i \left( -\sum_{\alpha=1}^r \beta_\alpha f_i^{(\alpha)} \right)$$

$$= -\sum_i p_i \ln p_i = \mathcal{S}[p]. \tag{5.21}$$

From the second line to the third we have used the equality of the averages of the $r$ functions $f_i^{(\alpha)}$ under $\{p_i\}$ and $\{q_i\}$, eqn (5.20). Thus

$$\mathcal{S}[q] \leqslant \mathcal{S}[p]. \tag{5.22}$$

Consistent with the method of Lagrange multipliers, we have thus found that *the entropy of the probability distribution $\{p_i\}$ having the form of eqn (5.17) and satisfying (5.20) cannot be smaller than that of any other $\{q_i\}$ also satisfying (5.20)*. (This is illustrated schematically in Fig. 5.2(b), where the point $P$ represents the entropy of the distribution (5.17), subject to the $r$ constraints (5.20).)

One particular case of this last result is when $\{q_i\}$ has the form

$$q_i = \exp\left( -\sum_{\alpha=1}^{r+1} \beta'_\alpha f_i^{(\alpha)} \right), \tag{5.23}$$

i.e., it is of maximum entropy with $r+1$ constraints, whereas $\{p_i\}$ is of maximum entropy with $r$ constraints, the first $r$ constraints having the same value for the two distributions, as indicated in eqn (5.20). (This is illustrated schematically in Fig. 5.2(a), where the point $Q$ represents the entropy of the distribution (5.23), subject to $r + 1$ constraints, as explained above.) The inequality (5.22) then states that

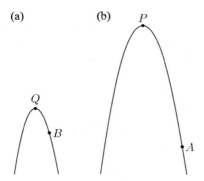

FIG. 5.2. Schematic representation of the entropy $\mathcal{S}$ for distributions with (a) $r+1$ constraints, and (b) $r$ constraints, *all* distributions in (a) and (b) satisfying eqn (5.20). The points $Q$ and $P$ represent the corresponding maximum-entropy distributions, eqns (5.23) and (5.17), respectively.

$$\mathcal{S}\left[\exp\left(-\sum_{\alpha=1}^{r+1}\beta'_\alpha f_i^{(\alpha)}\right)\right] \leqslant \mathcal{S}\left[\exp\left(-\sum_{\alpha=1}^{r}\beta_\alpha f_i^{(\alpha)}\right)\right], \qquad (5.24)$$

i.e., *lifting a constraint from a maximum-entropy distribution—without changing the values of the remaining constraints—cannot lower $\mathcal{S}$, provided that the new distribution is also of maximum entropy.* As illustrated schematically in Fig. 5.2, every point of (a), including its maximum $Q$, is lower than the maximum $P$ in (b). However, no definite statement can be made when comparing the entropies of two distributions subject to $r+1$ and $r$ constraints, but not necessarily of maximum entropy, as illustrated by the points $B$ and $A$ in Fig. 5.2.

### 5.1.2 Continuous random variables

Now consider an example with a continuous variable. Let $x \in (-\infty, +\infty)$. The probability density $p(x)$ is related to the differential probability by

$$dP(x) = p(x)\,dx, \qquad (5.25)$$

and the entropy is defined as

$$\mathcal{S} = -\int p(x) \ln p(x)\,dx. \qquad (5.26)$$

From the above discussion it is clear that among all the distributions that are normalized and have fixed first and second moments (equal to $\mu$ and $\mu^2 + \sigma^2$, respectively), the Gaussian

$$p(x) = \frac{e^{-(x-\mu)^2/\sigma^2}}{(2\pi\sigma^2)^{1/2}} = e^{-\alpha-\beta x-\gamma x^2} \qquad (5.27)$$

has the largest entropy.

Let us recall an interesting situation where a Gaussian distribution does arise. Consider a number $N$ of statistically independent variables with a common *arbitrary* (the actual restrictions are very mild!) distribution. The distribution of their sum becomes approximately Gaussian for $N \gg 1$ and, as $N \to \infty$ and with a proper rescaling, it tends to a zero-centered Gaussian with unit variance. This result is known as the *central-limit theorem* (CLT) of statistics [103, Vol. 1, p. 206]. We thus see that the resulting distribution is sensitive only to the centroid and width of the original one, *all other details* (e.g., higher-order moments) *being irrelevant* in the limit $N \to \infty$. This reduction is precisely reflected in the fact that the entropy carried by a Gaussian is the largest among all those distributions that have the same centroid and width. This interpretation has to be contrasted with the more subjective one of Jaynes [97], who would look for a distribution having maximum entropy, 'while agreeing with whatever information is *given*'. In this *universal* regime, our knowledge or ignorance of the particular details makes no difference to the limiting distribution. When we are slightly short of the universal limit, these particular details will surely show up as corrections. Far from the limiting regime, of course, our method will fail altogether.

As another example of a continuous variable, consider the angle $\theta \in (0, 2\pi)$, with the relation

$$\mathrm{d}P(\theta) = p(\theta)\,\mathrm{d}\theta \tag{5.28}$$

between the differential probability and the probability density. The entropy is defined as

$$\mathcal{S} = -\int p(\theta) \ln p(\theta)\,\mathrm{d}\theta. \tag{5.29}$$

With no other constraint except normalization, $\mathcal{S}$ is maximal for $p(\theta) = 1/2\pi$. On the other hand, with the constraint $\langle \cos\theta \rangle = \mu$, $\mathcal{S}$ is maximal for

$$p(\theta) = \mathrm{e}^{-\alpha - \beta \cos\theta}. \tag{5.30}$$

## 5.2 The role of symmetries in motivating a natural probability measure

We observe that, for a continuous variable, as distinct from the case of discrete variables, we have to specify how to factorize the differential probability into the product of a probability density and what has sometimes been called the 'prior' [117] ($\mathrm{d}x$ and $\mathrm{d}\theta$ in the above examples). The choice of the prior corresponds to making a definite postulate of 'equal-a-priori probabilities', as, after all, one always does in statistical studies. For a compact space, the prior is normalizable and defines the maximum-entropy distribution in the absence of constraints.

In many cases of interest the physical problem is invariant under a certain *symmetry operation*; there is then a 'natural' way to 'weigh' the physical quantities in question, provided by the *measure* that remains *invariant* under that symmetry operation. Let us make it clear, though, that it is the confrontation with experiment that finally decides whether the choice is a reasonable one or not.

Let us remark that choosing the prior as the invariant measure under a certain symmetry operation has the appealing consequence that the resulting entropy is also invariant under that operation [10]. As an illustration, in the above example of eqn (5.28), $d\theta$ remains invariant under the addition of an arbitrary constant phase $\theta_0$, i.e.,

$$\theta' = \theta + \theta_0, \quad d\theta = d\theta'. \tag{5.31}$$

Denoting by $p(\theta)$ and $q(\theta')$ the probability densities of the variables $\theta$ and $\theta'$, respectively, we have

$$p(\theta)\, d\theta = q(\theta')\, d\theta'. \tag{5.32}$$

Since $d\theta = d\theta'$ (eqn (5.31)), we have

$$q(\theta') = p(\theta(\theta')) = p(\theta' - \theta_0). \tag{5.33}$$

For the entropy we then have

$$\mathcal{S}[p(\theta)] = -\int p(\theta) \ln p(\theta)\, d\theta, \tag{5.34}$$

so that, applying eqns (5.32) and (5.33), we find

$$\mathcal{S}[p(\theta)] = -\int q(\theta') \ln q(\theta')\, d\theta'$$
$$= \mathcal{S}[q(\theta')]. \tag{5.35}$$

More elaborate symmetry considerations will be presented in later chapters.

## 5.3 Applications to equilibrium statistical mechanics

There is compelling experimental evidence that a reduced description of a thermodynamic system in equilibrium is possible in terms of a relatively small number of parameters, such as temperature, volume, order parameters, etc., which then play the role of the *relevant parameters* (thermodynamic variables), as discussed in the introduction to the present chapter; other numerous microscopic details become irrelevant in the thermodynamic limit. It thus seems appealing to seek the probability distribution for the microstates of the system by employing the maximum-entropy arguments presented above; the relevant parameters would then be the natural candidates for constraining the maximization of the information-theoretic entropy.

We would like to emphasize though that the information-theoretic entropy is essentially a mathematical concept, unlike the thermodynamic entropy, which originated in a physical context. The whole purpose of the discussion that follows is to establish a meaningful link between the two.

As examples, we discuss the situations that give rise to the classical microcanonical and canonical ensembles (see the footnote on p. 228), and to the quantum mechanical canonical ensemble.

### 5.3.1 The classical microcanonical ensemble

In classical statistical mechanics it is a *canonical transformation* that plays the role of the symmetry operation referred to earlier; a prior chosen as the Liouville measure $dp\,dq$, i.e., as the elementary extension in phase space ($p$ denotes the set of all momenta and $q$ the set of coordinates), is invariant under that operation.

Consider an *isolated* $N$-particle system confined to a volume $V$, the total energy of the system $E$ being fixed. According to the above ideas, we seek the probability distribution in phase space that maximizes the information-theoretic entropy $\mathcal{S}$, for fixed $N$, $V$ and $E$, with no further constraints. We find the differential probability

$$dP(p,q) \propto \delta(E - H(p,q))\,dp\,dq. \tag{5.36}$$

This is the *microcanonical ensemble*. It assigns equal probabilities to equal volumes in the phase space contained between two close surfaces of constant energies $E$ and $E + \delta E$. If that shell is divided up into $n$ equal volume elements, or cells, then the probability assigned to each one of them is

$$p_i = \frac{1}{n}. \tag{5.37}$$

Now imagine that, *in addition*, we impose on our system $r$ constraints of the type (5.16). For instance, the presence of partitions in the container could be expressed as in (5.16), using, for the functions $f_i^\alpha$ of (5.15), suitable functions (called 'indicator functions') which are defined to be equal to 1 in a certain region of space and 0 everywhere else. Also assume that, in *equilibrium*, we always assign to the set of $n$ volume elements in phase space described above a probability distribution of the type (5.17). Then, as discussed immediately after eqn (5.23), lifting a constraint, but retaining the same value for the remaining ones, cannot decrease $\mathcal{S}$ for the new *equilibrium* state; this property is like the one described by the second law for the thermodynamic entropy of the initial and final equilibrium states. Thus, having fixed $N$, $V$ and $E$, and in the absence of further constraints, looking for the distribution that maximizes $\mathcal{S}$ is simply in accordance with the second law. In this case the information-theoretic entropy $\mathcal{S}$ coincides, up to a multiplicative constant, with the thermodynamic entropy.

### 5.3.2 The classical canonical ensemble

Now suppose that our $N$-particle system, confined to a volume $V$, is in equilibrium with a thermal bath. The energy of the system is no longer fixed, but fluctuates, due to the interaction with the bath, around an average that we call $U$. If we have evidence that the physics of the problem can be expressed entirely in terms of these quantities, then we look for the probability distribution that, for *fixed* $N$ and $V$, maximizes the information entropy $\mathcal{S}$, under the following constraints:

(i) it be normalized, implying that

$$\langle 1 \rangle = 1; \tag{5.38a}$$

(ii) the *average* energy is $U$, i.e.,

$$\langle H(p,q)\rangle = U. \tag{5.38b}$$

Using the method of Lagrange multipliers, we find the differential probability

$$dP(p,q) = e^{-\alpha-\beta H(p,q)}\, dp\, dq, \tag{5.39}$$

where the Lagrange multipliers $\alpha$ and $\beta$ have to be found so as to fulfill the constraints of eqns (5.38). This is the *canonical ensemble*. It is well known that $\beta$ can be interpreted as the inverse *temperature*, i.e., $\beta = 1/k_B T$, with $k_B$ being Boltzmann's constant.

Note that the *fixed* nature of the parameters $N$ and $V$ is implied in the structure of the Hamiltonian $H(p,q)$ of the given system itself. On the other hand, the energy does fluctuate, but is constrained to be equal to $U$ on *average*.

In a discrete version of the above, if we divide up the phase space into cells of equal volume, then the canonical ensemble assigns to the $i$th cell the probability

$$p_i = e^{-\alpha-\beta E_i}, \tag{5.40}$$

where $E_i$ is the energy associated with that cell.

The fact that the canonical ensemble corresponds to a *fixed value of the temperature* rather than of the energy (which fluctuates) can be incorporated explicitly in the analysis as follows. We realize that, from the point of view of the method of Lagrange multipliers, the above derivation amounts to maximizing the new function

$$\begin{aligned}\mathcal{S}'_\beta[p] &= \mathcal{S}[p] - \beta U[p] \\ &= -\sum_i p_i \ln p_i - \beta \sum_i E_i p_i\end{aligned} \tag{5.41}$$

over the set $\{p_i\}$, for a *fixed* value of $\beta$, which can be identified with the inverse temperature of the thermal bath, the constraint now being that the probability distribution $\{p_i\}$ be normalized. The value of $U$, i.e.,

$$U = \sum_i E_i p_i, \tag{5.42}$$

is then determined as a result of the calculation. The new function $\mathcal{S}'_\beta[p]$ will be identified below with the negative of the free energy.

Let us now see how the above maximization procedure can be interpreted from the point of view of the second law. In preparation for this, we first make the following considerations.

Imagine that we impose on our system $r$ constraints of the type (5.16) (the normalization constraint will be counted among these, so that we are really considering $r-1$ *additional* constraints). Consider two probability distributions

$\{p_i\}$ and $\{q_i\}$, under which the $r$ functions (5.15) have the same average, as in eqn (5.20). Assume now that $\{p_i\}$ is of the form

$$p_i = \exp\left(-\beta E_i - \sum_{\alpha=1}^{r} \beta_\alpha f_i^{(\alpha)}\right). \tag{5.43}$$

Gibbs' inequality (5.18) then gives

$$-\sum_i q_i \ln q_i \leqslant -\sum_i q_i \left(-\beta E_i - \sum_{\alpha=1}^{r} \beta_\alpha f_i^{(\alpha)}\right)$$

$$= \beta \sum_i E_i q_i + \sum_{\alpha=1}^{r} \beta_\alpha \left\langle f^{(\alpha)} \right\rangle_q$$

$$= \beta \sum_i E_i q_i + \sum_{\alpha=1}^{r} \beta_\alpha \left\langle f^{(\alpha)} \right\rangle_p$$

$$= \beta \sum_i E_i q_i + \sum_i p_i \left(\sum_{\alpha=1}^{r} \beta_\alpha f_i^{(\alpha)}\right)$$

$$= \beta \sum_i E_i q_i - \sum_i p_i (\beta E_i + \ln p_i), \tag{5.44}$$

so that

$$-\sum_i q_i \ln q_i - \beta \sum_i E_i q_i \leqslant -\sum_i p_i \ln p_i - \beta \sum_i E_i p_i, \tag{5.45}$$

or

$$S'_\beta[q] \leqslant S'_\beta[p], \tag{5.46}$$

where $S'_\beta$ and $U$ are defined in eqns (5.41) and (5.42), respectively, for the probability distribution $\{p_i\}$, with similar definitions for the probability distribution $\{q_i\}$.

We thus find that *the quantity $S'_\beta[p]$ for a probability distribution $\{p_i\}$ having the form (5.43) and satisfying (5.20) cannot be smaller than the $S'_\beta[q]$ for any other distribution $\{q_i\}$ also satisfying (5.20) and for the same $\beta$.*

In particular, assume that $q_i$ is of the form

$$q_i = \exp\left(-\beta E_i - \sum_{\alpha=1}^{r+1} \beta'_\alpha f_i^{(\alpha)}\right), \tag{5.47}$$

that the requirements (5.20) are fulfilled, and that, in addition,

$$\left\langle f^{(r+1)} \right\rangle_q = C_{r+1} \tag{5.48}$$

has been specified. The probability distribution $\{q_i\}$ is of maximum $S'_\beta$ for fixed $\beta$ and $r+1$ constraints, whereas $\{p_i\}$ is of maximum $S'_\beta$ for the same $\beta$ and $r$

constraints, the first $r$ constraints having the same value for the two distributions, as in eqn (5.20). The inequality (5.46) then states that *lifting a constraint from a maximum-$S'_\beta$ distribution—without changing $\beta$ or the values of the remaining constraints—cannot lower $S'_\beta$*, provided that the new distribution is also of maximum $S'_\beta$.

We now return to our statistical mechanical problem. Suppose that to a system in *equilibrium*, with $N$, $V$ and $\beta$ *fixed*, and with $r$ constraints, we always assign a probability distribution of the form given in eqn (5.43). Then, as discussed above, every time we lift a constraint—retaining the same value for the remaining ones and for $\beta$—the quantity $S'_\beta$ for the final *equilibrium* state cannot be lower than that for the initial state. It is apt, at this point, to call $-S'_\beta/\beta = U - S/\beta \equiv F$ the *free energy*. We see that, every time we lift a constraint, $F$ for the final equilibrium state cannot be larger than that for the initial state. This property is precisely of the same nature as that described by the second law for the thermodynamic entropy and free energy. Thus, having fixed $N$, $V$ and $\beta$, and in the absence of further constraints, looking for the distribution with maximum $S'_\beta$, or minimum $F$, giving the result (5.40), is simply in accordance with the second law of thermodynamics. Thus, the maximum-entropy criterion not only conforms to the second law, but the method is so powerful that, in addition, it actually gives the statistical distribution of the dynamical variables in question—the Gibbsian distribution in this case.

### 5.3.3 *The quantum mechanical canonical ensemble*

Since the system interacts with a thermal bath (although we assume here no exchange of particles), it cannot be described by a wave function (corresponding to what is called a *pure state*), but can be described by a density matrix $\rho$ (corresponding to what is called a *mixed state*). The expectation value of an observable represented by the operator $O$ is then evaluated as

$$\langle O \rangle = \text{Tr}\,(\rho O). \tag{5.49}$$

In particular, the normalization requirement, $\langle 1 \rangle = 1$, fixes $\text{Tr}\,\rho$, i.e.,

$$\langle 1 \rangle = \text{Tr}\,\rho = 1. \tag{5.50a}$$

We also require the expectation value of the Hamiltonian to equal the internal energy $U$, i.e.,

$$\langle H \rangle = \text{Tr}\,(\rho H) = U. \tag{5.50b}$$

We now wish to maximize the entropy

$$\mathcal{S} = -\,\text{Tr}\,(\rho \ln \rho), \tag{5.51}$$

subject to the constraints (5.50). Notice that, in a representation in which $\rho$ is diagonal, $\mathcal{S}$ of eqn (5.51) has all the properties of the entropy discussed earlier.

In principle, $\rho$ and $H$ do not necessarily commute. Let us call $|i\rangle$ the basis in which the Hamiltonian $H$ is diagonal, with eigenvalues $E_i$. To be specific, suppose that our Hilbert space consists of $n$ states; the index $i$ then runs from 1 to $n$.

Our variational problem is now

$$\delta\left\{\mathrm{Tr}\left[-\rho\ln\rho - (\alpha - 1)\rho - \beta H\rho\right]\right\} = 0, \qquad (5.52)$$

where, for convenience, we have written the first multiplier as $\alpha - 1$. The variation of the last term is

$$\delta\,\mathrm{Tr}\,(H\rho) = \mathrm{Tr}\,(H\delta\rho). \qquad (5.53)$$

Next consider the variation of $\rho\ln\rho$. The variation of $\rho$ can be carried out in two steps, the first being a unitary transformation which preserves the trace and hence gives zero variation, and the second being a variation of the eigenvalues, so that

$$\delta\,\mathrm{Tr}\,(\rho\ln\rho) = \mathrm{Tr}\,(1 + \ln\rho)\,\delta\rho. \qquad (5.54)$$

Thus (5.52) gives

$$\mathrm{Tr}\left[(\ln\rho + \alpha + \beta H)\,\delta\rho\right] = 0. \qquad (5.55)$$

Written in the basis $|i\rangle$ that diagonalizes $H$, this last expression has the structure

$$\sum_{ij} A_{ij}(\delta\rho)_{ji} = 0. \qquad (5.56)$$

The density matrix $\rho$ is Hermitian. Since we accept only variations $\delta\rho$ that uphold this property, we have

$$\sum_{i} A_{ii}(\delta\rho)_{ii} + \sum_{i<j} A_{ij}(\delta\rho)_{ji} + \sum_{i<j} A_{ji}(\delta\rho)^*_{ji} = 0, \qquad (5.57)$$

and, since these variations can be taken to be arbitrary, we require that

$$A_{ii} = A_{ij}(i<j) = A_{ji}(i<j) = 0 \quad \Rightarrow \quad A = 0. \qquad (5.58)$$

Applying this result to our eqn (5.55), we have

$$\ln\rho + \alpha + \beta H = 0, \qquad (5.59)$$

or

$$\rho = e^{-\alpha - \beta H}, \qquad (5.60)$$

which tells us that $\rho$ and $H$ *are diagonal in the same basis*. This is the well-known result for the quantum mechanical canonical ensemble.

The above result can be readily generalized to the case of several constraints, by simply replacing $\beta H$ by $\sum_\alpha \beta_\alpha f^\alpha$, with $f^\alpha$ being Hermitian operators.

## 5.4 The maximum-entropy criterion in the context of statistical inference

For the sake of completeness, and also for that of contrast, we would like to conclude this chapter with a brief discussion of the maximum-entropy method (MEM), set generally in the context of mathematical statistics—of how to draw a unique inference rationally from a limited set of measured data which is insufficient, and also uncertain because of the random measurement errors, i.e., noise. We reserve the term maximum-entropy approach (MEA) to the context of the physical problems discussed in earlier sections and to its applications in the chapters that follow. Clearly, we have here an inverse problem, the so-called ill-posed problem, in that the number of unknowns (the possible conclusions, or the parameters that model a conclusion) exceeds, often by far, the number of knowns (the experimental data), thus making a solution by straightforward inversion non-unique. The MEM here helps us lift, or *regularize*, this degeneracy of possible solutions by selecting the one which incorporates and is consistent with all that is known, while remaining maximally non-committal with respect to the missing information—the maxim being 'whereof we do not know, thereof we must be silent!' The sense in which this is so is made precise by Bayes' theorem, *as reformulated by Laplace*, which we will now discuss below [47, 103].

We start from the well-known rules for composing probabilities, namely that the joint probability $P(CD)$ of two propositions $C$ and $D$, say, is related to their conditional probabilities $P(C|D)$ and $P(D|C)$, and the corresponding unconditional, or marginal, probabilities $P(C)$ and $P(D)$, by $P(CD) = P(C|D)P(D) = P(D|C)P(C)$. The Bayes theorem then states the obvious:

$$P(C|D) = \frac{P(D|C)P(C)}{P(D)} \propto P(D|C)P(C). \tag{5.61}$$

(The denominator $P(D)$ is an unimportant normalization factor here, and will be suppressed hereinafter, inasmuch as one is only interested in the relative probabilities of the different conclusions possible.)

Now, Bayes' theorem may be written in the following somewhat descriptive form which is highly suggestive and transparent:

P(conclusion | new data) $\propto$ P(new data | conclusion) $\times$ P(conclusion | old data).

This is then what provides the Bayesian update of the statistical inference, or conclusion, as the old data set is updated to the new set of data.

Accordingly, the propositions $C$ and $D$ have a very general connotation here. Thus, $P(C|D)$ may be, and indeed shall be, taken to be the probability a posteriori for the conclusion $C$ conditional on the new data. Next, $P(C)$ is the probability of the conclusion $C$ based on the old data—it is a priori, and is the hardest to get at. It is aptly called the Bayesian *prior* in statistics, and is often determined from some general considerations of what reasonably constitutes the *equally probable* elementary events for the problem in question. Finally, $P(D|C)$

is the probability of the new data $D$ conditioned on the conclusion. It is often referred to as the *likelihood function*, first introduced by R. A. Fisher in the mid-1950s [74] in his small-sample statistics. This is the part which is easiest to write down in practice. Now, one has to simply maximize, over the space of conclusions, the product of the two factors on the right-hand side of the sign of proportionality in eqn (5.61) to arrive at the best (most probable) conclusion. This inductive method for statistical inference, from the particular to the general, has found extensive and successful applications to problems in diverse fields, e.g., image enhancement in astronomy, X-ray structure determination in crystallography, nuclear and atomic spectra analysis, and correlation spectroscopy—all involving the statistical analysis of insufficient and uncertain data [47].

We will now complete our discussion of MEM in this Bayesian setting by considering a typical problem of statistical inference from a limited and noisy data set, somewhat analogous to the example treated in Section 5.1.

Let $N$ be the number of trials, each trial having $n$ possible elementary outcomes, with $N_i$ being the frequency of occurrence of the $i$th elementary possibility. (Thus, e.g., in a problem of imaging, $n$ may correspond to the number of pixels on a CCD (charged coupled device) camera, while $N$ is the total number of photons incident on the CCD, and $N_i$ is the number of photons hitting the $i$th pixel.) This gives $W = N!/\Pi_i N_i!$ configurations which are degenerate—all corresponding to the same image to within the resolution of the pixels. Let the $n$ elementary possibilities all be equally probable a priori. This is the well-known Bayes' hypothesis. With this, we at once have for the prior that

$$\text{P(conclusion} \,|\, \text{old data)} \propto W \propto e^{S'},$$

where $S' = -\sum_i N_i \ln N_i$ is the entropy. Here the factorials have been replaced by their Stirling approximation, as $N_i \gg 1$. Next, consider the likelihood function P(new data | conclusion). Assuming the measurement errors to be Gaussian, one has

$$\text{P(new data} \,|\, \text{conclusion)} \propto \exp\left[-\frac{\chi^2(\{N_i\})}{2}\right],$$

with

$$\chi^2(\{N_i\}) = \frac{\sum_i \left(N_i - N_i^{\text{expt}}\right)^2}{\sigma_i^2},$$

the $\sigma_i^2$ being the mean-squared error. Here, $\chi^2(\{N_i\})$ is the usual goodness-of-fit parameter (the *chi-squared* statistic or the so-called sufficient statistic) and $N_i^{\text{expt}}$ is the measured frequency of occurrence. All we have to do now is to maximize the product of the two factors on the right-hand side of eqn (5.61). Equivalently, we can maximize the exponent, where the entropy tends to make the distribution as broad as possible, while the $\chi^2(\{N_i\})$ tends to minimize the misfits $(N_i - N_i^{\text{expt}})^2$ overall. In practice, one maximizes

$$-\sum_i N_i \ln N_i - \frac{\lambda}{2}\chi^2(\{N_i\})$$

over $\{N_i\}$. This amounts to maximizing the entropy $-\sum_i N_i \ln N_i$, subject to the constraint $\chi^2(\{N_i\}) \sim N$, corresponding to the physical requirement that one can never fit the data better than the tolerance set by the noise. The Lagrange multiplier $\lambda$ ensures this constraint. There may, of course, be other constraints too.

Admittedly, the choice of the constraints and the question of their sufficiency remains the *bad problem* of MEM, inasmuch as the latter contains within it no systematic prescription for the constraints—these must be supplied from the outside. The MEM is subjective in the original sense of Jaynes [97], where whatever happens to be known is to be used as the constraint. This, however, is not the case with MEA. Quite on the contrary, we believe that there is a central-limit theorem which is valid in a physical context, as we have discussed earlier. Of course, an actual implementation of MEA is quite similar to that for the case of MEM. This does not, however, diminish the conceptual difference between the two viewpoints. Thus, as for the choice of the prior in MEA, we must treat all possibilities related by the symmetries underlying the physical problem as strictly equally probable. In MEM, on the other hand, it seems certainly most *natural, unbiased, least prejudiced* and most *uniform* to treat all the mutually exclusive and exhaustive possibilities about which we know nothing as equally probable—the principle of insufficient reason of Laplace [97]. Clearly, there is no way to prove such a Bayesian hypothesis. It is to be tested for validity only a posteriori. Logically, this is the best form of statistical inference by induction.

Finally, we provide some remarks on the use of the term *entropy*, as extended outside its original thermal context of equilibrium statistical mechanics, into the athermal context of statistics *without* underlying mechanics. The information-theoretic entropy function $\mathcal{S} = -\sum_i p_i \ln p_i$, introduced by Shannon [158,159], and called so at the suggestion of von Neumann, and finally used by Jaynes [97] in his subjective statistical mechanics, clearly measures, numerically, the uncertainty associated with a probability distribution $\{p_i\}$. (Thus, e.g., with a binary choice of base 2 for the logarithm, the entropy $\mathcal{S}$ roughly counts the number of questions of the yes/no type required to be asked to arrive at a reasonably accurate answer.) Indeed, Shannon used this information-theoretic entropy for the efficient encoding of information for error-free transmission over noisy channels. Moreover, as used in MEA, the information entropy certainly shares, at least formally, several features in common with the physical entropy of Planck and Gibbs. Still, it seems, however, that *uncertainty* would have perhaps been a more suitable, and acceptable, name for entropy in the context of MEA/MEM.

# 6

# ELECTRONIC TRANSPORT THROUGH OPEN CHAOTIC CAVITIES

This chapter is devoted to the study of quantum chaotic scattering, i.e., wave transport through open chaotic cavities. A cavity is said to be open if it is connected by means of ideal leads to asymptotically free regions far from the quasi-bound states of the cavity. Intuitively, the open cavity is said to be chaotic when the dynamics inside the corresponding closed cavity is chaotic in the classical limit, with the proviso that the leads are sufficiently thin. More precisely, some authors have defined classical scattering as chaotic if (i) the dynamics in the fraction of phase space which is trapped for an infinite time is chaotic in the closed-system sense, or if (ii) as one iterates the scattering map, the dynamics is chaotic in the closed-system sense as the number of iterations tends to infinity.

The quantities of interest here are, in general, the scattering matrix elements or, more specifically, the transmission and reflection amplitude coefficients and the associated time delays, although the latter will not be dealt with in what follows. On rather general grounds, these quantities are expected to, and are indeed known to, fluctuate strongly in an irregular manner as a function of the energy (frequency) of the incident wave. The fluctuations are sample specific and very much in the nature of a *reproducible noise*, the so-called Ericson fluctuations (when the resonance width exceeds their spacing) which are well known from nuclear physics. This very feature in fact motivates our present treatment of chaotic scattering—a treatment which is macroscopic and frankly statistical in nature. It is based on the information-theoretic maximum-entropy approach (MEA), as introduced in Chapter 5, applied to random-matrix theory (RMT). Clearly, in spirit as also in detail, it is quite different from, but complementary to, other approaches to quantum chaos (the study of how quantum properties depend on the nature of the classical dynamics in a system), such as microscopic dynamical treatments based on the semi-classical approximation, now extensively practiced by a whole community of physicists. In such approaches, the non-integrability of the underlying Hamiltonian enters explicitly. (A Hamiltonian system is said to be non-integrable when the number of its degrees of freedom exceeds the number of conserved quantities it has in involution, i.e., commuting with each other. Such a dynamical system is known to have, in general, phase space regions that are partially chaotic at the classical level.) More relevant to our present viewpoint, however, is this deterministic classical chaos that effectively introduces an element of randomness and the associated probability, which ultimately underlies the MEA. We believe that the present approach to quantum chaos is valid when the system is fully chaotic.

It is important to emphasize here that the motion through the interior of the cavity is taken to be ballistic—without any impurity scattering, elastic or inelastic. The only scattering in question is the elastic scattering from the boundary (walls) of the cavity. For such a cavity, non-integrability and, therefore, chaos is due entirely to a lack of those geometric symmetries that one would find, e.g., in cavities of cylindrical or spherical shape, which would generate conserved quantities making the motion regular. Open cavities now include a wide range of optical, microwave and electronic cavities as common laboratory examples. The simplest and the oldest known open cavity, of course, is the Fabry–Perot interferometer of optics. An open chaotic optical cavity can be realized as an asymmetric dielectric microstructure, wherein the light wave is totally internally reflected at the boundary with a rarer medium outside. Here openness is realized through the refractive escape, and chaos is ensured through the geometric asymmetry. The underlying classical dynamics here is that of the ray-optical trajectory (of geometrical optics), and similarly for the chaotic multi-port microwave cavity with waveguide leads attached to it, for all the coming to and passing away of the waves. Extremely interesting chaotic cavities are realized in the quantum dots and quantum wells in microelectronic heterostructures, where the shape and size of the cavity can be controlled at will by a patterned gate voltage. One could also alter the discrete symmetry under time reversal (TRI) by the application of an external magnetic field to the cavity, and thus study the different universality classes. In this chapter we will be dealing mostly with the electronic chaotic cavities with and without TRI, in the absence of spin; but, of course, the method readily generalizes to the cases in which spin is present.

In Section 6.1 we introduce the notion of an ensemble of $S$ matrices and propose an invariant measure to weigh these. In Sections 6.2 and 6.3 we study the one-channel and multichannel cases, respectively. Section 6.4 develops an important mathematical method of invariant integration to deal with averages over the invariant measure, corresponding, physically, to the absence of prompt processes; such processes are taken into account explicitly, for a one-channel case, in Section 6.5. Sections 6.6 and 6.7 are devoted to comparisons of the theory with numerical and real experiments, respectively.

## 6.1 Statistical ensembles of $S$ matrices: the invariant measure

Extensive numerical simulations of quantum chaotic scattering in two-dimensional cavities are now available. Various geometries have been simulated. Figure 6.1 shows some results of one such simulation for the transmission coefficient as a function of the incident electron energy. Strong fluctuations of the transmission coefficient are clearly seen. These arise from the overlapping resonances of the open cavity. Overlap of the resonances depends on the number of channels—it is moderate for a single channel, but becomes more pronounced as the number of channels is increased [14, 16, 130]. This is due ultimately to the escape time broadening of the otherwise sharp levels relative to the level spacing. In this

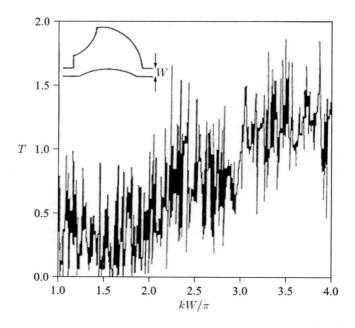

FIG. 6.1. Results of numerical solutions of the Schrödinger equation for a two-dimensional open chaotic cavity, shown in the inset in the upper left corner, giving sample-specific fluctuations of the transmission coefficient as a function of the wave number $k$. (From [130].)

book, resonance theory for one-dimensional systems and two-dimensional cavities is discussed in Sections 2.2 and 3.3, respectively, to which the interested reader is referred.

As indicated in Sections 1.1 and 2.2.8, we have to introduce for our statistical treatment the notion of an *ensemble*, and invoke *ergodicity* so as to relate what is calculated to what is actually measured. More explicitly, let us recall that, in nuclear physics and in most chaotic microwave cavities, experimental measurements have indeed been carried out on a *given* sample, e.g., a given chaotic cavity, where we sample the physical quantity of interest, e.g., the transmission coefficient, along the energy/frequency axis. In the quantum dot case, however, beautiful experiments have been performed, see [56], where an ensemble of dots with the same area but of different shapes has actually been constructed.

For a chaotic cavity these measured values fluctuate strongly, almost randomly, and so we can meaningfully generate statistics from these measured values on the *given* sample. We would now like to calculate these statistical fluctuations theoretically. To this end, we introduce an ensemble of macroscopically identical cavities represented by the corresponding ensemble of $S$ matrices, all at a *fixed* energy, whose statistical properties we now calculate based on our MEA. Next, we invoke *ergodicity*, namely that the measured statistics for a *given* cavity over the *energy axis* are the same as those calculated for the *ensemble of cavities* (S matrices) for a *fixed* energy. This is, of course, what we do in statistical mechanics, where we equate the time average of a physical quantity for a *given* sample to its phase space average at a fixed time.

However, before we proceed to implement the MEA, we must first decide on what we should really mean by *a priori equally probable* in the space of samples (i.e., in the space of $S$ matrices). This is much the same as the notion of *a priori equally probable* extension in phase space (the familiar $\mathrm{d}p\,\mathrm{d}q$) in statistical mechanics. It involves the idea of an invariant measure that defines equality-at-a-distance of volume elements centered at different points in the sample space of $S$ matrices. Again, we take our cue from statistical mechanics, where the well-known phase space measure arises from the requirement of an invariance—with respect to canonical transformations. The corresponding condition in the present case is to demand invariance of the measure under an automorphism of a given symmetry class of matrices into itself.

Keeping these general considerations in mind, we now proceed to determine the appropriate measure $\mathrm{d}\mu^{(\beta)}(S)$ for the ensemble belonging to the universality class $\beta$, where $\beta = 1, 2$ and 4 correspond to the orthogonal, unitary and symplectic ensembles, respectively. The invariance demanded is, for $\beta = 1$,

$$\mathrm{d}\mu^{(1)}(S) = \mathrm{d}\mu^{(1)}(S') \quad \text{with } S' = U_0 S U_0^\top, \tag{6.1}$$

where $U_0$ is an arbitrary, but fixed, unitary matrix, and the transformation $S$ to $S'$ is an automorphism of the set of unitary symmetric matrices into itself. Similarly, $\beta = 2$ corresponds to the unitary ensemble, and the invariance demanded is

$$\mathrm{d}\mu^{(2)}(S) = \mathrm{d}\mu^{(2)}(S') \quad \text{with } S' = U_0 S V_0, \tag{6.2}$$

where $U_0$ and $V_0$ are arbitrary, but fixed, unitary matrices. Both of these measures are known to be unique [67, 83, 176]; for $\beta = 2$, the result is the Haar measure of the unitary group. The well-known *circular ensembles* (CE), i.e., the *circular orthogonal ensemble* (COE) and the *circular unitary ensemble* (CUE) for $\beta = 1$ and 2, respectively, [67, 87] are defined by using the invariant measures of eqns (6.1) and (6.2) as the probability measures for the ensembles of $S$ matrices.

Before we specialize to any convenient representation of the invariant measure for the $N$-channel case of interest here, let us consider a general procedure for defining the invariant measure in a metric space, in which the differential line element, or the arc length, is given by the bilinear expression

$$\mathrm{d}s^2 = \sum_{ij} g_{ij}(x)\,\mathrm{d}x_i\,\mathrm{d}x_j, \tag{6.3}$$

where $g_{ij}$ is the metric tensor and $x_i$ are the coordinates (the independent variables). Invariance of the differential interval $\mathrm{d}s^2$ under the transformation $x_i \Rightarrow x'_i(x_1, x_2, \ldots)$ leads uniquely to the volume element

$$\mathrm{d}V = |\det g(x)|^{1/2} \prod_i \mathrm{d}x_i, \tag{6.4}$$

which is also invariant under the above transformation. This is then the invariant measure we were looking for. We will now apply this to the space of $S$ matrices.

The invariant differential arc length in the space of $S$ matrices in question can be taken to be

$$\mathrm{d}s^2 = \mathrm{Tr}\left(\mathrm{d}S^\dagger \, \mathrm{d}S\right), \tag{6.5}$$

which is clearly invariant under the automorphic transformations (6.1) and (6.2). In order to extract the metric tensor $g$, we need an explicit parametrization of the $S$ matrices. Of the several known explicit representations of the invariant measure, the one in terms of the eigenphases and eigenvectors of the $S$ matrix is the classic one. The polar representation of eqn (3.86) is of particular interest to us because of its connection to the conductance properties of the cavity, as shown in eqns (4.182)–(4.184). For the two-equal-lead case with $N_1 = N_2 = N$, we thus express the invariant measure explicitly in this parametrization and determine the measure from eqn (6.4). The algebra involved is heavy, but straightforward in principle, and has been carried out in Appendix D (see also [96]). We obtain

$$\mathrm{d}\mu^{(1)}(S) = P^{(1)}(\{\tau\}) \prod_{\alpha=1}^{N} \mathrm{d}\tau_a \prod_{i=1}^{2} \mathrm{d}\mu(v^{(i)}), \tag{6.6a}$$

$$P^{(1)}(\{\tau\}) = C_1 \prod_{a<b} |\tau_a - \tau_b| \prod_c \frac{1}{\sqrt{\tau_c}}, \tag{6.6b}$$

and

$$\mathrm{d}\mu^{(2)}(S) \, \mathrm{d}\mu(g) = P^{(2)}(\{\tau\}) \prod_{\alpha=1}^{N} \mathrm{d}\tau_a \prod_{i=1}^{4} \mathrm{d}\mu(v^{(i)}), \tag{6.7a}$$

$$P^{(2)}(\{\tau\}) = C_2 \prod_{a<b} |\tau_a - \tau_b|^2, \tag{6.7b}$$

with

$$\mathrm{d}\mu(g) = \frac{1}{(2\pi)^N} \prod_{a=1}^{N} \mathrm{d}\eta_a. \tag{6.8}$$

Here $\mathrm{d}\mu(v^{(i)})$ is the invariant (Haar) measure on the unitary group $U(N)$, and $C_1$ and $C_2$ are normalization constants.

Some observations are in order here. Notice that the forms of the joint distribution of $\{\tau\}$ in $P^{(1)}(\tau)$ and $P^{(2)}(\tau)$ give short-ranged repulsion of the $\tau$s. The $\mathrm{d}\mu(g)$ for the $\beta = 2$ case is given in terms of the phases $\eta_a$. The identification and counting of the independent parameters occurring in the polar representation (3.86) of the random $S$ matrices, chosen above in calculating the invariant measure, calls for some discussion. Recall that an $n$-dimensional unitary symmetric matrix has $n(n+1)/2$ independent real parameters. With $n = 2N$ in the present case of $2N \times 2N$ $S$ matrices, this gives $N(2N + 1)$ independent real parameters. In the polar representation of eqn (3.86), this is made up of the $N^2$ independent parameters coming from each of the $N$-dimensional matrices $v^{(1)}$ and $v^{(2)}$ and the $N$ independent $\tau_a$s for the case $\beta = 1$. For the case $\beta = 2$, however, we seem to have $N$ parameters too many—eqn (3.86) has $4N^2 + N$ independent real parameters, while an arbitrary $2N \times 2N$ unitary matrix only needs $4N^2$ independent real parameters. There is a parametric redundancy here. This is readily understood in terms of the invariance of the unitary $S$ matrix, as parametrized in eqn (3.86), under the transformation

$$v^{(1)} \to v^{(1)}g, \quad v^{(2)} \to v^{(2)}g, \quad v^{(3)} \to g^{-1}v^{(3)}, \quad v^{(4)} \to g^{-1}v^{(4)}, \qquad (6.9)$$

where $g = \mathrm{diag}\,(\mathrm{e}^{i\eta_1}, \ldots, \mathrm{e}^{i\eta_N})$ is the diagonal phase matrix. Thus, while in principle the $N$ redundancy parameters can be eliminated because of the invariance of the $S$ matrices under the transformation (6.9), we will conveniently continue to use, in the following, the parametrization (3.86) as such. We expect the $N$ phases $\eta_a$s to drop out from any physically meaningful probability distribution for the $S$ matrices.

## 6.2 The one-channel case

Let us note at the outset that the one-channel case turns out not to require the MEA, inasmuch as it happens to be a well-posed problem. It is included in this section as an example to contrast with the $N$-channel case, where the MEA is involved in an essential way.

The one-channel case is fully described by a $1 \times 1$ $S$ matrix, with $S(E) = \mathrm{e}^{i\theta(E)}$ as in Section 2.2.8, and physically corresponds to a cavity connected by a single, one-dimensional lead to the asymptotically free region. We recall that the phase $\theta(E)$ for reflection varies as we move along the real energy axis and we are looking for the relative frequency of occurrence, $\mathrm{d}P(\theta) = p(\theta)\,\mathrm{d}\theta$, of the elementary interval $\mathrm{d}\theta$ as the energy $E$ is varied over the entire energy axis $-\infty < E < +\infty$; the probability density $p(\theta)$ is given by the Poisson kernel of eqn (2.464).

We thus introduce a statistical model based on the idealization of *stationarity* and the assumption of *ergodicity*, as discussed at the end of Section 2.2.8. We first consider an ensemble of $S$ matrices supplied with a probability measure. We generate a *stationary random function* of energy $S(E)$ over the whole energy axis $-\infty < E < +\infty$ in the following way. We construct one member of the ensemble of $S$ matrices by picking a particular (infinite) set of resonance energies $E_\lambda$ and widths $\Gamma_\lambda$ for the $K$ matrix of eqn (2.452) (evaluated at the real energy $E$), the former being sampled from a properly 'unfolded' Gaussian ensemble [40]; different sampling then generates other members of the ensemble. Under the requirement of ergodicity, whose validity was discussed at the end of Section 2.2.8, energy averages should equal ensemble averages. Ergodicity, along with the analyticity property of the $S$ matrix, leads at once to the relation

$$\langle S^m \rangle = \langle S \rangle^m, \quad \ldots, \qquad (6.10)$$

known as the *analyticity–ergodicity* (AE) condition. Here $\langle \cdots \rangle$ denotes an ensemble average. Note that stationarity makes the ensemble averages energy independent. This fixes the ensemble measure *uniquely* in terms of the average $\langle S \rangle$. Thus, we obtain the following expression for the normalized probability density:

$$p_{\langle S \rangle}(\theta) = \frac{1}{2\pi} \frac{1 - |\langle S \rangle|^2}{|S - \langle S \rangle|^2}. \tag{6.11}$$

This shows a one-parameter universality in that the probability density involves the average $\langle S \rangle$ only. In point of fact, this result can be interpreted as arising from a generalized central-limit theorem (CLT), as discussed in [128]. It is quite remarkable that the above statistical model leads to a non-degenerate solution. Clearly, the AE condition along with stationarity has turned our problem into a well-posed one. This is, however, not to be expected in general, and, indeed, the $N$-channel problem turns out to be one such case of an ill-posed problem. We shall consider this case later.

It may be noted in passing that eqn (6.10) implies that a function $f(S)$ that is analytic in its argument, and can thus be expanded in a power series in $S$, must fulfill the *reproducing property* [87, 133]

$$f(\langle S \rangle) = \int f(S) \, \mathrm{d}P_{\langle S \rangle}(S). \tag{6.12}$$

It is because the probability measure appears as the kernel of this integral equation that it is called Poisson's kernel.

The average of the $S$ matrix, often called the *optical* $S$ matrix, has a very appealing physical significance [72]. So far we have considered the stationary solutions of the Schrödinger equation. If we construct a wave packet as a linear combination of eigenfunctions of the Hamiltonian with coefficients $A(E)$, then the time-dependent solution in the waveguide will have the form

$$\psi(x,t) = \int A(E) \mathrm{e}^{-\mathrm{i}(kx + (E/\hbar)t)} \, \mathrm{d}E + \int A(E) S(E) \mathrm{e}^{\mathrm{i}(kx - (E/\hbar)t)} \, \mathrm{d}E$$
$$= \psi_{\mathrm{in}}(x,t) + \psi_{\mathrm{sc}}(x,t), \tag{6.13}$$

where $\psi_{\mathrm{in}}(x,t)$ denotes the incident wave packet and $\psi_{\mathrm{sc}}(x,t)$ denotes the scattered one. As we have seen, $S(E)$ may show wild fluctuations as a function of energy. We separate out a smooth part $\bar{S}$, obtained by locally energy averaging $S(E)$, and write

$$S(E) = \bar{S} + S^{\mathrm{fl}}(E). \tag{6.14}$$

The scattered wave packet in the waveguide can thus be written as the following sum of two contributions:

$$\psi_{\mathrm{sc}}(x,t) \approx \bar{S} \int A(E) \mathrm{e}^{\mathrm{i}(kx - (E/\hbar)t)} \, \mathrm{d}E + \int A(E) S^{\mathrm{fl}}(E) \mathrm{e}^{\mathrm{i}(kx - (E/\hbar)t)} \, \mathrm{d}E. \tag{6.15}$$

In the first term we have taken $\bar{S}$ out of the integral sign, under the assumption of its slow energy variation. This term has the same structure as that of the incident wave packet, except that it travels away from the cavity and has the factor $\bar{S}$ in front. It thus represents the fraction of the wave packet that comes out promptly from the cavity. We say that this *prompt response* is due to *direct processes* taking place in the system. In contrast, the second term in eqn (6.15) represents the *delayed* part of the scattered packet. Thus, the single parameter that Poisson's kernel depends upon, i.e., the optical $S$ matrix $\bar{S} = \langle S \rangle$, is a measure of the prompt response from our scattering system.

## 6.3 The multichannel case

A scattering matrix of dimensionality $M$ describes, in general, a multi-lead problem with a total of $M$ channels, as explained in Section 3.2. Now the Argand diagram discussed in the previous section for $M = 1$ has to be generalized to include the axes $\operatorname{Re} S_{ab}$ and $\operatorname{Im} S_{ab}$, with $a, b = 1, \ldots, M$. We obtain a picture in which $S$ is restricted to move on the surface determined by unitarity and, for $\beta = 1$, symmetry ($S = S^{\mathsf{T}}$).

We have to assume $E$ to be far from thresholds, so that, locally, we again replace $k$ by $k_0$ in eqn (3.270) and end up with a meromorphic matrix function $S(E)$ which is *analytic* in the upper half of the complex-energy plane and has resonance poles in the lower half-plane. The study of the statistical properties of $S$ is again simplified by idealizing $S(E)$, for real $E$, as a *stationary random-matrix function* of $E$ satisfying the condition of *ergodicity*. The AE properties for the $M = 1$ case generalize to give

$$\langle (S_{a_1 b_1})^{m_1} \cdots (S_{a_k b_k})^{m_k} \rangle = \langle S_{a_1 b_1} \rangle^{m_1} \cdots \langle S_{a_k b_k} \rangle^{m_k} . \tag{6.16}$$

It is important to point out that this expression involves only $S$ matrix elements, whereas $S^*$ matrix elements are absent. For any function $f(S)$ that admits an expansion in non-negative powers of $S_{11}, \ldots, S_{MM}$ (analytic in $S$), the above eqn (6.16) again implies the reproducing property (6.12).

It is our aim to study those problems which can be characterized in terms of *two time-scales* only, arising from the prompt and equilibrated components, as explained in Section 1.1 and at the end of Section 6.2. We thus seek a probability distribution for $S$ which can be parametrized solely in terms of the optical matrix $\langle S \rangle$. We introduce the probability density $p_{\langle S \rangle}^{(\beta)}(S)$ through the equation

$$\mathrm{d}P_{\langle S \rangle}^{(\beta)}(S) = p_{\langle S \rangle}^{(\beta)}(S) \, \mathrm{d}\mu_\beta(S) \tag{6.17}$$

and demand the AE conditions for $S$, eqn (6.16), to be satisfied. Notice that the conditions (6.16) contain, on the right-hand side, the optical matrix $\langle S \rangle$ only, which is thus the only (matrix) parameter entering the problem.

In the above eqn (6.17), $\mathrm{d}\mu^{(\beta)}(S)$ is the invariant measure, defined *uniquely* by the definition introduced earlier. The average of $S$ evaluated with this measure vanishes (this intuitively clear result will be shown explicitly later), so that the prompt, or direct, components described earlier vanish. It is easy to check that the AE requirements (6.16) or, equivalently, the reproducing property (6.12), are satisfied exactly for the invariant measure. However, it was shown in [133] that, in contrast to the $M = 1$ case studied above, for $M > 1$ the AE conditions and reality of the answer are *not* enough to determine the probability distribution (6.17) uniquely. Remarkably, though, the probability density ($V_\beta$ is a normalization factor)

$$p_{\langle S \rangle}(S) = V_\beta^{-1} \frac{[\det(I - \langle S \rangle \langle S \rangle^\dagger)]^{(\beta M + 2 - \beta)/2}}{|\det(I - S\langle S \rangle^\dagger)|^{\beta M + 2 - \beta}}, \qquad (6.18)$$

known again as *Poisson's kernel*, not only satisfies the AE requirements (6.16) [87], but the *entropy* $\mathcal{S}$ associated with it (see Chapter 5), namely

$$\mathcal{S}[p] \equiv -\int p_{\langle S \rangle}(S) \ln p_{\langle S \rangle}(S) \, d\mu(S), \qquad (6.19)$$

is *greater than or equal to that of any other probability density satisfying the AE requirements for the same* $\langle S \rangle$ [133]. For $M = 1$, we recover the result for the one-channel case, eqn (6.11).

We end this section with some general remarks on the information-theoretic content of Poisson's kernel. Here we have to distinguish between *general properties*, like unitarity of the $S$ matrix (flux conservation), analyticity of $S(E)$ implied by causality, and the presence or absence of TRI (and spin–rotation symmetry when spin is taken into account)—which determine the universality class (orthogonal, unitary or symplectic)—and *particular properties* of the system parametrized by the ensemble average $\langle S \rangle$ ($= \overline{S}$ under ergodicity), which control the presence of *prompt*, or *direct processes* in the scattering problem. System-specific *details other than the optical S are assumed to be irrelevant*. The optical matrix $\langle S \rangle$ plays the role of the *physically relevant parameters* discussed in the previous chapter.

The fact that for $M > 1$ the conditions of reality and AE do not uniquely fix the distribution for $S$ is understandable, because, in general, there are time-scales other than those associated with the prompt and the equilibrated components [3]. This degeneracy is, however, removed by the maximum-entropy argument, where the prompt and equilibrated components and the associated optical $S$ are the only physically relevant quantities.

In addition to the completely general derivation of Poisson's kernel above, we present a concrete construction of this distribution following [41, 42]. For the equilibrated part of the response, suppose that there is an $S$ matrix $S_0$ which is distributed according to the circular ensemble. For the prompt response, imagine a scattering process $S_1$ occurring prior to the response $S_0$. The total scattering is the composition of these two parts. Specifically, imagine bunching the $L$ leads of the cavity into a 'superlead' containing $M$ incoming and $M$ outgoing waves. Along the superlead, between the cavity and infinity, we connect a scatterer (of the appropriate symmetry class) described by $S_1$. Since there are $M$ incoming and $M$ outgoing waves on either side of the scatterer, $S_1$ is $2M$-dimensional and can be written as

$$S_1 = \begin{bmatrix} r_1 & t_1' \\ t_1 & r_1' \end{bmatrix}. \tag{6.20}$$

The composition of the two scattering processes (see Section 3.1.4) yields the total $S$, i.e.,

$$S = r_1 + t_1'(1 - S_0 r_1')^{-1} S_0 t_1. \tag{6.21}$$

Equation (6.21) relates the problem with prompt processes to the one without these. One can show that the invariant measure for $S_0$ induces Poisson's measure (6.18) for $S$, with $\langle S \rangle = r_1$. Moreover, one can show [76,87] that the distribution is independent of the choice of $t_1$ and $t_1'$, as long as they belong to a unitary matrix $S_1$.

Note that throughout this work we use arguments which refer only to physical information expressible entirely in terms of the $S$ matrix. An alternative point of view is to express $S$ in terms of an underlying Hamiltonian, for which one makes statistical or information-theoretic assumptions. These two points of view give, in fact, the same results; one can prove [41, 42, 90, 120, 148] that, for $\langle S \rangle = 0$, a Gaussian ensemble for the underlying Hamiltonian gives a circular ensemble for the resulting $S$. The argument was extended to $\langle S \rangle \neq 0$ in [41, 42] using the transformation (6.21) above.

## 6.4 Absence of prompt (direct) processes

By definition, the optical matrix $\langle S \rangle$ vanishes in the absence of prompt processes, and the Poisson kernel (6.18) reduces to the invariant measure, i.e., the CE, so that

$$dP_{\langle S \rangle=0}^{(\beta)} = d\mu^{(\beta)}(S). \tag{6.22}$$

We now study some of the moments of $T$ and then its full distribution (see also [96]). It will turn out that averages over the invariant measure of products of $S$-matrix elements (invariant integration) can be evaluated using solely the properties of the measure, without performing any integration explicitly [127, 135].

### 6.4.1 *Averages of products of $S$: weak localization and conductance fluctuations*

*The unitary case ($\beta = 2$)*

We analyze the unitary case first, as it is simpler. We denote by $M$ the dimensionality of the $S$ matrix. To begin with, consider the average

$$\langle S_{a\alpha} \rangle_0^{(2)} = \int S_{a\alpha} \, d\mu^{(2)}(S), \tag{6.23}$$

which trivially vanishes, because of the equal weight given by the invariant measure to $S_{a\alpha}$ and to its negative. This procedure will then be generalized to more

complicated averages. If $U^0$ is an *arbitrary*, but *fixed*, unitary matrix, then we define the transformed $\widetilde{S}$ by
$$S = U^0 \widetilde{S}. \tag{6.24}$$
Introducing eqn (6.24) into (6.23), we have
$$\langle S_{a\alpha}\rangle_0^{(2)} = \sum_{a'} U^0_{aa'} \int \widetilde{S}_{a'\alpha}\, d\mu^{(2)}(\widetilde{S}) = \sum_{a'} U^0_{aa'} \langle S_{a'\alpha}\rangle_0^{(2)}, \tag{6.25}$$
where we have used the defining property (6.2) of the invariant measure and the definition of $\langle S_{a'\alpha}\rangle_0$. Thus, for the choice
$$U^0_{aa'} = e^{i\theta_a} \delta_{aa'}, \tag{6.26}$$
we find
$$\langle S_{a\alpha}\rangle_0 = e^{i\theta_a} \langle S_{a\alpha}\rangle_0. \tag{6.27}$$
As this expression should hold for arbitrary $\theta_a$, we conclude that
$$\langle S_{a\alpha}\rangle_0 = 0. \tag{6.28}$$
The above argument can now be generalized to prove that
$$\left\langle \left[S_{b_1\beta_1}\cdots S_{b_p\beta_p}\right]\left[S_{a_1\alpha_1}\cdots S_{a_q\alpha_q}\right]^*\right\rangle_0^{(2)} = 0 \tag{6.29}$$
for $p \neq q$, and, also, unless $\{a_1,\ldots,a_p\}$ and $\{b_1,\ldots,b_p\}$ constitute the same set of indices except for order; the same condition holds for the sets $\{\alpha_1,\ldots,\alpha_p\}$ and $\{\beta_1,\ldots,\beta_p\}$. In particular, consider $p = q = 1$. Using the same argument as above, we find
$$\langle S_{b\beta} S^*_{a\alpha}\rangle_0^{(2)} = \sum_{a'b'} U^0_{bb'} \left(U^0_{aa'}\right)^* \langle S_{b'\beta} S^*_{a'\alpha}\rangle_0^{(2)} \quad \text{for all } U^0. \tag{6.30}$$
For $U^0$ as given in eqn (6.26), we have
$$\langle S_{b\beta} S^*_{a\alpha}\rangle_0 = e^{i(\theta_b - \theta_a)} \langle S_{b\beta} S^*_{a\alpha}\rangle_0, \tag{6.31}$$
which vanishes unless $b = a$. A similar argument involving right multiplication implies that $\beta = \alpha$. The only non-vanishing possibility in eqn (6.30) is thus
$$\left\langle |S_{a\alpha}|^2\right\rangle_0 = \sum_{a'} |U^0_{aa'}|^2 \left\langle |S_{a'\alpha}|^2\right\rangle_0 \quad \text{for all } U^0. \tag{6.32}$$
For instance, the choice of the matrix $U^0$ that produces a permutation of the indices 1 and 2 gives
$$\left\langle |S_{1\alpha}|^2\right\rangle_0 = \left\langle |S_{2\alpha}|^2\right\rangle_0. \tag{6.33}$$

Similarly,
$$\left\langle |S_{1\alpha}|^2 \right\rangle_0 = \cdots = \left\langle |S_{M\alpha}|^2 \right\rangle_0. \qquad (6.34)$$

From unitarity we thus have
$$\sum_{a=1}^{M} \left\langle |S_{a\alpha}|^2 \right\rangle_0 = 1 \quad \Longrightarrow \quad \left\langle |S_{a\alpha}|^2 \right\rangle_0^{(2)} = \frac{1}{M}. \qquad (6.35)$$

As an application, we calculate the first and second moments of the conductance of a cavity having two connecting leads with $N_1$ and $N_2$ open channels (so that $M = N_1 + N_2$). For the first moment (average), we obtain

$$\langle T \rangle_0^{(2)} = \sum_{a=1}^{N_1} \sum_{b=1}^{N_2} \left\langle |t_{ab}|^2 \right\rangle_0^{(2)} = \frac{N_1 N_2}{N_1 + N_2} = \left[ \frac{1}{N_1} + \frac{1}{N_2} \right]^{-1}. \qquad (6.36)$$

This is clearly the classical series addition of the two dimensionless conductances $N_1$ and $N_2$. (The superscript on the angular brackets indicates $\beta = 2$.)

Similarly, one can obtain the following results [127]:

$$\left\langle |S_{12}|^2 |S_{34}|^2 \right\rangle_0^{(2)} = \frac{1}{M^2 - 1}, \qquad (6.37a)$$

$$\left\langle |S_{12}|^2 |S_{13}|^2 \right\rangle_0^{(2)} = \frac{1}{M(M+1)}, \qquad (6.37b)$$

$$\left\langle |S_{12}|^4 \right\rangle_0^{(2)} = \frac{2}{M(M+1)}. \qquad (6.37c)$$

Here 1, 2, 3 and 4 stand for any quartet of different indices, so that $n$ in each case must be large enough to accommodate as many indices as necessary. These expressions can be used to obtain the second moment of the conductance as

$$\langle T^2 \rangle_0^{(2)} = \sum_{a,c=1}^{N_1} \sum_{b,d=1}^{N_2} \left\langle |t_{ab}|^2 |t_{cd}|^2 \right\rangle_0^{(2)} = \frac{N_1^2 N_2^2}{(N_1 + N_2)^2 - 1}. \qquad (6.38)$$

The associated variance of the conductance is then given by

$$[\text{var } T]_0^{(2)} = \frac{N_1^2 N_2^2}{(N_1 + N_2)^2 \left[ (N_1 + N_2)^2 - 1 \right]}. \qquad (6.39)$$

It is interesting that the above expression directly gives the so-called *universal conductance fluctuations* (UCF), in that

$$[\text{var } T]_0^{(2)} \to \frac{K^2}{(K+1)^4}, \qquad (6.40)$$

in the limit as $N_1, N_2 \to \infty$ with $N_1/N_2 = K$ fixed. Equation (6.40) gives a constant which depends only on the ratio $N_1/N_2$, all other details of the cavity becoming irrelevant. Since this limit corresponds to increasing the width of the waveguides, and so also of the full system, the fact that the result is constant is the analog of the well-known UCF for quasi-one-dimensional disordered systems [7, 24, 126]. In particular, for $K = 1$, $\operatorname{var} T \to 1/16$, which is slightly less than the value $1/15$ for quasi-one-dimensional disordered conductors.

*The orthogonal case ($\beta = 1$)*

Again, $M$ will denote the dimensionality of the $S$ matrix. In the present case, the appropriate transformed $\widetilde{S}$ is

$$S = U^0 \widetilde{S} (U^0)^\top, \tag{6.41}$$

so that we can write the average $\langle S_{ab} \rangle_0^{(1)}$ as

$$\langle S_{ab} \rangle_0^{(1)} = \sum_{a'} U^0_{aa'} U^0_{bb'} \int \widetilde{S}_{a'b'} \, \mathrm{d}\mu^{(1)}(\widetilde{S}) = \sum_{a'} U^0_{aa'} U^0_{bb'} \langle S_{a'b'} \rangle_0^{(1)}. \tag{6.42}$$

Similar to the unitary case, we take as the arbitrary fixed matrix $U^0$ the one given by eqn (6.26) and obtain

$$\langle S_{ab} \rangle_0 = \mathrm{e}^{\mathrm{i}(\theta_a + \theta_b)} \langle S_{ab} \rangle_0 \quad \Longrightarrow \quad \langle S_{ab} \rangle_0 = 0, \tag{6.43}$$

since $\theta_a$ and $\theta_b$ are arbitrary.

Again, the above argument can be generalized to prove that

$$\left\langle \left[ S_{a_1 b_1} \cdots S_{a_p b_p} \right] \left[ S_{\alpha_1 \beta_1} \cdots S_{\alpha_q \beta_q} \right]^* \right\rangle_0^{(1)} = 0, \tag{6.44}$$

unless $p = q$, and unless $\{a_1, b_1, \ldots, a_p, b_p\}$ and $\{\alpha_1, \beta_1, \ldots, \alpha_p, \beta_p\}$ constitute the same set of indices except for order. As an example, for $p = q = 1$ we find

$$\langle S_{ab} S^*_{\alpha\beta} \rangle_0^{(1)} = \sum_{a'b'\alpha'\beta'} U^0_{aa'} U^0_{bb'} \left( U^0_{\alpha\alpha'} U^0_{\beta\beta'} \right)^* \langle S_{a'b'} S^*_{\alpha'\beta'} \rangle_0^{(1)} \quad \text{for all } U^0. \tag{6.45}$$

For the $U^0$ used above, eqn (6.26), we have

$$\langle S_{ab} S^*_{\alpha\beta} \rangle_0^{(1)} = \mathrm{e}^{\mathrm{i}(\theta_a + \theta_b - \theta_\alpha - \theta_\beta)} \langle S_{ab} S^*_{\alpha\beta} \rangle_0^{(1)}, \tag{6.46}$$

which vanishes unless $\{a, b\} = \{\alpha, \beta\}$ or $\{\beta, \alpha\}$. For $a = b = \alpha = \beta = 1$, and using as $U^0$ the matrix

$$U^0 = \begin{bmatrix} u^0 \, [2 \times 2] & 0 \, [2 \times (M-2)] \\ 0 \, [(M-2) \times 2] & I \, [(M-2) \times (M-2)] \end{bmatrix}, \tag{6.47a}$$

where the dimensionality of each submatrix is indicated in square brackets and the two-dimensional matrix $u^0$ is given by

## ABSENCE OF PROMPT (DIRECT) PROCESSES

$$U^0 = \begin{bmatrix} U^0_{11} & U^0_{12} \\ U^0_{21} & U^0_{22} \end{bmatrix}, \tag{6.47b}$$

we find

$$\left\langle |S_{11}|^2 \right\rangle_0 = |U^0_{11}|^4 \left\langle |S_{11}|^2 \right\rangle_0 + |U^0_{12}|^4 \left\langle |S_{22}|^2 \right\rangle_0 + 4 |U^0_{11}|^2 |U^0_{12}|^2 \left\langle |S_{12}|^2 \right\rangle_0. \tag{6.48}$$

If $U^0$ is taken to be the matrix that produces a permutation of the indices 1 and 2, we find

$$\left\langle |S_{11}|^2 \right\rangle_0 = \left\langle |S_{22}|^2 \right\rangle_0, \tag{6.49}$$

so that (6.48) gives

$$\left\langle |S_{11}|^2 \right\rangle_0 = \left[ |U^0_{11}|^4 + |U^0_{12}|^4 \right] \left\langle |S_{11}|^2 \right\rangle_0 + 4 |U^0_{11}|^2 |U^0_{12}|^2 \left\langle |S_{12}|^2 \right\rangle_0. \tag{6.50}$$

Using the unitarity of $U^0$, we finally obtain

$$\left\langle |S_{11}|^2 \right\rangle_0^{(1)} = 2 \left\langle |S_{12}|^2 \right\rangle_0^{(1)}. \tag{6.51}$$

This result is very important. It states that time-reversal invariance (TRI) has the consequence that *the average of the absolute value squared of a diagonal S-matrix element is twice as large as that of an off-diagonal one*, under the invariant measure. We call this doubling effect *coherent back-scattering enhancement* (see also Chapter 8). By unitarity, the value of each one of these averages is given by

$$\left\langle |S_{aa}|^2 \right\rangle_0^{(1)} = \frac{2}{M+1}, \tag{6.52a}$$

$$\left\langle |S_{a \neq b}|^2 \right\rangle_0^{(1)} = \frac{1}{M+1}. \tag{6.52b}$$

We can now calculate as follows the *average conductance* when the cavity is connected to two leads with $N_1$ and $N_2$ open channels ($M = N_1 + N_2$):

$$\langle T \rangle_0^{(1)} = \sum_{a=1}^{N_1} \sum_{b=1}^{N_2} \left\langle |t_{ab}|^2 \right\rangle_0^{(1)} = \frac{N_1 N_2}{N_1 + N_2 + 1}. \tag{6.53}$$

The extra '1' appearing in the denominator, as compared with eqn (6.36), is interpreted as the *weak-localization correction* (WLC), a symmetry effect resulting from TRI, as also explained in Chapter 8 for disordered systems. We can rewrite eqn (6.53), separating out the WLC term, as

$$\langle T \rangle_0^{(1)} = \frac{N_1 N_2}{N_1 + N_2} - \frac{N_1 N_2}{(N_1 + N_2)(N_1 + N_2 + 1)}. \tag{6.54}$$

In particular, for $N_1 = N_2 = N$ and for $N \to \infty$, corresponding to a large system, the WLC term tends to the *universal* number $-1/4$.

One can also prove the following results ($M$ being again the dimensionality of the $S$ matrix) [135]:

$$\left\langle |S_{12}|^2 |S_{34}|^2 \right\rangle_0^{(1)} = \frac{M+2}{M(M+1)(M+3)}, \tag{6.55a}$$

$$\left\langle |S_{12}|^2 |S_{13}|^2 \right\rangle_0^{(1)} = \frac{1}{M(M+3)}, \tag{6.55b}$$

$$\left\langle |S_{12}|^4 \right\rangle_0^{(1)} = \frac{2}{M(M+3)}. \tag{6.55c}$$

A comment like the one made immediately after eqn (6.37) also applies here. We now find the second moment of the conductance to be

$$\left\langle T^2 \right\rangle_0^{(1)} = \frac{N_1 N_2 \left[ N_1 N_2 (N_1 + N_2 + 2) + 2 \right]}{(N_1 + N_2)(N_1 + N_2 + 1)(N_1 + N_2 + 3)}, \tag{6.56}$$

and its variance is

$$[\text{var}\, T]_0^{(1)} = \frac{2 N_1 N_2 (N_1 + 1)(N_2 + 1)}{(N_1 + N_2)(N_1 + N_2 + 1)^2 (N_1 + N_2 + 3)} \to 2 \frac{K^2}{(K+1)^4}. \tag{6.57}$$

The limit in eqn (6.57) is as $N_1, N_2 \to \infty$ with $N_1/N_2 = K$ a fixed number. Note that in this *universal* limit the variance is now exactly twice as large as that for $\beta = 2$, eqn (6.40), as a consequence of time-reversal invariance. For the particular case $K = 1$, the limiting value of the variance is $1/8$, to be compared with the value $2/15$ for quasi-one-dimensional disordered conductors.

### 6.4.2 *The distribution of the conductance in the two-equal-lead case*

In the last section we calculated the first two moments of the statistical distribution of the dimensionless conductance for a chaotic cavity, in the absence of prompt processes. This suffices to characterize the full distribution, if the latter is Gaussian; in fact, this can be shown to be the case in the limit as $N \to \infty$ [151]. For finite $N$, however, deviations from a Gaussian distribution are appreciable, and make the problem even more appealing. To this interesting case we now turn our attention.

For this purpose, the polar representation of eqn (3.86) is particularly useful, since the dimensionless conductance is directly related to the $\tau_a$s, whose joint probability distribution we know. Specifically, the distribution of the transmission $T$ of eqn (4.184) can be obtained by direct integration of the $P^{(\beta)}(\{\tau\})$ of eqns (6.6b) and (6.7b) as follows:

$$w^{(\beta)}(T) = \int \delta(T - \sum_a \tau_a) P^{(\beta)}(\{\tau_a\}) \prod_a d\tau_a. \tag{6.58}$$

We now analyze a few examples.

## The case $N = 1$

Now we have only one $\tau_a \equiv \tau$, so that $0 \leq T = \tau \leq 1$ and eqns (6.6b) and (6.7b) give

$$w^{(1)}(T) = \frac{1}{2\sqrt{T}}, \tag{6.59a}$$

$$w^{(2)}(T) = 1. \tag{6.59b}$$

For $\beta = 1$, we thus have a higher probability of finding small $T$s than $T \sim 1$; this is clearly a symmetry effect, a result of TRI that favors back-scattering and hence low conductances.

## The case $N = 2$

Now $T = \tau_1 + \tau_2$, and $0 \leq T \leq 2$. In the two following exercises we prove that

$$w^{(1)}(T) = \begin{cases} \frac{3}{2}T, & 0 < T < 1, \\ \frac{3}{2}\left(T - 2\sqrt{T-1}\right), & 1 < T < 2, \end{cases} \tag{6.60a}$$

$$w^{(2)}(T) = 2\left(1 - |1 - T|\right)^3. \tag{6.60b}$$

For $\beta = 1$, we have a square-root cusp at $T = 1$. We find, once again, a higher probability for the occurrence of $T < 1$ than $T > 1$. On the other hand, for $\beta = 2$, $w(T)$ is again symmetric around $T = 1$.

**Exercise 6.1** Derive eqn (6.60b).

For the two-channel distribution we have to perform the integral (6.58) in the two-dimensional space $\tau_1, \tau_2$, inside the square $0 < \tau_1 < 1$, $0 < \tau_2 < 1$ indicated in Fig. 6.2. The change of variables

$$T = \tau_1 + \tau_2, \quad \tau = \tau_1 - \tau_2 \tag{6.61}$$

will be found to be advantageous. As shown in Fig. 6.2, when $0 < T < 1$ the variable $\tau$ varies in the interval $-T < \tau < T$, whereas when $1 < T < 2$ we have $T - 2 < \tau < 2 - T$.

From eqn (6.7b) for the invariant measure we find, for two channels,

$$P^{(2)}(\tau_1, \tau_2) = C\left(\tau_1 - \tau_2\right)^2. \tag{6.62}$$

The normalization constant is found from

$$1 = C \iint_0^1 (\tau_1 - \tau_2)^2 \, d\tau_1 \, d\tau_2$$
$$= \frac{C}{2}\left[\int_0^1 dT \int_{-T}^{T} d\tau \cdot \tau^2 + \int_1^2 dT \int_{T-2}^{2-T} d\tau \cdot \tau^2\right] = \frac{C}{6}, \tag{6.63}$$

so that $C = 6$.

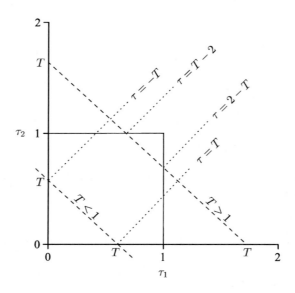

FIG. 6.2. Region of integration for finding the distribution $w(T)$ in the $N = 2$ case.

The distribution of the spinless conductance is thus

$$w^{(2)}(T) = 3\left[\int_0^1 dT'\,\delta(T-T')\int_{-T'}^{T'} d\tau\cdot\tau^2 + \int_1^2 dT'\,\delta(T-T')\int_{T'-2}^{2-T'} d\tau\cdot\tau^2\right], \quad (6.64)$$

so that
$$w^{(2)}(T<1) = 2T^3, \quad w^{(2)}(1<T<2) = 2(2-T)^3. \quad (6.65)$$
These two equations can be condensed into the result of eqn (6.60b).

**Exercise 6.2** Derive eqn (6.60a).

From eqn (6.6b) for the invariant measure we find, for two channels,

$$P^{(1)}(\tau_1,\tau_2) = C\frac{|\tau_1-\tau_2|}{\sqrt{\tau_1\tau_2}}. \quad (6.66)$$

The normalization constant is found from

$$\begin{aligned}
1 &= C\iint_0^1 \frac{|\tau_1-\tau_2|}{\sqrt{\tau_1\tau_2}}\,d\tau_1\,d\tau_2 \\
&= \frac{C}{2}\left[\int_0^1 dT\int_{-T}^{T} d\tau\,\frac{2|\tau|}{\sqrt{T^2-\tau^2}} + \int_1^2 dT\int_{T-2}^{2-T} d\tau\,\frac{2|\tau|}{\sqrt{T^2-\tau^2}}\right] \\
&= \frac{4}{3}C, \quad (6.67)
\end{aligned}$$

so that $C = 3/4$.

The distribution of the spinless conductance is thus

$$w^{(1)}(T) = \frac{3}{8} \left[ \int_0^1 dT' \, \delta(T-T') \int_{-T'}^{T'} d\tau \, \frac{2|\tau|}{\sqrt{T'^2 - \tau^2}} \right. \\ \left. + \int_1^2 dT' \, \delta(T-T') \int_{T'-2}^{2-T'} d\tau \, \frac{2|\tau|}{\sqrt{T'^2 - \tau^2}} \right], \quad (6.68)$$

so that

$$w^{(1)}(T<1) = \frac{3}{2}T, \quad w^{(1)}(1<T<2) = \frac{3}{2}\left(T - 2\sqrt{T-1}\right), \quad (6.69)$$

which is eqn (6.60a) which we were required to prove.

*The case $N = 3$*

Now $T = \tau_1 + \tau_2 + \tau_3$, and $0 \leqslant T \leqslant 3$. Proceeding as before, we find, for $\beta = 2$,

$$w^{(2)}(T) = \begin{cases} (9/42)T^8, & 0 < T < 1, \\ -(2781/14) + (6588/7)T - 1818T^2 + 1836T^3 \\ \quad - 1035T^4 + 324T^5 - 54T^6 + (36/7)T^7 - (3/7)T^8, & 1 < T < \frac{3}{2}, \end{cases} \quad (6.70)$$

and the distribution is symmetric about $T = 3/2$. As mentioned above, $w^{(\beta)}(T)$ gradually approaches a Gaussian distribution.

*Arbitrary $N$*

In this case it is straightforward to obtain the dependence of the tail of the distribution in the region $0 < T < 1$; a calculation presented in the following exercise shows that

$$w_N^{(\beta)}(T) \propto T^{\beta N^2/2 - 1}. \quad (6.71)$$

**Exercise 6.3** Derive eqn (6.71).

The integration region selected by the delta function in eqn (6.58) is $(N-1)$-dimensional. In the particular case of $T < 1$, each of the $\tau_a$s ($a = 1, \ldots, N$) varies, over that region, in the interval $0 < \tau_a < T$. This is clearly illustrated in Fig. 6.2 for $N = 2$, where the relevant region is the segment extending from $(\tau_1, \tau_2) = (0, T)$ to $(\tau_1, \tau_2) = (T, 0)$. For $T < 1$ the corners of the $N$-dimensional hypercube are not relevant; they become relevant for $T > 1$. We can thus write, for $T < 1$,

$$w_N^{(\beta)}(T<1) = C_N^{(\beta)} \int_0^T \cdots \int_0^T \delta\left(T - \sum_{a=1}^N \tau_a\right) \prod_{a<b} |\tau_a - \tau_b|^\beta \prod_c \tau_c^{(\beta-2)/2} \prod_i d\tau_i. \quad (6.72)$$

Introducing the new variables $\sigma_a = \tau_a/T$ ($a = 1, \ldots, N$) which vary in the interval $(0, 1)$, we have

$$w_N^{(\beta)}(T<1) = C_N^{(\beta)} \int_0^1 \cdots \int_0^1 \delta\left(T - T\sum_{a=1}^N \sigma_a\right) T^{(N(N-1)/2)\beta}$$
$$\times \prod_{a<b} |\sigma_a - \sigma_b|^\beta \, T^{N(\beta-2)/2} \prod_c \sigma_c^{(\beta-2)/2} T^N \prod_i d\sigma_i$$
$$= C_N^{(\beta)} T^{\beta N^2/2 - 1} \int_0^1 \cdots \int_0^1 \delta\left(1 - \sum_{a=1}^N \sigma_a\right)$$
$$\times \prod_{a<b} |\sigma_a - \sigma_b|^\beta \prod_c \sigma_c^{(\beta-2)/2} \prod_i d\sigma_i, \qquad (6.73)$$

which behaves as the power of $T$ indicated in eqn (6.71).

## 6.5 Presence of prompt (direct) processes

We now need Poisson's kernel (6.18) in its full generality. We discuss below some analytical results for the distribution of the spinless conductance $T$ in the case of a cavity connected to two one-open-channel leads ($N_1 = N_2 = 1$); the associated $S$ matrix is two-dimensional. There is only one $\tau$ in this case, and it is its distribution that we seek, as $T = \tau$. Some of the expressions that we derive are used for comparing with numerical results in Section 6.6, where plots of several examples are displayed.

We write the optical $S$ matrix $\overline{S}$, a subunitary matrix, as

$$\overline{S} = \begin{bmatrix} x & w \\ z & y \end{bmatrix}, \qquad (6.74)$$

where the entries are, in general, complex numbers.

### 6.5.1 The case $\beta = 2$

From eqn (6.18) we write the differential probability for the $S$ matrix as

$$dP_{\overline{S}}^{(2)}(S) = \frac{\left[\det\left(I - \overline{S}\,\overline{S}^\dagger\right)\right]^n}{\left|\det\left(I - S\overline{S}^\dagger\right)\right|^{2n}} d\mu_0^{(2)}(S), \qquad (6.75)$$

where

$$d\mu_0^{(2)}(S) = \frac{d\mu^{(2)}(S)}{V}, \quad \int d\mu_0^{(2)}(S) = 1. \qquad (6.76)$$

We are interested here in the case $n = 2$.

In the polar representation (3.86) with $n = 2$, the $S$ matrix has the form

$$S = \begin{bmatrix} e^{i\alpha} & 0 \\ 0 & e^{i\beta} \end{bmatrix} \begin{bmatrix} -\sqrt{1-\tau} & \sqrt{\tau} \\ \sqrt{\tau} & \sqrt{1-\tau} \end{bmatrix} \begin{bmatrix} e^{i\gamma} & 0 \\ 0 & e^{i\delta} \end{bmatrix}, \qquad (6.77)$$

the invariant measure (6.76) being (see eqns (6.7))

$$d\mu_0(S) = d\tau \, \frac{d\alpha \, d\beta \, d\gamma \, d\delta}{(2\pi)^4}. \tag{6.78}$$

We consider the following three cases.

(i) As a particular case, suppose that the optical $S$ matrix is diagonal, so that there is *only direct reflection* and no direct transmission; in eqn (6.74) we set $w = z = 0$.

Substituting eqns (6.74), (6.77) and (6.78) into (6.75), we find

$$dP^{(2)}_{x,y}(\tau, \varphi, \psi) = \frac{(1 - X^2)^2 (1 - Y^2)^2}{\left| \left(e^{-i\varphi} + x^*\sqrt{1-\tau}\right) \left(e^{-i\psi} - y^*\sqrt{1-\tau}\right) - x^* y^* \tau \right|^4} \frac{d\tau \, d\varphi \, d\psi}{(2\pi)^2}, \tag{6.79}$$

where $\varphi = \alpha + \gamma$, $\psi = \beta + \delta$, $X = |x|$ and $Y = |y|$. The distribution of the conductance $T$ is thus

$$w^{(2)}_{X,Y}(T) = (1 - X^2)^2 (1 - Y^2)^2$$
$$\times \left\langle \frac{1}{\left| \left(e^{-i\varphi} + X\sqrt{1-T}\right) \left(e^{-i\psi} - Y\sqrt{1-T}\right) - XYT \right|^4} \right\rangle_{\varphi, \psi}, \tag{6.80}$$

where $\langle \cdots \rangle_{\varphi, \psi}$ denotes an average over the variables $\varphi$ and $\psi$ in the interval $(0, 2\pi)$. The result is (recall that $0 < T < 1$)

$$w^{(2)}_{X,Y}(T) = K \frac{A - B(1-T) + C(1-T)^2 + D(1-T)^3}{[E - 2F(1-T) + G(1-T)^2]^{5/2}}, \tag{6.81}$$

where

$$K = (1 - X^2)^2 (1 - Y^2)^2, \tag{6.82a}$$
$$A = (1 - X^4 Y^4)(1 - X^2 Y^2), \tag{6.82b}$$
$$B = (X^2 + Y^2)(1 - 6X^2 Y^2 + X^4 Y^4) + 4X^2 Y^2 (1 + X^2 Y^2), \tag{6.82c}$$
$$C = (1 + X^2 Y^2)(6X^2 Y^2 - X^4 - Y^4) - 4X^2 Y^2 (X^2 + Y^2), \tag{6.82d}$$
$$D = (X^2 + Y^2)(X^2 - Y^2)^2, \tag{6.82e}$$
$$E = (1 - X^2 Y^2)^2, \tag{6.82f}$$
$$F = (1 + X^2 Y^2)(X^2 + Y^2) - 4X^2 Y^2, \tag{6.82g}$$
$$G = (X^2 - Y^2)^2. \tag{6.82h}$$

This result reduces to 1 when $X = Y = 0$, as expected. A particularly interesting case is that of 'equivalent channels', i.e., $X = Y$, for which the expression (6.81) reduces to

$$w^{(2)}_{X,X}(T) = (1 - X^2) \frac{(1 - X^4)^2 + 2X^2 (1 + X^4) T + 4X^4 T^2}{[(1 - X^2)^2 + 4X^2 T]^{5/2}}. \tag{6.83}$$

The structure of this result is clear if we notice that

$$w^{(2)}_{X,X}(0) = \left(\frac{1+X^2}{1-X^2}\right)^2 > 1, \quad w^{(2)}_{X,X}(1) = \frac{(1-X^2)(1+X^4)}{(1+X^2)^3} < 1,$$

and hence $w^{(2)}_{X,X}(0) > w^{(2)}_{X,X}(1)$, so that small conductances are emphasized, as expected, because of the presence of direct reflection and no direct transmission.

(ii) The case of *only direct transmission* and no direct reflection is obtained by setting $x = y = 0$ in eqn (6.74). The conductance distribution is obtained from eqn (6.81) with the replacements $X \to W = |w|$, $Y \to Z = |z|$ and $T \to 1 - T$. In the equivalent-channel case we now obtain a conductance distribution that emphasizes large conductances.

(iii) The case of a general optical $S$ matrix, eqn (6.74), has also been studied and the result expressed in terms of quadratures; because of its complexity, it will not be quoted here.

### 6.5.2 The case $\beta = 1$

This case is more complicated than that for $\beta = 2$, and we have only succeeded in treating analytically some particular cases. Take, for instance, a diagonal matrix $\overline{S}$, i.e., $w = z = 0$ in eqn (6.74). With the same notation as above, we find

$$w^{(1)}_{X,Y}(T) = \left|(1-X^2)(1-Y^2)\right|^{3/2} \frac{1}{2\sqrt{T}}$$
$$\times \left\langle \frac{1}{\left|(e^{-i\varphi} + X\sqrt{1-T})(e^{-i\psi} - Y\sqrt{1-T}) - XYT\right|^3} \right\rangle_{\varphi,\psi},$$
(6.84)

a result that has to be integrated numerically. When $X = Y = 0$, the distribution (6.84) reduces to $1/2\sqrt{T}$, as expected. It is interesting to notice that for $Y = 0$ the above result can be integrated analytically, to give

$$w^{(1)}_{X,0}(T) = \frac{(1-X^2)^{3/2}}{2\sqrt{T}} {}_2F_1\left(3/2, 3/2; 1; X^2(1-T)\right),$$
(6.85)

${}_2F_1$ being a hypergeometric function [2].

## 6.6 Numerical calculations and comparison with theory

As we mentioned at the beginning of this chapter, the maximum-entropy approach that we have been discussing is expected to be valid for cavities in which the classical dynamics is completely chaotic, a property that refers to the *long-time* behavior of the system. It is in such structures that the long-time response is equilibrated and classically ergodic, and therefore one can expect that maximum-entropy considerations will be relevant. In this section we examine particular cavities numerically in order to determine to what extent our approach really holds. The structures that we consider are all 'billiards'—they consist of hard walls surrounding a cavity—with two leads. We start by considering structures in which the absence of direct processes is assured, and then move on to structures that obviously support direct processes.

In the 'quantum chaos' literature, several billiards are used as standard examples of closed chaotic systems, of which the two most studied ones are the Sinai billiard (the region enclosed between a square and a circle centered in the square) and the stadium billiard (two half-circles joined by straight edges). The classical dynamics in these two billiards is known to be completely chaotic classically. To construct an open system to test our theory, we then take one of these billiards and attach two leads to it. The open-stadium billiard was studied previously precisely for this reason [12, 13, 65, 95, 98]. Here we directly compare numerical results for this system to the predictions of the maximum-entropy approach. The numerical methods employed in these calculations are covered in detail in [11].

6.6.1 *Absence of prompt (direct) processes*

To ensure the absence of obviously direct processes, we have introduced 'stoppers' into the stadium, to block both the direct and the whispering gallery trajectories (that hug the boundary of the cavity); examples are shown in Fig. 6.3. In order to study its statistical properties, the conductance is calculated as a function of energy.

A statistical ensemble was generated parametrically based on the following considerations. The energy variation of the conductance is on the scale of $\hbar\gamma_{esc}$ (the escape rate from the cavity) [29–31, 95], which is much smaller than the spacing between the modes in the leads ($\hbar v_F/W$). Thus, across the energy axis, many independent samplings of the conductance at a fixed number of modes may be obtained. In addition, we vary slightly the position of the stoppers so as to change the interference effects and collect better statistics. Also, for a nonzero magnetic field, two values of $B$ were used. The numerical results in Fig. 6.3 used fifty energies for each $N$ (all chosen away from the threshold for the modes) and the six different stopper configurations shown. We see that the agreement with the CE is very good for both the mean and the variance, both for $B = 0$ and for nonzero $B$.

A much more dramatic prediction of the model is the strongly non-Gaussian distribution of $T$ for a small number of modes. The analytic results for the CE were given in Section 6.4.2. These are compared to the numerical results for $N = 1$ and 2 in Fig. 6.4, using the same data as in Fig. 6.3. Note that the data is consistent with a square-root singularity in the case $N = 1$ and $B = 0$, and with cusps in the two $N = 2$ cases. Thus we see that, even for this much more stringent test, the agreement between the behavior of real cavities and the CE is excellent.

The above predictions for the variance of the conductance arising from the CE were used recently [180] in relation to the transport problem through graphene quantum dots.

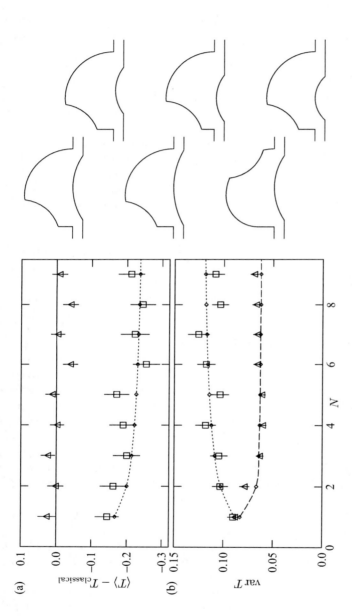

FIG. 6.3. The magnitude of (a) the quantum correction to the classical conductance, and (b) the conductance fluctuations as a function of the number of modes $N$ in the lead. The numerical results for $B = 0$ (squares with statistical error bars) agree with the prediction of the COE (points, dotted line), while those for $B \neq 0$ (triangles) agree with the CUE (points, dashed line). The six cavities shown on the right were used; the average of the numerical results is plotted. Note that each cavity has stoppers to block both the direct and whispering gallery trajectories. For a nonzero magnetic field, $BA/\phi_0 = 2$ and $4$, where $A$ is the area of the cavity. The classical transmission probabilities for these cavities ranged from 0.46 to 0.51, with a mean of 0.49. (From [131].)

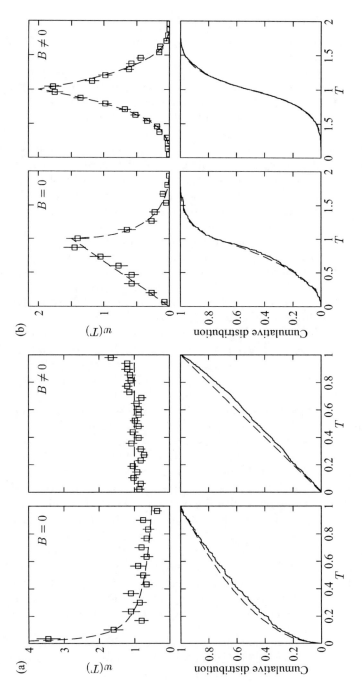

FIG. 6.4. The distribution of the transmission intensity at fixed (a) $N = 1$, or (b) $N = 2$, in both the absence and presence of a magnetic field, compared to the analytic COE and CUE results. The panels in the first row are histograms; those in the second row are cumulative distributions. Note both the strikingly non-Gaussian distributions and the good agreement between the numerical results and the CE in all cases. The cavities and energy sampling points used are the same as those in Fig. 6.3; for $B \neq 0$, the values $BA/\phi_0 = 2, 3, 4$ and $5$ were used. (From [131].)

### 6.6.2 Presence of prompt (direct) processes

We now consider more general structures than those used in the last section, i.e., we remove the stoppers that blocked short paths, and compare the numerical results with the predictions of Poisson's kernel, following [131] and [16]. We will actually consider structures where we expect *two* widely separated time-scales, a prompt response and an equilibrated one. Since we have only obtained explicit analytical results in the case $N = 1$ (see Section 6.5), we will only study this case numerically.

The conductance for several billiards, shown in Fig. 6.5, has been computed. Further, statistics were collected by (i) sampling in an energy window larger than the energy correlation length but smaller than the interval over which the prompt response changes, and (ii) using several slightly different structures. We emphasize that the stopper is used here to increase statistics, not block short (direct) paths. As in the absence of direct processes, since we are mostly averaging over energy, we rely on ergodicity to compare the numerical distributions to the ensemble averages of the maximum-entropy model. The optical $S$ matrix was extracted directly from the numerical data and used as $\langle S \rangle$ in Poisson's kernel; in this sense the theoretical curves shown below are *parameter free*.

The numerical simulations were carried out for two different values of the magnetic field intensity, as discussed below.

We first consider the billiard shown in the upper left part of Fig. 6.5, with no potential barriers at the openings and at a low magnetic field ($BA/\phi_0 = 2$, where $A$ is the area of the cavity; by a low magnetic field we mean that the cyclotron radius $r_c$ is much larger than the size of the cavity—$r_c = 55W$, shown to scale). In this case $w(T)$ is nearly uniform (Fig. 6.5(a)), and $\langle S \rangle$ is small because direct trajectories are negligible in this large structure. We thus obtain good agreement with the constant distribution (6.59b) predicted by the invariant measure.

In order to see the effect of direct processes we vary $\langle S \rangle$ by making the following changes: (i) introduce potential barriers at the openings of the leads into the cavity; (ii) increase the magnetic field; or (iii) extend the leads into the cavity. The barriers cause an immediate reflection and skew the distribution towards small $T$ (Fig. 6.5(b)). This is in contrast with the transmitted particles which are trapped for a much longer time; this provides the two very different times-scales referred to above. Secondly, the large magnetic field ($BA/\phi_0 = 80$) corresponds to $r_c$ being just larger than the width of the lead ($r_c = 1.4W$). Physically, the field increases one component of the direct transmission—the one corresponding to skipping orbits along the lower edge—and skews the distribution towards large $T$ (Fig. 6.5(c)). Thirdly, extending the leads into the cavity increases the direct transmission in both directions and also skews the distribution towards large $T$ (Fig. 6.5(d)).

In each of these cases, the numerical histogram is compared with the maximum-entropy model (solid lines) in which the numerically obtained $\overline{S}$ is inserted. In panels Fig. 6.5(b–d) the curve plotted is the analytic expression of eqn (6.81)

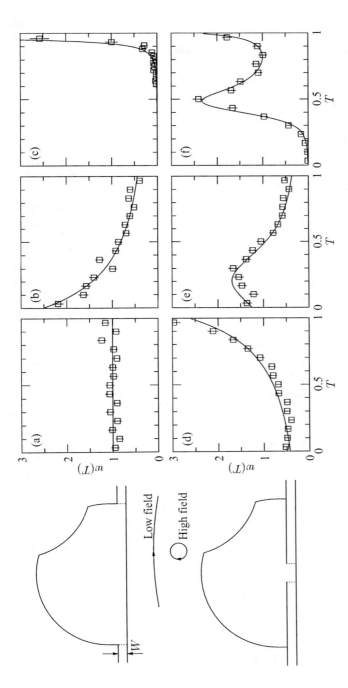

FIG. 6.5. The transmission coefficient distribution in a simple billiard (a–c) and a billiard with leads extended into the cavity (d–f) for $N = 1$. The magnitude of the magnetic field and the presence or absence of a potential barrier at the entrance to the leads (marked by dotted lines in the sketches of the structures) are as follows: (a,d) low field, no barrier; (b,e) low field, with barrier; (c) high field, no barrier; (f) high field, with barrier. Cyclotron orbits for both fields, drawn to scale, are shown on the left. The squares with statistical error bars are the numerical results; the solid lines are the predictions of the maximum-entropy model, parametrized by an optical $S$ matrix extracted from the numerical data. The agreement is good in all cases. (From [16, 131]).

and the corresponding one for direct transmission. Note the excellent agreement with the theoretical model.

Since the long-time classical dynamics in each of the three structures corresponding to Fig. 6.5(a,b,d) is chaotic, these results show that a wide variety of behavior is possible for chaotic scattering, the invariant measure description applying only when there is a single characteristic time-scale.

For the case shown in Fig. 6.5(c), the dynamics is not completely chaotic because of the small cyclotron radius, and so one would not expect the circular ensemble to apply. In [14] it was found that increasing the magnetic flux through the structure beyond a few flux quanta spoiled the agreement with the circular ensemble; we now know that a nonzero $\langle S \rangle$ is generated and that the present model describes the data very well.

By using extended leads with barriers at their ends, one can cause both prompt transmission and reflection; the result is shown in Fig. 6.5(e). Finally, increasing the magnetic field in this structure produces a large average transmission and a large average reflection. The resulting $w(T)$, Fig. 6.5(f), has a surprising two-peak structure; one peak near $T = 1$ is caused by the large direct transmission, and another peak is near $T = 1/2$. Even in these two unusual cases, the prediction of the maximum-entropy model is in excellent agreement with the numerical results.

## 6.7  Dephasing effects: comparison with experimental data

Quantum transport through ballistic chaotic cavities in terms of the random $S$-matrix theory was seen in the last section to be highly successful in explaining *numerical* simulations for structures in which the assumptions of the theory are expected to hold; this agreement includes the average conductance, its variance and probability density.

We now turn to actual experiments performed on *physical* systems. At first sight the disagreement between theory and experiment is striking, quantitatively for the WLC and the variance of the conductance, and qualitatively for the shape of the conductance distribution function [56, 89].

We now show that these discrepancies can be understood in terms of the decoherence of the electron wave function, which can be parametrized effectively in terms of a single phase-breaking length $l_\phi$; this decoherence effect has been neglected so far. Note, however, that, even if $l_\phi$ is larger than the linear size of the cavity, for sufficiently narrow leads the electron may spend enough time inside the sample to feel the effect of phase-breaking mechanisms. The discussion of these effects and their relevance to the description of the experimental data is the subject of the present section.

In order to simulate the presence of phase-breaking events, we invoke a model invented by M. Büttiker [48], where, in addition to the physical leads 1 and 2, a 'fake lead' 3 is inserted that connects the cavity to a phase-randomizing reservoir. In this phenomenological model, phase-breaking events in the cavity are simulated by the following mechanism: any particle which exits the cavity through lead 3 is replaced by a particle from the reservoir; since the replacement particle comes from a reservoir, it is incoherent with respect to the exiting particle, and hence the phase breaking.

As usual, leads 1 and 2 are attached to reservoirs at chemical potentials $\mu_1$ and $\mu_2$, respectively, whereas the chemical potential $\mu_3$ of the fake reservoir is fixed by demanding the net current in lead 3 to vanish. (We recall here that ideally the reservoirs cannot reflect an incoming electron wave; they can only absorb and re-emit as black bodies.) The two-terminal dimensionless conductance is then found to be

$$g \equiv \frac{G}{e^2/h} = 2\left(T^{21} + \frac{T^{23}T^{31}}{T^{32} + T^{31}}\right). \tag{6.86}$$

In this equation, the factor 2 accounts for spin explicitly and $T^{ij}$ is the transmission coefficient for 'spinless electrons' from lead $j$ to lead $i$. This expression for $g$ has a natural interpretation: the first term is the coherent transmission and the second term is the sequential transmission from 1 to 2 via 3. In terms of the $S$ matrix, $T^{ij}$ is

$$T^{ij} = \sum_{a_i=1}^{N_i} \sum_{b_j=1}^{N_j} \left|S^{ij}_{a_i b_j}\right|^2, \tag{6.87}$$

where, using the notation of Section 3.2.2, the indices $i$ and $j$ refer to the leads, and $a_i$ and $b_j$ refer to the channels in the corresponding leads. Accordingly, $N_i$ is the number of channels in lead $i$. The total number of channels in the physical and fake leads will be denoted by

$$M = \sum_i N_i, \tag{6.88}$$

as in eqn (3.234). We also introduce the number of *phase-breaking channels* $N_\phi \equiv N_3$. Physically, $N_\phi$ is related to the phase-breaking scattering rate $\gamma_\phi$ via the relation $N_\phi/(N_1 + N_2) \approx \gamma_\phi/\gamma_{\text{esc}}$, where $\gamma_{\text{esc}}$ is the escape rate from the cavity. This 'fake-lead' model has been used successfully for various studies of the effect of phase breaking in mesoscopic systems [59, 60, 62, 86, 105].

We now make the assumption that the *total* $M \times M$ scattering matrix $S$ obeys the distribution (6.22) given by the invariant measure [15, 43], i.e., all the leads are treated *at par*. Through this assumption, the effect of the third lead is somehow felt as being uniformly distributed in space rather than being lumped at any given point of the cavity. This even-handed statistical treatment makes the 'fake-lead' approach more physically reasonable. A further generalization to the case of non-ideal coupling of the fake lead to the cavity is studied in [44] but will not be presented here. In the following, we confine the analytical discussion to the large-$N_\phi$ limit, then present the results of numerical random-matrix theory calculations, and finally compare with experiments.

### 6.7.1 The limit of large $N_\phi$

We find below the average and variance of the conductance when $N_\phi \gg 1$ [15]. As we shall see, for this purpose we need the average and the covariance of the transmission coefficients $T^{ij}$ introduced above.

From eqns (6.35) and (6.52), the average of $T^{ij}$ ($i \neq j$) is given by (as usual, $\beta = 1, 2$)

$$\langle T^{ij} \rangle_0^{(\beta)} = \frac{N_i N_j}{M + \delta_{\beta 1}}, \tag{6.89}$$

so that

$$\langle T^{21} \rangle_0^{(\beta)} = \frac{N_2 N_1}{M + \delta_{\beta 1}} \approx \frac{N_2 N_1}{N_3} + O\left(\frac{1}{N_3^2}\right), \tag{6.90}$$

$$\langle T^{31} \rangle_0^{(\beta)} = \frac{N_3 N_1}{M + \delta_{\beta 1}} \approx N_1 \left[1 - \frac{N_1 + N_2 + \delta_{\beta 1}}{N_3} + O\left(\frac{1}{N_3^2}\right)\right], \tag{6.91}$$

$$\langle T^{32} \rangle_0^{(\beta)} = \frac{N_3 N_2}{M + \delta_{\beta 1}} \approx N_2 \left[1 - \frac{N_1 + N_2 + \delta_{\beta 1}}{N_3} + O\left(\frac{1}{N_3^2}\right)\right], \tag{6.92}$$

where the sign '$\approx$' refers to the situation $N_3 \gg 1$, while $N_1$ and $N_2$ are $O(1)$. Turning to the covariance, using eqns (6.37) and (6.89), one finds, for $\beta = 2$,

$$\langle \delta T^{ij} \delta T^{kl} \rangle_0^{(2)} = \frac{N_i N_j}{M^2 (M^2 - 1)} \left(M^2 \delta_{ik} \delta_{jl} - N_k M \delta_{jl} - N_l M \delta_{ik} + N_k N_l\right), \tag{6.93}$$

where we have written

$$T^{ij} = \overline{T}^{ij} + \delta T^{ij}. \tag{6.94}$$

For convenience, we use interchangeably a bar or a bracket $\langle \cdots \rangle_0^{(\beta)}$ to indicate an average over the invariant measure. Likewise, using eqns (6.55) for $\beta = 1$ yields

$$\begin{aligned}\langle \delta T^{ij} & \delta T^{kl} \rangle_0^{(1)} \\ &= \frac{1}{M(M+1)^2(M+3)} \Big\{ N_i N_j (M+1)(M+2)(\delta_{ik}\delta_{jl} + \delta_{il}\delta_{jk}) \\ &\quad + 2N_i N_k \delta_{ij}\delta_{kl} + 2N_i N_k N_l \delta_{ij} + 2N_i N_j N_k \delta_{kl} + 2N_i M(M+1)\delta_{ijkl} \\ &\quad + 2N_i N_j N_k N_l - (M+1)\big[2N_i N_l \delta_{ijk} + 2N_i N_k \delta_{ijl} \\ &\quad + 2N_i N_j (\delta_{ikl} + \delta_{jkl}) N_i N_j N_l (\delta_{ik} + \delta_{jk}) + N_i N_j N_k (\delta_{il} + \delta_{jl})\big] \Big\}.\end{aligned} \tag{6.95}$$

Here, a Kronecker delta with two or more indices vanishes unless all of its indices coincide, in which case its value is 1. These expressions are valid for arbitrary $N_i$ and $N_j$, and also for $i = j$ if $T^{ii}$ is interpreted as the reflection coefficient $R^{ii}$ from lead $i$ back to itself.

From eqns (6.93) and (6.95) we find, for the variance of $T_{ij}$ ($i \neq j$) ($\delta T^{ij} = T^{ij} - \overline{T}^{ij}$),

$$\left\langle \left(\delta T^{ij}\right)^2 \right\rangle_0^{(2)} = \frac{N_i N_j \left(M - N_i\right)\left(M - N_j\right)}{M^2 \left(M^2 - 1\right)} \tag{6.96}$$

and

$$\left\langle \left(\delta T^{ij}\right)^2 \right\rangle_0^{(1)} = \frac{N_i N_j \left[\left(M + 1 - N_i\right)\left(M + 1 - N_j\right) - \left(M + 1\right) + N_i N_j\right]}{M \left(M + 1\right)^2 \left(M + 3\right)}. \tag{6.97}$$

For two physical leads only, i.e., for $N_\phi = 0$, these last two expressions reduce to eqns (6.39) and (6.57), respectively. On the other hand, for $N_\phi \gg 1$ and $i = 1, 2$, we have

$$\frac{\left[\left\langle \left(\delta T^{i3}\right)^2 \right\rangle_0^{(\beta)}\right]^{1/2}}{\overline{T}^{i3}} \sim O\left(\frac{1}{N_\phi}\right). \tag{6.98}$$

Thus, $\delta T^{23}/\overline{T}^{23}$ and $\delta T^{13}/\overline{T}^{13}$ are small quantities and we can make an expansion in powers of them.

The average of the conductance of eqn (6.86) for $N_\phi \gg 1$ is obtained as follows, by using eqn (6.98):

$$\langle g \rangle_0^{(\beta)} = 2 \left\{ \overline{T}^{21} + \frac{\overline{T}^{23}\overline{T}^{31}}{\overline{T}^{32} + \overline{T}^{31}} \left[1 + O\left(\frac{1}{N_\phi^2}\right)\right] \right\}. \tag{6.99}$$

Substituting eqns (6.90)–(6.92) into (6.99), we have

$$\langle g \rangle_0^{(\beta)} = 2 \frac{N_2 N_1}{N_2 + N_1} \left(1 - \frac{\delta_{\beta 1}}{N_\phi}\right) + O\left(\frac{1}{N_\phi^2}\right), \tag{6.100}$$

which gives the weak-localization correction

$$\delta \langle g \rangle \equiv \langle g \rangle_0^{(1)} - \langle g \rangle_0^{(2)} = -2 \frac{N_2 N_1}{N_2 + N_1} \frac{1}{N_\phi} + O\left(\frac{1}{N_\phi^2}\right). \tag{6.101}$$

If we subtract the average conductance from $g$ and call $\delta g$ the difference, then we obtain, to lowest order in $1/N_\phi$,

$$\frac{1}{2}\delta g \equiv \frac{1}{2}\left[g - \langle g \rangle_0^{(\beta)}\right]$$

$$\approx \delta T^{21} + \frac{\overline{T}^{23}\overline{T}^{31}}{\overline{T}^{32} + \overline{T}^{31}} \left(\frac{\delta T^{23}}{\overline{T}^{23}} + \frac{\delta T^{31}}{\overline{T}^{31}} - \frac{\delta T^{32} + \delta T^{31}}{\overline{T}^{32} + \overline{T}^{31}}\right) \tag{6.102}$$

$$\approx \delta T^{21} + \frac{N_1}{N_2 + N_1}\delta T^{23} + \left(\frac{N_2}{N_2 + N_1}\right)^2 \delta T^{31} - \frac{N_2 N_1}{(N_2 + N_1)^2}\delta T^{32}. \tag{6.103}$$

Squaring this last expression and averaging, we obtain the variance of the conductance in terms of $\langle \delta T^{ij} \delta T^{kl} \rangle_0$. We must now substitute the variances and covariances given in eqns (6.93) and (6.95) to obtain the following final result:

$$[\text{var } g]_0^{(\beta)} = \left[\frac{2N_2 N_1}{(N_2 + N_1) N_\phi}\right]^2 \frac{2}{\beta} \left[1 + (2-\beta)\frac{N_2^3 + N_1^3}{N_2 N_1 (N_2 + N_1)^2}\right] + \cdots . \quad (6.104)$$

The ratio of the variances for $\beta = 1$ and $\beta = 2$ is

$$\frac{[\text{var } g]_0^{(1)}}{[\text{var } g]_0^{(2)}} = 2\left[1 + \frac{N_2^3 + N_1^3}{N_2 N_1 (N_2 + N_1)^2}\right] + \cdots . \quad (6.105)$$

We observe from this last equation that, first, this ratio is independent of $N_\phi$ to leading order, and that, second, the ratio of variances is larger than 2, and as high as 3 for $N_2 = N_1 = N = 1$. For comparison, in the absence of phase breaking and for $N_2 = N_1$, this ratio lies between 1 and 2.

### 6.7.2 Arbitrary $N_\phi$

In order to evaluate the effects of phase breaking when $N_\phi$ is not large, we numerically generated random $M \times M$ unitary or unitary symmetric matrices and computed $g$ from eqn (6.86). Fig. 6.6 shows the weak-localization correction and the variance of the conductance as the number of modes in the leads is varied for several fixed $N_\phi$. This result is relevant to experiments at fixed temperature in which the size of the opening to the cavity is varied. Though $\delta \langle g \rangle$ and var $g$ are nearly independent of $N$ in the perfectly coherent limit—the 'universality' discussed above—phase-breaking channels cause variation. Thus the universality can be seen only if $N_\phi \ll N$; otherwise, the behavior is approximately linear.

The limits of very large $N_\phi$ and $N_\phi = 0$ suggest, for WLC and var $g$, an approximate interpolation formula for arbitrary values of $N_\phi$, which, for the case $N_2 = N_1 = N$, turns out to be [15]

$$\delta \langle g \rangle \approx -\frac{N}{2N + N_\phi}, \quad (6.106)$$

$$[\text{var } g]_0^{(\beta)} \approx \left\{ \left([\text{var } g]_{0, N_\phi=0}^{(\beta)}\right)^{-1/2} + \left([\text{var } g]_{0, N_\phi \gg 1}^{(\beta)}\right)^{-1/2} \right\}^{-2}. \quad (6.107)$$

The agreement with the numerical simulations shown in Fig. 6.6 appears to be good, except for $N = 1$ and small $N_\phi$.

Figure 6.7 shows the conductance distribution in the weak phase-breaking regime with a small number of modes in the lead. From the variance and mean, we know that the phase breaking will narrow the distribution and make it more symmetric because the weak-localization correction tends to zero. Not surprisingly, the phase breaking in addition smooths the distribution, and the extreme non-Gaussian structure in $w(g)$ is washed out.

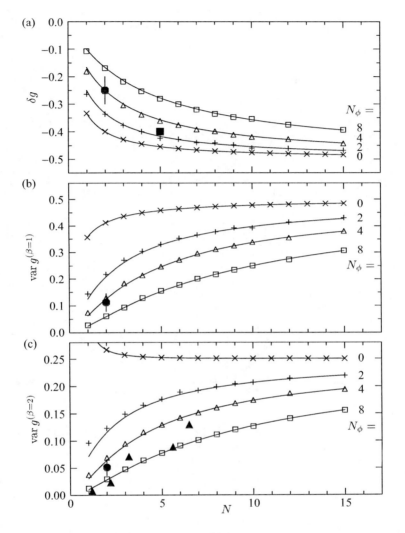

FIG. 6.6. The magnitude of quantum transport effects as a function of the number of channels in the leads, $N_1 = N_2 = N$, for $N_\phi = 0$, 2, 4 and 8. (a) The weak-localization correction. (b) The variance for the orthogonal case ($B = 0$). (c) The variance for the unitary case (nonzero $B$). The open symbols are numerical results (20 000 matrices used; statistical error is the symbol size). The solid lines are interpolation formulae. The solid symbols are experimental results of [58] (squares), [26,27] (triangles) and [56] (circles) corrected for thermal averaging. The introduction of phase breaking decreases the 'universality' of the results but leads to good agreement with experiment. (From [15].)

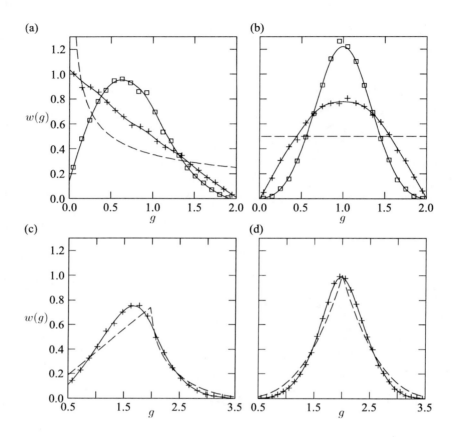

FIG. 6.7. The probability density of the conductance in the orthogonal (a,c) ($B = 0$) and unitary (b,d) ($B \neq 0$) cases for $N = 1$ (a,b) and $N = 2$ (c,d). Increasing the phase breaking from zero (dashed lines, analytic) to $N_\phi = 1$ ('+' symbols, numerical) to $N_\phi = 2$ (squares, numerical) smooths the distribution. (From [15].)

### 6.7.3 Physical experiments

In this section we compare the maximum-entropy model with real physical experiments, in which the mean, variance and distribution of the conductance are measured in the presence of phase breaking. Firstly, the solid symbols in Fig. 6.6 show results for the mean and variance of the conductance from three experiments [26, 27, 56, 58]. Since the theoretical results take into account, through $N_\phi$, the effect of finite temperature insofar as dephasing is concerned, without including thermal smearing, the experimental variance has been corrected to compensate for this latter effect.[2] In the case of [56], measurements of all three quantum transport quantities were made, and so this data can be used to test the consistency of the theory. Notice that a value of $N_\phi = 4$ to 8 allows for the simultaneous fitting of the WLC and the variance for both $\beta = 1$ and 2. This indicates that the theory and experiment are indeed in good agreement.

Secondly, the full distribution of the conductance has been measured in the $N = 1$ case and compared to the random-matrix theory including phase breaking [89]. The results are shown in Fig. 6.8. In the $B = 0$ case, the experimental distribution is clearly not Gaussian and is skewed towards small values of $g$, as expected. On the other hand, the distribution is not nearly as dramatic as the completely coherent theory suggests. An important recent development in relation to the fictitious-lead model described above has been carried out in [44], with the motivation of describing a more spatially uniform distribution of phase-breaking events; the fictitious lead is considered to support an infinite number of modes, each with vanishing transmission, allowing a continuous value for the dephasing rate. The experimental results are analyzed in [89] with this improved model, and the result is also shown in Fig. 6.8. It is seen that the shape of the conductance distribution is reproduced well with a dephasing rate obtained independently from the WLC. This comparison provides strong support for the validity of the theory. However, the numerical value of the variance at various temperatures shows a discrepancy. In particular, the observed ratio of variances for $\beta = 1$ and $\beta = 2$ is considerably larger than that given by the model; this is as yet unexplained.

---

[2] To compensate for the thermal smearing in the experiment, we estimate the energy correlation length $E_c$ from the data given and then multiply the experimental variance by $3.5\, kT/E_c$. For [56], this produces a factor of approximately 2.6–4.5.

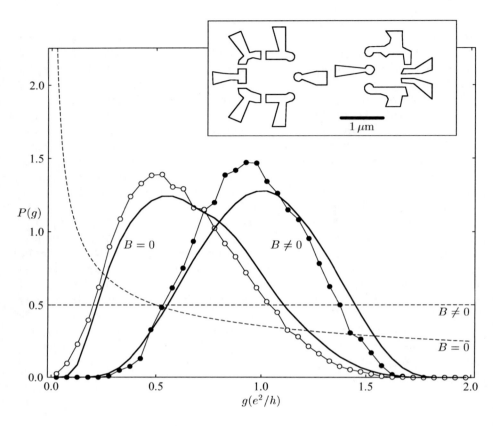

FIG. 6.8. Experimental conductance distributions for both $B = 0$ (open circles) and 40 mT (filled circles) for a $0.5\,\mu\mathrm{m}^2$ device at 100 mK with $N = 1$. These are compared to theoretical curves for both zero temperature (dashed lines, $\gamma_\phi = 0$) and nonzero temperature (solid lines, $kT/\Delta = 0.61$, $\gamma_\phi = 0.3$). Though the effect of nonzero temperature is substantial, the measured distribution at $B = 0$ is clearly not Gaussian. The inset is the pattern of gates defining the dot. (From [89].)

# 7
# ELECTRONIC TRANSPORT THROUGH QUASI-ONE-DIMENSIONAL DISORDERED SYSTEMS

In Chapter 3 we studied the scattering properties of quasi-one-dimensional systems. The potential indicated schematically in the shaded region of Fig. 3.1 is considered, in the present context, to be caused by impurities located at random inside the conductor. Clearly, a statistical description of the problem is called for. With this aim in mind, it is thus natural to contemplate a collection, or *ensemble*, of such conductors, all having the same macroscopic properties—such as the sample length $L$ and the mean free path $l_e$—but differing from one another in their sample-specific microscopic realization of disorder. This procedure parallels that followed in the previous chapter, in which we considered an ensemble of cavities having the same area but different shapes, or, under the assumed property of ergodicity, one single cavity sampled at various energies. In the present context of quasi-one-dimensional disordered conductors we consider sampling across disorder configurations. A more general parametric sampling, e.g., across various values of an applied magnetic field, brings up again the notion of ergodicity; such a general situation will not be touched upon here.

Just as in Chapter 6, the quantities of physical interest that we shall be concerned with here are, among others, as follows: the average, fluctuations about the average and the statistical distribution of the electrical conductance; the average and fluctuations of the transmission coefficients $T_{ab} = |t_{ab}|^2$ and of the reflection coefficients $R_{ab} = |r_{ab}|^2$, as well as the correlation between pairs of such coefficients. Again, these various quantities can be calculated, in principle, once we know the *probability distribution of the scattering matrix* $S$ (as in Chapter 6), or, equivalently, the probability distribution of *the transfer matrix* $M$ of the system. As it turns out, the $M$ matrix is particularly useful in the present context because of its multiplicative property that was studied earlier (see the end of Section 3.1.3 and eqn (3.110)).

The material presented in this chapter is, to a large extent, based on [126, 132, 137, 164], with substantial improvements. The chapter is organized as follows. In Section 7.1 we introduce the notion of ensemble for the $M$ matrices; for this purpose, needless to say, the concept of invariant measure is essential. Knowing the multiplicative combination law for the $M$ matrices of adjacent pieces of material, we then proceed to investigate the combination law for the associated probability distributions. In Section 7.2 we derive, for strictly one-dimensional disordered systems, the Fokker–Planck equation governing the evolution in sample length of the statistical distribution of the $M$ matrix. We first propose an ansatz for the statistical distribution of a thin piece of material, to be called a *building block* (BB). This ansatz contains the physical information relevant to the present problem and is obtained using, as a guideline, the general ideas from the maximum-entropy method discussed in Chapter 5. In Section 7.3 the analysis is extended to quasi-one-dimensional systems. In Section 7.4 we analyze, using the diffusion equation, the average conductance, its statistical fluctuations and make some considerations regarding its full statistical distribution. The exact solution of the diffusion equation for the unitary case $\beta = 2$ is also indicated. In Section 7.5 we analyze the correlations in the electronic transmission and reflection from disordered quasi-one-dimensional conductors.

Before developing this program, it is appropriate at this point to recall an important result obtained in [136]: there it was shown, for the orthogonal universality class $\beta = 1$, that a *limiting distribution* of physical quantities arises in the dense-weak-scattering limit (to be defined below, in Sections 7.2 and 7.3), within the particular class of models in which the unitary matrices $U$ and $V$ (see eqn (3.106)) defining the polar representation of the transfer matrices for the individual, microscopic, scattering units have an 'isotropic' distribution, i.e., given by their invariant measure. Only one *relevant physical parameter*, the mean free path, survives in the limit. The limiting distribution that was found constitutes a generalized *central-limit theorem*. The result obtained in [136] coincides with that of the maximum-entropy model that had been developed in [132], which gives the same diffusion equation found here, in eqn (7.137) below. That diffusion equation, at least for the isotropic case, can thus be interpreted as capturing the features arising from a central-limit theorem. We return briefly to these comments right after eqns (7.45) and (7.91).

A class of limiting distributions wider than that of [136] was studied in [138], in which the isotropy assumption of [136] was relaxed to a large extent. The motivation was to understand the influence, on the electronic transport, of the specific scattering properties of the various modes. A further generalization of the central-limit theorem was developed recently in [78].

The reader is referred to Section 1.3.3 for comments made in a wider perspective on the central-limit theorem.

## 7.1 Ensemble of transfer matrices; the invariant measure; the combination law and the Smoluchowski equation

Consider a quasi-one-dimensional disordered system of uniform cross-section, connected at both ends to waveguides, each supporting $N$ open channels. The scattering properties of our system will be described by means of its transfer matrix $M$, represented, in general, by eqn (3.95). Its polar representation is given in eqns (3.106) and (3.108).

It is important to identify the independent parameters occurring in the polar representation of eqn (3.106) which we shall be using. For $\beta = 1$, i.e., the orthogonal case, the unitary $N$-dimensional matrices $u^{(1)}$ and $v^{(1)}$ each contribute $N^2$ independent parameters, and the $\lambda_a$s, $a = 1, \ldots, N$, contribute $N$ additional ones, in agreement with the expected number of independent parameters (the same as for $n$-dimensional unitary symmetric $S$ matrices with $n = 2N$, i.e., $n(n+1)/2 = N(2N+1)$). However, for $\beta = 2$, eqn (3.106) contains $4N^2 + N$ parameters, i.e., $N$ more than is needed to specify an arbitrary matrix with $U(N, N)$ symmetry. This redundancy of the parametrization arises because a matrix $M$ parametrized as in eqn (3.106) is unchanged if the unitary matrices appearing there are subject to the transformation

$$u^{(1)} \to u^{(1)} g, \quad v^{(1)} \to g^{-1} v^{(1)},$$
$$u^{(2)} \to u^{(2)} g, \quad v^{(2)} \to g^{-1} v^{(2)}, \tag{7.1a}$$

where $g$ (reminiscent of a gauge freedom) is the diagonal matrix of phases

$$g = \begin{bmatrix} e^{i\eta_1} & \cdots & 0 \\ \vdots & \ddots & \vdots \\ 0 & \cdots & e^{i\eta_N} \end{bmatrix}. \tag{7.1b}$$

The invariance of $M$ under the transformation (7.1) could be used to eliminate $N$ parameters in the above unitary matrices. However, we shall choose the alternative of keeping the generality of eqn (3.106) and adding $N$ trivial phase parameters to those needed to specify an arbitrary matrix in $U(N, N)$. This is a convenient procedure, because any sensible probability distribution for the $M$ matrices must be invariant under the transformation (7.1), and therefore the $N$ additional phases appearing in (7.1) will drop out of the calculation of any statistical physical quantity, just through normalization of the distribution. Thus the parametrization (3.106) will be used systematically in what follows. This situation is similar to that encountered for the $S$ matrix in Section 6.1.

We start by noticing the elementary fact that, given a complex number $w$, one can extract its polar components, namely the magnitude and the phase. For instance, its magnitude, or radial component, which could be denoted by $\mathcal{R}(w)$, is obtained by performing the operation $\mathcal{R}(w) = +\sqrt{ww^*}$; when we write $w = re^{i\theta}$, $r$ being real and positive, this gives $r$. In a similar way, in what follows we shall need to extract from an $M$ matrix its polar components, to be denoted by

$$\mathcal{R}(M), \quad \mathcal{U}(M), \quad \mathcal{V}(M), \tag{7.2a}$$

which, when we write $M = U \Lambda V$ as in eqn (3.106), give

$$\lambda = \{\lambda_a\}, \quad u = \{u^{(i)}\}, \quad v = \{v^{(i)}\}, \tag{7.2b}$$

respectively. The explicit implementation of the three operations indicated symbolically in eqn (7.2) is performed in the following two exercises for $\beta = 1$ and $\beta = 2$.

**Exercise 7.1** Construct explicitly the operations indicated in eqns (7.2), intended to extract the polar components of a transfer matrix $M$ for the case $\beta = 1$.

From eqns (3.106) and (3.108), we write $M$ as

$$M = \begin{bmatrix} u\sqrt{1+\lambda}\,v & u\sqrt{\lambda}\,v^* \\ u^*\sqrt{\lambda}\,v & u^*\sqrt{1+\lambda}\,v^* \end{bmatrix} = \begin{bmatrix} \alpha & \beta \\ \beta^* & \alpha^* \end{bmatrix}. \quad (7.3)$$

We construct the Hermitian positive-definite matrix

$$h = \alpha\alpha^\dagger = u(1+\lambda)u^\dagger. \quad (7.4)$$

The eigenvalues of $h$ are $1+\lambda_a$, $a = 1,\ldots,N$, i.e., the radial coordinates we are looking for. Its $a$th eigenvector is the $a$th column of the matrix $u$, up to an overall phase factor $e^{i\phi_a}$. This means that, if we denote by $\hat{u}$ the matrix of eigenvectors, then we can extract $u$ up to a diagonal matrix of phases $d$, i.e.,

$$u = \hat{u}d, \quad (7.5)$$

where

$$d_{ab} = e^{i\phi_a}\delta_{ab}. \quad (7.6)$$

In fact, inserting eqn (7.5) into (7.4), we see that $h$ is independent of $d$. Similarly, if we construct the Hermitian matrix

$$k = \alpha^\dagger\alpha = v^\dagger(1+\lambda)v, \quad (7.7)$$

from its diagonalization we can extract $v$ up to a diagonal matrix of phases $e$, i.e.,

$$v = e\hat{v}, \quad (7.8)$$

where

$$e_{ab} = e^{i\chi_a}\delta_{ab}. \quad (7.9)$$

To find $d$ and $e$ we introduce eqns (7.5) and (7.8) into eqn (7.3) for $\alpha$ and $\beta$, to find

$$\alpha = \hat{u}d\sqrt{1+\lambda}\,e\hat{v} \quad\Rightarrow\quad (\hat{u}^\dagger\alpha\hat{v}^\dagger)_{ab} = (d\sqrt{1+\lambda}\,e)_{ab} = \sqrt{1+\lambda_a}\,e^{i(\phi_a+\chi_a)}\delta_{ab}, \quad (7.10a)$$

$$\beta = \hat{u}d\sqrt{\lambda}\,e^*\hat{v}^* \quad\Rightarrow\quad (\hat{u}^\dagger\alpha\hat{v}^\top)_{ab} = (d\sqrt{\lambda}\,e^*)_{ab} = \sqrt{\lambda_a}\,e^{i(\phi_a-\chi_a)}\delta_{ab}, \quad (7.10b)$$

from which we extract $e^{i(\phi_a+\chi_a)}$ and $e^{i(\phi_a-\chi_a)}$, and thereby $e^{i\phi_a}$ and $e^{i\chi_a}$. This solves the problem.

Notice that, knowing $u$ and $v$, we can construct the matrix $z = vu$ entering the block-diagonal matrix

$$Z = VU = \begin{bmatrix} z & 0 \\ 0 & z^* \end{bmatrix}$$

to be introduced later, eqn (7.14).

**Exercise 7.2** Extract the polar components of a transfer matrix $M$ for the case $\beta = 2$.

From eqn (3.106) we write $M$ as

$$M = \begin{bmatrix} u^{(1)}\sqrt{1+\lambda}\,v^{(1)} & u^{(1)}\sqrt{\lambda}\,v^{(2)} \\ u^{(2)}\sqrt{\lambda}\,v^{(1)} & u^{(2)}\sqrt{1+\lambda}\,v^{(2)} \end{bmatrix} = \begin{bmatrix} \alpha & \beta \\ \gamma & \delta \end{bmatrix}. \tag{7.11}$$

We construct the Hermitian matrices

$$\alpha\alpha^\dagger = u^{(1)}(1+\lambda)[u^{(1)}]^\dagger \;\Rightarrow\; u^{(1)} = \hat{u}^{(1)}d^{(1)},\; d^{(1)}_{ab} = e^{i\phi^{(1)}_a}\delta_{ab}, \tag{7.12a}$$

$$\alpha^\dagger\alpha = [v^{(1)}]^\dagger(1+\lambda)v^{(1)} \;\Rightarrow\; v^{(1)} = e^{(1)}\hat{v}^{(1)},\; e^{(1)}_{ab} = e^{i\chi^{(1)}_a}\delta_{ab}, \tag{7.12b}$$

$$\delta\delta^\dagger = u^{(2)}(1+\lambda)[u^{(2)}]^\dagger \;\Rightarrow\; u^{(2)} = \hat{u}^{(2)}d^{(2)},\; d^{(2)}_{ab} = e^{i\phi^{(2)}_a}\delta_{ab}, \tag{7.12c}$$

$$\delta^\dagger\delta = [v^{(2)}]^\dagger(1+\lambda)v^{(2)} \;\Rightarrow\; v^{(2)} = e^{(2)}\hat{v}^{(2)},\; e^{(2)}_{ab} = e^{i\chi^{(2)}_a}\delta_{ab}. \tag{7.12d}$$

The eigenvalues of these Hermitian matrices are $1+\lambda_a$, $a = 1, \ldots, N$. Their eigenvectors allow us to extract the unitary matrices $u^{(i)}$ and $v^{(i)}$ up to diagonal matrices of phases $d^{(i)}$ and $e^{(i)}$, as indicated. To find $d^{(i)}$ and $e^{(i)}$ we introduce the $u^{(i)}$ and $v^{(i)}$ of eqns (7.12) into eqn (7.11) for $\alpha$, $\beta$, $\gamma$ and $\delta$, to find

$$\alpha = \hat{u}^{(1)}d^{(1)}\sqrt{1+\lambda}\,e^{(1)}\hat{v}^{(1)} \;\Rightarrow\; ([\hat{u}^{(1)}]^\dagger \alpha [\hat{v}^{(1)}]^\dagger)_{ab} = \sqrt{1+\lambda_a}\,e^{i(\phi^{(1)}_a+\chi^{(1)}_a)}\delta_{ab}, \tag{7.13a}$$

$$\beta = \hat{u}^{(1)}d^{(1)}\sqrt{\lambda}\,e^{(2)}\hat{v}^{(2)} \;\Rightarrow\; ([\hat{u}^{(1)}]^\dagger \beta [\hat{v}^{(2)}]^\dagger)_{ab} = \sqrt{\lambda_a}\,e^{i(\phi^{(1)}_a+\chi^{(2)}_a)}\delta_{ab}, \tag{7.13b}$$

$$\gamma = \hat{u}^{(2)}d^{(2)}\sqrt{\lambda}\,e^{(1)}\hat{v}^{(1)} \;\Rightarrow\; ([\hat{u}^{(2)}]^\dagger \gamma [\hat{v}^{(1)}]^\dagger)_{ab} = \sqrt{\lambda_a}\,e^{i(\phi^{(2)}_a+\chi^{(1)}_a)}\delta_{ab}, \tag{7.13c}$$

$$\delta = \hat{u}^{(2)}d^{(2)}\sqrt{1+\lambda}\,e^{(2)}\hat{v}^{(2)} \;\Rightarrow\; ([\hat{u}^{(2)}]^\dagger \delta [\hat{v}^{(2)}]^\dagger)_{ab} = \sqrt{1+\lambda_a}\,e^{i(\phi^{(2)}_a+\chi^{(2)}_a)}\delta_{ab}, \tag{7.13d}$$

from which we obtain $e^{i(\phi^{(i)}_a+\chi^{(j)}_a)}$. The individual phases $\phi^{(i)}_a$ and $\chi^{(j)}_a$ can then be extracted, up an arbitrary set of $N$ phases $\eta_a$ for the former, and $-\eta_a$ for the latter, in agreement with the ambiguity discussed around eqn (7.1). This solves the problem.

Again, note that, knowing $u^{(i)}$ and $v^{(i)}$, we can construct the matrix $z^{(i)} = v^{(i)}u^{(i)}$ entering the block-diagonal matrix $Z$ to be introduced later, eqn (7.14).

We shall find it advantageous to use a variant of the polar representation of eqns (3.106) and (3.108). Since from eqn (3.106b) we can write

$$M = U(\Lambda Z)U^\dagger, \tag{7.14a}$$

where

$$Z = VU, \tag{7.14b}$$

we can parametrize a transfer matrix as follows:

$$M = \begin{bmatrix} u^{(1)} & 0 \\ 0 & u^{(2)} \end{bmatrix} \begin{bmatrix} \sqrt{1+\lambda}\,z^{(1)} & \sqrt{\lambda}\,z^{(2)} \\ \sqrt{\lambda}\,z^{(1)} & \sqrt{1+\lambda}\,z^{(2)} \end{bmatrix} \begin{bmatrix} [u^{(1)}]^\dagger & 0 \\ 0 & [u^{(2)}]^\dagger \end{bmatrix}. \tag{7.14c}$$

The $u^{(i)}$ and $z^{(i)}$ are arbitrary $N$-dimensional unitary matrices for $\beta = 2$, with the further restrictions

$$u^{(2)} = \left[u^{(1)}\right]^*, \quad z^{(2)} = \left[z^{(1)}\right]^* \tag{7.14d}$$

for $\beta = 1$. As the length of the quasi-one-dimensional system under consideration tends to zero, we have

$$\lambda \to 0, \tag{7.15a}$$

$$z^{(i)} \to I_N, \tag{7.15b}$$

while the $u^{(i)}$ remain arbitrary. We shall find the present representation somewhat more convenient than the original one (eqns (3.106) and (3.108)), for which, instead of (7.15b), we would have $v^{(i)} \to [u^{(i)}]^\dagger$. The $N \times N$ reflection and transmission matrices $r$ and $t$ of eqns (3.109) can now be written as

$$r = -u^{(2)}[z^{(2)}]^\dagger \sqrt{\frac{\lambda}{1+\lambda}} z^{(1)}[u^{(1)}]^\dagger, \tag{7.16a}$$

$$t = u^{(1)} \frac{1}{\sqrt{1+\lambda}} z^{(1)}[u^{(1)}]^\dagger. \tag{7.16b}$$

In the notation of eqn (7.2), we denote by

$$\mathcal{Z}(M) \tag{7.17a}$$

the operation that, when applied to $M$ as written in eqn (7.14a), gives

$$z = \{z^{(i)}\}. \tag{7.17b}$$

From eqns (7.14) we notice that multiplication of $M$ by arbitrary unitary matrices $U_0$ on the left and $V_0$ on the right does not change $\mathcal{R}(M)$, whereas multiplication by $U_0$ on the left and $U_0^\dagger$ on the right does not change $\mathcal{R}(M)$ or $\mathcal{Z}(M)$, i.e.,

$$\mathcal{R}(M) = \mathcal{R}(U_0 M V_0), \tag{7.18a}$$

$$\mathcal{R}(M) = \mathcal{R}(U_0 M U_0^\dagger), \quad \mathcal{Z}(M) = \mathcal{Z}(U_0 M U_0^\dagger), \tag{7.18b}$$

a property that we shall make use of later.

### 7.1.1 *The invariant measure*

The invariant (or Haar's) measure $d\mu(M)$ on the group of $M$ matrices is, by definition [83, 176], the measure $d\mu(M)$ that remains invariant when all the $M$s are multiplied by a single fixed $M_0$, i.e.,

$$d\mu(M) = d\mu(M_0 M). \tag{7.19}$$

In order to implement the program of calculating the invariant measure according to eqns (6.3) and (6.4), we define the differential arc element as

$$ds^2 = \text{Tr}\left[\Sigma_z \, dM^\dagger \, \Sigma_z \, dM\right]. \tag{7.20}$$

This expression remains invariant under the transformations (7.19). One can find the invariant measure in the polar representation of eqn (3.106) by substituting for $M$ in eqn (7.20) the form (3.106), extracting the metric tensor and then applying eqn (6.4). Following a procedure similar to the one used in Appendix D, [132, 137] obtained the result (for $\beta = 1$ and 2)

$$d\tilde{\mu}^{(\beta)}(M) = d\mu^{(\beta)}(M)\left[d\mu(g)\right]^{\beta-1} \tag{7.21a}$$

$$= J_\beta(\lambda) \prod_{a=1}^{N} d\lambda_a \prod_{i=1}^{\beta} d\mu(u^{(i)}) \prod_{i=1}^{\beta} d\mu(v^{(i)}) \tag{7.21b}$$

$$\equiv d\mu^{(\beta)}(\lambda) \, d\mu^{(\beta)}(u) \, d\mu^{(\beta)}(v), \tag{7.21c}$$

where

$$J_\beta(\lambda) = \prod_{a<b} |\lambda_a - \lambda_b|^\beta. \tag{7.22a}$$

The symbol $\lambda$ stands for $\{\lambda_a\}$, $a = 1, \ldots, N$, $u$ stands for $\{u^{(i)}\}$, $i = 1, \ldots, \beta$, and $v$ stands for $\{v^{(i)}\}$, $i = 1, \ldots, \beta$, so that $d\mu^{(\beta)}(\lambda)$, $d\mu^{(\beta)}(u)$ and $d\mu^{(\beta)}(v)$ represent the following:

$$d\mu^{(\beta)}(\lambda) = J_\beta(\lambda) \prod_{a=1}^{N} d\lambda_a, \tag{7.22b}$$

$$d\mu^{(\beta)}(u) = \prod_{i=1}^{\beta} d\mu(u^{(i)}), \tag{7.22c}$$

$$d\mu^{(\beta)}(v) = \prod_{i=1}^{\beta} d\mu(v^{(i)}). \tag{7.22d}$$

In eqn (7.21a), $d\mu(g)$ is relevant for $\beta = 2$ and, in terms of the phases of eqn (7.1), is given by

$$d\mu(g) = (2\pi)^{-N} \prod_{a=1}^{N} d\eta_a. \tag{7.23}$$

The invariant measure, which was written in eqn (7.21) in terms of $\lambda$, $u^{(i)}$ and $v^{(i)}$, can be easily found in the new representation of eqn (7.14) in terms of $\lambda$, $z^{(i)}$ and $u^{(i)}$, as we now show. The integral of an arbitrary function of $\lambda$, $u$ and $v$ over the invariant measure can be written as

$$I = \iiint \mathrm{d}\mu^{(\beta)}(\lambda) \, \mathrm{d}\mu^{(\beta)}(u) \, \mathrm{d}\mu^{(\beta)}(v) \, f(\lambda, u, v) \tag{7.24a}$$

$$= \int \mathrm{d}\mu^{(\beta)}(\lambda) \int \mathrm{d}\mu^{(\beta)}(u) \int \mathrm{d}\mu^{(\beta)}(v) \, f(\lambda, u, v) \tag{7.24b}$$

$$= \int \mathrm{d}\mu^{(\beta)}(\lambda) \int \mathrm{d}\mu^{(\beta)}(u) \int \mathrm{d}\mu^{(\beta)}(vu) \, f(\lambda, u, (vu)u^\dagger) \tag{7.24c}$$

$$= \int \mathrm{d}\mu^{(\beta)}(\lambda) \int \mathrm{d}\mu^{(\beta)}(u) \int \mathrm{d}\mu^{(\beta)}(z) \, F(\lambda, u, z). \tag{7.24d}$$

In eqn (7.24b) the integral over $v$ is performed first, $u$ being kept fixed; for this reason, $\mathrm{d}\mu^{(\beta)}(v)$ is written as $\mathrm{d}\mu^{(\beta)}(vu)$ in eqn (7.24c), using the defining property of Haar's measure on the unitary group. In eqn (7.24d), $z$ stands for $\{z^{(i)}\}$, $i = 1, \ldots, \beta$, with $z^{(i)} = v^{(i)} u^{(i)}$ (from the definition (7.14b)). The definition $F(\lambda, u, z) \equiv f(\lambda, u, (vu)u^\dagger)$ is also introduced. We thus find for the invariant measure that

$$\mathrm{d}\tilde{\mu}^{(\beta)}(M) = \mathrm{d}\mu^{(\beta)}(\lambda) \, \mathrm{d}\mu^{(\beta)}(u) \, \mathrm{d}\mu^{(\beta)}(z), \tag{7.25a}$$

with

$$\mathrm{d}\mu^{(\beta)}(z) = \prod_{i=1}^{\beta} \mathrm{d}\mu(z^{(i)}). \tag{7.25b}$$

### 7.1.2 The ensemble of transfer matrices

We now consider a collection or ensemble of random conductors of length $L$ and describe it in terms of an ensemble of $M$ matrices defined by means of a differential probability $\mathrm{d}\tilde{P}_L^\beta(M)$, which we split as

$$\mathrm{d}\tilde{P}_L^\beta(M) = p_L(M) \, \mathrm{d}\tilde{\mu}^{(\beta)}(M), \tag{7.26}$$

where $\mathrm{d}\tilde{\mu}^{(\beta)}(M)$ is given in eqns (7.21) and (7.25), and the *probability density* $p_L(M)$ is independent of the $\eta_a$s.

Suppose that we put together, from left to right, two wires of lengths $L''$ and $L'$, described by the transfer matrices $M''$ and $M'$, respectively. In order to find the transfer matrix $M$ of the combination, we run inevitably into the problem of evanescent modes, or closed channels, which was discussed in Section 3.1.7. However, we saw there that, if the adjacent ends of the two pieces being added are much farther apart than the characteristic decay lengths of the wave functions for the closed channels, then we can ignore the latter and multiply $M$ matrices associated with open channels only. Therefore, in order to proceed with the analysis, we shall assume, from now on, such a situation to hold, the limit to a continuous geometry only being taken later. We should point out, though, that numerical simulations [156], as well as some theoretical considerations, suggest that this situation can be relaxed to a large extent. For individual configurations, it appears that the inclusion of closed channels may be important. On the other hand, it is for averages over many configurations that the conditions do not seem to be so stringent.

For the combination of the two samples indicated in the previous paragraph, we thus write the resulting transfer matrix as the product of the individual open-channel transfer matrices, i.e.,

$$M = M'M''. \tag{7.27}$$

We assume the two segments to be *statistically independent* of each other and described by the differential probabilities

$$d\tilde{P}^{\beta}_{L''}(M'') = p_{L''}(M'') \, d\tilde{\mu}^{(\beta)}(M''), \tag{7.28a}$$

$$d\tilde{P}^{\beta}_{L'}(M') = p_{L'}(M') \, d\tilde{\mu}^{(\beta)}(M'), \tag{7.28b}$$

respectively. What is the probability distribution of the resulting transfer matrix $M$? To answer this question, we first construct the average of an arbitrary function of $M$ as follows:

$$\begin{aligned}\langle f(M)\rangle &= \iint f(M'M'') \, d\tilde{P}^{\beta}_{L'}(M') \, d\tilde{P}^{\beta}_{L''}(M'') \\ &= \int d\tilde{P}^{\beta}_{L'}(M') \int d\tilde{P}^{\beta}_{L''}(M'') \, f(M'M'') \\ &= \int d\tilde{\mu}^{(\beta)}(M') \, p_{L'}(M') \int d\tilde{\mu}^{(\beta)}(M'') \, p_{L''}(M'') f(M'M''). \end{aligned} \tag{7.29}$$

The last integration is performed over $M''$, keeping $M'$ fixed. We make the change of variables $M'' = M'^{-1}M$ (from eqn (7.27)) and integrate over $M$, for fixed $M'$. Since $d\tilde{\mu}^{(\beta)}(M'^{-1}M) = d\tilde{\mu}^{(\beta)}(M)$, because of the invariant property of the measure, we write the last integral in (7.29) as

$$\int d\tilde{\mu}^{(\beta)}(M) \, p_{L''}(M'^{-1}M) f(M), \tag{7.30}$$

and so

$$\langle f(M)\rangle = \int d\tilde{\mu}^{(\beta)}(M) \, f(M) \int d\tilde{\mu}^{(\beta)}(M') \, p_{L''}(M'^{-1}M) p_{L'}(M'). \tag{7.31}$$

The probability density of the resulting $M$ is thus given by

$$p_{L''+L'}(M) = \int d\tilde{\mu}^{(\beta)}(M') \, p_{L''}(M'^{-1}M) p_{L'}(M') \tag{7.32}$$

$$\equiv p_{L''} \otimes p_{L'}. \tag{7.33}$$

This operation, giving the combination law for the statistical distributions of two independent segments, will be referred to as a *convolution* over the group manifold of transfer matrices. It is clearly the generalization of the elementary concept of convolution that arises in the construction of the probability density of the sum of two statistically independent random variables.

Now suppose that we start with a system of length $L$ and enlarge it by adding, on its right-hand side, say, a thin slab of length $\delta L$, to be called a *building block* (BB). This is illustrated in Fig. 7.1. In eqn (7.33) we set $L'' = L$ and $L' = \delta L$, and obtain

$$p_{L+\delta L}(M) = \int \mathrm{d}\tilde{\mu}^{(\beta)}(M')\, p_L(M'^{-1}M) p_{\delta L}(M'). \tag{7.34a}$$

This result is analogous to the Smoluchowski equation for Markovian random processes [57, 166], where the role of the 'transition probability' is played here by the probability density $p_{\delta L}(M')$ for our BB. It is useful to write the last equation as

$$p_{L+\delta L}(M) = \left\langle p_L(M'^{-1}M) \right\rangle_{\delta L}, \tag{7.34b}$$

where $\langle \cdots \rangle_{\delta L}$ denotes an average over the building block variables, described by the probability density $p_{\delta L}(M')$. Thus, given the probability density $p_L(M'')$ for a system of length $L$ and that for the BB, $p_{\delta L}(M')$, eqn (7.34) allows us to find $p_{L+\delta L}(M)$ for the enlarged system of length $L + \delta L$. We shall find it convenient, in later sections, to follow a procedure familiar from the theory of Markovian processes to convert the integral equation (7.34) into a differential equation, known as the Fokker–Planck equation, involving $\partial p_L(M)/\partial L$.

Before tackling the general $N$-channel case, we illustrate in the next section some of the above relations, as well as the derivation of the diffusion equation, in a strictly one-dimensional problem described by only one channel.

## 7.2 The Fokker–Planck equation for a disordered one-dimensional conductor

Consider a strictly one-dimensional system, described in terms of just one channel and in the presence of TRI, i.e., $\beta = 1$. The various $M$ matrices are two-dimensional and can be parametrized in the polar representation of eqn (2.185) with the variant described above (starting from eqn (7.14)), i.e.,

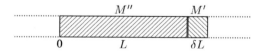

FIG. 7.1. A quasi-one-dimensional system of length $L$ is described by the transfer matrix $M''$ and the associated probability density $p_L(M'')$. The system is enlarged by the addition of a thin slab of length $\delta L$, called the building block (BB) in the text, which is described by the transfer matrix $M'$ and the associated probability density $p_{\delta L}(M')$. The total transfer matrix is $M = M'M''$ and the resulting probability density is the convolution of the individual probability densities, as given by eqn (7.34).

$$M = \begin{bmatrix} e^{i\mu} & 0 \\ 0 & e^{-i\mu} \end{bmatrix} \begin{bmatrix} \sqrt{1+\lambda}\,e^{i\theta} & \sqrt{\lambda}\,e^{-i\theta} \\ \sqrt{\lambda}\,e^{i\theta} & \sqrt{1+\lambda}\,e^{-i\theta} \end{bmatrix} \begin{bmatrix} e^{-i\mu} & 0 \\ 0 & e^{i\mu} \end{bmatrix} \quad (7.35\text{a})$$

$$= \begin{bmatrix} \sqrt{1+\lambda}\,e^{i\theta} & \sqrt{\lambda}\,e^{i(2\mu-\theta)} \\ \sqrt{\lambda}\,e^{i(\theta-2\mu)} & \sqrt{1+\lambda}\,e^{-i\theta} \end{bmatrix}. \quad (7.35\text{b})$$

We model our disordered system, including the BB, assuming that it contains a large number of very weak scatterers; this we call the *dense-weak-scattering limit*. We shall see below that, to convert the integral equation (7.34) into a differential equation, we need the BB to be shorter than the elastic mean free path, i.e., $\delta L \ll l_e$. We expect the BB to produce mainly forward scattering, giving a transfer matrix close to the unit matrix. In the polar representation of eqn (7.35) written for the BB (with primed variables, as in eqn (7.34)), we see that, as $\delta L \to 0$,

$$\lambda' \to 0, \quad \theta' \to 0. \quad (7.36)$$

Thus, as $\delta L \to 0$, the statistical distribution for the transfer matrix of the BB must develop appropriate delta functions to ensure the fulfillment of (7.36). A model for this statistical distribution is proposed below using a maximum-entropy criterion.

### 7.2.1 The maximum-entropy ansatz for the building block

First, suppose that we displace the BB of Fig. 7.1 to an $L$-independent reference position. Specifically, we center it around the origin of coordinates and call $\mathring{M}'$ the new transfer matrix, which is related to $M$ through eqn (2.214), i.e.,

$$\lambda' = \mathring{\lambda}', \quad (7.37\text{a})$$

$$\theta' = \mathring{\theta}', \quad (7.37\text{b})$$

$$\mu' = \mathring{\mu}' - kL, \quad (7.37\text{c})$$

where $k = \sqrt{2mE}/\hbar$, with $E$ being the energy of the incoming wave.

We now model the $\mathring{M}'$ probability distribution using a maximum-entropy ansatz, expecting that in the dense-weak-scattering limit the final results will be largely independent of the microscopic details. The information entropy associated with the probability density $p_{\delta L}(\mathring{M}')$ for the BB, defined as

$$\mathcal{S}[p_{\delta L}] = -\int p_{\delta L}(\mathring{M}') \ln\left[p_{\delta L}(\mathring{M}')\right] d\mu(\mathring{M}')$$

$$= -\int p_{\delta L}(\mathring{\lambda}', \mathring{\theta}', \mathring{\mu}') \ln\left[p_{\delta L}(\mathring{\lambda}', \mathring{\theta}', \mathring{\mu}')\right] d\mathring{\lambda}'\, d\mathring{\theta}'\, d\mathring{\mu}', \quad (7.38)$$

will be maximized subject to the following constraints.

(i) The normalization condition, i.e.,

$$\int p_{\delta L}(\mathring{\lambda}', \mathring{\theta}', \mathring{\mu}')\, d\mathring{\lambda}'\, d\mathring{\theta}'\, d\mathring{\mu}' = 1. \quad (7.39)$$

(ii) The average
$$\frac{\langle \mathring{\lambda}' \rangle_{\delta L}}{\delta L} \equiv \frac{1}{l_e} \tag{7.40}$$
will be assumed to be fixed. The meaning of this constraint is very physical. From eqns (2.187a) or (2.187b), the reflection coefficient associated with the BB is given by
$$\mathring{R}' = \frac{\mathring{\lambda}'}{1 + \mathring{\lambda}'}, \tag{7.41}$$
the prime again indicating the BB variables. Thus, for a short BB, the left-hand side of eqn (7.40) approximately represents the *average reflection probability per unit length*. In fact, this quantity will be identified later with the inverse of the *mean free path* $l_e$ for backward elastic scattering.

(iii) The average of a suitable $\lambda$-independent (periodic) function of $\mathring{\theta}'$,
$$\left\langle f(\mathring{\theta}') \right\rangle_{\delta L}, \tag{7.42}$$
will be assumed to be fixed for a given $\delta L$, and with the appropriate $\delta L$ dependence to fulfill the requirement described around eqn (7.36). We shall see that we need not be specific about the nature of the function $f(\mathring{\theta}')$. For instance, we may choose, as (7.42), the following:
$$\left\langle \left| e^{i\mathring{\theta}'} - 1 \right|^2 \right\rangle_{\delta L} = 4 \left\langle \sin^2 \frac{\mathring{\theta}'}{2} \right\rangle_{\delta L}, \tag{7.43}$$
a positive quantity depending on $\delta L$ which measures the spread of the $\mathring{\theta}'$ distribution around $\mathring{\theta}' = 0$.

The result of the maximization of the entropy subject to the above three constraints is
$$p_{\delta L}(\mathring{M}') = e^{\gamma_0 - \gamma_1 \mathring{\lambda}' - \gamma_2 f(\mathring{\theta}')}, \tag{7.44a}$$
the $\gamma_i$ being Lagrange multipliers needed to satisfy the three constraints, or
$$p_{\delta L}(\mathring{M}') = \frac{1}{\delta L / l_e} e^{-\mathring{\lambda}'/(\delta L / l_e)} p_{\delta L}(\mathring{\theta}'). \tag{7.44b}$$

Using the relations (7.37) between $M'$ and $\mathring{M}'$, we can write the probability density $p_{\delta L}(M')$ as
$$p_{\delta L}(M') = e^{\gamma_0 - \gamma_1 \lambda' - \gamma_2 f(\theta')} \tag{7.45a}$$
$$= \frac{1}{\delta L / l_e} e^{-\lambda'/(\delta L / l_e)} p_{\delta L}(\theta'). \tag{7.45b}$$

The MEA model of [126, 132, 137, 164] was obtained using only the constraints (i) and (ii) above for the BB. The result was an 'isotropic' probability distribution, i.e., one depending only on $\lambda$ and not on the phases of the transfer matrix.

In turn, [136] showed that the MEA selects the limiting distribution—in the sense of the *dense-weak-scattering limit*—within the class of models characterized by an isotropic distribution of the transfer matrix phases for the individual scattering units. It is expected, although it has not been shown, that the present MEA for the BB, which uses constraints (i), (ii) *and* (iii) above, will select the limiting distribution within a more realistic class of models for the individual scattering units. The fact that in both cases one ends up with the same diffusion equation for $p_L(\lambda)$ suggests that the latter equation is a result of a rather robust nature.

The BB distribution (7.45) *itself* (in terms of which the distribution for a system of finite length will be constructed below) is thus to be regarded as the limiting distribution of individual scattering-units distributions: obviously, the latter are no longer retrievable from the 'macroscopic' model (7.45). It is to be remarked though that the marginal distribution $p_{\delta L}(\lambda)$ for the BB (obtained by integrating (7.45) over $\theta'$) satisfies, to lowest order in $\delta L$, the final diffusion equation (7.67) that will be derived below.

### 7.2.2 Constructing the probability density for a system of finite length

We now investigate the structure of the probability density of a one-dimensional system of *finite* length $L$ that has been constructed using BBs with the properties described above. A first step towards this is given in the following exercise, which shows that *the convolution of two functions depending on $\lambda$ and $\theta$ depends again on $\lambda$ and $\theta$ only.* We say that this is a *persistent* property.

**Exercise 7.3** Prove that the convolution of a function $p_L(\lambda'',\theta'')$ with one of the form $p_{\delta L}(\lambda',\theta')$ (i.e., not necessarily factorizable as the MEA result (7.45)) is a function of $\lambda$ and $\theta$ only.

We use the polar representation of eqn (7.35). To prove the required statement, we start from the Smoluchowski equation (7.34), which, for the present one-dimensional case, reads

$$p_{L+\delta L}(\lambda,\theta,\mu) = \int p_L\left[\mathcal{R}\left(M'^{-1}M\right), \mathcal{Z}\left(M'^{-1}M\right)\right] p_{\delta L}(\lambda',\theta')\,d\lambda'\,\frac{d\theta'}{2\pi}\frac{d\mu'}{2\pi}, \quad (7.46)$$

where we have used the notation of eqns (7.2) and (7.17); the first argument of $p_L$ is the radial part of $M'' = M'^{-1}M$, i.e., $\lambda''$, and the second argument is the phase $\theta''$ thereof. From eqn (2.173) we write the inverse $M'^{-1}$ as

$$M'^{-1} = \sigma_z (M')^\dagger \sigma_z \qquad (7.47a)$$

$$= \begin{bmatrix} \sqrt{1+\lambda'}\,e^{-i\theta'} & -\sqrt{\lambda'}\,e^{i(2\mu'-\theta')} \\ -\sqrt{\lambda'}\,e^{i(\theta'-2\mu')} & \sqrt{1+\lambda'}\,e^{i\theta'} \end{bmatrix}, \qquad (7.47b)$$

which is again written in terms of the polar representation (7.35). The $(1,1)$th element of the product $M'^{-1}M = M''$ is thus

$$\alpha'' = \sqrt{1+\lambda''}\,e^{i\theta''}$$
$$= \sqrt{(1+\lambda')(1+\lambda)}\,e^{i(-\theta'+\theta)} - \sqrt{\lambda'\lambda}\,e^{i(2\mu'-\theta'+\theta-2\mu)}. \tag{7.48}$$

Squaring the modulus and subtracting 1, we obtain

$$\lambda'' = \lambda + \delta\lambda(\lambda, \lambda', \mu - \mu'), \tag{7.49a}$$

where

$$\delta\lambda(\lambda, \lambda', \mu - \mu') = \lambda'(2\lambda+1) - 2\sqrt{\lambda'(1+\lambda')}\sqrt{\lambda(1+\lambda)}\cos 2(\mu - \mu'). \tag{7.49b}$$

We divide $\alpha''$ of eqn (7.48) by its complex conjugate to obtain

$$e^{2i\theta''} = e^{2i(\theta-\theta')}\frac{\sqrt{(1+\lambda')(1+\lambda)} - \sqrt{\lambda'\lambda}\,e^{-2i(\mu-\mu')}}{\sqrt{(1+\lambda')(1+\lambda)} - \sqrt{\lambda'\lambda}\,e^{2i(\mu-\mu')}}, \tag{7.50a}$$

so that

$$\theta'' = \theta + \delta\theta, \tag{7.50b}$$
$$\delta\theta = -\theta' + \delta'\theta\left(\lambda, \lambda', \mu - \mu'\right), \tag{7.50c}$$

$2\delta'\theta$ being the phase of the fraction on the right-hand side of eqn (7.50a). We substitute the results (7.49a) and (7.50b,c) into eqn (7.46) to obtain

$$p_{L+\delta L}(\lambda, \theta, \mu)$$
$$= \langle p_L(\lambda + \delta\lambda, \theta + \delta\theta)\rangle_{\delta L} \tag{7.51a}$$
$$= \int p_L\left[\lambda + \delta\lambda(\lambda, \lambda', \mu - \mu'),\; \theta - \theta' + \delta'\theta\left(\lambda, \lambda', \mu - \mu'\right)\right] p_{\delta L}(\lambda', \theta')\,d\lambda'\,\frac{d\theta'}{2\pi}\,\frac{d\mu'}{2\pi}. \tag{7.51b}$$

In eqn (7.51a) we have used the notation introduced in eqn (7.34b); the expression is written more explicitly in eqn (7.51b). In the above integration, $\mu$ has to be regarded as a constant; with the change of variables $\mu' - \mu = \tilde{\mu}'$, which implies that $d\mu' = d\tilde{\mu}'$, we thus find

$$p_{L+\delta L}(\lambda, \theta, \mu)$$
$$= \int p_L\left[\lambda + \delta\lambda(\lambda, \lambda', -\tilde{\mu}'),\; \theta - \theta' + \delta'\theta\left(\lambda, \lambda', -\tilde{\mu}'\right)\right] p_{\delta L}(\lambda', \theta')\,d\lambda'\,\frac{d\theta'}{2\pi}\,\frac{d\tilde{\mu}'}{2\pi}$$
$$= p_{L+\delta L}(\lambda, \theta), \tag{7.52}$$

i.e., a function of $\lambda$ and $\theta$ only, as we were required to prove.

The combination of the $M$ matrices for the two segments illustrated in Fig. 7.1 is performed in the following exercise. Only the resulting $\lambda$ is considered, since the transmission coefficient is $T = 1/(1+\lambda)$ (see eqn (2.186)) and, by Landauer's formula, eqn (4.186), the dimensionless conductance is $g = 2T$.

**Exercise 7.4** Find the parameter $\lambda$ for the one-dimensional system of length $L + \delta L$ in Fig. 7.1 for the TRI case, in terms of the parameters $\lambda''$, $\theta''$ and $\mu''$ of the segment of length $L$, and $\lambda'$, $\theta'$ and $\mu'$ of the BB of length $\delta L$, in the polar representation of eqn (7.35).

Performing the matrix multiplication of eqn (7.27), we find the following expression for the resulting $(1,1)$th matrix element:

$$\alpha = \sqrt{1+\lambda}\,e^{i\theta} = \sqrt{(1+\lambda'')(1+\lambda')}\,e^{i(\theta''+\theta')} + \sqrt{\lambda''\lambda'}\,e^{i(\theta''-2\mu''-\theta'+2\mu')}. \quad (7.53)$$

Squaring the modulus and subtracting 1, we obtain

$$\lambda = \lambda'' + \lambda' + 2\lambda''\lambda' + 2\sqrt{\lambda''(1+\lambda'')\lambda'(1+\lambda')}\,\cos 2(\mu'' + \theta' - \mu'). \quad (7.54)$$

The statistical distribution of the variable $\lambda$ for the enlarged system of length $L + \delta L$ is the subject of the next two exercises.

**Exercise 7.5** Assume that the distribution of the transfer matrix for a segment of length $L$ and for the BB of length $\delta L$ to be added to it have the form $p_L(\lambda'', \theta'')$ and $p_{\delta L}(\lambda', \theta')$, respectively. As shown in Exercise 7.3, the distribution of the resulting transfer matrix for the enlarged system of length $L + \delta L$ has the form $p_{L+\delta L}(\lambda, \theta)$. Find the Fourier transform of the probability density (also known as the *characteristic function*, or *moment generating function*) of the variable $\lambda$ associated with the enlarged system.

By definition, the required characteristic function $\tilde{p}_{L+\delta L}(t)$ is given by

$$\tilde{p}_{L+\delta L}(t) = \left\langle e^{it\lambda(M',M'')} \right\rangle_{M',M''}. \quad (7.55)$$

Here, $\lambda(M', M'')$ is the resulting $\lambda$ given in eqn (7.54); it is a function of the parameters $\lambda'$, $\theta'$ and $\mu'$ of $M'$, and the parameters $\lambda''$ and $\mu''$ of $M''$. The notation $\langle \cdots \rangle_{M',M''}$ indicates an average over the probability distributions of $M'$ and $M''$, i.e.,

$$\tilde{p}_{L+\delta L}(t) = \int \cdots \int \exp\left\{it\left[\lambda'' + \lambda' + 2\lambda''\lambda'\right.\right.$$
$$\left.\left. + 2\sqrt{\lambda''(1+\lambda'')\lambda'(1+\lambda')}\,\cos 2(\mu'' + \theta' - \mu')\right]\right\}$$
$$\times p_L(\lambda'', \theta'')p_{\delta L}(\lambda', \theta')\,d\lambda'\,\frac{d\theta'}{2\pi}\,\frac{d\mu'}{2\pi}\,d\lambda''\,\frac{d\theta''}{2\pi}\,\frac{d\mu''}{2\pi}. \quad (7.56)$$

The integral of $p_L(\lambda'', \theta'')$ over $\theta''$ gives the marginal distribution $p_L(\lambda'')$. The integral over $\mu''$ is performed for fixed $\theta'$ and $\mu'$; we can thus make the change of variables $\mu'' + \theta' - \mu' = \tilde{\mu}''$ and realize that $d\mu'' = d\tilde{\mu}''$. We thus have

$$\tilde{p}_{L+\delta L}(t) = \int \cdots \int \exp\left\{\mathrm{it}\left[\lambda'' + \lambda' + 2\lambda''\lambda' + 2\sqrt{\lambda''(1+\lambda'')\lambda'(1+\lambda')}\cos 2\tilde{\mu}''\right]\right\}$$
$$\times p_L(\lambda'') p_{\delta L}(\lambda', \theta')\, \mathrm{d}\lambda'\, \frac{\mathrm{d}\theta'}{2\pi}\, \frac{\mathrm{d}\mu'}{2\pi}\, \mathrm{d}\lambda''\, \frac{\mathrm{d}\tilde{\mu}''}{2\pi}. \quad (7.57)$$

Now the phase $\mu'$ no longer appears in the integrand, so that the integral over the phase $\mu'$ is 1. The integral over the phase $\theta'$ leads to the marginal probability density $p_{\delta L}(\lambda')$ for the variable $\lambda'$ of the BB. We thus find

$$\tilde{p}_{L+\delta L}(t) = \iiint \exp\left\{\mathrm{it}\left[\lambda'' + \lambda' + 2\lambda''\lambda' + 2\sqrt{\lambda''(1+\lambda')\lambda'(1+\lambda'')}\cos 2\mu''\right]\right\}$$
$$\times p_L(\lambda'') p_{\delta L}(\lambda')\, \mathrm{d}\lambda'\, \mathrm{d}\lambda''\, \frac{\mathrm{d}\mu''}{2\pi}. \quad (7.58)$$

The actual probability density of the variable $\lambda$ for the enlarged segment of length $L + \delta L$ is found in the following exercise.

**Exercise 7.6** Find the probability density of the variable $\lambda$ for the enlarged segment of length $L + \delta L$, whose characteristic function was obtained in the previous exercise. Perform the calculation up to order $\langle \lambda' \rangle_{\delta L}$. The result will have an error of order $\langle (\lambda')^2 \rangle_{\delta L}$.

The required probability density $p_{L+\delta L}(\lambda)$ is the inverse Fourier transform of the $\tilde{p}_{\delta L}(t)$ given in eqn (7.58), i.e.,

$$p_{L+\delta L}(\lambda) = \frac{1}{2\pi}\int_{-\infty}^{+\infty} \mathrm{e}^{-\mathrm{it}\lambda}\, \tilde{p}_{L+\delta L}(t)\, \mathrm{d}t$$
$$= \frac{1}{2\pi}\int_{-\infty}^{+\infty} \mathrm{d}t \iiint \mathrm{e}^{\mathrm{it}(\lambda''-\lambda)} \mathrm{e}^{\mathrm{it}\lambda'(1+2\lambda'')} \mathrm{e}^{2\mathrm{it}\sqrt{\lambda''(1+\lambda')\lambda'(1+\lambda'')}\cos 2\mu''}$$
$$\times p_L(\lambda'') p_{\delta L}(\lambda')\, \mathrm{d}\lambda'\, \mathrm{d}\lambda''\, \frac{\mathrm{d}\mu''}{2\pi}. \quad (7.59)$$

We expand the second and third exponentials in the above integrand as follows:

$$p_{L+\delta L}(\lambda) = \iiint \mathrm{d}\lambda'\, \mathrm{d}\lambda''\, \frac{\mathrm{d}\mu''}{2\pi}\, \frac{1}{2\pi}\int_{-\infty}^{+\infty} \mathrm{d}t\, \mathrm{e}^{\mathrm{it}(\lambda''-\lambda)} p_L(\lambda'') p_{\delta L}(\lambda')$$
$$\times \left\{1 + \mathrm{it}\lambda'(1+2\lambda'') + \frac{(\mathrm{it})^2}{2!}[\lambda'(1+2\lambda'')]^2 + \cdots\right\}$$
$$\times \left\{1 + 2\mathrm{it}\sqrt{\lambda'(1+\lambda')\lambda''(1+\lambda'')}\cos 2\mu''\right.$$
$$\left. + \frac{(2\mathrm{it})^2}{2!}\lambda'(1+\lambda')\lambda''(1+\lambda'')\cos^2 2\mu'' + \cdots\right\}. \quad (7.60)$$

The integral over the phase $\mu''$ kills odd powers of $\cos 2\mu''$. Also, we make use of the identity

$$\frac{1}{2\pi}\int_{-\infty}^{+\infty}\mathrm{d}t\,(\mathrm{i}t)^k e^{\mathrm{i}t(\lambda''-\lambda)} = \delta^{(k)}(\lambda''-\lambda), \tag{7.61}$$

the right-hand side denoting the $k$th derivative of the delta function, to obtain

$$p_{L+\delta L}(\lambda) = \iint \mathrm{d}\lambda'\,\mathrm{d}\lambda''\,\Big[\delta(\lambda''-\lambda) + \delta''(\lambda''-\lambda)\lambda'(1+\lambda')\lambda''(1+\lambda'')$$
$$+ \delta'(\lambda''-\lambda)\lambda'(1+2\lambda'') + O(\lambda')^2\Big]p_L(\lambda'')p_{\delta L}(\lambda'), \tag{7.62}$$

or

$$p_{L+\delta L}(\lambda) = p_L(\lambda) + \frac{\partial^2}{\partial\lambda^2}\left[\lambda(1+\lambda)p_L(\lambda)\right]\left\langle\lambda'(1+\lambda')\right\rangle_{\delta L}$$
$$- \frac{\partial}{\partial\lambda}\left[(1+2\lambda)p_L(\lambda)\right]\langle\lambda'\rangle_{\delta L} + O\left\langle(\lambda')^2\right\rangle_{\delta L}. \tag{7.63}$$

The result (7.63) for $p_{L+\delta L}(\lambda)$ is expressed in terms of the moments $\langle(\lambda')^n\rangle_{\delta L}$ of the variable $\lambda'$ for the BB. These moments can now be calculated from our MEA for the BB, eqn (7.45). This is done in the next exercise, where a diffusion equation for the $\lambda$ probability density is also found.

**Exercise 7.7** From the MEA for the BB, eqn (7.45), find the marginal distribution $p_{\delta L}(\lambda')$. Find the moments of $\lambda'$ required in eqn (7.63), which gives the $\lambda$ probability density for the system of length $L+\delta L$. Take the continuum limit $\delta L \to 0$ to find a diffusion equation with increasing length for $p_L(\lambda)$.

We integrate eqn (7.45) over the phases to find the marginal distribution $p_{\delta L}(\lambda')$, i.e.,

$$p_{\delta L}(\lambda') = \frac{1}{\delta L/l_e}e^{-\lambda'/(\delta L/l_e)}, \tag{7.64}$$

which gives, for the $n$th moment of $\lambda'$,

$$\left\langle(\lambda')^n\right\rangle_{\delta L} = n!\left(\frac{\delta L}{l_e}\right)^n. \tag{7.65}$$

Substituting into eqn (7.63) we thus find

$$p_{L+\delta L}(\lambda) = p_L(\lambda) + \frac{\partial^2}{\partial\lambda^2}\left[\lambda(1+\lambda)p_L(\lambda)\right]\frac{\delta L}{l_e} - \frac{\partial}{\partial\lambda}\left[(1+2\lambda)p_L(\lambda)\right]\frac{\delta L}{l_e} + O\left(\frac{\delta L}{l_e}\right)^2$$
$$= p_L(\lambda) + \frac{\partial}{\partial\lambda}\left[\lambda(1+\lambda)\frac{\partial}{\partial\lambda}p_L(\lambda)\right]\frac{\delta L}{l_e} + O\left(\frac{\delta L}{l_e}\right)^2. \tag{7.66}$$

Taking $p_L(\lambda)$ to the left-hand side, dividing through by $\delta L$ and taking the continuum limit $\delta L \to 0$, we find

$$\frac{\partial p_s(\lambda)}{\partial s} = \frac{\partial}{\partial\lambda}\left[\lambda(1+\lambda)\frac{\partial p_s(\lambda)}{\partial\lambda}\right], \tag{7.67a}$$

where $s$ is the dimensionless length

$$s = \frac{L}{l_e}. \tag{7.67b}$$

We observe in passing that, with the simple substitution $x = 1 + 2\lambda$ and with

$$p_s(\lambda) = e^{s\lambda} f(x), \tag{7.68}$$

the function $f(x)$ satisfies Legendre's differential equation.

Equation (7.67) obtained in the last exercise is known as *Mel'nikov's equation* [139, 140]. It is a diffusion equation that describes the evolution with increasing length $L$ of the probability density $p_L(\lambda)$ for the variable $\lambda$. It has to be solved with the initial condition

$$w_{s=0}(\lambda) = \delta_+(\lambda), \tag{7.69}$$

where $\delta_+(\lambda)$ is the one-sided delta function, satisfying

$$\int_0^\infty \delta_+(\lambda)\,\mathrm{d}\lambda = 1. \tag{7.70}$$

An alternative way to derive the diffusion equation starts directly from the expression for the convolution, eqn (7.51), in a way to be described below. This is the procedure that will be generalized in the next section to the multichannel case.

We reproduce here eqns (7.51b) and (7.49b), namely

$$p_{L+\delta L}(\lambda, \theta) = \int p_L\left[\lambda + \delta\lambda(\lambda, \lambda', \mu - \mu'),\ \theta - \theta' + \delta'\theta(\lambda, \lambda', \mu - \mu')\right]$$
$$\times p_{\delta L}(\lambda', \theta')\,\mathrm{d}\lambda'\,\frac{\mathrm{d}\theta'}{2\pi}\,\frac{\mathrm{d}\mu'}{2\pi}, \tag{7.71}$$

$$\delta\lambda(\lambda, \lambda', \mu - \mu') = \lambda'(2\lambda + 1) - 2\sqrt{\lambda'(1+\lambda')}\sqrt{\lambda(1+\lambda)}\cos 2(\mu - \mu'). \tag{7.72}$$

We find the marginal distribution of the variable $\lambda$ by integrating over $\theta$, i.e.,

$$p_{L+\delta L}(\lambda) = \int p_{L+\delta L}(\lambda, \theta)\,\frac{\mathrm{d}\theta}{2\pi}$$
$$= \int \frac{\mathrm{d}\theta}{2\pi} \int p_L\left[\lambda + \delta\lambda(\lambda, \lambda', \mu - \mu'),\ \theta - \theta' + \delta'\theta(\lambda, \lambda', \mu - \mu')\right]$$
$$\times p_{\delta L}(\lambda', \theta')\,\mathrm{d}\lambda'\,\frac{\mathrm{d}\theta'}{2\pi}\,\frac{\mathrm{d}\mu'}{2\pi}. \tag{7.73}$$

The integral over $\theta$ can be performed, giving rise to the marginal distribution $p_L(\lambda'') = p_L(\lambda + \delta\lambda)$, i.e.,

$$p_{L+\delta L}(\lambda) = \int p_L\left[\lambda + \delta\lambda(\lambda, \lambda', \mu - \mu')\right] p_{\delta L}(\lambda', \theta')\,\mathrm{d}\lambda'\,\frac{\mathrm{d}\theta'}{2\pi}\,\frac{\mathrm{d}\mu'}{2\pi}, \tag{7.74a}$$

or
$$p_{L+\delta L}(\lambda) = \langle p_L(\lambda + \delta\lambda(\lambda, \lambda', \mu - \mu'))\rangle_{\delta L}. \tag{7.74b}$$

More explicitly, we make the change of variable $\tilde{\mu}' = \mu' - \mu$ and integrate over $\theta'$, to find

$$p_{L+\delta L}(\lambda) = \int p_L[\lambda + \delta\lambda(\lambda, \lambda', -\tilde{\mu}')]\, p_{\delta L}(\lambda')\, d\lambda'\, \frac{d\tilde{\mu}'}{2\pi}, \tag{7.74c}$$

in terms of the marginal distribution $p_{\delta L}(\lambda')$ of the variable $\lambda'$ for the BB.

The derivation of the diffusion equation starting from eqns (7.74) is carried out in the following exercise.

**Exercise 7.8** Expand the left-hand side of eqn (7.74b) in powers of $\delta L$ and its right-hand side in powers of $\delta\lambda$. Evaluate the averages $\langle\cdots\rangle_{\delta L}$ using, for the BB, the MEA expression given in eqn (7.45), and hence eqn (7.64) for the marginal probability distribution $p_{\delta L}(\lambda')$ of $\lambda'$. Take the continuum limit $\delta L \to 0$ to find Mel'nikov's equation.

The required expansion of eqn (7.74b) gives

$$p_L(\lambda) + \frac{\partial p_L(\lambda)}{\partial L}\delta L + \cdots = p_L(\lambda) + \frac{\partial p_L(\lambda)}{\partial \lambda}\langle \delta\lambda\rangle_{\delta L} + \frac{1}{2!}\frac{\partial^2 p_L(\lambda)}{\partial \lambda^2}\langle(\delta\lambda)^2\rangle_{\delta L} + \cdots. \tag{7.75}$$

From eqn (7.49b), the averages $\langle\delta\lambda\rangle_{\delta L}$ and $\langle(\delta\lambda)^2\rangle_{\delta L}$ are given by

$$\langle\delta\lambda\rangle_{\delta L} = (2\lambda + 1)\langle\lambda'\rangle_{\delta L} - 2\sqrt{\lambda(1+\lambda)}\left\langle\sqrt{\lambda'(1+\lambda')}\right\rangle_{\delta L}\langle\cos 2\tilde{\mu}'\rangle_{\delta L}$$
$$= (2\lambda + 1)\langle\lambda'\rangle_{\delta L} \tag{7.76}$$

and

$$\langle(\delta\lambda)^2\rangle_{\delta L} = 4\lambda(1+\lambda)\langle\lambda'(1+\lambda')\rangle_{\delta L}\langle\cos^2 2\tilde{\mu}'\rangle_{\delta L} + (2\lambda+1)^2\langle(\lambda')^2\rangle_{\delta L}$$
$$- 4(2\lambda+1)\sqrt{\lambda(1+\lambda)}\left\langle\lambda'\sqrt{\lambda'(1+\lambda')}\right\rangle_{\delta L}\langle\cos 2\tilde{\mu}'\rangle_{\delta L}$$
$$= 2\lambda(1+\lambda)\langle\lambda'(1+\lambda')\rangle_{\delta L} + (2\lambda+1)^2\langle(\lambda')^2\rangle_{\delta L}. \tag{7.77}$$

Using eqn (7.65) for the moments of the variable $\lambda'$, we find

$$\langle\delta\lambda\rangle_{\delta L} = (2\lambda+1)\frac{\delta L}{l_e}, \tag{7.78}$$

and

$$\langle(\delta\lambda)^2\rangle_{\delta L} = 2\lambda(1+\lambda)\frac{\delta L}{l_e} + O\left(\frac{\delta L}{l_e}\right)^2. \tag{7.79}$$

Higher powers of $\delta\lambda$ average to powers of $\delta L/l_e$ higher than unity. Substituting into eqn (7.75), we have

$$\frac{\partial p_L(\lambda)}{\partial L}\delta L + \cdots = (2\lambda+1)\frac{\partial p_L(\lambda)}{\partial \lambda}\frac{\delta L}{l_e} + \lambda(1+\lambda)\frac{\partial^2 p_L(\lambda)}{\partial \lambda^2}\frac{\delta L}{l_e} + \cdots. \tag{7.80}$$

Dividing through by $\delta L$ and taking the limit as $\delta L \to 0$ we find Mel'nikov's equation (7.67).

## 7.3 The Fokker–Planck equation for a quasi-one-dimensional multichannel disordered conductor

In this section we generalize to a quasi-one-dimensional multichannel disordered conductor the statistical model introduced earlier for one-channel systems.

We start with a discussion of the BB, to be described again in the dense-weak-scattering limit, in terms of which we build up the full conductor. Our BB, which is assumed to be a slab thinner than the elastic mean free path, $\delta L \ll l_e$, produces mainly forward scattering and its transfer matrix is close to the unit matrix. In the polar representation of eqn (7.14), the statistical distribution for the BB must develop, as $\delta L \to 0$, the appropriate delta functions to ensure the fulfillment of (7.15).

Just as we did in the one-channel case, we first displace the BB of Fig. 7.1 to the origin. We call $\mathring{M}'$ the new transfer matrix, which is related to $M'$ by

$$M' = D^{-1}(\mathbf{k}L)\mathring{M}'D(\mathbf{k}L), \tag{7.81}$$

where the unitary displacement matrix $D(\mathbf{k}L)$ is given in eqns (3.114) and (3.115). In the polar representation (7.14) we have

$$U'\Lambda'Z'U'^{\dagger} = \left[D^{-1}(\mathbf{k}L)\mathring{U}'\right]\mathring{\Lambda}'\mathring{Z}'\left[\mathring{U}'^{\dagger}D(\mathbf{k}L)\right], \tag{7.82}$$

so that

$$\Lambda' = \mathring{\Lambda}' \quad \Rightarrow \quad \lambda' = \mathring{\lambda}', \tag{7.83a}$$

$$Z' = \mathring{Z}' \quad \Rightarrow \quad z'^{(i)} = \mathring{z}'^{(i)}, \tag{7.83b}$$

$$U' = D^{-1}(\mathbf{k}L)\mathring{U}' \quad \Rightarrow \quad \begin{cases} u'^{(1)} = T^{-1}(\mathbf{k}L)\,\mathring{u}'^{(1)}, \\ u'^{(2)} = T(\mathbf{k}L)\,\mathring{u}'^{(2)}, \end{cases} \tag{7.83c}$$

$T(\mathbf{k}L)$ being given in eqn (3.115).

### 7.3.1 The maximum-entropy ansatz for the building block

We now use a maximum-entropy ansatz to model the probability distribution of the BB, expecting that for our dense collection of random weak scatterers the final results will be largely independent of the microscopic details. The information entropy associated with the probability density $p_{\delta L}(M')$ for the BB, defined as

$$\mathcal{S}[p_{\delta L}] = -\int p_{\delta L}(\mathring{M}') \ln\left[p_{\delta L}(\mathring{M}')\right] d\tilde{\mu}(\mathring{M}'), \tag{7.84}$$

will be maximized subject to the following constraints.

(i) The normalization condition, i.e.,

$$\int p_{\delta L}(\mathring{M}')\,d\tilde{\mu}(\mathring{M}') = 1. \tag{7.85}$$

(ii) The average

$$\frac{N^{-1}\left\langle \operatorname{tr} \overset{\circ}{\lambda}' \right\rangle_{\delta L}}{\delta L} \equiv \frac{1}{l_e} \tag{7.86}$$

will be assumed to be fixed. The meaning of this constraint is clear if we realize, from eqn (7.16a), that the total reflection coefficient arising from the BB is

$$R' = \operatorname{tr}(\overset{\circ}{r}\overset{\circ}{r}^\dagger) = \operatorname{tr}\left[\frac{\overset{\circ}{\lambda}'}{1+\overset{\circ}{\lambda}'}\right]. \tag{7.87}$$

Thus, for a thin BB, $N^{-1}\operatorname{tr}\overset{\circ}{\lambda}'$ is approximately the total reflection coefficient (summed over the exit channels and averaged over the incident ones), and the left-hand side of eqn (7.86) represents the reflection probability per unit length, which will be identified later with the inverse of the mean free path $l_e$ for backward scattering.

(iii) The average of a number of suitable functions (which we may assume to be complex, for generality) of the matrices $\overset{\circ}{z}'^{(i)}$, namely

$$\left\langle f_n(\overset{\circ}{z}'^{(i)}) \right\rangle_{\delta L}, \tag{7.88}$$

will be assumed to be fixed for a given $\delta L$, and with the appropriate $\delta L$ dependence to fulfill the requirement described around eqn (7.15). We shall see that we need not be specific about the nature of these functions. For instance, we may choose, as (7.88),

$$\left\langle \sum_i \operatorname{tr}\left(\overset{\circ}{z}'^{(i)} - I_N\right)\left(\overset{\circ}{z}'^{(i)} - I_N\right)^\dagger \right\rangle_{\delta L}, \tag{7.89}$$

a positive quantity depending on $\delta L$ that represents the deviation from the fulfillment of the condition (7.15b).

The result of the maximization of the entropy subject to the above constraints is

$$p_{\delta L}(\overset{\circ}{M}') = \exp\left[\gamma_0 - \gamma_1 \operatorname{tr}\overset{\circ}{\lambda}' - \operatorname{Re}\sum_{n=1}\gamma_{n+1}f_n(\overset{\circ}{z}'^{(i)})\right] \tag{7.90a}$$

$$= e^{\gamma_0 - \gamma_1 \operatorname{tr}\overset{\circ}{\lambda}'} p_{\delta L}(\overset{\circ}{z}'^{(i)}), \tag{7.90b}$$

the $\gamma_n$ being Lagrange multipliers needed to satisfy the above constraints.

Using the relations (7.83), we can write the BB probability density $p_{\delta L}(M')$ as

$$p_{\delta L}(M') = \exp\left[\gamma_0 - \gamma_1 \operatorname{tr}\lambda' - \operatorname{Re}\sum_n \gamma_{n+1}f_n(z'^{(i)})\right] \tag{7.91a}$$

$$= e^{\gamma_0 - \gamma_1 \operatorname{tr}\lambda'} p_{\delta L}(z'^{(i)}) \tag{7.91b}$$

$$= p_{\delta L}(\lambda') p_{\delta L}(z'^{(i)}). \tag{7.91c}$$

The comments made in the two paragraphs after eqn (7.45), with the corresponding extension to arbitrary $N$, also hold here.

We analyze some of the consequences of our ansatz (7.91), before using it in connection with the Smoluchowski equation (7.34). In particular, from eqn (7.91), the expressions (7.21) and (7.25) for the invariant measure and eqn (7.26), we can write the marginal joint probability density of the $\{\lambda_1', \ldots, \lambda_N'\} \equiv \lambda'$ as

$$w_{\delta L}^{(\beta)}(\lambda') = A e^{-\gamma_1 \sum_a \lambda_a'} J_\beta(\lambda'). \tag{7.92}$$

It is convenient to define the 'partition function'

$$Z(\gamma_1) = \int_0^\infty \cdots \int_0^\infty e^{-\gamma_1 \sum_a \lambda_a'} \prod_{b<c} |\lambda_b' - \lambda_c'|^\beta \prod_e d\lambda_e', \tag{7.93}$$

in terms of which the normalization constant $A$ can be written as

$$A = \frac{1}{Z(\gamma_1)} \tag{7.94}$$

and the average of $\sum_a \lambda_a'$ is

$$\left\langle \sum_a \lambda_a' \right\rangle = -\frac{\partial \ln Z(\gamma_1)}{\partial \gamma_1}. \tag{7.95}$$

With the change of variables $\mu_a = \gamma_1 \lambda_a'$, we can write the partition function as

$$Z(\gamma_1) = \gamma_1^{-N[(N-1)\beta/2+1]} \int_0^\infty \cdots \int_0^\infty e^{-\sum_a \mu_a} \prod_{b<c} |\mu_b - \mu_c|^\beta \prod_e d\mu_e$$
$$= \gamma_1^{-N[(N-1)\beta/2+1]} C_0^{(\beta)}(N), \tag{7.96}$$

where $C_0^{(\beta)}(N)$ depends on $N$ and $\beta$, but not on $\gamma_1$. From eqns (7.95) and (7.96), we write the average of $\sum_a \lambda_a'$ as

$$\left\langle \sum_a \lambda_a' \right\rangle = N \left( \frac{N-1}{2} \beta + 1 \right) \frac{1}{\gamma_1} = N \frac{\delta L}{l_e}, \tag{7.97}$$

where the last equality was obtained from the constraint (7.86). The Lagrange multiplier $\gamma_1$ is thus given by

$$\gamma_1 = \frac{(N-1)\beta/2 + 1}{\delta L/l_e} = \begin{cases} (N+1)/(2\delta L/l_e), & \beta = 1, \\ N/(\delta L/l_e), & \beta = 2. \end{cases} \tag{7.98}$$

Substituting $A$ and $\gamma_1$ into eqn (7.92), we thus find

$$w_{\delta L}^{(\beta)}(\lambda') = \frac{1}{C_0^{(\beta)}(N)} \left( \frac{(N-1)\beta/2 + 1}{\delta L/l_e} \right)^{N[(N-1)\beta/2+1]}$$
$$\times \exp\left[ -\frac{(N-1)\beta/2 + 1}{\delta L/l_e} \sum_a \lambda'_a \right] \prod_{b<c} |\lambda'_b - \lambda'_c|^\beta . \qquad (7.99)$$

We now construct the following average, which will be required later:

$$\left\langle \lambda'_{a_1} \cdots \lambda'_{a_p} \right\rangle_{\delta L}$$
$$= \frac{1}{Z} \int_0^\infty \cdots \int_0^\infty \lambda'_{a_1} \cdots \lambda'_{a_p} e^{-\gamma_1 \sum_a \lambda'_a} \prod_{b<c} |\lambda'_b - \lambda'_c|^\beta \prod_e d\lambda'_e$$
$$= \frac{1}{C_0^{(\beta)}(N)\gamma_1^p} \int_0^\infty \cdots \int_0^\infty \mu_{a_1} \cdots \mu_{a_p} e^{-\sum_a \mu_a} \prod_{b<c} |\mu_b - \mu_c|^\beta \prod_e d\mu_e$$
$$= \frac{C_{a_1 \cdots a_p}^{(\beta)}(N)}{C_0^{(\beta)}(N)\gamma_1^p}$$
$$= \frac{C_{a_1 \cdots a_p}^{(\beta)}(N)}{C_0^{(\beta)}(N)} \left( \frac{\delta L/l_e}{(N-1)\beta/2 + 1} \right)^p , \qquad (7.100a)$$

where $C_{a_1 \cdots a_p}^{(\beta)}(N)$ does not depend on $\gamma_1$. Thus, for a fixed $N$, the result (7.100a) behaves like $(\delta L/l_e)^p$. Similarly, we find that

$$\left\langle \lambda'^q_a \right\rangle_{\delta L} \sim \left( \frac{\delta L}{l_e} \right)^q , \qquad (7.100b)$$

with $q$ not necessarily an integer.

### 7.3.2 *Constructing the probability density for a system of finite length*

We now investigate the structure of the probability distribution of a quasi-one-dimensional disordered conductor constructed by putting together BBs with the probability density of eqn (7.91). We extend to the quasi-one-dimensional case the *persistent* property studied in Section 7.2.2 for one dimension by proving that *the convolution* (as in eqn (7.33)) *of two functions depending on $\lambda$ and $z = \{z^{(i)}\}$ depends again on $\lambda$ and $z$ only*. We start with the Smoluchowski equation (7.34), i.e.,

$$p_{L+\delta L}(\lambda, z, u) = \int p_L \left[ \mathcal{R}\left( M'^{-1} M \right), \mathcal{Z}\left( M'^{-1} M \right) \right]$$
$$\times p_{\delta L}(\lambda', z') d\mu(\lambda') d\mu(z') d\mu(u'), \qquad (7.101)$$

where we have used the notation of eqns (7.2) and (7.17); the first argument of $p_L$ is the radial part of $M'' = M'^{-1} M$, i.e., $\lambda''$, and the second argument is the 'angular' matrices $z''$ thereof. Using eqn (7.14), we write the combination $M'^{-1}M$ as

$$M'^{-1}M = U'(Z')^\dagger \Lambda'^{-1}(U')^\dagger U\Lambda Z U^\dagger. \tag{7.102a}$$

From the properties (7.18), we can write the $\mathcal{R}$ and $\mathcal{Z}$ parts of $M'^{-1}M$ as

$$\mathcal{R}\left(M'^{-1}M\right) = \mathcal{R}\left(\Lambda'^{-1}(U^\dagger U')^\dagger \Lambda\right), \tag{7.102b}$$

$$\mathcal{Z}\left(M'^{-1}M\right) = \mathcal{Z}\left(\Lambda'^{-1}(U^\dagger U')^\dagger \Lambda Z(U^\dagger U')(Z')^\dagger\right). \tag{7.102c}$$

Substituting these last expressions into the combination law, eqn (7.101), we have

$$p_{L+\delta L}(\lambda, z, u) = \int p_L\left[\mathcal{R}\left(\Lambda'^{-1}(U^\dagger U')^\dagger \Lambda\right), \mathcal{Z}\left(\Lambda'^{-1}(U^\dagger U')^\dagger \Lambda Z(U^\dagger U')(Z')^\dagger\right)\right]$$
$$\times p_{\delta L}(\lambda', z')\, d\mu(\lambda')\, d\mu(z')\, d\mu(u'). \tag{7.103}$$

The integration over $u'$ is performed for fixed $u$. We can thus make the change of variables

$$\tilde{u}' = u^\dagger u', \tag{7.104a}$$

which stands for

$$\tilde{u}'^{(i)} = (u^{(i)})^\dagger u'^{(i)}, \tag{7.104b}$$

and use the invariance property of the measure $d\mu(u')$ to write

$$d\mu(\tilde{u}') = d\mu(u^{-1}u') = d\mu(u'). \tag{7.105}$$

Then (7.103) becomes

$$p_{L+\delta L}(\lambda, z, u) = \int p_L\left[\mathcal{R}\left(\Lambda'^{-1}(\tilde{U}')^\dagger \Lambda\right), \mathcal{Z}\left(\Lambda'^{-1}(\tilde{U}')^\dagger \Lambda Z\tilde{U}'(Z')^\dagger\right)\right]$$
$$\times p_{\delta L}(\lambda', z')\, d\mu(\lambda')\, d\mu(z')\, d\mu(\tilde{u}')$$
$$= p_{L+\delta L}(\lambda, z), \tag{7.106}$$

which is a function of $\lambda$ and $z$ only; this is the persistent property we wanted to show.

### 7.3.3 *The diffusion equation for the orthogonal universality class, $\beta = 1$*

We need a more explicit evaluation of the polar parameters of $M'^{-1}M$ appearing in eqns (7.101) and (7.106) in terms of those of $M$ and $M'$. This is done in part in the following exercise, using the procedure developed in Exercise 7.1.

**Exercise 7.9** Find, for $\beta = 1$, the polar parameters of $M'' = M'^{-1}M$ in terms of those of $M$ and $M'$.

From eqn (3.96) we write the inverse $M'^{-1}$ as

$$M'^{-1} = \Sigma_z (M')^\dagger \Sigma_z \tag{7.107a}$$

$$= \begin{bmatrix} (v')^\dagger \sqrt{1+\lambda'}\,(u')^\dagger & -(v')^\dagger \sqrt{\lambda'}\,(u')^\top \\ -(v')^\top \sqrt{\lambda'}\,(u')^\dagger & (v')^\top \sqrt{1+\lambda'}\,(u')^\top \end{bmatrix}. \tag{7.107b}$$

Thus

$$M'' = M'^{-1} M$$

$$= \begin{bmatrix} \alpha'' & \beta'' \\ (\beta'')^* & (\alpha'')^* \end{bmatrix} = \begin{bmatrix} u''\sqrt{1+\lambda''}\,v'' & u''\sqrt{\lambda''}\,(v'')^* \\ (u'')^*\sqrt{\lambda''}\,v'' & (u'')^*\sqrt{1+\lambda''}\,(v'')^* \end{bmatrix} \tag{7.108a}$$

$$= \begin{bmatrix} (v')^\dagger \sqrt{1+\lambda'}\,(u')^\dagger u\sqrt{1+\lambda}\,v & (v')^\dagger \sqrt{1+\lambda'}\,(u')^\dagger u\sqrt{\lambda}\,v^* \\ \quad -(v')^\dagger \sqrt{\lambda'}\,(u')^\top u^*\sqrt{\lambda}\,v & \quad -(v')^\dagger \sqrt{\lambda'}\,(u')^\top u^*\sqrt{1+\lambda}\,v^* \\ -(v')^\top \sqrt{\lambda'}\,(u')^\dagger u\sqrt{1+\lambda}\,v & -(v')^\top \sqrt{\lambda'}\,(u')^\dagger u\sqrt{\lambda}\,v^* \\ \quad +(v')^\top \sqrt{1+\lambda'}\,(u')^\top u^*\sqrt{\lambda}\,v & \quad +(v')^\top \sqrt{1+\lambda'}\,(u')^\top u^*\sqrt{1+\lambda}\,v^* \end{bmatrix}. \tag{7.108b}$$

Following the procedure developed in Exercise 7.1, we first construct the matrix

$$h'' = \alpha''(\alpha'')^\dagger = u''(1+\lambda'')(u'')^\dagger$$
$$= (v')^\dagger \sqrt{1+\lambda'}\,(u')^\dagger u(1+\lambda) u^\dagger u' \sqrt{1+\lambda'}\,v'$$
$$\quad - (v')^\dagger \sqrt{1+\lambda'}\,(u')^\dagger u\sqrt{\lambda(1+\lambda)}\,u^\top (u')^* \sqrt{\lambda'}\,v'$$
$$\quad - (v')^\dagger \sqrt{\lambda'}\,(u')^\top u^* \sqrt{\lambda(1+\lambda)}\,u^\dagger u' \sqrt{1+\lambda'}\,v'$$
$$\quad + (v')^\dagger \sqrt{\lambda'}\,(u')^\top u^* \lambda u^\top (u')^* \sqrt{\lambda'}\,v'. \tag{7.109}$$

Notice that the matrix $h''$ does not depend on $v$. The eigenvalues of $h''$, i.e., $1 + \lambda''_a$, $a = 1, \ldots, N$, depend on the matrices $\lambda$, $\lambda'$ and $u^\dagger u'$. Its eigenvectors form a matrix $\hat{u}''$, which may differ from $u''$ by a diagonal matrix of phases $d''$, i.e.,

$$u'' = \hat{u}'' d''. \tag{7.110a}$$

The matrix $\hat{u}''$ depends on the matrices $\lambda$, $\lambda'$, $u^\dagger u'$ and $v'$; from eqn (7.109) we see that $\hat{u}''$ depends on $v'$ in the simple way

$$\hat{u}'' = (v')^\dagger \hat{\hat{u}}'', \tag{7.110b}$$

where $\hat{\hat{u}}''$ depends on the matrices $\lambda$, $\lambda'$ and $u^\dagger u'$. We can thus write $u''$ as

$$u'' = (v')^\dagger \hat{\hat{u}}'' d''. \tag{7.110c}$$

Similarly, we construct the matrix

$$\begin{aligned}
k'' &= (\alpha'')^\dagger \alpha'' = (v'')^\dagger (1+\lambda'') v'' \\
&= v^\dagger \sqrt{1+\lambda}\, u^\dagger u'(1+\lambda')(u')^\dagger u\sqrt{1+\lambda}\, v \\
&\quad - v^\dagger \sqrt{1+\lambda}\, u^\dagger u' \sqrt{\lambda'(1+\lambda')}\, (u')^\top u^* \sqrt{\lambda}\, v \\
&\quad - v^\dagger \sqrt{\lambda}\, u^\top (u')^* \sqrt{\lambda'(1+\lambda')}\, (u')^\dagger u\sqrt{1+\lambda}\, v \\
&\quad + v^\dagger \sqrt{\lambda}\, u^\top (u')^* \lambda' (u')^\top u^* \sqrt{\lambda}\, v.
\end{aligned} \quad (7.111)$$

*Notice that the matrix $k''$ does not depend on $v'$. The eigenvalues of $k''$, i.e., $1+\lambda''_a$, $a = 1, \ldots, N$, depend on the matrices $\lambda$, $\lambda'$ and $u^\dagger u'$. Its eigenvectors form a matrix $(\hat{v}'')^\dagger$, where $\hat{v}''$ may differ from $v''$ by a diagonal matrix of phases $e''$*, i.e.,

$$v'' = e'' \hat{v}''. \quad (7.112a)$$

The matrix $\hat{v}''$ depends on the matrices $\lambda$, $\lambda'$, $u^\dagger u'$ and $v$; from eqn (7.111) we see that $\hat{v}''$ depends on $v$ in the simple way

$$\hat{v}'' = \hat{\hat{v}}'' v, \quad (7.112b)$$

where $\hat{\hat{v}}''$ depends on the matrices $\lambda$, $\lambda'$ and $u^\dagger u'$. We can thus write $v''$ as

$$v'' = e'' \hat{\hat{v}}'' v. \quad (7.112c)$$

To find $d''$ and $e''$ we substitute $u''$ and $v''$ from eqns (7.110) and (7.112) into eqn (7.108a) for $\alpha''$ and $\beta''$, to find

$$\alpha'' = (v')^\dagger \hat{\hat{u}}'' d'' \sqrt{1+\lambda''}\, e'' \hat{\hat{v}}'' v, \quad (7.113a)$$

$$\beta'' = (v')^\dagger \hat{\hat{u}}'' d'' \sqrt{\lambda''}\, (e'')^* (\hat{\hat{v}}'')^* v^*, \quad (7.113b)$$

and, using $\alpha''$ and $\beta''$ from eqn (7.108b), we have

$$d'' \sqrt{1+\lambda''}\, e'' = (\hat{\hat{u}}'')^\dagger \left[ \sqrt{1+\lambda'}\, (u^\dagger u')^\dagger \sqrt{1+\lambda} - \sqrt{\lambda'}\, (u^\dagger u')^\top \sqrt{\lambda} \right] (\hat{\hat{v}}'')^\dagger, \quad (7.114a)$$

$$d'' \sqrt{\lambda''}\, (e'')^* = (\hat{\hat{u}}'')^\dagger \left[ \sqrt{1+\lambda'}\, (u^\dagger u')^\dagger \sqrt{\lambda} - \sqrt{\lambda'}\, (u^\dagger u')^\top \sqrt{1+\lambda} \right] (\hat{\hat{v}}'')^\top. \quad (7.114b)$$

Assuming that we have already found $\lambda''$, we can find $d''$ and $e''$ from the last two equations in terms of $\lambda$, $\lambda'$ and $u^\dagger u'$.

In conclusion, we have found the polar parameters of $M'' = M'^{-1} M$ as follows.

(i) We have $\lambda''$, which depends only upon $\lambda$, $\lambda'$ and $u^\dagger u'$ and is independent of $v$ and $v'$. This parallels eqn (7.49) for the one-dimensional case.

(ii) We also have $u''$ and $v''$ from eqns (7.110c) and (7.112c), respectively, where the dependence on $v$ and $v'$ is given explicitly, and $\hat{\hat{u}}'' d''$ and $e'' \hat{\hat{v}}''$ depend only on $\lambda$, $\lambda'$ and $u^\dagger u'$.

From eqns (7.14) we can thus write the matrix $z''$ as

$$\begin{aligned}z'' &= v''u'' \\ &= e''\hat{v}''v(v')^\dagger \hat{u}''d'' \\ &= e''\hat{v}''(vu)u^\dagger u'(v'u')^\dagger \hat{u}''d'' \\ &= (e''\hat{v}'')z\left[(u^\dagger u')(z')^\dagger \hat{u}''d''\right] \\ &= u'''zu'''',\end{aligned} \qquad (7.115)$$

where $u'''$ depends on $\lambda$, $\lambda'$ and $u^\dagger u'$, and $u''''$ depends on the same quantities and on $z'$. This generalizes the result (7.50) that was deduced for the one-dimensional case. The result (i) of Exercise 7.9 for $\lambda''$ and the result (7.115) for $z''$ are to be substituted for the $\mathcal{R}$ and $\mathcal{Z}$, respectively, that appear in eqn (7.106). The linear relation between $z''$ and $z$ that is shown explicitly in eqn (7.115) will be very important in what follows. We find

$$p_{L+\delta L}(\lambda, z) = \int p_L\left[\lambda''(\lambda, \lambda', \tilde{u}'),\ u'''(\lambda, \lambda', \tilde{u}')z\ u''''(\lambda, \lambda', \tilde{u}', z')\right] \\ \times p_{\delta L}(\lambda', z')\,\mathrm{d}\mu(\lambda')\,\mathrm{d}\mu(z')\,\mathrm{d}\mu(\tilde{u}'). \qquad (7.116)$$

This equation is the $N$-channel generalization of eqn (7.51) for the one-dimensional case. We now integrate both sides over $z$. On the left-hand side we obtain the marginal distribution $p_{L+\delta L}(\lambda)$. On the right-hand side we integrate over $z$ first, keeping $\lambda'$, $z'$ and $\tilde{u}'$ fixed; we make the change of variables (see eqn (7.115))

$$z'' = u'''zu'''', \qquad (7.117)$$

and write

$$p_{L+\delta L}(\lambda) = \int \left\{\int p_L\left[\lambda''(\lambda, \lambda', \tilde{u}'),\ z''\right]\mathrm{d}\mu(z'')\right\} p_{\delta L}(\lambda', z')\,\mathrm{d}\mu(\lambda')\,\mathrm{d}\mu(z')\,\mathrm{d}\mu(\tilde{u}'). \qquad (7.118)$$

The integral over $z''$ gives the marginal distribution $p_L(\lambda'')$, and the integral over $z'$ gives the marginal distribution $p_{\delta L}(\lambda')$. We finally write

$$p_{L+\delta L}(\lambda) = \int p_L\left[\lambda''(\lambda, \lambda', \tilde{u}')\right]p_{\delta L}(\lambda')\,\mathrm{d}\mu(\lambda')\,\mathrm{d}\mu(\tilde{u}'). \qquad (7.119)$$

We rewrite the above equation in a more compact way as in eqn (7.34b), i.e.,

$$p_{L+\delta L}(\lambda) = \langle p_L(\lambda + \delta\lambda)\rangle_{\delta L}, \qquad (7.120)$$

where we have written $\lambda'' = \lambda + \delta\lambda$ and $\delta\lambda = \delta\lambda(\lambda, \lambda', \tilde{u}')$. The notation $\langle \cdots \rangle_{\delta L}$ denotes an average over the BB variables; since only $\lambda'$ and $\tilde{u}'$ occur, this average is performed using the distribution $p_{\delta L}(\lambda')$ of eqn (7.91c) for $\lambda'$ and the invariant measure $\mathrm{d}\mu(\tilde{u}')$ for $\tilde{u}'$.

Expanding the left-hand side of eqn (7.120) in powers of $\delta L$ and its right-hand side in powers of the variables $\delta \lambda_a$, $a = 1, \ldots, N$, we find

$$p_L(\lambda) + \frac{\partial p_L(\lambda)}{\partial L} \delta L + \cdots$$

$$= p_L(\lambda) + \sum_{a=1}^{N} \frac{\partial p_L(\lambda)}{\partial \lambda_a} \langle \delta \lambda_a \rangle_{\delta L} + \frac{1}{2!} \sum_{a,b} \frac{\partial^2 p_L(\lambda)}{\partial \lambda_a \partial \lambda_b} \langle \delta \lambda_a \delta \lambda_b \rangle_{\delta L} + \cdots .$$
(7.121)

In the multichannel case we do not have a simple, closed expression for the $\delta \lambda_a = \delta \lambda_a (\lambda, \lambda', \tilde{u}')$, $a = 1, \ldots, N$, in contrast to the $N = 1$ case, eqn (7.49b). Thus our immediate task is to analyze these quantities. For simplicity of notation, we shall write $u'$ for $\tilde{u}'$ in what follows.

We notice that the required information lies in the eigenvalues $1 + \lambda''_a$ of the matrix $k''$ given in eqn (7.111), which can be written as

$$k'' = 1 + v^\dagger (\lambda + w) v,$$
(7.122)

where the matrix $w$ is given by

$$w = \sqrt{1+\lambda}\, u' \lambda' (u')^\dagger \sqrt{1+\lambda} - \sqrt{1+\lambda}\, u' \sqrt{\lambda'(1+\lambda')}\, (u')^\top \sqrt{\lambda}$$
$$- \sqrt{\lambda}\, (u')^* \sqrt{\lambda'(1+\lambda')}\, (u')^\dagger \sqrt{1+\lambda} + \sqrt{\lambda}\, (u')^* \lambda' (u')^\top \sqrt{\lambda}.$$
(7.123)

The eigenvalues of the matrix $\lambda + w$ are the quantities $\lambda_a + \delta \lambda_a$ which we are looking for. The matrix $\lambda$ is diagonal, with matrix elements $\lambda_a$, $a = 1, \ldots, N$. The Hermitian matrix $w$ can be considered to be a perturbation; the resulting 'corrections' $\delta \lambda_a$ to the eigenvalues, to be inserted into eqn (7.121), will be calculated using perturbation theory, i.e.,

$$\delta \lambda_a = w_{aa} + \sum_{b(\neq a)} \frac{w_{ab} w_{ba}}{\lambda_a - \lambda_b} + \cdots$$
(7.124a)

$$= \delta \lambda_a^{(1)} + \delta \lambda_a^{(2)} + \cdots ,$$
(7.124b)

where the matrix elements $w_{ab}$ of the perturbation matrix $w$ of eqn (7.123) are given by

$$w_{ab} = \sum_c \Big\{ \sqrt{1+\lambda_a}\, [u'_{ac} \lambda'_c (u'_{bc})^*]\, \sqrt{1+\lambda_b}$$
$$- \sqrt{1+\lambda_a}\, \Big[u'_{ac} \sqrt{\lambda'_c(1+\lambda'_c)}\, u'_{bc}\Big] \sqrt{\lambda_b}$$
$$- \sqrt{\lambda_a}\, \Big[(u'_{ac})^* \sqrt{\lambda'_c(1+\lambda'_c)}\, (u'_{bc})^*\Big] \sqrt{1+\lambda_b}$$
$$+ \sqrt{\lambda_a}\, [(u'_{ac})^* \lambda'_c u'_{bc}]\, \sqrt{\lambda_b} \Big\}.$$
(7.125)

Before proceeding, we notice that, for the calculation of the expectation values occurring in eqn (7.121), we need the expectation values of powers of matrix elements of the unitary matrix $u'$ over the invariant measure of the $N$-dimensional unitary group (to be indicated by $\langle \cdots \rangle_0$). Some of these expectation values were calculated in Section 6.4.1 (see, in particular, eqns (6.29), (6.35) and (6.37)). The results which are relevant here are repeated, for convenience, in the following equations:

$$\langle u'_{ab} \rangle_0 = \langle u'_{ab} u'_{cd} \rangle_0$$
$$= \langle u'_{ab} u'_{cd} u'_{ef} (u'_{gh})^* \rangle_0 = 0, \quad (7.126a)$$

$$\langle u'_{ab} (u'_{cd})^* \rangle_0 = \frac{\delta_{ac} \delta_{bd}}{N}, \quad (7.126b)$$

$$\langle u'_{ac} u'_{bc} (u'_{ad} u'_{bd})^* \rangle_0 = \frac{1 + \delta_{ab}}{N(N+1)} \delta_{cd}. \quad (7.126c)$$

The expectation value of the first-order correction in eqn (7.124) is now

$$\left\langle \delta \lambda_a^{(1)} \right\rangle_{\delta L} = \langle w_{aa} \rangle_{\delta L}$$
$$= (1 + \lambda_a) \sum_c \langle u'_{ac} \lambda'_c (u'_{ac})^* \rangle_{\delta L} + \lambda_a \sum_c \langle (u'_{ac})^* \lambda'_c u'_{ac} \rangle_{\delta L}$$
$$= (1 + 2\lambda_a) \sum_c \langle \lambda'_c \rangle_{\delta L} \left\langle |u'_{ac}|^2 \right\rangle_0$$
$$= (1 + 2\lambda_a) \frac{1}{N} \sum_c \langle \lambda'_c \rangle_{\delta L}$$
$$= (1 + 2\lambda_a) \frac{\delta L}{l_e}. \quad (7.127)$$

We have used eqns (7.126a,b) and the constraint (7.86). The expectation value of the second-order correction in eqn (7.124) is

$$\left\langle \delta \lambda_a^{(2)} \right\rangle_{\delta L} = \sum_{b(\neq a)} \frac{\langle w_{ab} w_{ba} \rangle_{\delta L}}{\lambda_a - \lambda_b}$$
$$= \sum_{b(\neq a)} \frac{1}{\lambda_a - \lambda_b}$$
$$\times \sum_{cd} \Big\langle \Big\{ \sqrt{1 + \lambda_a} \left[ u'_{ac} \lambda'_c (u'_{bc})^* \right] \sqrt{1 + \lambda_b}$$
$$- \sqrt{1 + \lambda_a} \left[ u'_{ac} \sqrt{\lambda'_c (1 + \lambda'_c)} u'_{bc} \right] \sqrt{\lambda_b}$$
$$- \sqrt{\lambda_a} \left[ (u'_{ac})^* \sqrt{\lambda'_c (1 + \lambda'_c)} (u'_{bc})^* \right] \sqrt{1 + \lambda_b}$$
$$+ \sqrt{\lambda_a} \left[ (u'_{ac})^* \lambda'_c u'_{bc} \right] \sqrt{\lambda_b} \Big\}$$

$$\times \left\{ \sqrt{1+\lambda_b} \left[ u'_{bd} \lambda'_d (u'_{ad})^* \right] \sqrt{1+\lambda_a} \right.$$
$$- \sqrt{1+\lambda_b} \left[ u'_{bd} \sqrt{\lambda'_d(1+\lambda'_d)}\, u'_{ad} \right] \sqrt{\lambda_a}$$
$$- \sqrt{\lambda_b} \left[ (u'_{bd})^* \sqrt{\lambda'_d(1+\lambda'_d)}\, (u'_{ad})^* \right] \sqrt{1+\lambda_a}$$
$$\left. + \sqrt{\lambda_b} \left[ (u'_{bd})^* \lambda'_d u'_{ad} \right] \sqrt{\lambda_a} \right\}_{\delta L}. \tag{7.128}$$

In the above product we retain only terms that are linear in $\lambda'$, which, from eqn (7.100), contribute linear terms in $\delta L$ in eqn (7.121). Notice that terms that behave as $(\lambda')^{3/2}$ do not contribute anyway, on account of the 'angular' integral (see eqn (7.126a)). We thus have

$$\left\langle \delta\lambda_a^{(2)} \right\rangle_{\delta L} \approx \sum_{b(\neq a)} \frac{1}{\lambda_a - \lambda_b}$$
$$\times \sum_{cd} \left[ (1+\lambda_a)\lambda_b \left\langle \sqrt{\lambda'_c(1+\lambda'_c)\lambda'_d(1+\lambda'_d)} \right\rangle_{\delta L} \langle u'_{ac} u'_{bc} (u'_{ad} u'_{bd})^* \rangle_0 \right.$$
$$\left. + (1+\lambda_b)\lambda_a \left\langle \sqrt{\lambda'_c(1+\lambda'_c)\lambda'_d(1+\lambda'_d)} \right\rangle_{\delta L} \langle (u'_{ac} u'_{bc})^* u'_{ad} u'_{bd} \rangle_0 \right]$$
$$\approx \sum_{b(\neq a)} \frac{\lambda_a + \lambda_b + 2\lambda_a\lambda_b}{\lambda_a - \lambda_b} \sum_c \langle \lambda'_c \rangle_{\delta L} \frac{1+\delta_{ab}}{N(N+1)}$$
$$= \frac{1}{N+1} \frac{\delta L}{l_e} \sum_{b(\neq a)} \frac{\lambda_a + \lambda_b + 2\lambda_a\lambda_b}{\lambda_a - \lambda_b}. \tag{7.129}$$

We have used eqn (7.126c) for the average of the unitary matrices, and the constraint (7.86). Before proceeding, we express the summation appearing in the last equation in a more convenient way. We write

$$\sum_{b(\neq a)} \frac{\lambda_a + \lambda_b + 2\lambda_a\lambda_b}{\lambda_a - \lambda_b} = -(N-1)(1+2\lambda_a) + 2\lambda_a(1+\lambda_a) \sum_{b\neq a} \frac{1}{\lambda_a - \lambda_b}$$
$$= -(N-1)(1+2\lambda_a) + 2\lambda_a(1+\lambda_a) \frac{\partial \ln J^{(1)}(\lambda)}{\partial \lambda_a}, \tag{7.130}$$

where $J^{(1)}(\lambda)$ is the Jacobian defined in eqn (7.22a). Substituting eqn (7.130) into (7.129), we thus have

$$\left\langle \delta\lambda_a^{(2)} \right\rangle_{\delta L} = \frac{1}{N+1} \frac{\delta L}{l_e} \left[ -(N-1)(1+2\lambda_a) + 2\lambda_a(1+\lambda_a) \frac{\partial \ln J^{(1)}(\lambda)}{\partial \lambda_a} \right]. \tag{7.131}$$

The quadratic term $\langle \delta\lambda_a \delta\lambda_b \rangle_{\delta L}$ in eqn (7.121) contains, in principle, the contribution from the various orders in perturbation theory, i.e.,

$$\langle \delta\lambda_a \delta\lambda_b \rangle_{\delta L} = \left\langle \left[ \delta\lambda_a^{(1)} + \delta\lambda_a^{(2)} + \cdots \right] \left[ \delta\lambda_b^{(1)} + \delta\lambda_b^{(2)} + \cdots \right] \right\rangle_{\delta L}. \quad (7.132)$$

In this expansion we only retain terms which are linear in $\lambda'$. We have

$$\langle \delta\lambda_a \delta\lambda_b \rangle_{\delta L} \approx \left\langle \left\langle \delta\lambda_a^{(1)} \delta\lambda_b^{(1)} \right\rangle_{\delta L} \right\rangle_{\delta L} = \langle w_{aa} w_{bb} \rangle_{\delta L}$$

$$= \left\langle \sum_{cd} \left\{ (1+\lambda_a) \left[ u'_{ac} \lambda'_c (u'_{ac})^* \right] \right.\right.$$
$$- \sqrt{\lambda_a(1+\lambda_a)} \left[ u'_{ac} \sqrt{\lambda'_c(1+\lambda'_c)}\, u'_{ac} \right]$$
$$- \sqrt{\lambda_a(1+\lambda_a)} \left[ (u'_{ac})^* \sqrt{\lambda'_c(1+\lambda'_c)}\, (u'_{ac})^* \right]$$
$$\left. + \lambda_a \left[ (u'_{ac})^* \lambda'_c u'_{ac} \right] \right\}$$

$$\times \left\{ (1+\lambda_b) \left[ u'_{bd} \lambda'_d (u'_{bd})^* \right] \right.$$
$$- \sqrt{\lambda_b(1+\lambda_b)} \left[ u'_{bd} \sqrt{\lambda'_d(1+\lambda'_d)}\, u'_{bd} \right]$$
$$- \sqrt{\lambda_b(1+\lambda_b)} \left[ (u'_{bd})^* \sqrt{\lambda'_d(1+\lambda'_d)}\, (u'_{bd})^* \right]$$
$$\left.\left. + \lambda_b \left[ (u'_{bd})^* \lambda'_d u'_{bd} \right] \right\} \right\rangle_{\delta L}$$

$$\approx 2 \sum_{cd} \sqrt{\lambda_a(1+\lambda_a)\lambda_b(1+\lambda_b)} \left\langle \sqrt{\lambda'_c \lambda'_d} \right\rangle_{\delta L} \langle u'_{ac} u'_{ac} (u'_{bd} u'_{bd})^* \rangle_0$$

$$= 2 \sum_{cd} \sqrt{\lambda_a(1+\lambda_a)\lambda_b(1+\lambda_b)} \left\langle \sqrt{\lambda'_c \lambda'_d} \right\rangle_{\delta L} \frac{2\delta_{ab}\delta_{cd}}{N(N+1)}$$

$$= \frac{4\delta_{ab}}{N(N+1)} \lambda_a(1+\lambda_a) \sum_c \langle \lambda'_c \rangle_{\delta L}$$

$$= \frac{4\delta_{ab}}{N+1} \lambda_a(1+\lambda_a) \frac{\delta L}{l_e}. \quad (7.133)$$

We have made use of eqn (7.126c). We now substitute eqns (7.127), (7.131) and (7.133) into the expansion (7.121), to obtain (the superscript '(1)' indicates that we are dealing with the orthogonal case)

$$\frac{\partial p_s^{(1)}(\lambda)}{\partial s} = \sum_a \left\{ (1+2\lambda_a) \frac{\partial p_s^{(1)}(\lambda)}{\partial \lambda_a} \right.$$

$$+ \frac{1}{N+1}\left[-(N-1)(1+2\lambda_a) + 2\lambda_a(1+\lambda_a)\frac{\partial \ln J(\lambda)}{\partial \lambda_a}\right]\frac{\partial p_s^{(1)}(\lambda)}{\partial \lambda_a}$$

$$+ \frac{2}{N+1}\lambda_a(1+\lambda_a)\frac{\partial^2 p_s^{(1)}(\lambda)}{\partial \lambda_a^2}\Bigg\}$$

$$= \frac{2}{N+1}\frac{1}{J(\lambda)}\sum_a \frac{\partial}{\partial \lambda_a}\left[\lambda_a(1+\lambda_a)J(\lambda)\frac{\partial p_s^{(1)}(\lambda)}{\partial \lambda_a}\right]. \tag{7.134}$$

From eqns (7.26) and (7.21) we see that the joint probability density of the $\lambda_a$s, $a = 1, \ldots, N$, is given by

$$w_s^{(1)}(\lambda) = p_s^{(1)}(\lambda)J^{(1)}(\lambda). \tag{7.135}$$

Note the normalization

$$\int_0^\infty w_s^{(1)}(\lambda)\,d\lambda_1 \cdots d\lambda_N = 1. \tag{7.136}$$

Again, $s = L/l_e$ is the dimensionless length, just as in eqn (7.67b). We finally write eqn (7.134) in terms of the *joint probability density* $w_s^{(1)}(\lambda)$ as

$$\frac{\partial w_s^{(1)}(\lambda)}{\partial s} = \frac{2}{N+1}\sum_a \frac{\partial}{\partial \lambda_a}\left[\lambda_a(1+\lambda_a)J^{(1)}(\lambda)\frac{\partial}{\partial \lambda_a}\frac{w_s^{(1)}(\lambda)}{J^{(1)}(\lambda)}\right]. \tag{7.137}$$

This is the Fokker–Planck equation governing, for $\beta = 1$, the evolution with the conductor length of the joint probability density $w_s^{(1)}(\lambda)$ of the $\lambda_a$s, $a = 1, \ldots, N$. This diffusion equation has to be solved with the initial condition

$$w_{s=0}^{(1)}(\lambda_1, \ldots, \lambda_N) = \prod_{a=1}^N \delta_+(\lambda_a), \tag{7.138}$$

which generalizes the initial condition of eqn (7.69) for the one-dimensional case, with $\delta_+(\lambda_a)$ again denoting the one-sided delta function defined in eqn (7.70).

One can show that, for $L \ll l_e$ (i.e., $s \ll 1$), the maximum-entropy ansatz for $p_{\delta L}(\lambda)$ given in eqn (7.91c) indeed satisfies, to lowest order in $\delta L$, the above diffusion equation.

### 7.3.4 *The diffusion equation for the unitary universality class, $\beta = 2$*

Following Exercise 7.2, we can extract the polar parameters of $M'^{-1}M$ in terms of those of $M$ and $M'$, as they are needed in eqns (7.101) and (7.106). This is done in the following exercise.

**Exercise 7.10** Find, for $\beta = 2$, the polar parameters of $M'' = M'^{-1}M$ in terms of those of $M$ and $M'$.

From eqn (3.96) we write the inverse $M'^{-1}$ as

$$M'^{-1} = \Sigma_z (M')^\dagger \Sigma_z \tag{7.139a}$$

$$= \begin{bmatrix} (v'^{(1)})^\dagger \sqrt{1+\lambda'}\,(u'^{(1)})^\dagger & -(v'^{(1)})^\dagger \sqrt{\lambda'}\,(u'^{(2)})^\dagger \\ -(v'^{(2)})^\dagger \sqrt{\lambda'}\,(u'^{(1)})^\dagger & (v'^{(2)})^\dagger \sqrt{1+\lambda'}\,(u'^{(2)})^\dagger \end{bmatrix}. \tag{7.139b}$$

Thus

$$M'' = M'^{-1} M$$

$$= \begin{bmatrix} \alpha'' & \beta'' \\ \gamma'' & \delta'' \end{bmatrix} = \begin{bmatrix} u''^{(1)} \sqrt{1+\lambda''}\,v''^{(1)} & u''^{(1)} \sqrt{\lambda''}\,v''^{(2)} \\ u''^{(2)} \sqrt{\lambda''}\,v''^{(1)} & u''^{(2)} \sqrt{1+\lambda''}\,v''^{(2)} \end{bmatrix} \tag{7.140a}$$

$$= \begin{bmatrix} \begin{array}{l}(v'^{(1)})^\dagger \sqrt{1+\lambda'}\,(u'^{(1)})^\dagger u^{(1)} \sqrt{1+\lambda}\,v^{(1)} \\ \quad - (v'^{(1)})^\dagger \sqrt{\lambda'}\,(u'^{(2)})^\dagger u^{(2)} \sqrt{\lambda}\,v^{(1)} \\ -(v'^{(2)})^\dagger \sqrt{\lambda'}\,(u'^{(1)})^\dagger u^{(1)} \sqrt{1+\lambda}\,v^{(1)} \\ \quad + (v'^{(2)})^\dagger \sqrt{1+\lambda'}\,(u'^{(2)})^\dagger u^{(2)} \sqrt{\lambda}\,v^{(1)} \end{array} & \begin{array}{l} (v'^{(1)})^\dagger \sqrt{1+\lambda'}\,(u'^{(1)})^\dagger u^{(1)} \sqrt{\lambda}\,v^{(2)} \\ \quad - (v'^{(1)})^\dagger \sqrt{\lambda'}\,(u'^{(2)})^\dagger u^{(2)} \sqrt{1+\lambda}\,v^{(2)} \\ -(v'^{(2)})^\dagger \sqrt{\lambda'}\,(u'^{(1)})^\dagger u^{(1)} \sqrt{\lambda}\,v^{(2)} \\ \quad + (v'^{(2)})^\dagger \sqrt{1+\lambda'}\,(u'^{(2)})^\dagger u^{(2)} \sqrt{1+\lambda}\,v^{(2)} \end{array} \end{bmatrix}. \tag{7.140b}$$

Following the procedure developed in Exercise 7.2, we first construct the matrix

$$h'' = \alpha''(\alpha'')^\dagger = u''^{(1)}(1+\lambda'')(u''^{(1)})^\dagger$$
$$= (v'^{(1)})^\dagger \sqrt{1+\lambda'}\,(u'^{(1)})^\dagger u^{(1)} (1+\lambda)(u^{(1)})^\dagger u'^{(1)} \sqrt{1+\lambda'}\,v'^{(1)}$$
$$\quad - (v'^{(1)})^\dagger \sqrt{1+\lambda'}\,(u'^{(1)})^\dagger u^{(1)} \sqrt{\lambda(1+\lambda)}\,(u^{(2)})^\dagger (u'^{(2)}) \sqrt{\lambda'}\,v'^{(1)}$$
$$\quad - (v'^{(1)})^\dagger \sqrt{\lambda'}\,(u'^{(2)})^\dagger u^{(2)} \sqrt{\lambda(1+\lambda)}\,(u^{(1)})^\dagger u'^{(1)} \sqrt{1+\lambda'}\,v'^{(1)}$$
$$\quad + (v'^{(1)})^\dagger \sqrt{\lambda'}\,(u'^{(2)})^\dagger u^{(2)} \lambda (u^{(2)})^\dagger u'^{(2)} \sqrt{\lambda'}\,v'^{(1)}. \tag{7.141}$$

Notice that the matrix $h''$ does not depend on $v^{(1)}$. The eigenvalues of $h''$, i.e., $1+\lambda''_a$, $a = 1, \ldots, N$, depend on the matrices

$$\lambda, \quad \lambda', \tag{7.142a}$$

$$(u^{(1)})^\dagger u'^{(1)}, \quad (u^{(2)})^\dagger u'^{(2)}. \tag{7.142b}$$

Its eigenvectors form a matrix $\hat{u}''^{(1)}$, which may differ from $u''^{(1)}$ by a diagonal matrix of phases $d''^{(1)}$, i.e.,

$$u''^{(1)} = \hat{u}''^{(1)} d''^{(1)} = (v'^{(1)})^\dagger \hat{\tilde{u}}''^{(1)} d''^{(1)}, \tag{7.143}$$

where $\hat{\tilde{u}}''^{(1)}$ depends on the matrices of (7.142).

Similarly, we construct the matrix

$$\begin{aligned}k'' &= (\alpha'')^\dagger \alpha'' = (v''^{(1)})^\dagger (1 + \lambda'') v''^{(1)} \\
&= (v^{(1)})^\dagger \sqrt{1+\lambda}\,(u^{(1)})^\dagger u'^{(1)}(1+\lambda')(u'^{(1)})^\dagger u^{(1)} \sqrt{1+\lambda}\, v^{(1)} \\
&\quad - (v^{(1)})^\dagger \sqrt{1+\lambda}\,(u^{(1)})^\dagger u'^{(1)} \sqrt{\lambda'(1+\lambda')}\,(u'^{(2)})^\dagger (u^{(2)}) \sqrt{\lambda}\, v^{(1)} \\
&\quad - (v^{(1)})^\dagger \sqrt{\lambda}\,(u^{(2)})^\dagger u'^{(2)} \sqrt{\lambda'(1+\lambda')}\,(u'^{(1)})^\dagger u^{(1)} \sqrt{1+\lambda}\, v^{(1)} \\
&\quad + (v^{(1)})^\dagger \sqrt{\lambda}\,(u^{(2)})^\dagger u'^{(2)} \lambda' (u'^{(2)})^\dagger u^{(2)} \sqrt{\lambda}\, v^{(1)}.\end{aligned} \quad (7.144)$$

Notice that the matrix $k''$ does not depend on $v'^{(1)}$. The eigenvalues of $k''$, i.e., $1 + \lambda''_a$, $a = 1, \ldots, N$, depend on the matrices of (7.142). Its eigenvectors form a matrix $(\hat{v}''^{(1)})^\dagger$, where $\hat{v}''^{(1)}$ may differ from $v''^{(1)}$ by a diagonal matrix of phases $e''^{(1)}$, i.e.,

$$v''^{(1)} = e''^{(1)} \hat{v}''^{(1)} = e''^{(1)} \hat{\hat{v}}''^{(1)} v^{(1)}, \quad (7.145)$$

where $\hat{\hat{v}}''^{(1)}$ depends on the matrices of (7.142).

In a similar fashion, we can construct $\delta(\delta'')^\dagger$ and $(\delta'')^\dagger \delta$, from which we find

$$u''^{(2)} = \hat{u}''^{(2)} d''^{(2)} = (v'^{(2)})^\dagger \hat{\hat{u}}''^{(2)} d''^{(2)}, \quad (7.146)$$

$$v''^{(2)} = e''^{(2)} \hat{v}''^{(2)} = e''^{(2)} \hat{\hat{v}}''^{(2)} v^{(2)}, \quad (7.147)$$

where $\hat{\hat{u}}''^{(2)}$ and $\hat{\hat{v}}''^{(2)}$ again depend on the matrices of (7.142).

To find $d''^{(1)}$, $e''^{(1)}$, $d''^{(2)}$ and $e''^{(2)}$ we first substitute $u''^{(1)}$, $v''^{(1)}$, $u''^{(2)}$ and $v''^{(2)}$ from eqns (7.143), (7.145), (7.146) and (7.147), respectively, into eqn (7.140), which results in the following equations:

$$d''^{(1)} \sqrt{1+\lambda''}\, e''^{(1)}$$
$$= (\hat{\hat{u}}''^{(1)})^\dagger \left[\sqrt{1+\lambda'}\,(u'^{(1)})^\dagger u^{(1)} \sqrt{1+\lambda} - \sqrt{\lambda'}\,(u'^{(2)})^\dagger u^{(2)} \sqrt{\lambda}\right] (\hat{\hat{v}}''^{(1)})^\dagger, \quad (7.148\text{a})$$

$$d''^{(1)} \sqrt{\lambda''}\,(e''^{(2)})$$
$$= (\hat{\hat{u}}''^{(1)})^\dagger \left[\sqrt{1+\lambda'}\,(u'^{(1)})^\dagger u^{(1)} \sqrt{\lambda} - \sqrt{\lambda'}\,(u'^{(2)})^\dagger u^{(2)} \sqrt{1+\lambda}\right] (\hat{\hat{v}}''^{(2)})^\dagger, \quad (7.148\text{b})$$

$$d''^{(2)} \sqrt{\lambda''}\, e''^{(1)}$$
$$= (\hat{\hat{u}}''^{(2)})^\dagger \left[\sqrt{1+\lambda'}\,(u'^{(2)})^\dagger u^{(2)} \sqrt{\lambda} - \sqrt{\lambda'}\,(u'^{(1)})^\dagger u^{(1)} \sqrt{1+\lambda}\right] (\hat{\hat{v}}''^{(1)})^\dagger, \quad (7.148\text{c})$$

$$d''^{(2)} \sqrt{1+\lambda''}\, e''^{(2)}$$
$$= (\hat{\hat{u}}''^{(2)})^\dagger \left[\sqrt{1+\lambda'}\,(u'^{(2)})^\dagger u^{(2)} \sqrt{1+\lambda} - \sqrt{\lambda'}\,(u'^{(1)})^\dagger u^{(1)} \sqrt{\lambda}\right] (\hat{\hat{v}}''^{(2)})^\dagger. \quad (7.148\text{d})$$

Assuming that we have already found $\lambda''$, from the last four equations we can find $d''^{(1)} e''^{(1)}$, $d''^{(1)} e''^{(2)}$, $d''^{(2)} e''^{(1)}$ and $d''^{(2)} e''^{(2)}$ in terms of the matrices of (7.142). The matrices $d''^{(i)}$ can thus be found up to an arbitrary set of $N$ phases $\eta_a$, and the matrices $e''^{(i)}$ can be found up to the $N$ phases $-\eta_a$ (here $i=1,2$).

In conclusion, we have found the polar parameters of $M'' = M'^{-1}M$ as follows.

(i) We have $\lambda''$, which depends only upon the matrices of (7.142).

(ii) We also have $u''^{(i)}$ and $v''^{(i)}$ as in eqns (7.143), (7.145), (7.146) and (7.147), where the dependence on $v^{(i)}$ and $v'^{(i)}$ is given explicitly, and $\hat{u}''^{(i)} d''^{(i)}$ and $e''^{(i)} \hat{\hat{v}}''^{(i)}$ depend only upon the matrices of (7.142) (here $i=1,2$).

From eqns (7.14) we can thus write the matrix $z''^{(i)}$ ($i=1,2$) as

$$\begin{aligned} z''^{(i)} &= v''^{(i)} u''^{(i)} \\ &= e''^{(i)} \hat{\hat{v}}''^{(i)} v^{(i)} (v'^{(i)})^\dagger \hat{u}''^{(i)} d''^{(i)} \\ &= e''^{(i)} \hat{\hat{v}}''^{(i)} (v^{(i)} u^{(i)}) u^{(i)\dagger} u'^{(i)} (v'^{(i)} u'^{(i)})^\dagger \hat{u}''^{(i)} d''^{(i)} \\ &= (e''^{(i)} \hat{\hat{v}}''^{(i)}) z^{(i)} \left[ (u^{(i)\dagger} u'^{(i)}) (z'^{(i)})^\dagger \hat{u}''^{(i)} d''^{(i)} \right] \\ &= u'''^{(i)} z^{(i)} u''''^{(i)}, \end{aligned} \qquad (7.149)$$

where $u'''^{(i)}$ depends on the matrices of (7.142), and $u''''$ depends on the same quantities and on $z'^{(i)}$.

Substituting these results into the combination eqn (7.106) and making the change of variables (7.104), we find

$$\begin{aligned} p_{L+\delta L}(\lambda, z) = \int p_L \big[ &\lambda''(\lambda, \lambda', \tilde{u}'^{(1)}, \tilde{u}'^{(2)}), \\ &u'''^{(1)}(\lambda, \lambda', \tilde{u}'^{(1)}, \tilde{u}'^{(2)}) z^{(1)} u''''^{(1)}(\lambda, \lambda', \tilde{u}'^{(1)}, \tilde{u}'^{(2)}, z'^{(1)}), \\ &u'''^{(2)}(\lambda, \lambda', \tilde{u}'^{(1)}, \tilde{u}'^{(2)}) z^{(2)} u''''^{(2)}(\lambda, \lambda', \tilde{u}'^{(1)}, \tilde{u}'^{(2)}, z'^{(2)}) \big] \\ &\times p_{\delta L}(\lambda', z'^{(1)}, z'^{(2)}) \, d\mu(\lambda') \, d\mu(z') \, d\mu(\tilde{u}'). \end{aligned} \qquad (7.150)$$

We now integrate both sides with respect to $z$. On the left-hand side we obtain the marginal distribution $p_{L+\delta L}(\lambda)$. On the right-hand side we integrate over $z$ first, keeping $\lambda'$, $z'^{(i)}$ and $\tilde{u}'^{(i)}$ fixed; we make the change of variables (see eqn (7.149))

$$z''^{(i)} = u'''^{(i)} z^{(i)} u''''^{(i)} \qquad (7.151)$$

and write

$$\begin{aligned} p_{L+\delta L}(\lambda) = \int & d\mu(\lambda') \, d\mu(z') \, d\mu(\tilde{u}') \, p_{\delta L}(\lambda', z'^{(1)}, z'^{(2)}) \\ & \times \int d\mu(z'') \, p_L[\lambda''(\lambda, \lambda', \tilde{u}'^{(1)}, \tilde{u}'^{(2)}), z''^{(1)}, z''^{(2)}]. \end{aligned} \qquad (7.152)$$

The integral over $z''$ gives the marginal distribution $p_L(\lambda'')$, and the integral over $z'$ gives the marginal distribution $p_{\delta L}(\lambda')$. We finally find

$$p_{L+\delta L}(\lambda) = \int p_L[\lambda''(\lambda, \lambda', \tilde{u}'^{(1)}, \tilde{u}'^{(2)})] p_{\delta L}(\lambda') \, d\mu(\lambda') \, d\mu(\tilde{u}'), \tag{7.153}$$

which is, for $\beta = 2$, the equivalent of eqn (7.119), that was written for $\beta = 1$. We now proceed in a way similar to that for $\beta = 1$. Again, we write eqn (7.153) in a more compact way, as in eqn (7.120), i.e.,

$$p_{L+\delta L}(\lambda) = \langle p_L(\lambda + \delta\lambda) \rangle_{\delta L}, \tag{7.154}$$

with the understanding that $\lambda'' = \lambda + \delta\lambda$ and $\delta\lambda = \delta\lambda(\lambda, \lambda', \tilde{u}'^{(1)}, \tilde{u}'^{(2)})$. The notation $\langle \cdots \rangle_{\delta L}$ denotes an average over the BB variables; since only $\lambda'$, $\tilde{u}'^{(1)}$ and $\tilde{u}'^{(2)}$ occur, this average is performed using the distribution $p_{\delta L}(\lambda')$ of eqn (7.91c) for $\lambda'$ and the invariant measure $d\mu(\tilde{u}'^{(i)})$ for $\tilde{u}'^{(i)}$. From now on we omit the tilde, for simplicity.

An expansion like that of eqn (7.121) and the perturbation expansion of eqn (7.124) also apply here. Again, we notice that we need the eigenvalues $1 + \lambda_a''$ of the matrix $k''$ of eqn (7.144), which can be written as

$$k'' = 1 + (v^{(1)})^\dagger (\lambda + w) v^{(1)}, \tag{7.155}$$

where the matrix $w$ is now given by

$$w = \sqrt{1+\lambda} \left[ u'^{(1)} \lambda' (u'^{(1)})^\dagger \right] \sqrt{1+\lambda}$$
$$- \sqrt{1+\lambda} \left[ u'^{(1)} \sqrt{\lambda'(1+\lambda')} (u'^{(2)})^\dagger \right] \sqrt{\lambda}$$
$$- \sqrt{\lambda} \left[ u'^{(2)} \sqrt{\lambda'(1+\lambda')} (u'^{(1)})^\dagger \right] \sqrt{1+\lambda}$$
$$+ \sqrt{\lambda} \left[ u'^{(2)} \lambda' (u'^{(2)})^\dagger \right] \sqrt{\lambda}. \tag{7.156}$$

The matrix $\lambda$ is diagonal, with matrix elements $\lambda_a$, $a = 1, \ldots, N$. The Hermitian matrix $w$ can be considered to be a perturbation; the resulting 'corrections' $\delta\lambda_a$ to the eigenvalues are the quantities to be inserted into eqn (7.121). These corrections will again be calculated in perturbation theory, as in eqn (7.124), where the matrix elements $w_{ab}$ of the perturbation matrix $w$ are now given by

$$w_{ab} = \sum_c \left\{ \sqrt{1+\lambda_a} \left[ u'^{(1)}_{ac} \lambda'_c (u'^{(1)}_{bc})^* \right] \sqrt{1+\lambda_b} \right.$$
$$- \sqrt{1+\lambda_a} \left[ u'^{(1)}_{ac} \sqrt{\lambda'_c(1+\lambda'_c)} (u'^{(2)}_{bc})^* \right] \sqrt{\lambda_b}$$
$$- \sqrt{\lambda_a} \left[ u'^{(2)}_{ac} \sqrt{\lambda'_c(1+\lambda'_c)} (u'^{(1)}_{bc})^* \right] \sqrt{1+\lambda_b}$$
$$\left. + \sqrt{\lambda_a} \left[ u'^{(2)}_{ac} \lambda'_c (u'^{(2)}_{bc})^* \right] \sqrt{\lambda_b} \right\}. \tag{7.157}$$

The expectation value of the first-order correction in eqn (7.124) is then

$$\begin{aligned}
\left\langle \delta\lambda_a^{(1)} \right\rangle_{\delta L} &= \langle \delta w_{aa} \rangle_{\delta L} \\
&= (1+\lambda_a) \sum_c \left\langle u'^{(1)}_{ac} \lambda'_c (u'^{(1)}_{ac})^* \right\rangle_{\delta L} + \lambda_a \sum_c \left\langle (u'^{(2)}_{ac})^* \lambda'_c u'^{(2)}_{ac} \right\rangle_{\delta L} \\
&= (1+2\lambda_a) \sum_c \left\langle u'_{ac} \lambda'_c (u'_{ac})^* \right\rangle_{\delta L} \\
&= (1+2\lambda_a) \sum_c \langle \lambda'_c \rangle_{\delta L} \left\langle |u'_{ac}|^2 \right\rangle_0 \\
&= (1+2\lambda_a) \frac{1}{N} \sum_c \langle \lambda'_c \rangle_{\delta L} \\
&= (1+2\lambda_a) \frac{\delta L}{l_e}. \quad (7.158)
\end{aligned}$$

We have used eqns (7.126a,b) and the constraint (7.86). The expectation value of the second-order correction in eqn (7.124) is

$$\begin{aligned}
\left\langle \delta\lambda_a^{(2)} \right\rangle_{\delta L} &= \sum_{b(\neq a)} \frac{\langle w_{ab} w_{ba} \rangle_{\delta L}}{\lambda_a - \lambda_b} \\
&= \sum_{b(\neq a)} \frac{1}{\lambda_a - \lambda_b} \\
&\quad \times \sum_{cd} \Big\langle \Big\{ \sqrt{1+\lambda_a} \left[ u'^{(1)}_{ac} \lambda'_c (u'^{(1)}_{bc})^* \right] \sqrt{1+\lambda_b} \\
&\qquad - \sqrt{1+\lambda_a} \left[ u'^{(1)}_{ac} \sqrt{\lambda'_c(1+\lambda'_c)} (u'^{(2)}_{bc})^* \right] \sqrt{\lambda_b} \\
&\qquad - \sqrt{\lambda_a} \left[ u'^{(2)}_{ac} \sqrt{\lambda'_c(1+\lambda'_c)} (u'^{(1)}_{bc})^* \right] \sqrt{1+\lambda_b} \\
&\qquad + \sqrt{\lambda_a} \left[ u'^{(2)}_{ac} \lambda'_c (u'^{(2)}_{bc})^* \right] \sqrt{\lambda_b} \Big\} \\
&\quad \times \Big\{ \sqrt{1+\lambda_b} \left[ u'^{(1)}_{bd} \lambda'_d (u'^{(1)}_{ad})^* \right] \sqrt{1+\lambda_a} \\
&\qquad - \sqrt{1+\lambda_b} \left[ u'^{(1)}_{bd} \sqrt{\lambda'_d(1+\lambda'_d)} (u'^{(2)}_{ad})^* \right] \sqrt{\lambda_a} \\
&\qquad - \sqrt{\lambda_b} \left[ u'^{(2)}_{bd} \sqrt{\lambda'_d(1+\lambda'_d)} (u'^{(1)}_{ad})^* \right] \sqrt{1+\lambda_a} \\
&\qquad + \sqrt{\lambda_b} \left[ u'^{(2)}_{bd} \lambda'_d (u'^{(2)}_{ad})^* \right] \sqrt{\lambda_a} \Big\} \Big\rangle_{\delta L}. \quad (7.159)
\end{aligned}$$

We only retain terms which are linear in $\lambda'$, and therefore obtain

$$\left\langle \delta\lambda_a^{(2)} \right\rangle_{\delta L} \approx \sum_{b(\neq a)} \frac{1}{\lambda_a - \lambda_b}$$

$$\times \sum_{cd} \left[ (1+\lambda_a)\lambda_b \left\langle \sqrt{\lambda'_c \lambda'_d} \right\rangle_{\delta L} \left\langle u'^{(1)}_{ac}(u'^{(1)}_{ad})^* \right\rangle_0 \left\langle (u'^{(2)}_{bc})^* u'^{(2)}_{bd} \right\rangle_0 \right.$$

$$\left. + (1+\lambda_b)\lambda_a \left\langle \sqrt{\lambda'_c \lambda'_d} \right\rangle_{\delta L} \left\langle (u'^{(1)}_{bc})^* u'^{(1)}_{bd} \right\rangle_0 \left\langle u'^{(2)}_{ac}(u'^{(2)}_{ad})^* \right\rangle_0 \right]$$

$$= \sum_{b(\neq a)} \frac{1}{\lambda_a - \lambda_b}$$

$$\times \sum_{cd} \left[ (1+\lambda_a)\lambda_b \left\langle \sqrt{\lambda'_c \lambda'_d} \right\rangle_{\delta L} \frac{\delta_{cd}}{N} \frac{\delta_{cd}}{N} \right.$$

$$\left. + (1+\lambda_b)\lambda_a \left\langle \sqrt{\lambda'_c \lambda'_d} \right\rangle_{\delta L} \frac{\delta_{cd}}{N} \frac{\delta_{cd}}{N} \right]$$

$$= \frac{1}{N} \frac{\delta L}{l_e} \sum_{b(\neq a)} \frac{\lambda_a + \lambda_b + 2\lambda_a \lambda_b}{\lambda_a - \lambda_b}. \qquad (7.160)$$

We have used eqn (7.126b) for the average of the unitary matrices, and the constraint (7.86). We make use of eqn (7.130) and eqn (7.22a) for the Jacobian $J^{(2)}(\lambda)$ for $\beta = 2$ to write

$$\sum_{b(\neq a)} \frac{\lambda_a + \lambda_b + 2\lambda_a \lambda_b}{\lambda_a - \lambda_b} = -(N-1)(1+2\lambda_a) + \lambda_a(1+\lambda_a) \frac{\partial \ln J^{(2)}(\lambda)}{\partial \lambda_a}. \qquad (7.161)$$

Substituting into eqn (7.160), we thus have

$$\left\langle \delta\lambda_a^{(2)} \right\rangle_{\delta L} = \frac{1}{N} \frac{\delta L}{l_e} \left[ -(N-1)(1+2\lambda_a) + \lambda_a(1+\lambda_a) \frac{\partial \ln J^{(2)}(\lambda)}{\partial \lambda_a} \right]. \qquad (7.162)$$

In the quadratic term $\langle \delta\lambda_a \delta\lambda_b \rangle_{\delta L}$ in eqn (7.121) we retain only terms which are linear in $\lambda'$. We have

$$\langle \delta\lambda_a \delta\lambda_b \rangle_{\delta L} \approx \left\langle \delta\lambda_a^{(1)} \delta\lambda_b^{(1)} \right\rangle_{\delta L} = \langle w_{aa} w_{bb} \rangle_{\delta L}$$

$$= \left\langle \sum_{cd} \left\{ (1+\lambda_a) \left[ u'^{(1)}_{ac} \lambda'_c (u'^{(1)}_{ac})^* \right] \right. \right.$$

$$- \sqrt{\lambda_a(1+\lambda_a)} \left[ u'^{(1)}_{ac} \sqrt{\lambda'_c(1+\lambda'_c)} (u'^{(2)}_{ac})^* \right]$$

$$- \sqrt{\lambda_a(1+\lambda_a)} \left[ u'^{(2)}_{ac} \sqrt{\lambda'_c(1+\lambda'_c)} (u'^{(1)}_{ac})^* \right]$$

$$\left. + \lambda_a \left[ (u'^{(2)}_{ac})^* \lambda'_c u'^{(2)}_{ac} \right] \right\}$$

$$\times \left\{ (1+\lambda_b) \left[ u'^{(1)}_{bd} \lambda'_d (u'^{(1)}_{bd})^* \right] \right.$$

$$\left. - \sqrt{\lambda_b(1+\lambda_b)} \left[ u'^{(1)}_{bd} \sqrt{\lambda'_d(1+\lambda'_d)} (u'^{(2)}_{bd})^* \right] \right.$$

$$-\sqrt{\lambda_b(1+\lambda_b)}\left[u_{bd}^{\prime(2)}\sqrt{\lambda_d'(1+\lambda_d')}\left(u_{bd}^{\prime(1)}\right)^*\right.$$
$$\left.+\lambda_b\left[\left(u_{bd}^{\prime(2)}\right)^*\lambda_d'u_{bd}^{\prime(2)}\right]\right]\Big\rangle_{\delta L}$$
$$\approx \sum_{cd}\sqrt{\lambda_a(1+\lambda_a)\lambda_b(1+\lambda_b)}\left\langle\sqrt{\lambda_c'\lambda_d'}\right\rangle_{\delta L}$$
$$\times\left[\left\langle u_{ac}^{\prime(1)}(u_{bd}^{\prime(1)})^*\right\rangle_0\left\langle (u_{ac}^{\prime(2)})^*u_{bd}^{\prime(2)}\right\rangle_0\right.$$
$$\left.+\left\langle u_{ac}^{\prime(2)}(u_{bd}^{\prime(2)})^*\right\rangle_0\left\langle(u_{ac}^{\prime(1)})^*u_{bd}^{\prime(1)}\right\rangle_0\right]$$
$$=\sum_{cd}\sqrt{\lambda_a(1+\lambda_a)\lambda_b(1+\lambda_b)}\left\langle\sqrt{\lambda_c'\lambda_d'}\right\rangle_{\delta L}2\frac{\delta_{ab}\delta_{cd}}{N^2}$$
$$=2\frac{\delta_{ab}}{N}\lambda_a(1+\lambda_a)\frac{1}{N}\sum_c\langle\lambda_c'\rangle_{\delta L}$$
$$=2\frac{\delta_{ab}}{N}\lambda_a(1+\lambda_a)\frac{\delta L}{l_e}. \tag{7.163}$$

We now substitute eqns (7.158), (7.162) and (7.163) into the expansion (7.121), to obtain (the superscript '(2)' indicates that we are dealing with the unitary case)

$$\frac{\partial p_s^{(2)}(\lambda)}{\partial s}=\sum_a\left\{\left[(1+2\lambda_a)-\frac{N-1}{N}(1+2\lambda_a)+\frac{1}{N}\lambda_a(1+\lambda_a)\frac{\partial\ln J^{(2)}(\lambda)}{\partial\lambda_a}\right]\right.$$
$$\left.\times\frac{\partial p_s^{(2)}(\lambda)}{\partial\lambda_a}+\frac{1}{N}\lambda_a(1+\lambda_a)\frac{\partial^2 p_s^{(2)}(\lambda)}{\partial\lambda_a^2}\right\}$$
$$=\frac{1}{N}\frac{1}{J^{(2)}(\lambda)}\sum_a\frac{\partial}{\partial\lambda_a}\left[\lambda_a(1+\lambda_a)J^{(2)}(\lambda)\frac{\partial p_s^{(2)}(\lambda)}{\partial\lambda_a}\right]. \tag{7.164}$$

The dimensionless length $s$ was defined in eqn (7.67b). We notice again from eqns (7.26) and (7.21) that the joint probability density of the $\lambda_a$s, $a=1,\ldots,N$, is given by

$$w_s^{(2)}(\lambda)=p_s^{(2)}(\lambda)J^{(2)}(\lambda). \tag{7.165}$$

We finally write eqn (7.164) in terms of the joint probability density $w_L(\lambda)$ as

$$\frac{\partial w_s^{(2)}(\lambda)}{\partial s}=\frac{1}{N}\sum_a\frac{\partial}{\partial\lambda_a}\left[\lambda_a(1+\lambda_a)J^{(2)}(\lambda)\frac{\partial}{\partial\lambda_a}\frac{w_s^{(2)}(\lambda)}{J^{(2)}(\lambda)}\right]. \tag{7.166}$$

This is the diffusion equation governing, for $\beta=2$, the evolution with the conductor length of the joint probability density $w_s^{(2)}(\lambda)$ of the $\lambda_a$s, $a=1,\ldots,N$; it has to be solved with the initial condition indicated in eqn (7.138).

## 7.4 A unified form of the diffusion equation for the various universality classes describing quasi-one-dimensional disordered conductors: calculation of expectation values

The diffusion equation found in eqns (7.137) and (7.166) for the orthogonal and the unitary universality classes, $\beta = 1$ and $\beta = 2$, respectively, can be written in a unified way as

$$\frac{\partial w_s^{(\beta)}(\lambda)}{\partial s} = \frac{2}{\beta N + 2 - \beta} \sum_a \frac{\partial}{\partial \lambda_a} \left[ \lambda_a(1+\lambda_a) J^{(\beta)}(\lambda) \frac{\partial}{\partial \lambda_a} \frac{w_s^{(\beta)}(\lambda)}{J^{(\beta)}(\lambda)} \right]. \quad (7.167)$$

This equation can be shown to describe the symplectic universality class, $\beta = 4$, as well, although this latter case will not be considered here. We remind the reader that $s$ is the dimensionless length (see eqn (7.67b))

$$s = \frac{L}{l_e}, \quad (7.168)$$

i.e., the length measured in units of the elastic mean free path, and that the above diffusion equation has to be solved with the initial condition (see eqn (7.138))

$$w_{s=0}^{(\beta)}(\lambda_1, \ldots, \lambda_N) = \prod_{a=1}^{N} \delta_+(\lambda_a). \quad (7.169)$$

Our diffusion equation (7.167) is now widely referred to as the DMPK equation (after Dorokhov [64] and Mello, Pereyra and Kumar [132]) in extensive literature on quantum transport through a quasi-1-dimensional (multi-channel) disordered conductor.

In what follows we shall be more interested in the evolution with length of the expectation value of the functions $F(\lambda) \equiv F(\lambda_1, \ldots, \lambda_N)$ (for example, the dimensionless conductance $g = T = \sum_a 1/(1+\lambda_a)$), rather than in the joint distribution $w_s^{(\beta)}(\lambda)$ itself. Such an evolution is certainly governed by the above diffusion equation as well. Multiplying both sides of eqn (7.167) by $F(\lambda)$ and integrating over $\{\lambda_a\}$, we find, for the expectation value

$$\langle F \rangle_s^{(\beta)} = \int F(\lambda) w_s^{(\beta)}(\lambda) \prod_{a=1}^{N} d\lambda_a, \quad (7.170)$$

the evolution equation

$$\frac{1}{2}(\beta N + 2 - \beta) \partial_s \langle F \rangle_s^{(\beta)}$$
$$= \left\langle \sum_a \left[ \lambda_a(1+\lambda_a) \frac{\partial^2 F}{\partial \lambda_a^2} + (1+2\lambda_a) \frac{\partial F}{\partial \lambda_a} \right] \right\rangle$$

$$+ \frac{\beta}{2} \sum_{a \neq b} \frac{\lambda_a(1+\lambda_a)\partial F/\partial \lambda_a - \lambda_b(1+\lambda_b)\partial F/\partial \lambda_b}{\lambda_a - \lambda_b} \Bigg\rangle_s^{(\beta)}.$$
(7.171)

### 7.4.1 The moments of the conductance

An important quantity which we shall examine is the $p$th moment of the dimensionless conductance $T$. Setting $F = T^p$ in eqn (7.171) and defining

$$T_k = \sum_a \frac{1}{(1+\lambda_a)^k}, \tag{7.172}$$

we find the evolution equation

$$(2 - \beta + \beta N)\partial_s \langle T^p \rangle_s^{(\beta)}$$
$$= \langle -\beta p T^{p+1} - (2-\beta)p T^{p-1} T_2 + 2p(p-1)T^{p-2}(T_2 - T_3) \rangle_s^{(\beta)}.$$
(7.173a)

We notice the important feature that, on the right-hand side of eqn (7.173a), there appear other quantities in addition to the moments $\langle T^q \rangle_s^{(\beta)}$ of the conductance that we are interested in; in this sense we say that the moments $\langle T^p \rangle_s^{(\beta)}$ *do not form a closed set*. We thus need to find the evolution equations for these other quantities. The calculation is carried out in [126, 137], which results in the following equations:

$$(2 - \beta + \beta N)\partial_s \langle T^p T_2 \rangle_s^{(\beta)}$$
$$= \langle [2\beta T^{p+2} - \beta(p+4)T^{p+1}T_2 + 2(2-\beta)T^p T_2 - 4(2-\beta)T^p T_3$$
$$- (2-\beta)pT^{p-1}T_2^2] + 8pT^{p-1}(T_3 - T_4) + 2p(p-1)T^{p-2}T_2(T_2 - T_3) \rangle_s^{(\beta)},$$
(7.173b)

$$(2 - \beta + \beta N)\partial_s \langle T^p T_3 \rangle_s^{(\beta)}$$
$$= \langle [-\beta(p+6)T^{p+1}T_3 + 6\beta T^{p+1}T_2 - 3\beta T^p T_2^2] + 6(2-\beta)T^p T_3$$
$$- 9(2-\beta)T^p T_4 - (2-\beta)pT^{p-1}T_2 T_3 + 12pT^{p-1}(T_4 - T_5)$$
$$+ 2p(p-1)T^{p-2}T_3(T_2 - T_3) \rangle_s^{(\beta)},$$
(7.173c)

$$(2 - \beta + \beta N)\partial_s \langle T^p T_2^2 \rangle_s^{(\beta)}$$
$$= \langle [-\beta(p+8)T^{p+1}T_2^2 + 4\beta T^{p+2}T_2] - (2-\beta)pT^{p-1}T_2^3$$
$$+ 4(2-\beta)T^p T_2^2 - 8(2-\beta)T^p T_2 T_3 + 2p(p-1)T^{p-2}(T_2 - T_3)T_2^2$$
$$+ 16T^p(T_4 - T_5) + 16pT^{p-1}T_2(T_3 - T_4) \rangle_s^{(\beta)}.$$
(7.173d)

We notice that the evolution equations for the quantities other than the moments of $T$ appearing in eqn (7.173a) contain a hierarchy of further new quantities which had not appeared before. It is thus clear that finding an exact solution for the moments of the conductance as a function of length is unfeasible, at least within the present method. However, let us point out that in many cases of actual physical interest the number of channels $N$ is very large, $N \gg 1$. In this limit,

$$l_e \ll Nl_e \approx \xi_{1D}, \tag{7.174}$$

and the mean free path is well separated from the *quasi-one-dimensional localization length*, which we have denoted by $\xi_{1D}$. When the actual length $L$ of our quasi-one-dimensional system is smaller than the mean free path, i.e., $L < l_e$, the system is in the *ballistic regime*. When $L$ is in the range defined by the mean free path and the quasi-one-dimensional localization length, i.e.,

$$l_e \ll L \ll Nl_e, \tag{7.175a}$$

or, in terms of the dimensionless length $s = L/l_e$,

$$1 \ll s \ll N, \tag{7.175b}$$

we shall find the *diffusive*, or *metallic regime*, with a crossover to the *localized regime*, that occurs when $L \gg Nl_e$. We shall see in what follows that, for a large number of channels, $N \gg 1$, we can construct a meaningful approximation scheme to solve the above equations, which covers the ballistic and metallic regimes, i.e.,

$$0 < L \ll Nl_e, \tag{7.176a}$$

or

$$0 < s \ll N. \tag{7.176b}$$

The essential mathematical point is that, for every order $N^k$, we shall encounter a finite 'closed set' (in the sense used above) of coupled equations which we shall be able to solve.

We propose the following expansions in decreasing powers of $N$:

$$\langle T^p \rangle_s^{(\beta)} = N^p f_{p,0}(s) + N^{p-1} f_{p,1}(s) + N^{p-2} f_{p,2}(s) + \cdots, \tag{7.177a}$$

$$\langle T^p T_2 \rangle_s^{(\beta)} = N^{p+1} g_{p+1,0}(s) + N^p g_{p+1,1}(s) + N^{p-1} g_{p+1,2}(s) + \cdots, \tag{7.177b}$$

$$\langle T^p T_3 \rangle_s^{(\beta)} = N^{p+1} h_{p+1,0}(s) + N^p h_{p+1,1}(s) + N^{p-1} h_{p+1,2}(s) + \cdots, \tag{7.177c}$$

$$\langle T^p T_2^2 \rangle_s^{(\beta)} = N^{p+2} l_{p+2,0}(s) + N^{p+1} l_{p+2,1}(s) + N^p l_{p+2,2}(s) + \cdots. \tag{7.177d}$$

The various functions $f_{p,m}(s), g_{p,m}(s), \ldots$ must satisfy the initial conditions

$$f_{p,m}(s=0) = g_{p,m}(s=0) = \cdots = \delta_{m0}. \tag{7.178}$$

We introduce these expansions into eqn (7.173a) for $\langle T^p \rangle_s^{(\beta)}$ and find

$$(2-\beta)\left[N^p f'_{p,0}(s) + N^{p-1} f'_{p,1}(s) + N^{p-2} f'_{p,2}(s) + \cdots\right]$$
$$+ \beta \left[N^{p+1} f'_{p,0}(s) + N^p f'_{p,1}(s) + N^{p-1} f'_{p,2}(s) + \cdots\right]$$
$$= -\beta p \left[N^{p+1} f_{p+1,0}(s) + N^p f_{p+1,1}(s) + N^{p-1} f_{p+1,2}(s) + \cdots\right]$$
$$- (2-\beta) p \left[N^p g_{p,0}(s) + N^{p-1} g_{p,1}(s) + N^{p-2} g_{p,2}(s) + \cdots\right]$$
$$+ 2p(p-1)\left[N^{p-1} g_{p-1,0}(s) + N^{p-2} g_{p-1,1}(s) + \cdots\right]$$
$$- 2p(p-1)\left[N^{p-1} h_{p-1,0}(s) + N^{p-2} h_{p-1,1}(s) + \cdots\right]. \tag{7.179}$$

Equating the coefficients of $N^{p+1}$ in the above equation, we find the following recursive differential equations for the functions $f_{p,0}(s)$:

$$f'_{p,0}(s) + p f_{p+1,0}(s) = 0, \tag{7.180}$$

whose solution, with the initial condition (7.178), i.e.,

$$f_{p,0}(0) = 1, \quad \forall p, \tag{7.181}$$

is

$$f_{p,0}(s) = \frac{1}{(1+s)^p}. \tag{7.182}$$

Notice that it is essential, for the consistency of the above method, to have $s \ll N$, as indicated in eqn (7.176). For instance, should $s \sim N$, then $f_{p,0}(s)$ would no longer be independent of $N$, but rather $f_{p,0}(s) \sim 1/N^p$, and the above identification of powers of $N$ would be wrong.

We equate the coefficients of $N^p$ in eqn (7.179); we also equate the coefficients of $N^{p+2}$ in the equation obtained by substituting the expansions (7.177) into eqn (7.173b) for $\langle T^p T_2\rangle_s^{(\beta)}$. As a result, we obtain the following pair of recursive differential equations:

$$f'_{p,1}(s) + p f_{p+1,1}(s) = -\frac{2-\beta}{\beta}\left[f'_{p,0}(s) + p g_{p,0}(s)\right],$$
$$g'_{p,0}(s) + (p+3) g_{p+1,0}(s) = 2 f_{p+1,0}(s) \tag{7.183}$$

for $f_{p,1}(s)$ and $g_{p,0}(s)$, to be solved with the initial conditions

$$f_{p,1}(0) = 0, \quad g_{p,0} = 1, \quad \forall p; \tag{7.184}$$

use has to be made of the expression for $f_{p,0}(s)$ given in eqn (7.182). The result is

$$f_{p,1}(s) = -\frac{2-\beta}{\beta} \frac{p s^3}{3(1+s)^{p+2}}, \tag{7.185a}$$

$$g_{p,0}(s) = \frac{2s^3 + 6s^2 + 6s + 3}{3(1+s)^{p+3}}. \tag{7.185b}$$

We can thus proceed systematically to obtain results to any desired order in our $N$ expansion. For instance, for the $p$th moment of the conductance we find ($\beta = 1, 2$)

$$\langle T^p \rangle_s^{(\beta)} = \frac{N^p}{(1+s)^p} - \frac{2-\beta}{\beta} \frac{ps^3}{3(1+s)^{p+2}} N^{p-1}$$
$$+ \frac{p}{90} \left[ \left( \frac{2-\beta}{\beta} \right)^2 \frac{(11p-9)s^8 + \cdots}{(1+s)^{p+6}} \right.$$
$$\left. + \frac{2(\beta-1)(4-\beta)}{\beta^2} \frac{(3p-5)s^6 + \cdots}{(1+s)^{p+4}} \right] N^{p-2} + \cdots . \quad (7.186)$$

The two numerators inside the square bracket in the last equation are polynomials in $s$, given explicitly in [137] and [126].

*The average conductance*

Setting $p = 1$ in eqn (7.186) we find, in the metallic regime (7.175),

$$\langle T \rangle_s^{(\beta)} = \frac{N}{s} - \frac{1}{3} \frac{2-\beta}{\beta} + \cdots \quad (7.187a)$$
$$= \frac{N l_e}{L} - \frac{1}{3} \frac{2-\beta}{\beta} + \cdots . \quad (7.187b)$$

For each order in $N \gg 1$ we have kept the leading term in $s \gg 1$. This is equivalent to taking the limits $N \to \infty$ and $s \to \infty$, with $N/s \equiv g_0$ fixed. The first term in the above equation is essentially Drude's law, once we recall that the number of channels $N$ is proportional to the transverse cross-section of the wire. The next term has a quantum mechanical origin. It is a negative correction for the orthogonal case, $\beta = 1$, just as it occurs in *weak-localization* theory (to be discussed in more detail in the next chapter), which we also encountered in the previous chapter for cavities and for the same universality class. This term is absent for the unitary case, $\beta = 2$. As it happens, eqn (7.187) is valid for the symplectic class, $\beta = 4$, as well, in which case it gives a positive anti-weak-localization correction.

*The variance of the conductance*

Setting $p = 2$ in eqn (7.186) and working again in the metallic regime (7.175), we can find the second moment of the conductance and, using the above result for the average, we can extract the variance. We discover that both the terms which are quadratic and linear in $N$ cancel, the first non-vanishing term in the series being the one proportional to $N^0$, i.e.,

$$(\text{var } T)_s^{(\beta)} = \frac{2}{15\beta} + \cdots . \quad (7.188)$$

The first significant term is thus independent of the number of channels $N$ ($\gg 1$), the length of the system $L$ and the degree of disorder embodied in the mean free path $l_e$ (although we know that $L$ and $l_e$ occur in the dimensionless ratio $s$). Just as in the previous chapter for cavities, eqns (6.40) and (6.57), we speak of *universal conductance fluctuations* (UCF). For the cavity case, when the number of channels is the same in the entrance and exit waveguides ($K = 1$) and very large, one finds $(\mathrm{var}\, T)^{(\beta)} = 1/8\beta$, a result which is remarkably close, but not identical, to that of eqn (7.188) for quasi-one-dimensional systems in the metallic regime.

*The distribution of the conductance*

In terms of the variables $\{x_i\}$, related to the $N$ transmission eigenvalues $T_i = 1/(1+\lambda_i)$ by $T_i \equiv 1/\cosh^2 x_i$, Beenakker and Rejaei [22, 23] showed that the exact solution of the diffusion equation (7.167) for $\beta = 2$ is given by

$$\tilde{w}_s^{(2)}(\{x_i\}) = C(s) \prod_{i<j}(\sinh^2 x_j - \sinh^2 x_i) \prod_i \sinh 2x_i$$

$$\times \mathrm{Det}\left\{\int_0^\infty \mathrm{d}k\, k^{2m-1} \mathrm{e}^{-k^2 s/4N} \tanh\left(\frac{\pi k}{2}\right) P_{(ik-1)/2}(\cosh 2x_n)\right\}, \tag{7.189}$$

where, as usual, $s$ is the dimensionless length defined in eqn (7.168), $C(s)$ is an $x$-independent normalization factor, $P_{(ik-1)/2}$ are Legendre ('toroidal') functions and $\mathrm{Det}\{a_{nm}\}$ denotes the determinant of the $N \times N$ matrix with elements $a_{nm}$.

For $\beta = 1$ there is no known exact solution of the diffusion equation. However, in the metallic regime ($1 \ll s \ll N$) the results for both $\beta$ values can be written [21–23] in the general form of a Gibbs distribution $\tilde{w}_s^{(2)}(\{x_i\}) \propto \mathrm{e}^{-\beta H(\{x_i\})}$, where

$$H(x) = \sum_{i<j} U(x_i, x_j) + \sum_i V(x_i), \tag{7.190a}$$

with

$$U(x_i, x_j) = -\frac{1}{2}\left(\ln|\sinh^2 x_j - \sinh^2 x_i| + \ln|x_j^2 - x_i^2|\right), \tag{7.190b}$$

$$V(x) = \frac{\gamma}{2\beta s}x^2 - \frac{1}{2\beta}\ln|x \sinh 2x|, \tag{7.190c}$$

and $\gamma = \beta(N-1) + 2$. The function $H(x_1, \ldots, x_N)$ may be interpreted as the Hamiltonian function of $N$ interacting particles at positions $x_i$ in one dimension, $U(x_i, x_j)$ being the interaction potential and $V(x)$ an effective confining potential [21].

The distribution $P(g)$ of the conductance is known to evolve from a Gaussian (deep in the diffusive, metallic regime) to a log-normal distribution (deep in the localized, insulating regime) [150]. Although $P(g)$ cannot be easily obtained algebraically from the above expressions, various approximations show that this behavior is well described by the diffusion equation. In the crossover regime, [77] found the main statistical properties of $P(g)$ arising from the diffusion equation using a classical Monte Carlo calculation [28,99]. The calculations were first performed for $\beta = 2$ using the full exact solution (eqn (7.189)). Calculations for $\beta = 2$ using the approximate expressions given in eqns (7.190) are indistinguishable from the exact solution, all the way from the ballistic to the localized regime. The results for $\beta = 1$ were obtained only from the approximate eqn (7.190), but, as for $\beta = 2$, one expects them to be a very good approximation everywhere. The results are shown in Fig. 7.2, where $P(g)$ is plotted for both $\beta = 1$ and 2 and for different values of $\langle g \rangle$. The Monte Carlo results (solid lines in Fig. 7.2) are plotted together with the results of extensive quasi-one-dimensional

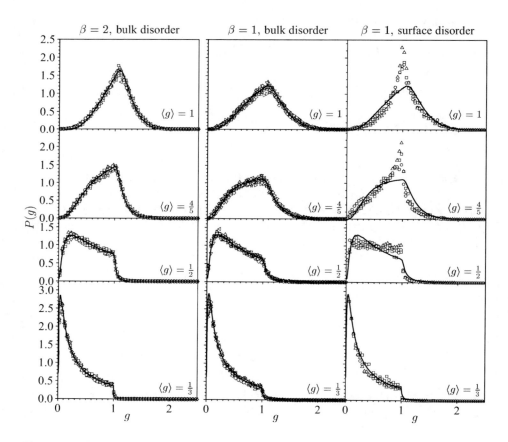

FIG. 7.2. Conductance distribution $P(g)$ for bulk-disordered wires for $\beta = 1$ and 2. The solid lines represent the Monte Carlo solution of the diffusion equation for $N = 6$, as discussed in the text. The symbols represent the numerical calculations based on the Anderson model for different configurations [77]. Results for surface-disordered wires are also shown for comparison [79].

tight-binding-model calculations (symbols), carried out for bulk disorder. As can be seen in Fig. 7.2, the diffusion equation results are in full agreement with the numerical distributions obtained from the Anderson model, for both $\beta = 1$ and 2.

In contrast, the results for bulk disorder are clearly different from those obtained for surface disorder and $\beta = 1$ [79], which are also shown in Fig. 7.2 for comparison.

These results motivated the study of [78], where a potential model is developed that describes in a better way than DMPK the results for bulk disorder and also for surface disorder. It represents an improvement on the model of [138] (for instance, the energy appears explicitly in the formalism, which is not the case for previous models). The model of [138] emerges in the short-wavelength approximation; DMPK, in turn, is a particular case of the results of [138], when all the channels are assumed to contribute on an equal footing.

## 7.5 The correlations in the electronic transmission and reflection from disordered quasi-one-dimensional conductors

The diffusion equation, eqn (7.167), describes the evolution with the length of the quasi-one-dimensional system of the joint probability distribution of the parameters $\{\lambda_a\}$. As a consequence, it governs the evolution of the statistical properties of those quantities which are expressible in terms of these parameters, such as the conductance $T$ studied in the previous section.

From eqns (7.16) and (3.109) it is clear that the study of the statistical properties of the *individual* transmission and reflection coefficients $T_{ab}$ and $R_{ab}$, respectively, would involve the knowledge of the joint probability distribution of *all* the polar parameters defining the transfer matrix $M$, i.e., the 'radial' parameters $\{\lambda_a\}$ and the matrices $u^{(i)}$ and $v^{(i)}$, or, equivalently, $u^{(i)}$ and $z^{(i)}$. An evolution equation providing this description is, unfortunately, not available to us at this moment. However, it seems reasonable to assume that, for a wide class of initial distributions for the phases of the BB, a large number of convolutions will eventually fill out the available compact space of phases with uniform density. This is at least the behavior that has been observed when multiplying many matrices belonging to a *compact* group, such as the unitary group [125]. As already pointed out immediately after eqn (7.45), in previous publications on the subject [126, 132, 137, 164] the model was such that the various unitary matrices entering the polar decomposition of $M$ were distributed according to the invariant measure ('isotropy') for any length. This condition has been relaxed in the above presentation. However, we make the plausible assumption that, under the matrix multiplication of transfer matrices $M$ (which belong to a non-compact group), the phases, which are described by unitary matrices, will eventually (i.e., in the *metallic regime*) be described by the corresponding invariant measure.

The calculation of the averages and correlations of individual transmission and reflection coefficients given below follows the presentation of [127, 129, 137]. The averages of products of elements of unitary matrices over the Haar measure, or invariant measure, of the unitary group will form an essential ingredient for the calculations mentioned above. As in [127], we define

$$Q_{b_1\beta_1,\ldots,b_m\beta_m}^{a_1\alpha_1,\ldots,a_l\alpha_l} = \langle (u_{b_1\beta_1}\cdots u_{b_m\beta_m})(u_{a_1\alpha_1}\cdots u_{a_m\alpha_m})^*\rangle_0, \qquad (7.191)$$

where $u_{a\alpha}$ are elements of an $N$-dimensional unitary matrix and $\langle\cdots\rangle_0$ indicates an average over Haar's measure of the group $U(N)$. Some of these quantities were evaluated in Section 6.4.1 when the case $\beta = 2$ was considered. More general expressions can be found in [127].

In terms of this notation, we can write the average and second moments of the individual transmission coefficients as

$$\langle T_{ab}\rangle_s^\beta = \sum_{\alpha\beta} Q_{a\beta}^{a\alpha} Q_{\beta b}^{\alpha b} \langle \sqrt{\tau_\alpha \tau_\beta}\rangle_s^\beta, \qquad (7.192)$$

$$\langle T_{ab}T_{a'b'}\rangle_s^\beta = \sum_{\alpha,\beta,\alpha',\beta'} Q_{a\beta,a'\beta'}^{a\alpha,a'\alpha'} Q_{\beta b,\beta'b'}^{\alpha b,\alpha'b'} \langle \sqrt{\tau_\alpha \tau_\beta \tau_{\alpha'} \tau_{\beta'}}\rangle_s^\beta. \qquad (7.193)$$

The quantities $\tau_a$ are the radial polar parameters of the $S$ matrix that were introduced in eqn (3.86) and which were used extensively in the previous chapter; they are related through eqn (3.107) to the parameters $\lambda_a$ that have been used more systematically in the present chapter.

In [127] one finds, for some of the $Q$ coefficients,

$$Q_{b\beta}^{a\alpha} = \frac{\delta_b^a \delta_\beta^\alpha}{N}, \qquad (7.194)$$

$$Q_{b\beta,b'\beta'}^{a\alpha,a'\alpha'} = \frac{\delta_b^a \delta_{b'}^{a'} \delta_\beta^\alpha \delta_{\beta'}^{\alpha'} + \delta_b^{a'} \delta_{b'}^a \delta_\beta^{\alpha'} \delta_{\beta'}^\alpha}{N^2 - 1} - \frac{\delta_b^a \delta_{b'}^{a'} \delta_\beta^{\alpha'} \delta_{\beta'}^\alpha + \delta_b^{a'} \delta_{b'}^a \delta_\beta^\alpha \delta_{\beta'}^{\alpha'}}{N(N^2 - 1)}. \qquad (7.195)$$

From eqns (7.194) and (7.192), we find

$$\langle T_{ab}\rangle_s^\beta = \frac{\langle T\rangle_s^\beta}{N^2}. \qquad (7.196)$$

Using eqn (7.195) we can find the second moment of eqn (7.193), and the covariance is

$$C_s^\beta(T_{ab}, T_{a'b'})$$
$$= \langle T_{ab}T_{a'b'}\rangle_s^\beta - \langle T_{ab}\rangle_s^\beta \langle T_{a'b'}\rangle_s^\beta \qquad (7.197a)$$
$$= \frac{1}{(N^2-1)^2}\left\{\left[\left(1+\frac{1}{N^2}\right)\langle T^2\rangle_s^\beta - \frac{2}{N}\langle T_2\rangle_s^\beta\right]\delta_{aa'}\delta_{bb'}\right.$$
$$+ \left[\left(1+\frac{1}{N^2}\right)\langle T_2\rangle_s^\beta - \frac{2}{N}\langle T^2\rangle_s^\beta\right](\delta_{aa'}+\delta_{bb'})$$
$$\left.+ \left(1+\frac{1}{N^2}\right)\langle T^2\rangle_s^\beta - \frac{2}{N}\langle T_2\rangle_s^\beta - \frac{(N^2-1)^2}{N^4}[\langle T\rangle_s^\beta]^2\right\}. \qquad (7.197b)$$

The above equation is exact; as a check, we can easily verify that its sum over the indices $a$, $b$, $a'$ and $b'$ gives precisely $\operatorname{var} T$.

So far, only isotropy of the distribution, as defined above, has been used. As we have already mentioned, the various averages appearing in eqn (7.197) can be obtained from the solution of the diffusion equation as an expansion in decreasing powers of $N$. For instance, from eqn (7.186) for $p = 2$, from eqn (7.187) and from a similar expansion for $\langle T_2 \rangle_s^{(\beta)}$ we find, in the diffusive regime,

$$\langle T \rangle_s^{(\beta)} = \frac{N}{s} - \frac{1}{3}\frac{2-\beta}{\beta} + \cdots, \tag{7.198}$$

$$\langle T^2 \rangle_s^{(\beta)} = \frac{N^2}{s^2} - \frac{2-\beta}{\beta}\frac{2}{3}\frac{N}{s} + \cdots, \tag{7.199}$$

$$\langle T_2 \rangle_s^{(\beta)} = \frac{2N}{3s} - \frac{2-\beta}{\beta}\frac{4}{45} + \cdots, \tag{7.200}$$

$$\langle T_{ab} \rangle_s^{(\beta)} = \frac{\langle T \rangle_s^{(\beta)}}{N^2} = \frac{1}{Ns} - \frac{2-\beta}{3N^2\beta} + \cdots, \tag{7.201}$$

$$\langle T^2 \rangle_s^{(\beta)} = [\langle T \rangle_s^{(\beta)}]^2 + \operatorname{var} T$$
$$= N^4 \langle T_{ab} \rangle_s^{(\beta)} \langle T_{a'b'} \rangle_s^{(\beta)} + \frac{2}{15\beta} + O\left(\frac{1}{N}\right). \tag{7.202}$$

Again, for each order in $N \gg 1$ we have kept the leading term in $s \gg 1$.

We denote the three terms occurring in $C_s^\beta(T_{ab}, T_{a'b'})$, eqn (7.197b), by $C^{(1)}$, $C^{(2)}$ and $C^{(3)}$, respectively, and realize that, with regards to their $N$ dependence, they have the structure

$$C_s^\beta(T_{ab}, T_{a'b'}) = \left[\frac{A}{N^2} + \frac{B}{N^3} + \frac{C}{N^4} + \cdots\right] \delta_{aa'}\delta_{bb'} \\ + \left[\frac{D}{N^3} + \frac{E}{N^4} + \cdots\right](\delta_{aa'} + \delta_{bb'}) + \left[\frac{F}{N^4} + \cdots\right]. \tag{7.203}$$

Take, for instance, $C_s^\beta(T_{12}, T_{12})$, i.e., $a = a'$ and $b = b'$; then $C^{(1)}$, $C^{(2)}$ and $C^{(3)}$ contribute and $A/N^2$ is the leading term. Now take $C_s^\beta(T_{12}, T_{13})$, i.e., $a = a'$ and $b \neq b'$; then $C^{(2)}$ and $C^{(3)}$ contribute and $D/N^3$ is the leading term. For $C_s^\beta(T_{13}, T_{24})$, i.e., $a \neq a'$ and $b \neq b'$, only $C^{(3)}$ contributes and $F/N^4$ is the leading term. We thus see that only $A$, $D$ and $F$ are needed if all we want is the leading term in each case. Our covariance can be written in the form

$$C_s^\beta(T_{ab}, T_{a'b'})$$
$$= \langle T_{ab} \rangle_s^{(\beta)} \langle T_{a'b'} \rangle_s^{(\beta)} \left[\delta_{aa'}\delta_{bb'} + \frac{2}{3 \langle T \rangle_s^{(\beta)}}(\delta_{aa'} + \delta_{bb'}) + \frac{2}{15\beta[\langle T \rangle_s^{(\beta)}]^2}\right]. \tag{7.204}$$

From the previous expression we can obtain, for instance, the variance of the individual transmission coefficient $T_{ab}$ as

$$\operatorname{var} T_{ab} = \left[\langle T_{ab}\rangle_s^{(\beta)}\right]^2 \left[1 + \frac{4s}{3N} + \frac{2s^2}{15\beta N^2}\right]$$
$$\approx \left[\langle T_{ab}\rangle_s^{(\beta)}\right]^2, \qquad (7.205)$$

which is consistent with a Rayleigh distribution for $T_{ab}$.

We can find the variance of the transmission coefficient $T_b = \sum_a T_{ab}$, summed over all the final channels $a$ for a given incident channel $b$, by setting, in eqn (7.197), $b = b'$ and summing over $a$ and $a'$, with the result that

$$\operatorname{var} T_b = \sum_{aa'} C_{ab,a'b}$$
$$= \frac{1}{N(N+1)} \left[\langle T^2\rangle + \langle T_2\rangle\right] - \frac{\langle T\rangle^2}{N^2}. \qquad (7.206)$$

Substituting from eqns (7.198)–(7.200), we find

$$\operatorname{var} T_b = \frac{2}{3Ns} + O(N^0), \qquad (7.207)$$

where, in the first term, we have kept the leading contribution in $s$.

Finally, we turn to the reflection coefficients. From eqn (3.109a) we write the reflection coefficient $R_{ab}$ as

$$R_{ab} = \sum_{\alpha\beta} [v^{(2)}_{\alpha a}]^* v^{(2)}_{\beta a} v^{(1)}_{\alpha b} [v^{(1)}_{\beta b}]^* \left[\frac{\lambda_\alpha \lambda_\beta}{(1+\lambda_\alpha)(1+\lambda_\beta)}\right]^{1/2}. \qquad (7.208)$$

In what follows we restrict ourselves to the calculation of the average reflection coefficients. The correlation between reflection coefficients can be found in the references given above.

For the orthogonal case, $\beta = 1$, we have the restriction $v^{(2)} = [v^{(1)}]^*$, so that

$$\langle R_{ab}\rangle_s^{(\beta=1)} = \sum_{\alpha\beta} Q^{\beta a,\beta b}_{\alpha a,\alpha b} \left\langle \left[\frac{\lambda_\alpha \lambda_\beta}{(1+\lambda_\alpha)(1+\lambda_\beta)}\right]^{1/2}\right\rangle_s^{(\beta=1)}. \qquad (7.209)$$

Using eqn (7.195) for the $Q$ coefficient, we can write

$$\langle R_{ab}\rangle_s^{(\beta=1)} = \frac{1+\delta_{ab}}{N(N+1)} \langle R\rangle_s^{(\beta=1)}. \qquad (7.210)$$

This result means that backward scattering to the same channel is enhanced by a factor of 2, as compared with the scattering to any other channel. A similar phenomenon has already been noticed in the previous chapter, in eqns (6.52). This is precisely the prediction of *weak-localization theory*—explained in more detail in the next chapter—where the argument is that the various paths contribute with random phases, except for a path and its time-reversed counterpart, which contribute coherently and give rise to a factor of 2 in the backward direction (*coherent back-scattering* (CBS)). The same argument predicts that, when a magnetic field is present and hence time-reversal symmetry is destroyed, the above-mentioned enhancement is absent.

We can easily check the prediction of our model for the unitary case, $\beta = 2$. Averaging eqn (7.208) we find

$$\langle R_{ab} \rangle_s^{(\beta=2)} = \sum_{\alpha\beta} Q_{\beta a}^{\alpha a} Q_{\alpha b}^{\beta b} \left\langle \left[ \frac{\lambda_\alpha \lambda_\beta}{(1+\lambda_\alpha)(1+\lambda_\beta)} \right]^{1/2} \right\rangle_s^{(\beta=2)}. \tag{7.211}$$

Using eqn (7.194) for the $Q$ coefficient, we find

$$\langle R_{ab} \rangle_s^{(\beta=2)} = \frac{\langle R \rangle_s^{(\beta=2)}}{N^2}, \tag{7.212}$$

which indeed shows that the backward enhancement factor is absent.

The DMPK equation has found applications for a number of interesting nanoscopic conductors: for example, carbon nanotubes [55, 167], as well as mesoscopic superconducting wires [91]. The DMPK equation has also been generalized to other universality classes for discussing the universal transport properties of disordered quasi-1-D conductors [45, 46, 92, 144, 145]. Generalization beyond quasi-1-dimension has also been proposed [146].

8

# AN INTRODUCTION TO LOCALIZATION THEORY

This chapter is a general introduction to the subject of disorder-induced localization and the metal–insulator transition that follows from it. Verily, together they constitute a paradigmatic example of complex wave interference that dominates the physics of phase-coherent transport. This is true of the transition regime of disordered metals and impure semiconductors at sufficiently low temperatures when the scattering is elastic and the mean free path is small and comparable with the electron wavelength. Then the usual approximation, namely that between successive scatterings the free electron follows a classical trajectory, breaks down. Now, there are two distinct but closely related aspects to it. One is about the nature of the one-electron eigenfunction, namely the possibility of its spatial localization by a potential that is constant in time but varies randomly in space, on a length-scale which is microscopic. This is the so-called *strong localization* (SL), due originally to P. W. Anderson, who studied it in the context of the possible absence of electronic diffusion on certain random lattices which are disordered sufficiently strongly and microscopically—hence Anderson localization [1,9]. The other aspect concerns the electrical transport. It calculates the scale-dependent quantum correction to the classical, or Drude conductivity arising from a certain subtle feature of wave interference which is constructive and undamped despite disorder, even because of it—the so-called *coherent back-scattering* (CBS) [25]. It dominates quantum transport in the absence of, but close to, the conditions of strong localization. This has been termed *weak localization* (WL) [25,115]. In this chapter, strong localization will be described briefly and in physical terms only, while the weak localization will be treated analytically and in relatively greater detail. This is simply because the latter has greater relevance to the mesoscopic conductors of interest here. We will also discuss the basic ideas of the scaling theory of localization that provides a smooth connection between the SL and the WL regimes. Throughout this chapter we will, however, assume the electrons to be non-interacting—with one another and with other excitations, e.g., phonons. This is necessary for maintaining the condition of phase coherence over the whole sample. This condition has been realized in practice at sufficiently low temperatures, typically at and below the temperature of liquid helium and down to millikelvins. A canonical example of such a disordered system is an impurity-band semiconductor in which the randomly distributed impurity centers introduce orbitals that overlap weakly to form a disordered impurity band. Strong disorder and a relatively much weaker electron–electron interaction can then be realized through compensation, i.e., by adding $p$-type impurities (e.g., boron) to an $n$-type (e.g., P-doped) semiconductor (e.g., silicon), thereby creating a low concentration of mobile carriers but a high concentration of the charged scattering centers, namely the positively charged donors ($P^+$) and the negatively charged acceptors ($B^-$), that do scatter strongly [143].

Now, what could possibly be the rationale for including a whole chapter on localization in a book stated to be devoted to the complex wave interference that gives mesoscopic fluctuations of the sample-specific conductances and other quantities of interest, such as the scattering phase shifts? A partial answer to this would be that almost any discussion of phase-coherent transport in a scattering system involves physical quantities, concepts and defining terms that derive naturally from the original context of Anderson localization. The latter provides a general perspective on the whole discussion here. In addition, weak localization is of direct relevance to interesting phase-sensitive effects observed in mesoscopic systems, e.g., chaotic cavities, where surface scattering dominates, as discussed in Chapter 6, and quasi-one-dimensional disordered conductors, as discussed in Chapter 7. The material presented in Sections 8.5 and 8.6 will enable the reader to contrast the *diagrammatic perturbative* method with the *non-perturbative* one presented in Chapters 6 and 7, based on random-matrix theory. This was another motivation for including this chapter.

Although the material presented in this chapter is covered comprehensively in a number of rather specialized review articles and monographs, for instance [6, 25, 115, 143], we thought it apt to include this rather short chapter in order to make the book reasonably self-contained.

Finally, it is to be noted that the subject of Anderson localization has grown immensely in the last two decades, theoretically as well as in relation to experimental facts. Still, a number of questions remain unanswered and there are open problems. Thus, while WL is essentially well understood, at least at the semi-classical level of CBS, its development into SL with increasing disorder is still far from being so. Furthermore, there is the unresolved problem of the interplay of disorder and the electron–electron interaction present in all real systems. Indeed, one would expect the electron–electron interaction to grow stronger as we approach localization—after all, the localized electrons do not screen well. This continues to be a recurrent theme of serious discussions. Any kind of completeness is, therefore, clearly out of the question. None is intended.

## 8.1 Strong localization

Consider a *tight-binding* one-electron Hamiltonian

$$H = \sum_{j=1}^{N} \varepsilon_j |j\rangle\langle j| - t \sum_{\langle j,l \rangle}^{N} \left( |j\rangle\langle l| + |l\rangle\langle j| \right) \tag{8.1}$$

with one non-degenerate orbital $|j\rangle$ per lattice site $j$ ($= 1, 2, \ldots, N$) of site energy $\varepsilon_j$, and with the transfer matrix element $t$ ($> 0$) connecting the nearest-neighboring sites, $\langle j, l \rangle$. The case of all site energies being equal corresponds to a perfect crystal. Then the Bloch theorem holds and the eigenfunctions are complex running (modulated plane) waves $|\phi_{\bm k}\rangle = N^{-1/2} \sum_j \mathrm{e}^{i{\bm k}\cdot{\bm x}_j}|j\rangle$, labeled by the good quantum number ${\bm k}$ (the wave vector), and extending over the entire lattice. These states are normalized as $\langle\phi_k|\phi_{k'}\rangle = \delta_{{\bm k},{\bm k}'}$, with ${\bm k}$ running over the first Brillouin zone. (Notice the slight change in notation for the wave vector label ${\bm k}$ with respect to that used in earlier chapters.) The energy eigenvalues $\epsilon_k$ form a band of width $B$ ($\equiv 2zt$, where $z$ is the coordination number, equal to $2d$ for a $d$-dimensional simple cubic lattice) that contains sharp features, e.g., the well-known van Hove singularities in the density of states. Periodic boundary conditions are assumed here. (Another boundary condition giving real standing waves instead of the complex running waves may be chosen, but in the limit as $N \to \infty$, as is always implicit here, the difference is inconsequential.)

Disorder can now be introduced simply by, e.g., making the site energies $\{\varepsilon_j\}$ random while keeping the transfer matrix elements non-random. This is the so-called site-diagonal disorder. More specifically, we can take $\{\varepsilon_j\}$ to be a set of uncorrelated, identically-distributed random variables drawn from a rectangular probability distribution over the range $-W/2 < \varepsilon_j < +W/2$. This is consistent with the microscopic disorder having macroscopic homogeneity. We could equally well consider a Gaussian distribution with statistical averages $\langle \varepsilon_j \rangle = 0$, and $\langle \varepsilon_i \varepsilon_j \rangle = \delta_{ij} W^2$, say. Clearly, $W$ is a measure of the site energy mismatch that acts as a potential barrier disfavoring propagation (delocalization) from one site to the other. The bandwidth $B$, on the other hand, favors propagation (delocalization). The dimensionless ratio $W/B$ is then a measure of the degree or the strength of disorder. (It turns out to be essentially the same as the disorder parameter $\eta \equiv 1/k_F l_e$, where $k_F$ is the wave vector magnitude for the electrons at the Fermi energy, and $l_e$ is the elastic transport mean free path due to scattering by the static (quenched) disorder.) It may be recalled here that, for isotropic scatterers, the transport mean free path equals the scattering mean free path. This will be assumed throughout.

For the above tight-binding lattice model Hamiltonian for a disordered system, referred to as the Anderson Hamiltonian, it is now known, as originally demonstrated by P. W. Anderson [9], that in three dimensions there exists a critical value of the disorder parameter, $(W/B)_c$, such that, for $W/B > (W/B)_c$, all eigenstates of the system get localized exponentially and the system turns insulator—the Anderson insulator. Exponential localization means that the envelope of the exact energy eigenstate $|\psi_\alpha\rangle$ decays exponentially with distance far away from the position $x_\alpha$, where it peaks, i.e.,

$$|\psi_\alpha\rangle = \sum_{j=1}^{N} a_j^\alpha |j\rangle, \qquad (8.2\mathrm{a})$$

with

$$|a_j^\alpha| \sim \mathrm{e}^{-|x_j - x_\alpha|/\xi_\alpha}. \qquad (8.2\mathrm{b})$$

As depicted in Fig. 8.1, this defines the localization length $\xi_\alpha$, the spatial extent of the localized eigenfunction $\psi_\alpha$. Of course, the complex amplitudes $a_j^\alpha$ must oscillate with site $j$ so as to ensure orthogonality. Also, the wave function $\psi_\alpha$ is expected to peak at a site $j$ with which it is energetically in near-resonance, i.e., for which the energy eigenvalue $E_\alpha \approx \varepsilon_j$.

A general proof of the above localization theorem is outside the scope of this chapter. A heuristic, but physically robust, argument for the existence of the critical disorder $(W/B)_c$ can, however, be readily given as follows. In the limit of infinitely strong disorder $(B/W = 0)$, clearly all states are localized as site orbitals on the respective lattice sites. Now consider a small but finite value of $B/W$ that can be treated perturbatively. The effective transfer matrix element between any two sites which are $n$ sites apart will then be scaled down by a factor $\sim (B/W)^n$. But the number of such distant neighbors scales up by a factor $\sim n^d$. We now effectively have a new scaled lattice with the single sites replaced by blocks of sites (each containing $\sim n^d$ original sites), having a coordination number $\sim n^d$ and an inter-block transition matrix element $\sim B(B/W)^n$. The resulting bandwidth $\sim n^d B(B/W)^n$ tends to zero as $n \to \infty$. This implies localization. Thus, not only for $B/W = 0$, but also for a small but nonzero range of values of $B/W$, the system remains localized. In the opposite limit of large $B/W$ (weak disorder), however, the usual treatment in terms of the scattering of plane waves with an elastic mean free path $l_e$ must hold, implying the usual metallic conduction, and hence delocalization. (The above argument holds with the proviso that a metallic state exists at all in the limit of weak disorder,

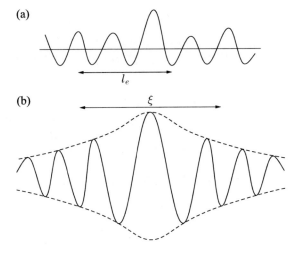

FIG. 8.1. (a) Extended eigenstate showing spatial phase randomization over the elastic mean free path $l_e$, and (b) localized eigenstate with amplitude envelope decaying exponentially over the localization length $\xi$ (schematic).

which we now know to be the case for $d > 2$.) Thus, a transition—the metal to insulator transition—between these two extremes is clearly indicated for some value of $B/W$. Of course, the nature of this transition, whether discontinuous, continuous, or even a mere crossover, cannot be ascertained from this heuristic argument.

The role of $W/B$, or $W/2zt$, the reciprocal of the so-called Thouless ratio, in determining the degree of localization, can be discerned from an examination of the simple case of $N = 2$, a lattice with just two sites 1 and 2, with the site energy difference $\varepsilon_2 - \varepsilon_1 = 2W$ ($> 0$, say). Elementary quantum mechanics of a two-level system tells us that the exact eigenstates tend to get localized on one or the other of the two sites as $W/t \to \infty$. More specifically, with $\varepsilon_1 + \varepsilon_2 = 0$, the lower-lying (bonding) eigenstate $\psi_+$ of energy $E_+ = -\sqrt{W^2 + t^2}$ localizes preferentially on site 1, with squared-modulus of the overlap $|\langle 1|\psi_+\rangle|^2 = \frac{1}{2}(1 + W/\sqrt{t^2 + W^2})$, while the higher-lying (anti-bonding) eigenstate $\psi_-$ of energy $E_- = +\sqrt{W^2 + t^2}$ localizes preferentially on site 2. In the limit as $W/t \to \infty$, the localization is complete, while the limit $W/t \to 0$ corresponds to complete delocalization—with equal sital overlap for both the eigenstates. This at the very least brings out the opposing effect of the site energy mismatch $W$ and the inter-site coupling $-t$. The more subtle effect of wave interference is, however, missed out in this two-site problem.

An important and numerically very meaningful measure of the degree of localization of an eigenstate $\psi_\alpha$ is the *inverse participation* ratio

$$p_\alpha = \frac{\sum_{j=1}^{N} |a_j^\alpha|^4}{\left(\sum_{j=1}^{N} |a_j^\alpha|^2\right)^2}. \tag{8.3}$$

It is at once verified that, for an extended state, $p_\alpha = O(1/N)$ as $N \to \infty$, while for a localized state $p_\alpha = O(1)$. Clearly, for an eigenstate localized on a single site $p_\alpha = 1$, while for a running (Bloch) wave $p_\alpha = 1/N$. For a three-dimensional exponentially-localized state, using (8.2b) and replacing the summation by an integral, we get $p_\alpha \sim (a/\xi_\alpha)^3$, where $a$ is the lattice constant. Thus, $p_\alpha$ goes from unity to zero as we go from the extreme localized to the extreme delocalized limit.

Closely associated with localization is the idea of *local density of states* (or the spectral function) $\nu_j(E)$, as distinct from the *global density of states*, $\nu(E)$. We have

$$\nu_j(E) = \sum_\alpha |\langle j|\alpha\rangle|^2 \delta(E - E_\alpha) \tag{8.4a}$$

and

$$\nu(E) = \sum_\alpha \delta(E - E_\alpha). \tag{8.4b}$$

Clearly, $\nu_j(E)$ counts the eigenstates $\psi_\alpha$ of energy $E_\alpha = E$ having appreciable overlap with (or equivalently, localized close to) the site $j$. The global density of states $\nu(E)$, on the other hand, counts all the eigenstates at the energy $E$ irrespective of their location in space. Thus, $\nu(E)$ is an extensive, additive quantity.

For a random system it coincides with its ensemble average—we say that it is *self-averaging* in the limit as $N \to \infty$ [35]. It is a featureless smooth function of $E$. The local density of states $\nu_j(E)$, on the other hand, will have a granular structure in the localized regime, as it counts the relatively small number of localized states that overlap appreciably with the site $j$. It will fluctuate from one microscopic realization of disorder to another. A subtle feature underlying $\nu_j(E)$ is the level-crossing avoidance of the spatially-localized nearby eigenfunctions. The localized eigenfunctions $|\alpha\rangle$ and $|\beta\rangle$, which are proximate in energy ($E_\alpha \approx E_\beta$), must necessarily be distant in space, $|x_\alpha - x_\beta| \gg \xi_{\alpha,\beta}$. This statistical correlation between eigenfunctions in respect of their separations in energy and in real space is an essential aspect of the localization induced by disorder [35]. This makes $\nu_j(E)$ granular for the localized states.

## 8.2  Mobility edge

Now knowing that for $W/B > (W/B)_c$ all states are localized, a question naturally arises as to what happens for a subcritical disorder. The answer, which was first given by N. F. Mott [143] and which has since been well substantiated, is that we then have both the extended as well as the localized states, but the two sets are separated by a sharply defined characteristic energy $E_c$ (or energies $E_{c1}$ and $E_{c2}$) such that the eigenstates lying toward the band edges are localized, while those toward the band center remain extended. Accordingly, the system undergoes a metal–insulator (M–I) transition as the Fermi level $E_F$ is tuned across this characteristic energy—very aptly called the mobility edge (see Fig. 8.2). It is important to note that the total density of states $\nu(E)$, however, remains continuous across the mobility edge [35]. As the disorder increases, the mobility edges $E_{c1}$ and $E_{c2}$ close in and eventually meet at the critical disorder $(W/B)_c$ when all states get localized. That the states nearer the band edges are the first to get localized is readily understood in terms of our earlier argument—

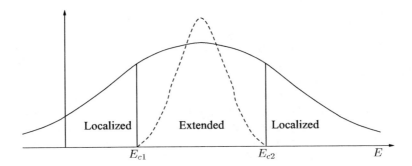

FIG. 8.2. Phase diagram for disordered systems showing the density of localized and extended states $\nu(E)$ (solid line), the mobility edges $E_{c1}$ and $E_{c2}$, and the mobility $\mu(E)$ (dashed line) as a function of the energy $E$ (schematic).

it is just that the relatively low density of states there makes it that much more improbable to find two sites close by in energy as well as in space.

The existence of a *sharply* defined transition energy, the *mobility edge* $E_c$, in a *one-electron* problem with quenched *randomness* is a real surprise of physics. (Again, the $N \to \infty$ limit is understood here.) A question closely associated with the sharpness of the mobility edge is that of the nature of the conductivity transition—is it of the first order, thereby implying a discontinuous jump in the conductivity $\sigma(E_F)$ at $E_F = E_c$, or is it of the second order, implying a continuous vanishing of $\sigma(E_F)$ at $E_c$ as $\sigma(E_F) \propto |E_F - E_c|^s$, which defines a conductivity critical exponent $s$. One is, of course, merely using the language of phase transitions, but it must be clearly understood that here we do not have a truly equilibrium thermodynamic transition—rather only a parametric transport transition. Also, at the absolute zero of temperature, the conductivity $\sigma$ is determined only by the nature of the eigenstates at the Fermi level, and hence we speak of $\sigma(E_F)$. Equivalently, we can speak of the electron mobility $\mu(E_F)$, with $\sigma(E_F) \approx ne\mu(E_F)$.

This brings us to the all-important point that marks the logical beginning of the new physics at and about the mobility edge. To appreciate this basic point, let us approach the mobility edge from the metallic side, where transport is best described in terms of the scattering of plane waves that randomizes the momentum. The electrical conductivity $\sigma(E_F)$ at zero temperature can then be expressed in terms of an elastic transport mean free path $l_e$ as $\sigma(E_F) = ne^2\tau/m$, namely the classical Drude formula, with $l_e = v_F\tau$. Here $e$ and $m$ are, respectively, the magnitude of the electronic charge and the mass, $n$ is the electron number density, $\tau$ is the mean free time and $v_F$ is the Fermi speed. Now, and this is the main point, for metallic conduction one would expect the mean free path $l_e \geqslant a$, where $a$ is the lattice spacing. This makes physical sense inasmuch as the electron can hardly be scattered before it traverses a lattice spacing. This minimum value of $l_e$ translates into a minimum value $\sigma_{\min}$ for the metallic conductivity [115, 143] (see Fig. 8.3). For a half-filled band one obtains, for $d = 3$, $\sigma_{\min}^{(3)} = e^2/3\pi^2\hbar a$, while for $d = 2$, $\sigma_{\min}^{(2)} \simeq 0.1\, e^2/\hbar$, a universal quantity. (In one dimension, of course, there is a well-known and rigorous result that all states are exponentially localized for an arbitrarily small disorder and, therefore, there is no truly metallic state to begin with.) Thus, a discontinuous conductivity jump is to be expected at the mobility edge, and was so predicted. Early measurements indeed indicated such a trend. This, however, now seems not to be supported by the experiments performed at very low temperatures and much closer to the mobility edge—much lower values for the metallic conductivity than $\sigma_{\min}$ have been reported [115]. Also, in two dimensions there is no evidence for a truly metallic state in the absence of the electron–electron interaction. As we will see later, the classical Drude-like expression misses out on a non-trivial and subtle quantum correction arising from a certain feature of the random wave interference that makes $\sigma_{\min}$ vanish, thus making the conductivity (or the metal–insulator) transition at the mobility edge a continuous one. Aside from this fine

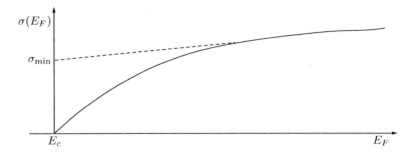

FIG. 8.3. Critical behavior of the conductivity $\sigma(E_F)$ as the Fermi energy $E_F$ approaches the mobility edge $E_c$ (solid line). The dashed line indicates the minimum metallic conductivity $\sigma_{\min}$ (schematic).

structure of the mobility edge, in the rather narrow critical regime, however, the $\sigma_{\min}$ of Mott remains an experimentally useful and robust quantity that operationally separates the good metallic conduction from insulation.

## 8.3  Coherent back-scattering (CBS)

One may ask why we have a transition and not just a crossover. The essential argument is that localized states cannot coexist with extended states at the same energy—quantum mechanical tunneling would admix them and destabilize the localized states. Thus, the localized states must be separated energy-wise from the continuum of extended states, without implying any gap in the density of states, of course. Hence the transition, rather than a crossover, at $E_c$. The discussion of the delocalization–localization transition (or the metal–insulator transition) has so far been based on the physically reasonable idea of the localizing effect of the energy mismatch $W$ due to disorder, opposing the delocalizing effect of the inter-site coupling matrix element ($-t$, or equivalently, the bandwidth $B$). This is highly reminiscent of the classical picture of competition between the kinetic and the potential energies which determines the phase transition from the mobile fluid state to the immobile solid state. Such a *classical* picture, however, misses the phenomenon of *random wave interference*. There is a subtle *quantum* feature, namely the *coherent back-scattering* (CBS), which begins to dominate transport as we approach the mobility edge from the delocalized side. Let us see how [25].

This is best seen by considering in perturbation theory the multiple scattering of an electron from an initial state $|\bm{k}\rangle$ with the wave vector $\bm{k}$ into the final state $|-\bm{k}\rangle$ with the wave vector $-\bm{k}$, directed oppositely to the initial wave vector $\bm{k}$. Such a back-scattering, when present, must clearly contribute most effectively to the resistivity, i.e., to localization. To see this, consider one general term in the $n$th-order Born scattering involving the process $P \equiv (\bm{k}_i \equiv \bm{k} \to \bm{k}_1 \to \bm{k}_2 \to \cdots \to \bm{k}_{n-2} \to \bm{k}_{n-1} \to -\bm{k} \equiv \bm{k}_f)$ that takes the initial state $|\bm{k}\rangle$ to the final state $|-\boldsymbol{k}\rangle$ via $n$ intermediate states. Now, to any such given process $P$ there corresponds a unique process $\bar{P} \equiv (\boldsymbol{k}_i \equiv \boldsymbol{k} \to -\boldsymbol{k}_{n-1} \to -\boldsymbol{k}_{n-2} \to \cdots \to -\boldsymbol{k}_2 \to -\boldsymbol{k}_1 \to -\boldsymbol{k} \equiv \boldsymbol{k}_f)$ that also takes the initial state $|\boldsymbol{k}\rangle$ to the final state $|-\boldsymbol{k}\rangle$, and does this via the same set of intermediate states but now visited in the time-reversed sequence. Clearly, $P$ and $\bar{P}$ constitute a time-reversed pair of alternatives which will now be shown to interfere constructively. The corresponding transition amplitudes $T(\boldsymbol{k} \to -\boldsymbol{k})$ and $\bar{T}(\boldsymbol{k} \to -\boldsymbol{k})$ in the $n$th-order Born approximation are

$$T(\boldsymbol{k} \to -\boldsymbol{k}) = V_{-\boldsymbol{k}\,\boldsymbol{k}_{n-1}} \frac{V_{\boldsymbol{k}_{n-1}\boldsymbol{k}_{n-2}}}{(\epsilon_{\boldsymbol{k}} - \epsilon_{\boldsymbol{k}_{n-1}})} \cdots \frac{V_{\boldsymbol{k}_2 \boldsymbol{k}_1}}{(\epsilon_{\boldsymbol{k}} - \epsilon_{\boldsymbol{k}_2})} \frac{V_{\boldsymbol{k}_1 \boldsymbol{k}}}{(\epsilon_{\boldsymbol{k}} - \epsilon_{\boldsymbol{k}_1})},$$

$$\bar{T}(\boldsymbol{k} \to -\boldsymbol{k}) = V_{-\boldsymbol{k}\,-\boldsymbol{k}_1} \frac{V_{-\boldsymbol{k}_1 -\boldsymbol{k}_2}}{(\epsilon_{\boldsymbol{k}} - \epsilon_{-\boldsymbol{k}_1})} \cdots \frac{V_{-\boldsymbol{k}_{n-2} -\boldsymbol{k}_{n-1}}}{(\epsilon_{\boldsymbol{k}} - \epsilon_{-\boldsymbol{k}_{n-2}})} \frac{V_{-\boldsymbol{k}_{n-1} \boldsymbol{k}}}{(\epsilon_{\boldsymbol{k}} - \epsilon_{-\boldsymbol{k}_{n-1}})}. \tag{8.5}$$

This corresponds to the three-dimensional generalization of the expansion given in eqn (2.121); here we have considered only one of the $n$th-order terms (together with its time-reversed process) obtained through the $n$th iteration of that equation. Here we have to identify $s$ and $s'$ of Chapter 2 with the direction of $\boldsymbol{k}$ and $\boldsymbol{k}'$, and $E$ with $\epsilon_{\boldsymbol{k}}$. Also, because of the normalization chosen for the wave function for the tight-binding Hamiltonian, the transition amplitudes in (8.5) have the dimension of energy.

Now, for a system which is symmetric under time-reversal (TRI, i.e., in the absence of any external magnetic field), the application of the anti-unitary operator $\theta$ for time inversion (as we did in connection with eqn (2.118)) shows that the matrix elements must have the symmetry $V_{\boldsymbol{k}'\boldsymbol{k}} \equiv \langle \boldsymbol{k}'|V|\boldsymbol{k}\rangle = \langle -\boldsymbol{k}|V|-\boldsymbol{k}'\rangle \equiv V_{-\boldsymbol{k}\,-\boldsymbol{k}'}$. Also, $\epsilon_{\boldsymbol{k}} = \epsilon_{-\boldsymbol{k}}$. Thus, the two partial amplitudes in eqn (8.5) are equal, $T(\boldsymbol{k} \to -\boldsymbol{k}) = \bar{T}(\boldsymbol{k} \to -\boldsymbol{k})$. This is seen from a quick inspection of the two right-hand side expressions in eqn (8.5) which are now identical except for the ordering of the factors in the numerator and in the denominator.

The above result, of course, holds exactly to all Born orders and for any real scattering potential, no matter how random, so long as the system fulfills TRI. Thus, we have a doubling of the return amplitude (i.e., $T + \bar{T} = 2T$), or quadrupling of the return probability (i.e., $|T + \bar{T}|^2 = 4|T|^2$) for coherent back-scattering, $\boldsymbol{k}_i \equiv \boldsymbol{k} \to -\boldsymbol{k} \equiv \boldsymbol{k}_f$. For scattering in any other direction ($\boldsymbol{k}_i \to \boldsymbol{k}_f \neq -\boldsymbol{k}_i$), there is no such systematic constructive enhancement. In general, the alternative paths $P_1$ and $P_2$, say, connecting the initial and the final states accumulate different phases, $\theta_1$ and $\theta_2$, at random, and the resulting amplitudes $T_1 = |T_1|e^{i\theta_1}$ and $T_2 = |T_2|e^{i\theta_2}$ interfere randomly due to the underlying potential disorder (see Fig. 8.4). One then has to simply add the probabilities, i.e., $|T_1 + T_2|^2 = |T_1|^2 + |T_2|^2 + 2|T_1||T_2|\cos(\theta_1 - \theta_2)$; however, the cross-term describing interference vanishes on ensemble averaging, i.e., $\langle\cos(\theta_1 - \theta_2)\rangle = 0$. This refocussing of the electron wave preferentially, and constructively, into a scattering direction opposite to that of incidence in a random medium is called coherent back-scattering (CBS). The simple kinetic picture of transport based on a transport mean free path $l_e$ alone misses this subtle effect inasmuch as one assumes

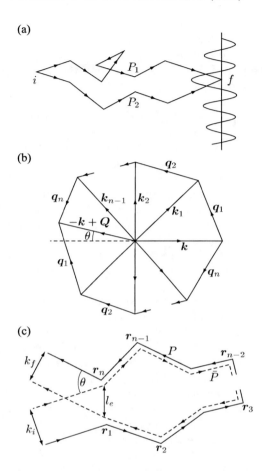

FIG. 8.4. (a) Interference of the partial amplitudes $T_1$ and $T_2$ traversing alternative paths $P_1$ and $P_2$ from the initial point $i$ to the final point $f$. (b) Coherent back-scattering from $\bm{k}_i = \bm{k}$ to $\bm{k}_f = -\bm{k}+\bm{Q}$, showing two partial amplitudes traversing time-reversed paths $P$ and $\bar{P}$ in the $\bm{k}$-space. (c) Coherent back-scattering shown in real space. The solid and dashed lines depict the pair of time-reversed paths acquiring a relative phase shift of $(\bm{k}_i + \bm{k}_f) \cdot (\bm{r}_n - \bm{r}_1)$. Exact CBS corresponds to $\theta = 0$, and $\bm{k}_f = -\bm{k}_i$.

complete randomization of the velocity (i.e., vanishing of the velocity–velocity correlation) on the length-scale $\sim l_e$, while the CBS gives a negative velocity–velocity correlation due to the refocussing, as discussed above. This in turn gives a negative *quantum* correction to the conductivity, the *weak-localization* (WL) correction, which we will evaluate later in this chapter. Recall that the WLC to the conductance was studied in detail in Chapter 6 for the case of ballistic chaotic cavities, and in Chapter 7 for quasi-one-dimensional disordered conductors. The weak-localization effect does, however, offend our sense of *naïve* realism—the transport mean free path becoming smaller than the lattice constant, the step-length itself, of course only close to and above the mobility edge!

Some remarks are in order before we close this section. Clearly, the phenomenon of CBS is true of any wave propagating in a random medium, including the matter waves in nuclear scattering (the elastic enhancement factor), or the light waves of optics. While the maximal CBS effect (doubling of the return wave amplitude) is true only for $\boldsymbol{k}_i = -\boldsymbol{k}_f$, a small deviation from the exact opposition would still give a pronounced refocussing (see Fig. 8.4(c); in fact, there is a cone of CBS with an opening angle of order $\lambda_F/2\pi l_e$, which is typically a few milliradians wide in the case of light waves and is directly measurable). It opens out as we approach the mobility edge from the conducting side. Also, any external magnetic field that destroys the time-reversal invariance will introduce a random relative phase shift between the alternatives of a time-reversed pair, $P$ and $\bar{P}$, and hence cut off the magnitude of the CBS effect for long return paths, as will be discussed later.

Coherent back-scattering is also implicated in the so-called *glory* effect—the halo seen around the edges of a shadow cast onto a distant scattering medium such as a cloud. (Here the shadow simply helps you look into the right direction so as to receive the back-scattered light.) In fact, enhanced back-scattering has been known for a long time in optics [4]. It can result from causes other than interference, e.g., from shadow working, lens action or from retro-reflection. The CBS, however, refers to enhanced back-scattering that results strictly from interference. Its signature is the inverse proportionality between the mean free path and the width of the CBS cone angle (i.e., $\propto 1/k_f l_e$). The connection between CBS and WL, however, went unnoticed until it was first pointed out by P. E. Wolf and G. Maret [179]. It turns out, however, that the CBS is only a semi-classical part of the full quantum mechanical effect giving WL. (By semi-classical here we mean that the amplitudes for the motion along the classical trajectories only are considered along with the appropriate phases, and added before squaring the resultant modulus.) There are quantum excursions from the classical trajectories that get neglected in the semi-classical treatment of CBS [84]. Finally, and most importantly, we should note that the CBS makes the effect of localization felt even before SL sets in by making the conductivity $\sigma$ a length-scale dependent quantity in the expression for the conductance, $G(L) = \sigma(L)L^{D-2}$.

## 8.4 Scaling theory

A treatment of the localization–delocalization transition across the mobility edge directly in terms of the changes in the nature of the individual eigenstates of a disordered system as a function of the energy is much too fine-grained and difficult. Besides, one hardly ever probes the wave functions! A closely related but directly measurable physical quantity of interest, however, is the sample conductance that effectively summarizes, at zero temperature, the nature of the eigenfunctions at the Fermi level as a whole. Let us, therefore, shift the emphasis from the wave functions to the *conductance*, and discuss the conductor–insulator transition as a function of the Fermi energy $E_F$. Of course, the critical nature of this transition as $E_F$ crosses the mobility edge $E_c$ can be revealed only in the infinite sample-size limit. This is much the same as in the case of the conventional phase transition that presupposes the thermodynamic limit. A finite-size system can merely show an analytic crossover rather than a singular (critical) behavior. The scaling theory of localization [1,115] reformulates this problem as that of the finite-size $L$-dependence of the sample conductance $G(L)$ for a given strength of disorder $(1/k_F l_e)$ and the value of the electron concentration (or of the Fermi energy $E_F$).

Consider a finite sample, a $d$-dimensional cube measuring $L$ on the side at absolute-zero temperature, i.e., $T = 0$. We now ask whether the sample is a conductor or an insulator. Inasmuch as a finite-sized system must necessarily have a nonzero conductance, no matter how strong the disorder may be, a meaningful answer to this question is not possible *per se*. Nevertheless, as the sample size $L$ is increased to infinity, it shall clearly evolve into either a conductor with a nonzero conductivity, or an insulator with zero conductivity. Which of the two it becomes must depend on the initial conditions defining the state of the sample at the initial length-scale, $L_0$, say. A large initial conductance $G(L_0)$ would scale up to the conducting state as $L \to \infty$, while a low initial conductance $G(L_0)$ would scale down to an insulating state. Indeed, the size-dependence of $G(L)$ gives away implicitly the conductor-*versus*-insulator nature of even the finite-size sample.

It is now possible to make these statements more precise, so as to make calculable predictions of a rather general validity. Let the initial size $L_0$ be chosen to be large compared to any microscopic length-scales present in the sample, e.g., the lattice spacing $a$ or the Fermi wavelength $\lambda_F$. We could take $L_0 \gtrsim l_e$, the elastic mean free path. We recall that $l_e$ measures the length-scale over which an electron wave of a definite wave vector randomizes its direction of motion. One must, however, have $a < l_e < \xi$; over the length-scale $L > l_e$, but $L < \xi$, the motion is always diffusive. Inasmuch as the conductance measures the coupling between the opposite hypercubic faces across the sample, one can make a highly plausible *ansatz*, namely that the conductance $G(L)$ is the only *relevant* scaling variable—in the sense that the conductance $G(\alpha L)$ on the length-scale $\alpha L$ ($\alpha > 1$) depends only on the conductance $G(L)$ at the length-scale $L$ and on the scale parameter $\alpha$, i.e.,

$$g(\alpha L) = f(g(L), \alpha) \tag{8.6}$$

or, equivalently,

$$g(L') = f\left(g(L), \frac{L'}{L}\right) \quad \text{with } \alpha = \frac{L'}{L}.$$

Here we have introduced the dimensionless conductance $g = G/(e^2/2\pi\hbar)$, the Thouless number, in terms of the natural quantum of conductance $e^2/2\pi\hbar$ (see eqn (4.179)). (Note that $\pi\hbar/e^2 \approx 12 k\Omega$.)

On expanding about $\alpha = 1$ (or $L' = L$), we obtain from eqn (8.6) that

$$\frac{\partial \ln g}{\partial \ln L} = \beta(g), \tag{8.7}$$

with

$$\beta(g) = \frac{1}{g}\left(\frac{\partial f(g,\alpha)}{\partial \alpha}\right)\bigg|_{\alpha=1},$$

which is a *universal* function of $g$ alone.

Equation (8.7) is central to the scaling theory of localization. It is a one-parameter scaling *ansatz*, and can be traced to a fundamental argument due originally to Thouless [168] relating the dimensionless conductance $g(L)$ of a mesoscopic $D$-dimensional hypercubic sample of size $L$ to the ratio $\delta/\Delta$, where $\Delta$ is the typical nearest-neighbor level spacing at the Fermi level $E_F$, while $\delta$ is the level-shift of an eigenstate at $E_F$ due to a change of the boundary condition, from periodic to anti-periodic across the sample, say. Clearly, $\delta$ measures the coupling (phase stiffness) across the length-scale $L$, while $\Delta$ measures the granularity (the typical energy mismatch). This naturally controls the conductance, as discussed earlier. All microscopic details enter *holomorphically* through this single ratio. Hence the rationale for the one-parameter scaling.

The dimensionality-dependent but otherwise universal *beta function* $\beta(g)$ is, of course, unknown. The predictive power of the scaling *ansatz*, however, comes from the fact that $\beta(g)$ can be calculated perturbatively in the asymptotic limits of $g \to 0$ (highly insulating) as well as $g \to \infty$ (highly conducting). A smooth (analytic and monotonic) interpolation between these two limits gives definite results, quite independently of the details of the interpolation. Finite size here ensures the assumed analyticity of $\beta(g)$. It turns out that the weak-localization correction to the conductivity due to CBS is indeed scale dependent, and provides a perturbative basis for the scaling theory. In the following sections we will present an impurity-diagrammatic calculation of the weak-localization correction to the classical (Drude) conductivity, and use the latter to discuss localization.

## 8.5 Weak localization: quantum correction to the conductivity

In this section we will calculate the weak-localization correction to the electrical conductivity that arises from the coherent back-scattering due to a random potential, as discussed earlier in physical terms. The calculation will be based on the Kubo linear response formula [63, 107, 108, 122] for the conductivity (see Chapter 4). It turns out, however, that a convergent perturbation expansion in powers of the impurity concentration, or the disorder parameter $(1/k_F l_e)$, does not exist [113]. Indeed, the CBS turns out to correspond to a subset of diagrams, the so-called maximally-crossed diagrams, (see Section 8.6.2) that must be summed to infinite order. This partial summation will be done using the standard impurity diagram technique. The latter enables us to perform ensemble averaging of the various quantities of interest in a systematic manner. The entire treatment will be for a disordered system of *non-interacting* electrons at zero temperature. In the absence of any electron–electron, inelastic and spin–flip scattering, the problem reduces to an essentially spinless one-electron problem where the fermionic nature of the electron, the Pauli exclusion principle, enters only through the fact that one has to consider the electrons at the Fermi energy only. Indeed, the exclusion principle does not enter into the calculation of the scattering processes, e.g., those involving the occupied states below the Fermi level as intermediate states. More explicitly, this is so because a process of scattering from the initial state $|\mathbf{k}_i\rangle$ to the final state $|\mathbf{k}_f\rangle$ via an occupied intermediate state $|\mathbf{k}\rangle$ with $\epsilon_{k_i} = \epsilon_{k_f} > E_F > \epsilon_k$ may be viewed to be so ordered that the electron occupying the intermediate state $|\mathbf{k}\rangle$ is first scattered out to the empty final state $|\mathbf{k}_f\rangle$, and then the electron in the initial state $|\mathbf{k}_i\rangle$ drops into the intermediate hole so created without violating the exclusion principle. This reduction leads to a great simplification of the standard many-body technique for calculating the conductivity using the Kubo linear response theory. Technically speaking, for this case the Kubo conductivity formula involves only the one-particle Green functions, as we will see below, and was indeed seen explicitly in Chapter 4.

### 8.5.1 *The Hamiltonian and the Green function*

The disordered system is well described by the Hamiltonian

$$H = -\frac{\hbar^2}{2m}\nabla^2 + V(\mathbf{r}), \qquad (8.8)$$

where $V(\mathbf{r})$, as in eqns (2.1) and (3.2), is now a one-body *random* potential. It is simpler for analytical purposes to use this continuum Hamiltonian rather than the tight-binding Anderson Hamiltonian on a lattice introduced earlier. We will take $V(\mathbf{r})$ to be a Gaussian random variable. Further simplification results from assuming $V(\mathbf{r})$ to have zero mean and a delta-function auto-correlation, i.e.,

$$\langle V(\mathbf{r})\rangle = 0, \qquad (8.9a)$$
$$\langle V(\mathbf{r})V(\mathbf{r}')\rangle \equiv W(\mathbf{r},\mathbf{r}') = \gamma\delta(\mathbf{r}-\mathbf{r}'). \qquad (8.9b)$$

The angular bracket $\langle\cdots\rangle$ here denotes the ensemble average. The Gaussian assumption introduces great simplifications, in that the probability distribution function for $V(\mathbf{r})$ involved in $\langle\cdots\rangle$ then depends only on the mean $\langle V(\mathbf{r})\rangle$ (now assumed to be zero) and the auto-correlation $\langle V(\mathbf{r})V(\mathbf{r}')\rangle$ (now taken to be delta-correlated). Moreover, the averages, or the expectation values of the products of Gaussian random variables involve just these two quantities. In particular, we have for the moments that

$$\langle V(\mathbf{r}_1)V(\mathbf{r}_2)\cdots V(\mathbf{r}_{2k+1})\rangle = 0, \qquad (8.10a)$$

$$\langle V(\mathbf{r}_1)V(\mathbf{r}_2)\cdots V(\mathbf{r}_{2k})\rangle$$
$$= \langle V(\mathbf{r}_1)V(\mathbf{r}_2)\rangle\langle V(\mathbf{r}_3)V(\mathbf{r}_4)\rangle\cdots\langle V(\mathbf{r}_{2k-1})V(\mathbf{r}_{2k})\rangle \qquad (8.10b)$$
$$+ \text{ all possible pairings.}$$

Thus, the expectation $\langle \cdots \rangle$ of a product of an odd number $2k+1$ of Gaussian variables vanishes, while the expectation of a product of any even number $2k$ of Gaussian variables is a sum of $(2k)!/k!2^k$ terms, with one term for each way that the $2k$ variables can be paired up. For example, for $2k = 4$ we have the three terms

$$\langle V(\boldsymbol{r}_1)V(\boldsymbol{r}_2)V(\boldsymbol{r}_3)V(\boldsymbol{r}_4)\rangle = W_{12}W_{34} + W_{13}W_{24} + W_{14}W_{23}, \tag{8.11}$$

where $W_{ij} = \langle V(\boldsymbol{r}_i)V(\boldsymbol{r}_j)\rangle$.

We note in passing that by $V(\boldsymbol{r})$ denoting a random variable we really imply infinitely many random variables corresponding to the set of potentials $\{V(\boldsymbol{r}_k)\}$ at the infinitely many points $\{\boldsymbol{r}_k\}$. One may visualize this as a limit of the Gaussian joint probability density function $P_k(V_1, V_2, \ldots, V_k)$ as $k \to \infty$. The above properties then follow readily from the Gaussian nature of $P_k$, namely

$$P_k(V_1, \ldots, V_k) = (2\pi)^{-k/2}(\text{Det}\,\mu)^{1/2}\exp\left[-\frac{1}{2}\sum_{jl}\mu_{jl}V_jV_l\right], \tag{8.12}$$

where the matrix $\mu$ is the inverse of the correlation matrix $W$ above, i.e., $\sum_l \mu_{jl}W_{lm} = \delta_{jm}$ (of course, in the case $\langle V_j\rangle \neq 0$ we must replace $V_j$ by $V_j - \langle V_j\rangle$ in the above). A more formal treatment requires the notion of probability density functionals.

In a real disordered system, such as an impurity semiconductor, the above Gaussian white-noise limit is realized approximately in the following manner. Let

$$V(\boldsymbol{r}) = \sum_{i=1}^{N} v(\boldsymbol{r} - \boldsymbol{R}_i), \tag{8.13}$$

where $v(\boldsymbol{r} - \boldsymbol{R}_i)$ is the potential at $\boldsymbol{r}$ due to an impurity located at $\boldsymbol{R}_i$. Consider the limit of weak disorder—a random dense weak-scattering limit (see also the paragraph immediately after eqn (7.35) and the second paragraph of Section 7.3)—as follows. Let $n_i$ be the mean concentration of the impurity centers, with $n_i^{-1/D}$ being the mean spacing between them, and $r_c$ the correlation length of the random potential. Then $\lambda_F \gg r_c \gg n_i^{-1/D}$ gives us the Gaussian white-noise limit. For the short-ranged impurity potential, we can take $v(\boldsymbol{r} - \boldsymbol{R}_i) = U_i\delta(\boldsymbol{r} - \boldsymbol{R}_i)$ with $U_i$ random, $\langle U_i\rangle = 0$ and $\langle U_iU_j\rangle = U^2\delta_{ij}$, and we have $W(\boldsymbol{r} - \boldsymbol{r}') = \gamma\delta(\boldsymbol{r} - \boldsymbol{r}')$ in the limit as $n_i \to \infty$ (dense), $U \to 0$ (weak), keeping $n_iU^2 \to \gamma$ (constant).

Next, we introduce the single-particle Green functions for the random Hamiltonian in eqn (8.8), in terms of which the weak-localization correction to the conductivity will be calculated [6,63].

The single-particle retarded Green function $G^{(+)}(\boldsymbol{r}, \boldsymbol{r}'; t)$ for a given realization of the random potential $V(\boldsymbol{r})$ is defined by

$$\left(i\hbar\frac{\partial}{\partial t} - H(\boldsymbol{r})\right) G^{(+)}(\boldsymbol{r},\boldsymbol{r}';t) = \delta(t)\delta(\boldsymbol{r}-\boldsymbol{r}'), \tag{8.14a}$$

with
$$G^{(+)}(\boldsymbol{r},\boldsymbol{r}';t) = 0 \quad \text{for } t \leqslant 0. \tag{8.14b}$$

We introduce the Fourier transform and its inverse as follows:

$$G^{(+)}(\boldsymbol{r},\boldsymbol{r}';E) = \int_{-\infty}^{+\infty} G^{(+)}(\boldsymbol{r},\boldsymbol{r}';t) e^{iEt/\hbar}\, dt, \tag{8.15a}$$

$$G^{(+)}(\boldsymbol{r},\boldsymbol{r}';t) = \int_{-\infty}^{+\infty} G^{(+)}(\boldsymbol{r},\boldsymbol{r}';E) e^{-iEt/\hbar}\, \frac{dE}{2\pi\hbar}. \tag{8.15b}$$

(Note the slight change of notation for the argument of the Green function, with respect to the one used in Chapters 2 and 3.) We can then express $G^{(+)}(\boldsymbol{r},\boldsymbol{r}';E)$ in terms of the eigenfunctions $\phi_m(\boldsymbol{r})$ and eigenvalues $E_m$ of the Hamiltonian $H(\boldsymbol{r})$ as

$$G^{(+)}(\boldsymbol{r},\boldsymbol{r}';E) = \lim_{\eta \to 0^+} \sum_m \frac{\phi_m(\boldsymbol{r})\phi_m^*(\boldsymbol{r}')}{E - E_m + i\eta}. \tag{8.16}$$

Note that this is the same as the Green function introduced in Section 2.1.3 for one dimension. It is readily verified that $G^{(+)}(\boldsymbol{r},\boldsymbol{r}';E)$ is analytic in the upper half of the complex-$E$ plane, consistent with its retarded nature (8.14b).

Similarly, we define the advanced Green function $G^{(-)}(\boldsymbol{r},\boldsymbol{r}';t)$ satisfying (8.14a), but now with the boundary condition $G^{(-)}(\boldsymbol{r},\boldsymbol{r}';t) = 0$ for time $t \geqslant 0$. Accordingly, $G^{(-)}(\boldsymbol{r},\boldsymbol{r}';E)$ is analytic in the lower half of the complex-energy plane, and is given by eqn (8.16), but with $i\eta$ replaced by $-i\eta$.

For a system of non-interacting particles, such as the one described by our one-body Hamiltonian $H$, the physical quantities of interest, e.g., the density of states $\nu$ and the conductivity $\sigma$, can be expressed in terms of the single-particle Green functions $G^{(+)}$ and $G^{(-)}$. We have, for instance,

$$\nu(E) = \sum_m \delta(E - E_m) = -\frac{1}{\pi} \operatorname{Tr} \operatorname{Im} G^{(+)}(\boldsymbol{r},\boldsymbol{r}';E)$$
$$= -\frac{1}{\pi} \int d\boldsymbol{r}\, \operatorname{Im} G^{(+)}(\boldsymbol{r},\boldsymbol{r};E), \tag{8.17}$$

where we have used the Dirac identity

$$\lim_{\eta \to 0^+} \frac{1}{x + i\eta} = P\frac{1}{x} - i\pi\delta(x). \tag{8.18}$$

As noted before, for a macroscopic sample, i.e., in the thermodynamic limit $\Omega \to \infty$, we expect many additive physical quantities such as $\nu(E)$ to self-average. In practice, therefore, we will be interested in calculating the ensemble average $\langle \cdots \rangle$ over all possible realizations of the disorder—of the random potential $V$. This requires calculating $\langle G^{(\pm)} \rangle$ in the present case. To this end, we will now introduce a powerful averaging technique—the impurity diagram technique, due originally to S. F. Edwards [69].

As a first step toward the systematic averaging of $G^{(+)}$ over the impurity potential $V(\boldsymbol{r})$, we expand $G^{(+)}$ in powers of $V$, treating the latter as a perturbation. The defining equation for $G^{(+)}(\boldsymbol{r}, \boldsymbol{r}'; E)$, namely

$$(E + i\eta - H)G^{(+)}(\boldsymbol{r}, \boldsymbol{r}'; E + i\eta) = \delta(\boldsymbol{r} - \boldsymbol{r}'), \tag{8.19}$$

can be conveniently rewritten as the matrix equation

$$(E + i\eta - H)G^{(+)} = 1, \tag{8.20a}$$

which can be formally solved to give

$$G^{(+)} = \frac{1}{\left(G_0^{(+)}\right)^{-1} - V}. \tag{8.20b}$$

Here we have introduced $G_0^{(+)}$, the unperturbed or *bare* Green's function, i.e., the one with $V = 0$, just as in Chapter 2. Using the well-known operator identity

$$\frac{1}{A - B} = \frac{1}{A} + \frac{1}{A}B\frac{1}{A} + \frac{1}{A}B\frac{1}{A}B\frac{1}{A} + \cdots, \tag{8.21}$$

we obtain, from eqn (8.20b),

$$G^{(+)} = G_0^{(+)} + G_0^{(+)} \sum_{n=1}^{\infty} \left(V G_0^{(+)}\right)^n, \tag{8.22}$$

just as in eqn (2.54).

Now recalling the general property of Gaussian random variables, eqn (8.10), and specifically for the delta-function-correlated impurity potential $V$, we can perform the impurity averaging of eqn (8.22) to give (with $\langle V \rangle = 0$) the equation shown in Fig. 8.5.

In Fig. 8.5 we have written out all the terms which are of second order in $V$ (there being only one such term, the second term on the right-hand side) and those of fourth order in $V$ (there being three such terms, the third, fourth and the fifth terms on the right-hand side). The over-arching dashed lines denote all possible pairwise contractions of the impurity potential, according to the Gaussian moment property, eqn (8.10). The equation of Fig. 8.5 can now be represented graphically, as shown in Fig. 8.6. Using

$$\langle V(\boldsymbol{r}_1)V(\boldsymbol{r}_2)\rangle = \gamma \delta(\boldsymbol{r}_1 - \boldsymbol{r}_2), \tag{8.23}$$

we write, in the coordinate representation, the equation in Fig. 8.5 in a more explicit fashion as

# WEAK LOCALIZATION CORRECTION TO CONDUCTIVITY

$$\langle G^{(+)} \rangle = G_0^{(+)} + G_0^{(+)} \overset{\frown}{V G_0^{(+)} V} G_0^{(+)}$$

$$+ G_0^{(+)} \overset{\frown}{V G_0^{(+)} V} G_0^{(+)} \overset{\frown}{V G_0^{(+)} V} G_0^{(+)}$$

$$+ G_0^{(+)} \overset{\frown}{V G_0^{(+)} \overset{\frown}{V G_0^{(+)} V} G_0^{(+)} V} G_0^{(+)}$$

$$+ G_0^{(+)} \overset{\frown}{V G_0^{(+)} \overset{\frown}{V G_0^{(+)} V} G_0^{(+)} V} G_0^{(+)} + \cdots$$

FIG. 8.5. The impurity averaging of the retarded Green function for a Gaussian random potential.

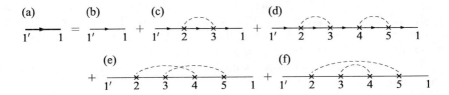

FIG. 8.6. Impurity-diagrammatic representation of the ensemble-averaged exact Green function $\langle G^{(+)} \rangle$ (thick solid line) in terms of the bare Green function $G_0^{(+)}$ (thin solid lines). Impurity lines (dashed) denote pairwise random-potential contractions. All intermediate coordinates are to be integrated over.

$$\langle G^{(+)}(\boldsymbol{r}, \boldsymbol{r}'; E) \rangle$$
$$= G_0^{(+)}(\boldsymbol{r}, \boldsymbol{r}'; E) + \iint d\boldsymbol{r}_1 \, d\boldsymbol{r}_2 \, G_0^{(+)}(\boldsymbol{r}, \boldsymbol{r}_1; E)$$
$$\times \langle V(\boldsymbol{r}_1) G_0^{(+)}(\boldsymbol{r}_1, \boldsymbol{r}_2; E) \cdot V(\boldsymbol{r}_2) \rangle \cdot G_0^{(+)}(\boldsymbol{r}_2, \boldsymbol{r}'; E) + \cdots$$
$$= G_0^{(+)}(\boldsymbol{r}, \boldsymbol{r}'; E)$$
$$+ \gamma \int d\boldsymbol{r}_1 \, G_0^{(+)}(\boldsymbol{r}, \boldsymbol{r}_1; E) G_0^{(+)}(\boldsymbol{r}_1, \boldsymbol{r}_1; E) \cdot G_0^{(+)}(\boldsymbol{r}_1, \boldsymbol{r}'; E) + \cdots .$$
(8.24)

Summing up the entire perturbation series in eqn (8.24) is clearly out of the question. (Indeed, a geometric series is about the only series that can be summed up in terms of elementary functions!) Truncating the series at a finite order so as to include the first few terms, on the other hand, would miss the singular aspect of the localization problem which is essentially of a non-perturbative nature. It is, however, possible to perform a partial summation, i.e., to sum up certain infinite subseries of important terms, or the *principal diagrams* contained in eqn (8.24), as follows. From an inspection of the structure of the diagrammatic expansion, eqn (8.24), it turns out that all the terms of the series can be generated iteratively from the generating Dyson equation [63]

$$\langle G^{(+)} \rangle = G_0^{(+)} + G_0^{(+)} \Sigma^{(+)} \langle G^{(+)} \rangle, \tag{8.25}$$

with the total irreducible self-energy part

$$\Sigma^{(+)} = \sum_i \Sigma_{(i)}^{(+)}, \tag{8.26}$$

where $\Sigma_{(i)}^{(+)}$ represents the $i$th self-energy diagram, irreducible in a precise sense, namely that it cannot be divided into two disjointed parts by cutting only one bare $G_0^{(+)}$ line. Figure 8.7 illustrates this diagrammatic construction. Thus, the diagrams in Fig. 8.7(b) are irreducible self-energy parts. These are built into the Green function diagrams given in Fig. 8.6(c,e,f). The diagram in Fig. 8.6(d) is, however, clearly reducible. It gets automatically generated iteratively by including the irreducible self-energy diagram of Fig. 8.7(b(i)) in the Dyson equation for $\langle G^{(+)} \rangle$.

Finally, we should note that this diagrammatic construction of the perturbation series is *modular* in the sense that each single line representing the bare $G_0^{(+)}$ in $\Sigma_{(i)}^{(+)}$ can itself be replaced by the exact $\langle G^{(+)} \rangle$. It is this inward inheritability of the diagrammatic rules that makes the technique so powerful.

Armed with this impurity-diagrammatic technique, we now proceed to calculate $\langle \nu(E) \rangle$ and, most importantly, $\langle \sigma(E_F) \rangle$.

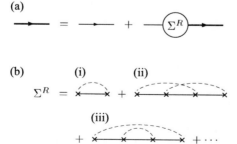

FIG. 8.7. (a) Diagrammatic representation of Dyson's equation for $\langle G^{(+)} \rangle$ in terms of the total irreducible self-energy $\Sigma^{(+)}$. (b) Total irreducible self-energy as the sum of the irreducible self-energy diagrams (i), (ii) and (iii). The superscript '$R$' denotes retarded, and is used interchangeably with the superscript '(+)'.

### 8.5.2 Ensemble-averaged Green's function in the self-consistent Born approximation

We will now calculate the ensemble-averaged Green function in what is essentially the self-consistent Born approximation to $\Sigma$ (Fig. 8.8), valid in the limit of weak delta-correlated disorder. We approximate the retarded irreducible self-energy $\Sigma^R (\equiv \Sigma^{(+)})$ by

$$\Sigma^{(+)}(\boldsymbol{k}; E) = \gamma \int \frac{d\boldsymbol{q}}{(2\pi)^d} G_0^{(+)}(\boldsymbol{k} - \boldsymbol{q}; E), \qquad (8.27a)$$

giving

$$\Sigma^{(+)}(\boldsymbol{k}; E) = -\frac{i\hbar}{2\tau}, \quad \text{with } \tau = \frac{\hbar}{2\pi\gamma v_F}. \qquad (8.27b)$$

Here we have introduced the Green function $G_0^{(+)}(\boldsymbol{k}; E)$, which is the bare Green function in the momentum $\boldsymbol{k}$ representation. Thus, we obtain

$$\langle G^{(+)}(\boldsymbol{k}; \epsilon) \rangle = \frac{1}{\epsilon - \xi_k + i\hbar/2\tau}, \qquad (8.28)$$

where we have measured the energy from the Fermi level as $\epsilon = E - E_F$, and $\xi_k = \epsilon_k - E_F$. The real part of the self-energy has been trivially absorbed as a shift of the Fermi level (chemical potential at zero temperature). Similarly, for the advanced Green function we have

$$\langle G^{(-)}(\boldsymbol{k}; \epsilon) \rangle = \frac{1}{\epsilon - \xi_k - i\hbar/2\tau}. \qquad (8.29)$$

These ensemble-averaged Green functions will now be used in the calculation of the electrical conductivity of a weakly-disordered system. We emphasize once again that calculating $\Sigma^{(+)}$ in this lowest approximation implies for $\langle G^{(+)} \rangle$ a summation over an infinite perturbation series.

We note in passing that, by Fourier transforming to direct space, the ensemble-averaged Green function can be readily shown to be short-ranged (decaying exponentially in space) with a range $v_F \tau = l_e$, where $l_e$ is the elastic mean free path and $\tau$ is the corresponding mean free lifetime for the particle in the plane-wave state. These are the typical values for scattering out of an initially prepared plane-wave state.

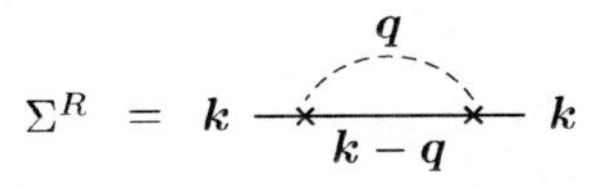

$$\Sigma^R = \boldsymbol{k} \xrightarrow{\quad \boldsymbol{q} \quad} \boldsymbol{k}$$
$$\phantom{\Sigma^R = \boldsymbol{k}} \boldsymbol{k} - \boldsymbol{q}$$

FIG. 8.8. Irreducible retarded self-energy diagram in the self-consistent Born approximation. In determining $\Sigma_{(i)}^+$, the two bare $G_0^{(+)}$ lines at the two ends, with which it is connected to the rest of the diagram, are to be discounted.

The corresponding ensemble-averaged density of states (DOS) is

$$\langle \nu(\epsilon) \rangle = -\frac{1}{\pi} \frac{1}{L^d} \mathrm{Tr}\, \mathrm{Im}\, G^{(+)} = \frac{1}{\pi} \int \frac{d\mathbf{k}}{(2\pi)^d} \frac{1/2\tau}{(\epsilon - \xi_k)^2 + (\hbar/2\tau)^2}$$
$$= \nu_F. \qquad (8.30)$$

Thus, if the unperturbed density of states is flat ($= \nu_F$) at the Fermi level, then so is the ensemble-averaged DOS, $\langle \nu(\epsilon) \rangle = \nu_F$.

## 8.6 Electrical conductivity of a disordered metal and quantum corrections: weak localization

In this section, which is heavily based on [6,61,115], we will be mainly concerned with wave interference effects in electrical conduction through a disordered conductor, e.g., a metal containing randomly distributed impurities that scatter the conduction electrons elastically. More specifically, therefore, we will derive an expression for the quantum correction to the classical, or the Drude, conductivity. These corrections arise, at least in part, from the coherent back-scattering—the so-called weak-localization (WL) effect—as discussed earlier in physical terms for a system with TRI. We will consider here the simplest model, namely that of a microscopically disordered conductor which is assumed to be statistically isotropic and, therefore, homogeneous, and without the electron–electron interaction. Further, we will restrict ourselves to the d.c. conductivity $\sigma_{\alpha\beta}(\omega \to 0)$ evaluated at the absolute zero of temperature.

As noted earlier, in the absence of the electron–electron interaction, the many-body problem essentially reduces to a one-body problem in that the (Kubo) conductivity can be expressed in terms of the product of single-particle retarded and advanced Green functions. Indeed, one can quite meaningfully define the conductivity for an electron in a given one-electron quantum state, and arrive at the total conductivity as the thermally weighted sum over these occupied states. What is central to the problem, however, is the question of averaging this sample-specific conductivity over the ensemble of configurations of the randomly distributed impurities.

In point of fact, the three-dimensional conductivity, unlike the conductance, turns out to be a self-averaging quantity—different parts of the macroscopic sample act as different instances of the sample, i.e., different members of the random ensemble. Thus, the conductivity measured on a given macroscopic sample, as is almost always the case, essentially equals the conductivity averaged over the ensemble of random impurity configurations. This justifies, in the thermodynamic limit, the mathematical operation of impurity averaging of the sample-specific Kubo expression for the conductivity obtained formally for a given realization (i.e., the given sample) drawn at random. The impurity averaging is readily carried out diagrammatically using the averaging law for the products of Gaussian random variables (potentials), as discussed earlier in this chapter. The fact that the quantum WL correction, admittedly a coherent wave-interference effect, is *not* washed out by this impurity-averaging procedure is a remarkable consequence that follows directly from CBS under TRI.

We could start directly with the general Kubo formula derived in Chapter 4 on LRT. Since, however, we will be ignoring the electron–electron interaction and confining ourselves only to the zero-frequency (d.c.) limit $\sigma_{\alpha\beta}^{\omega=0}$, which is real and, therefore, purely absorptive, it is simpler and, indeed, instructive to follow a somewhat different procedure based on the power dissipated in the sample, as described below. Consider a $d$-dimensional hypercubic sample measuring $L$ on the side, and with the cube edges conveniently aligned with the Cartesian axes. We will assume the thermodynamic limit $L \to \infty$ to be taken before the zero-frequency limit $\omega \to 0$. This ordering of limits is mathematically required to ensure a quasi-continuum for the electron energy levels, i.e., with the energy spacings (granularity) much smaller than $\hbar\omega$, a condition necessary for absorptive response, as discussed in Section 4.3.1.

Let the externally applied electric field be spatially uniform and time-harmonic, i.e.,

$$E_\alpha(t) = \frac{1}{2}\left(E_\alpha^\omega e^{-i\omega t} + E_\alpha^{-\omega} e^{i\omega t}\right) \equiv \operatorname{Re} E_\alpha^\omega e^{-i\omega t}, \qquad (8.31)$$

where $\alpha$ labels the Cartesian components.

This corresponds to a vector potential (see also Chapter 4 for a discussion of the gauge)

$$A_\alpha(t) = \frac{1}{2}\left(A_\alpha^\omega e^{-i\omega t} + A_\alpha^{-\omega} e^{i\omega t}\right) \equiv \operatorname{Re} A_\alpha^\omega e^{-i\omega t}, \qquad (8.32)$$

with

$$A_\alpha^\omega = \left(\frac{-\mathrm{i}c}{\omega}\right) E_\alpha^\omega,$$

as in eqn (4.30). The reality of $E$ and $A$ demands that $E_\alpha^{-\omega} = (E_\alpha^\omega)^*$ and $A_\alpha^{-\omega} = (A_\alpha^\omega)^*$.

Now, within the assumed linearity of response, the current density caused by the externally impressed electric field is in general given by

$$J_\alpha(\boldsymbol{r},t) = \sum_\beta \int_\Omega \mathrm{d}\boldsymbol{r}' \int_{-\infty}^t \mathrm{d}t'\, K_{\alpha\beta}(\boldsymbol{r},\boldsymbol{r}';t,t') E_\beta(\boldsymbol{r}',t'). \qquad (8.33)$$

Note the obvious connection between this last equation and eqn (4.114), namely that in the frequency domain the two kernels are related by the factor $-\mathrm{i}c/\omega$. The response function $K_{\alpha\beta}(\boldsymbol{r},\boldsymbol{r}';t,t')$ is non-local in space and time. As explained in Chapter 4, for a time-translationally invariant system, as is the case here, it is a function of the time difference $t - t'$. Further, causality demands that $K_{\alpha\beta}(\boldsymbol{r},\boldsymbol{r}';t,t') = 0$ for $t < t'$, and, therefore, the time integral in eqn (8.33) can be extended to $+\infty$ at the upper limit. The power $P$ absorbed in the system is then given by

$$P = \lim_{T\to\infty} \frac{1}{T} \int_{-T/2}^{+T/2} \sum_\alpha \int_\Omega J_\alpha(\boldsymbol{r},t) E_\alpha(\boldsymbol{r},t)\, \mathrm{d}\boldsymbol{r}\, \mathrm{d}t. \qquad (8.34)$$

Here we have averaged the power absorbed over a time interval $T \gg 2\pi/\omega$, the period of the time-harmonic field. Substituting from eqn (8.33) into eqn (8.34), we obtain

$$P = \frac{1}{4} \sum_{\alpha\beta} \int d\mathbf{r} \int d\mathbf{r}' \left[ E_\alpha^\omega E_\beta^{-\omega} K_{\alpha\beta}^\omega(\mathbf{r}, \mathbf{r}') + \text{c. c.} \right]$$

$$\equiv \frac{\Omega}{2} |\mathbf{E}^\omega|^2 \operatorname{Re} \sigma(\omega), \tag{8.35}$$

where we have defined $\sigma(\omega)$ through

$$\frac{1}{\Omega} \int d\mathbf{r} \int d\mathbf{r}' \, K_{\alpha\beta}^\omega(\mathbf{r}, \mathbf{r}') = \delta_{\alpha\beta} \sigma(\omega). \tag{8.36}$$

Here $\Omega = L^d$ is the volume of the $d$-dimensional macroscopic disordered sample. Note that we have used the fact that, for a statistically homogeneous and isotropic random sample of macroscopic size, the integrations over $\mathbf{r}$ and $\mathbf{r}'$ on the left-hand side of eqn (8.35) give an isotropic quantity $\sigma(\omega)$. This quantity is readily identified with the measured conductivity of the disordered system. This analysis is to be contrasted with the one followed in connection with the mesoscopic geometry in Chapter 4.

The next step now is to recalculate the power $P$ dissipated in the sample quantum mechanically using the well-known Fermi golden rule, and then equate the two expressions for $P$. We have from the golden rule that

$$P = \hbar\omega \left(\frac{2\pi}{\hbar}\right) \sum_{lm} \left(\frac{e}{mc}\right)^2 |\langle m|\mathbf{p} \cdot \mathbf{A}^\omega|l\rangle|^2 (f_m - f_l) \delta(\epsilon_l - \epsilon_m - \hbar\omega), \tag{8.37}$$

where $\mathbf{p} = -i\hbar\nabla$ is the electron momentum operator, and $|l\rangle$ and $|m\rangle$ are the single-particle eigenstates for the one-body impurity Hamiltonian with $\epsilon_l$ and $\epsilon_m$ being the corresponding energy eigenvalues. The coupling $(-e/mc)\mathbf{p} \cdot \mathbf{A}$ between the electron and the externally applied field has been treated in the usual electric-dipole approximation. The thermal weights $f_m$ and $f_l$ are the associated Fermi functions. From eqns (8.32), (8.35) and (8.37), we obtain

$$\operatorname{Re} \sigma(\omega) = \frac{\pi e^2}{\Omega m^2 \omega} \sum_{ml} |\langle m|p_x|l\rangle|^2 \cdot (f_m - f_l) \delta(\epsilon_l - \epsilon_m - \hbar\omega). \tag{8.38}$$

Here we have taken, without loss of generality, the external field to be along the $x$-axis, say. Thus, we have a microscopic expression for the real part of the conductivity. One can, of course, use the general analytic property of the causal response (i.e., the Kramers–Kronig dispersion relation) to obtain $\operatorname{Im} \sigma(\omega)$ from the $\operatorname{Re} \sigma(\omega)$. We are, however, interested in the d.c. conductivity only, which is always real (dissipative).

The right-hand side of eqn (8.38) is, of course, sample-specific. For a macroscopic sample (assumed to be statistically homogeneous and isotropic), however, the expression is self-averaging and thus allows us to meaningfully ensemble average it over the random impurity configurations. To this end, we find it apt to re-express the right-hand side of eqn (8.38) in terms of the single-particle retarded and advanced Green functions, and use the diagrammatic impurity-averaging technique introduced earlier in this chapter.

Writing out explicitly the matrix elements of the momentum operator $\boldsymbol{p}$ connecting the single-particle eigenstates, and making use of the relations (see eqn (B.28))

$$G^{(+)}(\boldsymbol{r}_1, \boldsymbol{r}_2; \epsilon) - G^{(-)}(\boldsymbol{r}_1, \boldsymbol{r}_2; \epsilon) = -2\pi \mathrm{i} \sum_m \phi_m(\boldsymbol{r}_1)\phi_m^*(\boldsymbol{r}_2)\delta(\epsilon - \epsilon_m),$$

$$\delta(\epsilon_l - \epsilon_m - \hbar\omega)(f_m - f_l) = -\iint \mathrm{d}\epsilon_1\, \mathrm{d}\epsilon_2\, \delta(\epsilon_1 - \epsilon_2 - \hbar\omega)$$
$$\times [f(\epsilon_1) - f(\epsilon_2)]\delta(\epsilon_1 - \epsilon_l)\delta(\epsilon_2 - \epsilon_m), \tag{8.39}$$

we obtain

$$\mathrm{Re}\, \sigma(\omega) = \frac{\pi e^2}{m^2 \Omega \omega} \sum_{ml} \int \frac{\mathrm{d}\epsilon}{(2\pi)^2} \iint \mathrm{d}\boldsymbol{r}_1\, \mathrm{d}\boldsymbol{r}_2$$
$$\times \left\{ p_{2x} \left[ G^{(+)}(\boldsymbol{r}_2, \boldsymbol{r}_1; \epsilon) - G^{(-)}(\boldsymbol{r}_2, \boldsymbol{r}_1; \epsilon) \right] \right\}$$
$$\times \left\{ p_{1x} \left[ G^{(+)}(\boldsymbol{r}_1, \boldsymbol{r}_2; \epsilon + \hbar\omega) - G^{(-)}(\boldsymbol{r}_1, \boldsymbol{r}_2; \epsilon + \hbar\omega) \right] \right\}$$
$$\times [f(\epsilon + \hbar\omega) - f(\epsilon)], \tag{8.40}$$

where we have integrated out one of the energy variables. We now proceed to the zero-frequency (d.c.) limit with the proviso that the thermodynamic limit ($\Omega \to \infty$) is understood to have preceded the $\omega \to 0$ limit, so as to ensure a quasi-continuum of energy levels, i.e., with the level spacings $\Delta \ll \hbar\omega$, as was discussed earlier. Further, we specialize to the zero-temperature $T$ limit. Then, noting that

$$\lim_{T \to 0} \lim_{\omega \to 0} \frac{f(\epsilon + \hbar\omega) - f(\epsilon)}{\hbar\omega} = -\delta(\epsilon - \epsilon_F), \tag{8.41}$$

we obtain, for the d.c. conductivity at zero temperature,

$$\sigma(0) = \frac{e^2 \hbar^3}{4\pi m^2} \iint \frac{\mathrm{d}\boldsymbol{r}_1\, \mathrm{d}\boldsymbol{r}_2}{\Omega} \partial_{2x}[G_F^{(+)}(\boldsymbol{r}_2, \boldsymbol{r}_1) - G_F^{(-)}(\boldsymbol{r}_2, \boldsymbol{r}_1)]$$
$$\times \partial_{1x}[G_F^{(+)}(\boldsymbol{r}_1, \boldsymbol{r}_2) - G_F^{(-)}(\boldsymbol{r}_1, \boldsymbol{r}_2)], \tag{8.42}$$

with the $\Omega \to \infty$ limit assumed implicitly (see also eqn (B.41)). We have used $p_{2x} = -i\hbar\partial_{2x}$, etc. It is convenient now to go over to the wave vector $\boldsymbol{k}$ representation through the Fourier transforms

$$G_F^{(\pm)}(\boldsymbol{r}_1, \boldsymbol{r}_2) = \iint G_F^{(\pm)}(\boldsymbol{k}_1, \boldsymbol{k}_2) e^{i(\boldsymbol{k}_1 \cdot \boldsymbol{r}_1 - \boldsymbol{k}_2 \cdot \boldsymbol{r}_2)} \frac{d\boldsymbol{k}_1}{(2\pi)^d} \frac{d\boldsymbol{k}_2}{(2\pi)^d},$$

$$G_F^{(\pm)}(\boldsymbol{k}_1, \boldsymbol{k}_2) = \iint G_F^{(\pm)}(\boldsymbol{r}_1, \boldsymbol{r}_2) e^{-i(\boldsymbol{k}_1 \cdot \boldsymbol{r}_1 - \boldsymbol{k}_2 \cdot \boldsymbol{r}_2)} d\boldsymbol{r}_1 \, d\boldsymbol{r}_2, \qquad (8.43)$$

with $G_F^{(+)}(\boldsymbol{k}_1, \boldsymbol{k}_2) = [G_F^{(-)}(\boldsymbol{k}_2, \boldsymbol{k}_1)]^*$, which follows from TRI (see Section 2.1.3). Further, as we will presently see, the dominant contribution to the right-hand side of eqn (8.42) comes only from the cross-terms involving products of the type $G^{(+)}G^{(-)}$. With this, the expression for the zero-temperature d.c. conductivity simplifies to

$$\sigma(0) = \frac{e^2 \hbar^3}{2\pi m^2 \Omega} \iint \frac{d\boldsymbol{k}_1}{(2\pi)^d} \frac{d\boldsymbol{k}_2}{(2\pi)^d} \, k_{1x} k_{2x} \left[ G_F^{(+)}(\boldsymbol{k}_1, \boldsymbol{k}_2) G_F^{(-)}(\boldsymbol{k}_2, \boldsymbol{k}_1) \right]. \qquad (8.44)$$

The expression on the right-hand side of eqn (8.44) is, of course, sample-specific. But, as discussed earlier, we can now conveniently replace this by its ensemble average $\langle \cdots \rangle$ over the random impurity configurations, giving

$$\sigma(0) = \frac{e^2 \hbar^3}{2\pi m^2 \Omega} \int \frac{d\boldsymbol{k}_1}{(2\pi)^d} \frac{d\boldsymbol{k}_2}{(2\pi)^d} \, k_{1x} k_{2x} \left\langle G_F^{(+)}(\boldsymbol{k}_1, \boldsymbol{k}_2) G_F^{(-)}(\boldsymbol{k}_2, \boldsymbol{k}_1) \right\rangle. \qquad (8.45)$$

Equation (8.45) is central to our further derivations—of the classical conductivity and the quantum (WL) correction to it. Note that at zero temperature, as is the case considered here and in all what follows, all the quantities in eqn (8.45) are to be evaluated at the Fermi surface, i.e., for $|\boldsymbol{k}_1| = |\boldsymbol{k}_2| = k_F$; hence the subscript $F$. The product $G_F^{(+)} G_F^{(-)}$ within the angular bracket in eqn (8.45) can be represented diagrammatically, as shown in Fig. 8.9—as the Kubo conductivity bubble. This is now to be averaged using the impurity-diagrammatic technique.

### 8.6.1 Classical (Drude) conductivity

The impurity averaging involved in eqn (8.45) can now be carried out diagrammatically, as discussed earlier in this chapter (see Fig. 8.9). At the simplest level of averaging, we replace the average of the product by the product of the averages, i.e.,

$$\left\langle G_F^{(+)}(\boldsymbol{k}_1, \boldsymbol{k}_2) G_F^{(-)}(\boldsymbol{k}_2, \boldsymbol{k}_1) \right\rangle \approx \left\langle G_F^{(+)}(\boldsymbol{k}_1, \boldsymbol{k}_2) \right\rangle \left\langle G_F^{(-)}(\boldsymbol{k}_2, \boldsymbol{k}_1) \right\rangle, \qquad (8.46)$$

where

$$\left\langle G_F^{(+)}(\boldsymbol{k}_1, \boldsymbol{k}_2) \right\rangle \equiv (2\pi)^d \delta(\boldsymbol{k}_1 - \boldsymbol{k}_2) \left\langle G_F^{(+)}(\boldsymbol{k}_1) \right\rangle,$$

$$\left\langle G_F^{(-)}(\boldsymbol{k}_2, \boldsymbol{k}_1) \right\rangle = (2\pi)^d \delta(\boldsymbol{k}_1 - \boldsymbol{k}_2) \left\langle G_F^{(-)}(\boldsymbol{k}_1) \right\rangle,$$

with, as before,

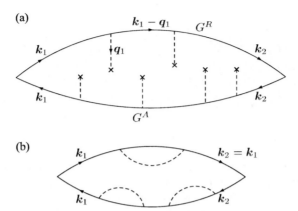

FIG. 8.9. (a) Typical conductivity bubble before impurity averaging. The dashed lines denote scattering by the impurities (crosses), and the solid lines denote the Green functions $G_0^{(+)}$ and $G_0^{(-)}$. (b) Typical non-vanishing diagram after impurity averaging and contributing to the classical (Drude) conductivity.

$$\left\langle G_F^{(+)}(\boldsymbol{k}) \right\rangle = \frac{1}{\epsilon_F - \epsilon_k + i\hbar/2\tau},$$

$$\left\langle G_F^{(-)}(\boldsymbol{k}) \right\rangle = \frac{1}{\epsilon_F - \epsilon_k - i\hbar/2\tau},$$

and

$$\tau = \frac{1}{2\pi\nu_F\gamma},$$

as in eqns (8.27b), (8.28) and (8.29). Substituting from eqn (8.46) into eqn (8.45), we obtain

$$\sigma(0) = \frac{e^2\hbar^3}{2\pi m^2} \int \frac{d\boldsymbol{k}}{(2\pi)^d}\, k_x^2 \left\langle G_F^{(+)}(\boldsymbol{k}) \right\rangle \left\langle G_F^{(-)}(\boldsymbol{k}) \right\rangle. \qquad (8.47)$$

Here we have made use of the expression for the square of the delta function $(\delta(\boldsymbol{k}))^2 = \delta(\boldsymbol{k})\Omega/(2\pi)^d$. Of course, we could always use interchangeably the quasi-continuous $\boldsymbol{k}$ (and the Kronecker delta $\delta_{\boldsymbol{k},\boldsymbol{k}'}$) and the continuous $\boldsymbol{k}$ (and the Dirac delta function $\delta(\boldsymbol{k} - \boldsymbol{k}')$), with

$$\sum_{\boldsymbol{k}} = \frac{\Omega}{(2\pi)^d} \int d\boldsymbol{k},$$

$$\delta_{\boldsymbol{k},\boldsymbol{k}'} = \frac{(2\pi)^d \delta(\boldsymbol{k} - \boldsymbol{k}')}{\Omega}.$$

Defining the energy variable $\xi = \epsilon_k - \epsilon_F$, measured from the Fermi level, replacing $k_\alpha^2$ by its angular average $k_F^2/d$, and transforming the integral over the momentum $\boldsymbol{k}$ into an energy ($\xi$) integral, we obtain

$$\sigma(0) = \frac{ne^2}{m}\frac{\hbar}{2\pi}\int_{-\infty}^{+\infty}\frac{1}{\xi - i\hbar/2\tau}\frac{1}{\xi + i\hbar/2\tau}\,\mathrm{d}\xi, \qquad (8.48)$$

where we have extended the limits of the $\xi$-integration to $\pm\infty$, recalling that the main contribution to the integral comes from within $\pm\hbar/2\tau$ of the Fermi level for weak scattering, $\hbar/2\tau \ll \epsilon_F$. Straightforward contour integration then yields

$$\sigma_{\text{Drude}}(0) = \frac{ne^2\tau}{m}. \qquad (8.49)$$

Here $n$ is the electron number density. This is the well-known classical Drude expression for the d.c. conductivity at zero temperature. We can clearly see now that the ignored terms of the type $G^{(-)}G^{(-)}$ and $G^{(+)}G^{(+)}$ in eqn (8.42) would have finally made no contribution as these products are analytic in the lower and the upper half-planes, respectively, permitting an appropriate choice of the contour that makes their contributions vanish identically.

It can be readily seen, diagrammatically, that the Drude expression for the conductivity given by eqn (8.49) represents the crudest approximation in which the impurity lines interconnecting the two propagators $G^{(+)}$ and $G^{(-)}$ have been totally neglected—leading to the simplifying factorization (8.46). Only impurity lines terminating on the same individual propagator line have been considered, giving the product of their averages (see Fig. 8.9(b)).

Let us, therefore, pause here, and reconsider our neglect of such scattering processes. First, take the so-called ladder diagrams, wherein the (dotted) impurity lines do connect the (solid) propagator lines for $G_F^{(+)}$ and $G_F^{(-)}$, but they do so without crossing one another—hence the name *ladder* (*L*) *diagrams*—as shown in Fig. 8.10. Now, for the isotropic, i.e., short-ranged or point-like impurity scatterers (as is implicit in our delta-correlated Gaussian random potential giving a momentum-independent scattering strength parameter $\gamma$), the contributions of these ladder diagrams to the conductivity vanish identically. This is

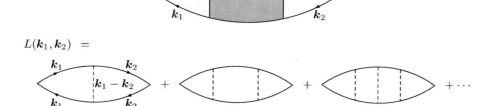

FIG. 8.10. Ladder diagrams in which the impurity lines (dashed) connect the Green function lines (solid) in the conductivity bubble, without crossing one another.

because of the vector character of the conductivity vertex, namely the occurrence of the factor $\boldsymbol{k}_{1x}\boldsymbol{k}_{2x}$ in the expression (8.45) for the conductivity. To see this more explicitly, consider the contribution of a ladder diagram to $\sigma$ with any number of impurity lines. This involves a factor of the type

$$\gamma \int \frac{\mathrm{d}\boldsymbol{k}_2 \, \boldsymbol{k}_{2x}}{(2\pi)^d} G_{0F}^{(+)}(\boldsymbol{k}_2) G_{0F}^{(-)}(\boldsymbol{k}_2), \qquad (8.50)$$

which vanishes because of the $\boldsymbol{k}_{2x}$ in the integrand. Note that the subscript '$0F$' here denotes the unperturbed (0) Green function evaluated at the Fermi energy ($F$). Also notice that the propagators here involve the absolute magnitude $|\boldsymbol{k}_2|$, and that the interaction parameter $\gamma$ is independent of the momentum transfer (because of the short-ranged scatterers, and hence the $s$-wave-like scattering assumed here). This then justifies our neglect of the ladder impurity diagrams. It may be noted in passing that, for anisotropic scattering, on the other hand, the ladder diagrams do contribute to the conductivity. Indeed, it is precisely these diagrams that would contribute the factor $1 - \cos\theta_{\boldsymbol{k}_1,\boldsymbol{k}_2}$, well known from the Boltzmann kinetic transport theory, that enhances the relative effectiveness of large momentum transfer processes. The ladder diagrams give the classical diffusion that underlies conduction, and are hence called the *diffusons*.

Having thus disposed of the ladder ($L$) diagrams with non-crossing impurity lines, we now turn our attention to the other class of diagrams, namely the so-called maximally-crossed, or fan diagrams ($X$), as depicted in Fig. 8.11. Semi-classically, these correspond to the coherent back-scattering (CBS), as discussed earlier in qualitative terms, and give the leading quantum (WL) correction to the classical (Drude) conductivity. The correction terms turn out to have a singular length-scale and dimensionality dependence which is all-important in the context of quantum transport.

### 8.6.2 *Weak localization (WL) and quantum correction to the classical (Drude) conductivity: the maximally-crossed diagrams*

The maximally-crossed diagrams, as shown in Fig. 8.11, involve crossed impurity lines in the conductivity bubble connecting the Green function lines $G_F^{(+)}$ and $G_F^{(-)}$ as the latter traverse the same set of scatterers, but in the time-reversed order, giving constructive interference under TRI. In order to calculate their contribution to the conductivity, it is necessary to calculate the kernel $X$ shown in Fig. 8.11. Now, a maximally-crossed diagram for the kernel $X$ can be conveniently turned around into an uncrossed ladder diagram just by spatially reversing one of the Green function lines, for $G^{(-)}$, say. Very properly, these crossed diagrams are now called the *Cooperons*. The iterative nature of the resulting ladder diagrams enables us to perform the partial summation to all orders as an infinite geometric series. We obtain

$$X(\boldsymbol{k}_1, \boldsymbol{k}_2) = \frac{\gamma}{\Omega} \frac{\Lambda(\boldsymbol{k}_1, \boldsymbol{k}_2)}{1 - \Lambda(\boldsymbol{k}_1, \boldsymbol{k}_2)}, \qquad (8.51)$$

where

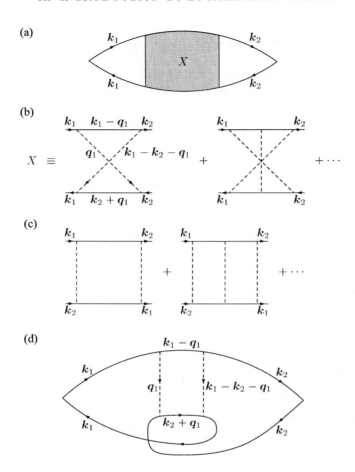

FIG. 8.11. (a) Conductivity bubble with maximally-crossed impurity lines ($X$). (b) Maximally-crossed (fan) diagrams comprising $X$. (c) The unfolded fan diagram $X$ with one propagator line spatially reversed. (d) Conductivity bubble with the fan diagram $X$ unfolded into a ladder diagram.

$$\Lambda(\boldsymbol{k}_1, \boldsymbol{k}_2) = \gamma \int \frac{\mathrm{d}\boldsymbol{q}}{(2\pi)^d} \left( \frac{1}{\epsilon_{\boldsymbol{k}_1+\boldsymbol{q}} - \epsilon_F - \mathrm{i}\eta} \right) \left( \frac{1}{\epsilon_{\boldsymbol{k}_2-\boldsymbol{q}} - \epsilon_F + \mathrm{i}\eta} \right).$$

We now make an important observation, namely that $\Lambda(\boldsymbol{k}_1, \boldsymbol{k}_2 = -\boldsymbol{k}_1) = 1$, making $X(\boldsymbol{k}_1, \boldsymbol{k}_2)$ divergent for $\boldsymbol{k}_1 + \boldsymbol{k}_2 = 0$. This is readily verified by replacing the $\boldsymbol{q}$-integral by the energy ($\epsilon_k$) integral via the density of states $\nu_F$, performing the $\epsilon_k$ contour integral by the method of residues, and noting that $\eta = \pi\gamma\nu_F$ (see eqn (8.27b)). Here we have extended the integration limits of $\epsilon_k$ to $\pm\infty$, which is a good approximation for $k_F l_e \gg 1$, or $\eta \ll E_F$, as is assumed throughout, i.e., the weak-scattering limit. This prompts us to introduce the total momentum

$Q = k_1 + k_2$, thereby obtaining

$$\Lambda(k_1, k_2) = \pi i \gamma \nu_F \left\langle \frac{1}{-(\hbar^2/m) k \cdot Q + 2i\eta} \right\rangle_{\text{angular}}$$

$$= 1 - \frac{\hbar^2 v_F^2 Q^2}{4\eta^2 d} + \cdots, \qquad (8.52)$$

where the subscript 'angular' denotes an average over the angle between $k$ and $Q$ in $d$ dimensions, and $k = (k_1 - k_2)/2$. From eqn (8.51) we have

$$X(k_1, k_2) = \frac{\gamma}{\Omega} \frac{4\eta^2 d}{\hbar^2 v_F^2} \frac{1}{Q^2}$$

$$\equiv X(Q). \qquad (8.53)$$

All we have to do now is substitute $X(k_1, k_2)$ from eqn (8.53) into the expression (8.45) for the conductivity $\sigma$, retaining only the maximally-crossed diagrams as in Fig. 8.11. Thus we obtain

$$\delta\sigma \equiv \sigma(0) - \sigma_{\text{Drude}}(0) = -\frac{2e^2}{\pi\hbar} \eta^2 \gamma \frac{1}{\Omega} \sum_k \left[\langle G_F^R(k) \rangle\right]^2 \left[\langle G_F^A(-k) \rangle\right]^2 \frac{1}{\Omega} \sum_Q \frac{1}{Q^2}. \qquad (8.54)$$

In arriving at eqn (8.54), we have made use of the fact noted earlier that the dominant contribution to the right-hand side of eqn (8.45) comes from the small-$Q$ ($\equiv k_1 + k_2$) region. Accordingly, we have set $k_1 = -k_2 \equiv k$ in all of the non-singular factors on the right-hand side of eqn (8.52). Straightforward energy-contour integration then gives

$$\frac{1}{\Omega} \sum_k \left[G_F^{(+)}(k)\right]^2 \left[G_F^{(-)}(-k)\right]^2 = \nu_F \int_{-\infty}^{+\infty} \left(\frac{1}{\epsilon - \epsilon_F - i\eta}\right)^2 \left(\frac{1}{\epsilon - \epsilon_F + i\eta}\right)^2 d\epsilon$$

$$= \frac{\pi \nu_F}{2\eta^3}, \qquad (8.55)$$

and

$$\frac{1}{\Omega} \sum_Q \frac{1}{Q^2} = \frac{1}{(2\pi)^d} \frac{2(\pi)^{d/2}}{\Gamma(d/2)} \int_{Q_{\min}}^{Q_{\max}} Q^{d-3} dQ, \qquad (8.56)$$

with $Q_{\max} = 1/l_e$ and $Q_{\min} = 1/l_\phi$, where $l_e$ and $l_\phi$ are, respectively, the elastic mean free path and the phase-breaking length. Here we have introduced the wave vector cut-offs $Q_{\max}$ and $Q_{\min}$, or the corresponding cut-off length-scales $l_e$ and $l_\phi$, in the otherwise divergent momentum integral, consistently with the conditions for phase-coherent transport. We will return to this point shortly. Substituting from eqns (8.55) and (8.56) into eqn (8.54), we get

$$\delta\sigma = -\frac{e^2}{\pi\hbar} \left(\frac{1}{2\pi}\right)^d \frac{2\pi^{d/2}}{\Gamma(d/2)} \int_{1/l_\phi}^{1/l_e} Q^{d-3} dQ. \qquad (8.57)$$

As there is no spin–flip scattering here, we have multiplied the conductivity expression by a factor of 2 to account for the two independent channels (the two spin projections for the spin-1/2 electrons) adding up in parallel. Thus, finally, we obtain the following expressions for the dimensionality $d$-dependent quantum corrections to the conductivity:

$$\delta\sigma = \begin{cases} -(1/\pi)\left(e^2/\hbar\right)(l_\phi - l_e), & d=1, \\ -(1/\pi^2)\left(e^2/\hbar\right)\ln(l_\phi/l_e), & d=2, \\ -(1/\pi^3)\left(e^2/\hbar\right)(1/l_e - 1/l_\phi), & d=3. \end{cases} \quad (8.58)$$

The negative sign of these quantum corrections for all three dimensionalities clearly reflects the physics of coherent back-scattering, namely that it enhances the *return* probability and hence reduces the conductivity.

### 8.6.3 *Scale dependence of the conductivity*

The wave vector cut-offs, or the corresponding reciprocal length-scales $l_e$ and $l_\phi$ introduced in the expressions above, have a direct physical significance. The upper cut-off $1/l_e$ for the wave vector $\mathbf{Q}$, of course, simply measures the effective angular width ($\sim 1/k_F l_e$) of the coherent back-scattering cone—the diffraction-limited width corresponding to an aperture of linear dimension $l_e$. The lower limit $1/l_\phi$ is, however, different, and is determined by the condition of phase-coherent transport, namely that the electron suffers no phase-breaking inelastic scattering from, or entanglement with, the environment, e.g., the phonons. Now, the phase-coherent volume has a linear dimension of the order of the Thouless length $L_T$ (see the second paragraph of Section 1.3), or the sample size $L$, whichever is the smaller of the two. This gives the condition $l_\phi = \min\{L, L_T\}$. Aside from phase breaking, however, one may also have to consider dephasing due to the presence of a magnetic field, if any, that breaks TRI. Here the two alternative return paths of the CBS that would normally interfere constructively will now acquire equal but opposite phase shifts $\pm 2\pi\phi/\phi_0$, where $\phi$ is the magnetic flux linking the closed return paths and $\phi_0 = \hbar c/e$ is the quantum of magnetic flux. Inasmuch as the projected area of the diffusive closed return path normal to the magnetic field is a random quantity, we expect a dephasing of the CBS amplitudes over the magnetic length-scale $L_B \sim (\hbar c/2eB)^{1/2}$. Thus, the cut-off length-scale is $l_\phi = \min\{L, L_T, L_B\}$. Recalling that $L_T$ is a function of temperature (and that at sufficiently low temperatures the latter enters the physics of the problem through $L_T$ only), the weak-localization correction provides a convenient, in fact the best, probe with which to measure the phase-coherence length by measuring the temperature dependence or the magnetic field $B$ dependence of the conductance at low temperatures. Thus we get a calculable and measurable negative temperature coefficient of resistance and negative magneto-resistance. This has indeed led to a whole new field of weak-localization study of the phase-sensitive effects, particularly in two-dimensional disordered systems—thin films or interfaces. It is apt to remark here that the above diagrammatics is extendable, and has indeed been extended to other kinds of scattering, the most interesting of all being the scattering involving spin–orbit coupling which dominates for heavy metal impurities, e.g., Au in two-dimensional Mg thin films [25]. The spin–orbit interaction maintains the TRI but breaks spin–rotational invariance; for a sufficiently strong spin–orbit interaction we end up in the so-called *symplectic* case— sometimes mentioned in the previous chapters (see, for instance, Section 1.3.3 and Section 7.4, immediately after eqn (7.167))—where we have *anti-localization*, instead of localization, and a change in sign of the magneto-resistance. Here the electron spin axis changes randomly as the electron traverses the CBS alternatives, and thus accumulates equal and opposite geometric phases along the two alternative return paths. The randomness of this relative phase shift leads to dephasing that tends to suppress the normal WL effect. Clearly, aligning the electron spins with a strong magnetic field suppresses the accumulation of the above random geometric phases (dephasing) and hence the anti-localization effect of spin–orbit coupling and the concomitant positive magneto-resistance, as is indeed observed. Weak localization realizes, in effect, a time-of-flight experiment with the conduction electrons in a disordered metal. It is perhaps best viewed as a multi-slit Young experiment with wave interference from the randomly located scatterers which it, therefore, probes.

The length-scale-dependent conductivity given by eqn (8.58) implies a one-parameter scaling law for the conductance referred to in Section 8.4. Indeed, one can at once write down the following expression for the beta function:

$$\beta(g) \equiv \frac{\partial \ln g}{\partial \ln L} = d - 2 - \frac{1}{2\pi^2}\frac{1}{g}, \qquad (8.59)$$

with the convenient redefinition $g = \sigma(L\pi)^{d-2}$. Thus, for $d = 1$ and 2 we have $\beta(g) < 0$, and the conductance scales down to zero as $L \to \infty$, thereby implying the absence of a truly metallic state for $d \leqslant 2$. For $d = 3$, however, the beta function has a zero at $g = g_c$ ($= 1/2\pi^2$). Here $g_c$ is an unstable fixed point, so that for $g > g_c$ the conductance scales up to that of a true metal, while for $g < g_c$ the conductance scales down to an insulating state. Clearly, $g = g_c$ is to be identified with the mobility edge.

Many of the features predicted by the scaling theory are in qualitative agreement with experiments on the magnetic field and the temperature dependence of the electrical transport at low temperatures. The neglect of the electron–electron interaction, however, continues to be a serious and strongly debated limitation of the physical domain of validity of the one-parameter scaling theory of localization, and of the associated disorder-induced metal–insulator transition that it claims to describe.

Finally, it may be noted that the above impurity-diagrammatic treatment can be, and indeed has been, applied to calculate the variance of the statistical fluctuations of the conductance [116]. This has led to the so-called universal conductance fluctuations (UCF), which are observed in the good metallic regime and which were encountered earlier in Chapters 6 and 7.

# APPENDIX A

## THE THEOREM OF KANE–SEROTA–LEE

Equation (4.124a) gives the conductivity tensor in RPA, which we reproduce here as follows:

$$\sigma^\omega_{\alpha\beta}(\boldsymbol{r},\boldsymbol{r}') = \frac{1}{\hbar\omega}\lim_{\eta\to 0^+}\int_0^\infty d\tau\, e^{i\omega\tau-\eta\tau}\left\langle\left[\widetilde{j}_\alpha(\boldsymbol{r},\tau),\widetilde{j}_\beta(\boldsymbol{r}',0)\right]\right\rangle_{\rm sc} \\ + i\frac{e^2 n_{\rm sc}(\boldsymbol{r})}{m\omega}\delta(\boldsymbol{r}-\boldsymbol{r}')\delta_{\alpha\beta}. \quad (\text{A.1})$$

The Schrödinger equation arising from the Hamiltonian $H_{\rm sc}$ is

$$(E_{NM} - H_{\rm sc})|NM\rangle = 0, \quad (\text{A.2})$$

where $|NM\rangle$ denotes the $M$th eigenstate with $N$ electrons and $E_{NM}$ is the corresponding energy. As explained in the text, the thermodynamic limit has to be taken (in the present case, this means making our system infinitely long in the $x$ direction) before the limit as $\eta \to 0$ is evaluated. In that limit, $M$ becomes a continuous index (perhaps in combination with other discrete indices); for instance, for plane-wave states $M$ would represent the collection of continuous indices $\{\boldsymbol{k}_1, \boldsymbol{k}_2, \ldots, \boldsymbol{k}_N\}$. For simplicity of notation, however, and with the above understanding, in what follows *we write summations over the index $M$ instead of the corresponding integrals.*

Taking the divergence on both sides of eqn (A.1) and using the continuity equation (4.127), we have

$$\partial_\alpha \sigma^\omega_{\alpha\beta}(\boldsymbol{r},\boldsymbol{r}') = \frac{1}{\hbar\omega}\lim_{\eta\to 0^+}\int_0^\infty d\tau\, e^{i\omega\tau-\eta\tau}(-1)\frac{\partial}{\partial\tau}\left\langle\left[\widetilde{\rho}(\boldsymbol{r},\tau),\widetilde{j}_\beta(\boldsymbol{r}',0)\right]\right\rangle_{\rm sc} \\ + i\frac{e^2}{m\omega}\partial_\beta\left[n_{\rm sc}(\boldsymbol{r})\delta(\boldsymbol{r}-\boldsymbol{r}')\right]. \quad (\text{A.3})$$

Let $A$ denote the integral in the first line of this last equation. Integrating by parts, we find

$$A = \left\langle\left[\widetilde{\rho}(\boldsymbol{r},0),\widetilde{j}_\beta(\boldsymbol{r}',0)\right]\right\rangle_{\rm sc} + (i\omega-\eta)\int_0^\infty d\tau\, e^{i\omega\tau-\eta\tau}\left\langle\left[\widetilde{\rho}(\boldsymbol{r},\tau),\widetilde{j}_\beta(\boldsymbol{r}',0)\right]\right\rangle_{\rm sc}. \quad (\text{A.4})$$

The commutator in the first line gives

$$\left[\widetilde{\rho}(\boldsymbol{r},0),\widetilde{j}_\beta(\boldsymbol{r}',0)\right] = -i\hbar\frac{e^2 \hat{n}(\boldsymbol{r}')}{m}\frac{\partial}{\partial x_\beta}\delta(\boldsymbol{r}-\boldsymbol{r}'). \quad (\text{A.5})$$

The equality (A.5) is proved as follows. At $t = 0$ the interaction and Schrödinger representations coincide, so that

$$\frac{2m}{e^2}\left[\widetilde{\rho}(\boldsymbol{r},0),\widetilde{j}_\beta(\boldsymbol{r}',0)\right]$$
$$= \frac{2m}{e^2}\left[\hat{\rho}(\boldsymbol{r}),\hat{j}_\beta(\boldsymbol{r}')\right]$$
$$= \sum_{ij}\left[\delta\left(\boldsymbol{r}-\boldsymbol{r}_i\right),\delta\left(\boldsymbol{r}'-\boldsymbol{r}_j\right)p_{j\beta}+p_{j\beta}\delta\left(\boldsymbol{r}'-\boldsymbol{r}_j\right)\right]$$
$$= \sum_{ij}\left\{\delta\left(\boldsymbol{r}'-\boldsymbol{r}_j\right)\left[\delta\left(\boldsymbol{r}-\boldsymbol{r}_i\right),p_{j\beta}\right]+\left[\delta\left(\boldsymbol{r}-\boldsymbol{r}_i\right),p_{j\beta}\right]\delta\left(\boldsymbol{r}'-\boldsymbol{r}_j\right)\right\}.$$

The commutator in the last line is

$$[\delta\left(\boldsymbol{r}-\boldsymbol{r}_i\right),p_{j\beta}] = -[p_{j\beta},\delta\left(\boldsymbol{r}-\boldsymbol{r}_i\right)]$$
$$= i\hbar\frac{\partial}{\partial x_{j\beta}}\delta(\boldsymbol{r}-\boldsymbol{r}_i) = i\hbar\delta_{ij}\frac{\partial}{\partial x_{i\beta}}\delta(\boldsymbol{r}-\boldsymbol{r}_i)$$
$$= -i\hbar\delta_{ij}\frac{\partial}{\partial x_\beta}\delta(\boldsymbol{r}-\boldsymbol{r}_i),$$

so that

$$\frac{2m}{e^2}\left[\widetilde{\rho}(\boldsymbol{r},0),\widetilde{j}_\beta(\boldsymbol{r}',0)\right] = -i\hbar\sum_i\left[\delta(\boldsymbol{r}'-\boldsymbol{r}_i)\frac{\partial\delta(\boldsymbol{r}-\boldsymbol{r}_i)}{\partial x_\beta}+\frac{\partial\delta(\boldsymbol{r}-\boldsymbol{r}_i)}{\partial x_\beta}\delta(\boldsymbol{r}'-\boldsymbol{r}_i)\right]$$
$$= -2i\hbar\left[\sum_i\delta(\boldsymbol{r}'-\boldsymbol{r}_i)\right]\frac{\partial\delta(\boldsymbol{r}-\boldsymbol{r}')}{\partial x_\beta},$$

from which the desired equality (A.5) follows.

The right-hand side of eqn (A.5), and thus the first term of (A.4), is purely imaginary, so that taking the real part of eqn (A.3) we find

$$\mathrm{Re}\left[\partial_\alpha\sigma^\omega_{\alpha\beta}\left(\boldsymbol{r},\boldsymbol{r}'\right)\right] = \mathrm{Re}\left[\lim_{\eta\to 0^+}\frac{i\omega-\eta}{\hbar\omega}\int_0^\infty d\tau\, e^{i\omega\tau-\eta\tau}\left\langle\left[\widetilde{\rho}(\boldsymbol{r},\tau),\widetilde{j}_\beta(\boldsymbol{r}',0)\right]\right\rangle_{\mathrm{sc}}\right]. \tag{A.6}$$

We now calculate the commutator occurring in eqn (A.6) to be

$$\left\langle\left[\widetilde{\rho}(\boldsymbol{r},\tau),\widetilde{j}_\beta(\boldsymbol{r}',0)\right]\right\rangle_{\mathrm{sc}}$$
$$= \left\langle\widetilde{\rho}(\boldsymbol{r},\tau)\widetilde{j}_\beta(\boldsymbol{r}',0)-\widetilde{j}_\beta(\boldsymbol{r}',0)\widetilde{\rho}(\boldsymbol{r},\tau)\right\rangle_{\mathrm{sc}}$$
$$= \sum_{NMN'M'}P(N,E_{NM})$$
$$\times\left[\left\langle NM\left|e^{(i/\hbar)H_{\mathrm{sc}}\tau}\hat{\rho}(\boldsymbol{r})e^{-(i/\hbar)H_{\mathrm{sc}}\tau}\right|N'M'\right\rangle\left\langle N'M'\left|\hat{j}_\beta(\boldsymbol{r}')\right|NM\right\rangle\right.$$
$$\left.-\left\langle NM\left|\hat{j}_\beta(\boldsymbol{r}')\right|N'M'\right\rangle\left\langle N'M'\left|e^{(i/\hbar)H_{\mathrm{sc}}\tau}\hat{\rho}(\boldsymbol{r})e^{-(i/\hbar)H_{\mathrm{sc}}\tau}\right|NM\right\rangle\right]$$

$$= \sum_{NMM'} P(N, E_{NM})$$
$$\times \left[ e^{(i/\hbar)(E_{NM}-E_{NM'})\tau} \langle NM|\hat{\rho}(\boldsymbol{r})|NM'\rangle \langle NM'|\hat{j}_\beta(\boldsymbol{r}')|NM\rangle \right.$$
$$\left. - e^{(i/\hbar)(E_{NM'}-E_{NM})\tau} \langle NM|\hat{j}_\beta(\boldsymbol{r}')|NM'\rangle \langle NM'|\hat{\rho}(\boldsymbol{r})|NM\rangle \right],$$
(A.7)

where
$$P(N, E_{NM}) = \frac{e^{-\beta(E_{NM}-\mu N)}}{\mathcal{Z}(\beta,\mu)} \quad \text{(A.8)}$$

denotes the statistical mechanical weight of the state $|NM\rangle$. We have used the fact that $\hat{\rho}(\boldsymbol{r})$ and $\hat{j}_\beta(\boldsymbol{r}')$ conserve the particle number. In eqn (A.6) we also need an integral of the type
$$\int_0^\infty d\tau\, e^{i\Omega\tau - \eta\tau} = \frac{i}{\Omega + i\eta}.$$

Substituting this last equation and eqn (A.7) into (A.6), we find

$$\operatorname{Re}\left[\partial_\alpha \sigma^\omega_{\alpha\beta}(\boldsymbol{r},\boldsymbol{r}')\right]$$
$$= \operatorname{Re}\left\{ \lim_{\eta\to 0^+} \frac{i}{\hbar\omega}(i\omega - \eta) \sum_N \sum_{MM'} P(N, E_{NM}) \right.$$
$$\times \left[ \frac{\langle NM|\hat{\rho}(\boldsymbol{r})|NM'\rangle \langle NM'|\hat{j}_\beta(\boldsymbol{r}')|NM\rangle}{\omega + (1/\hbar)(E_{NM}-E_{NM'}) + i\eta} \right.$$
$$\left.\left. - \frac{\langle NM|\hat{j}_\beta(\boldsymbol{r}')|NM'\rangle \langle NM'|\hat{\rho}(\boldsymbol{r})|NM\rangle}{\omega + (1/\hbar)(E_{NM'}-E_{NM}) + i\eta} \right] \right\}$$
$$= -\operatorname{Re}\left\{ \lim_{\eta\to 0^+} \frac{\omega+i\eta}{\omega} \sum_N \sum_{MM'} [P(N,E_{NM}) - P(N,E_{NM'})] \right.$$
$$\left. \times \frac{\langle NM|\hat{\rho}(\boldsymbol{r})|NM'\rangle \langle NM'|\hat{j}_\beta(\boldsymbol{r}')|NM\rangle}{\hbar\omega + E_{NM} - E_{NM'} + i\hbar\eta} \right\}.$$
(A.9)

We will find it convenient to introduce, in the right-hand side of this last equation, two energy integrals of the form

$$\text{Re}\left[\partial_\alpha \sigma_{\alpha\beta}^\omega (\boldsymbol{r}, \boldsymbol{r}')\right]$$
$$= -\text{Re}\left\{\lim_{\eta \to 0^+} \frac{\omega + i\eta}{\omega} \sum_N \sum_{MM'} \iint dE\, dE'\, \delta(E - E_{NM})\, \delta(E' - E_{NM'})\right.$$
$$\left. \times \frac{P(N, E) - P(N, E')}{\hbar\omega + E - E' + i\hbar\eta} \langle NM | \hat{\rho}(\boldsymbol{r}) | NM' \rangle \left\langle NM' \left| \hat{j}_\beta(\boldsymbol{r}') \right| NM \right\rangle \right\}. \tag{A.10}$$

We need a limit of the type
$$\lim_{\eta \to 0^+} \frac{\omega + i\eta}{\Omega + i\eta} = \lim_{\eta \to 0^+} (\omega + i\eta) \frac{\Omega - i\eta}{\Omega^2 + \eta^2}$$
$$= \lim_{\eta \to 0^+} (\omega + i\eta) \left[\frac{\Omega}{\Omega^2 + \eta^2} - i\pi \frac{\eta/\pi}{\Omega^2 + \eta^2}\right]$$
$$= \omega \left[\mathcal{P}\frac{1}{\Omega} - i\pi\delta(\Omega)\right]. \tag{A.11}$$

Thus
$$\text{Re}\left[\partial_\alpha \sigma_{\alpha\beta}^\omega (\boldsymbol{r}, \boldsymbol{r}')\right]$$
$$= -\text{Re} \sum_N \sum_{MM'} \iint dE\, dE'\, \delta(E - E_{NM})\, \delta(E' - E_{NM'})$$
$$\times \left[P(N, E) - P(N, E')\right] \left[\frac{\mathcal{P}}{\hbar\omega + E - E'} - i\pi\delta(\hbar\omega + E - E')\right]$$
$$\times \langle NM | \hat{\rho}(\boldsymbol{r}) | NM' \rangle \left\langle NM' \left| \hat{j}_\beta(\boldsymbol{r}') \right| NM \right\rangle. \tag{A.12}$$

In the d.c. limit, i.e., as $\omega \to 0$ (with the understanding that the thermodynamic limit has been taken first, even before the $\eta \to 0^+$ limit, so that, according to the comment made immediately after eqn (A.2), the summation $\sum_{MM'}$ really stands for an integral over continuous indices), we have

$$\lim_{\omega \to 0} \text{Re}\left[\partial_\alpha \sigma_{\alpha\beta}^\omega (\boldsymbol{r}, \boldsymbol{r}')\right]$$
$$= -\text{Re} \sum_N \sum_{MM'} \iint dE\, dE'\, \delta(E - E_{NM})\, \delta(E' - E_{NM'})$$
$$\times \mathcal{P}\frac{P(N, E) - P(N, E')}{E - E'} \langle NM | \hat{\rho}(\boldsymbol{r}) | NM' \rangle \left\langle NM' \left| \hat{j}_\beta(\boldsymbol{r}') \right| NM \right\rangle$$
$$+ \text{Re} \sum_N \sum_{MM'} i\pi \iint dE\, dE'\, \delta(E - E_{NM})\, \delta(E' - E_{NM'})\, \delta(E - E')$$
$$\times \left[P(N, E) - P(N, E')\right] \langle NM | \hat{\rho}(\boldsymbol{r}) | NM' \rangle \left\langle NM' \left| \hat{j}_\beta(\boldsymbol{r}') \right| NM \right\rangle. \tag{A.13}$$

The last two lines in the above equation do not contribute, due to the delta function $\delta(E - E')$, so that

$$\lim_{\omega \to 0} \text{Re}\left[\partial_\alpha \sigma^\omega_{\alpha\beta}(\boldsymbol{r}, \boldsymbol{r}')\right]$$

$$= -\text{Re} \sum_N \sum_{MM'} \iint dE\, dE'\, \delta(E - E_{NM})\, \delta(E' - E_{NM'})$$

$$\times \mathcal{P} \frac{P(N, E) - P(N, E')}{E - E'} \langle NM | \hat{\rho}(\boldsymbol{r}) | NM' \rangle \langle NM' | \hat{j}_\beta(\boldsymbol{r}') | NM \rangle. \tag{A.14}$$

The matrix element of the electron density operator $\hat{\rho}(\boldsymbol{r})$ between the states $\Phi(\boldsymbol{r}_1, \ldots, \boldsymbol{r}_N)$ and $\Psi(\boldsymbol{r}_1, \ldots, \boldsymbol{r}_N)$ is

$$(\Phi, \hat{\rho}(\boldsymbol{r})\Psi) = \sum_{i=1}^N \int d^3\boldsymbol{r}_1 \cdots \int d^3\boldsymbol{r}_N\, \delta(\boldsymbol{r}_i - \boldsymbol{r})\, \Phi^*(\boldsymbol{r}_1, \ldots, \boldsymbol{r}_N) \Psi(\boldsymbol{r}_1, \ldots, \boldsymbol{r}_N). \tag{A.15}$$

The matrix element of the current operator $\hat{j}_\beta(\boldsymbol{r})$ between the same two states is given by eqn (4.10) as

$$\left(\Phi, \hat{j}_\beta(\boldsymbol{r})\Psi\right) = -e \frac{\hbar}{2im} \sum_{i=1}^N \int d^3\boldsymbol{r}_1 \cdots \int d^3\boldsymbol{r}_N\, \delta(\boldsymbol{r}_i - \boldsymbol{r})$$

$$\times \left\{ \Phi^*(\boldsymbol{r}_1, \ldots, \boldsymbol{r}_N) \left[\frac{\partial}{\partial x_{i\beta}} \Psi(\boldsymbol{r}_1, \ldots, \boldsymbol{r}_N)\right] \right.$$

$$\left. - \left[\frac{\partial}{\partial x_{i\beta}} \Phi^*(\boldsymbol{r}_1, \ldots, \boldsymbol{r}_N)\right] \Psi(\boldsymbol{r}_1, \ldots, \boldsymbol{r}_N) \right\}. \tag{A.16}$$

We now assume that the system under study is *time-reversal invariant*. Thus, we can always find a basis of real eigenfunctions of the Hamiltonian $H_{\text{sc}}$. Equations (A.15) and (A.16) then show that the matrix elements of the density and the current operators between two such eigenstates are real and purely imaginary, respectively. Thus the right-hand side of eqn (A.14) vanishes and, as a result,

$$\lim_{\omega \to 0} \text{Re}\left[\partial_\alpha \sigma^\omega_{\alpha\beta}(\boldsymbol{r}, \boldsymbol{r}')\right] = 0. \tag{A.17}$$

Similarly, one finds

$$\lim_{\omega \to 0} \text{Re}\left[\partial'_\beta \sigma^\omega_{\alpha\beta}(\boldsymbol{r}, \boldsymbol{r}')\right] = 0, \tag{A.18}$$

which is eqn (4.144) in the text (see [100]).

# APPENDIX B

## THE CONDUCTIVITY TENSOR IN RPA

We shall be dealing with the conductivity tensor of eqn (4.124). We denote by $C_{\alpha\beta}(\boldsymbol{r},\boldsymbol{r}',\tau)$ the expectation value of the commutator appearing in that equation, i.e.,

$$C_{\alpha\beta}(\boldsymbol{r},\boldsymbol{r}',\tau) = \left\langle e^{(i/\hbar)H_{\rm sc}\tau}\hat{j}_\alpha(\boldsymbol{r})e^{-(i/\hbar)H_{\rm sc}\tau}\hat{j}_\beta(\boldsymbol{r}')\right\rangle_{\rm sc} \\ - \left\langle \hat{j}_\beta(\boldsymbol{r}')e^{(i/\hbar)H_{\rm sc}\tau}\hat{j}_\alpha(\boldsymbol{r})e^{-(i/\hbar)H_{\rm sc}\tau}\right\rangle_{\rm sc}, \qquad (\text{B.1})$$

and compute it separately. The angular brackets $\langle\cdots\rangle_{\rm sc}$ indicate the average (4.89) over the grand canonical ensemble associated with the Hamiltonian $H_{\rm sc}$ of eqn (4.85).

Denoting the $M$th $N$-particle eigenstate of $H_{\rm sc}$ by $|N;M\rangle$, its energy by $E_{NM}$ and its statistical weight by $P(N;M)$, we have

$$C_{\alpha\beta}(\boldsymbol{r},\boldsymbol{r}',\tau) \\ = \sum_{NN'}\sum_{MM'} P(N;M) \\ \times \Big[\left\langle N;M\left|e^{(i/\hbar)H_{\rm sc}\tau}\hat{j}_\alpha(\boldsymbol{r})e^{-(i/\hbar)H_{\rm sc}\tau}\right|N';M'\right\rangle\left\langle N';M'\left|\hat{j}_\beta(\boldsymbol{r}')\right|N;M\right\rangle \\ - \left\langle N;M\left|\hat{j}_\beta(\boldsymbol{r}')\right|N';M'\right\rangle\left\langle N';M'\left|e^{(i/\hbar)H_{\rm sc}\tau}\hat{j}_\alpha(\boldsymbol{r})e^{-(i/\hbar)H_{\rm sc}\tau}\right|N;M\right\rangle\Big], \\ (\text{B.2})$$

or

$$C_{\alpha\beta}(\boldsymbol{r},\boldsymbol{r}',\tau) \\ = \sum_{NN'}\sum_{MM'} P(N;M) \\ \times \Big[e^{(i/\hbar)(E_{NM}-E_{N'M'})\tau}\left\langle N;M\left|\hat{j}_\alpha(\boldsymbol{r})\right|N';M'\right\rangle\left\langle N';M'\left|\hat{j}_\beta(\boldsymbol{r}')\right|N;M\right\rangle \\ - e^{(i/\hbar)(E_{N'M'}-E_{NM})\tau}\left\langle N;M\left|\hat{j}_\beta(\boldsymbol{r}')\right|N';M'\right\rangle\left\langle N';M'\left|\hat{j}_\alpha(\boldsymbol{r})\right|N;M\right\rangle\Big]. \\ (\text{B.3})$$

We can also write the above equation as

$$C_{\alpha\beta}(\boldsymbol{r},\boldsymbol{r}',\tau) = \sum_{N}\sum_{M\neq M'} [P(N;M) - P(N;M')] e^{(i/\hbar)(E_{NM}-E_{NM'})\tau} \\ \times \left\langle N;M\left|\hat{j}_\alpha(\boldsymbol{r})\right|N;M'\right\rangle\left\langle N;M'\left|\hat{j}_\beta(\boldsymbol{r}')\right|N;M\right\rangle, \qquad (\text{B.4})$$

where we have made use of the fact that the current operator conserves the particle number $N$. Notice that only states with $M \neq M'$ enter into the summation.

We write the state $|N;M\rangle$ in second-quantization language as

$$|N;M\rangle = \prod_{\mu}{}' c_{\mu}^{\dagger} |0\rangle, \qquad (B.5)$$

where $c_{\mu}^{\dagger}$ is the creation operator for an electron in the $\mu$th single-particle state $\psi_{\mu}(\boldsymbol{r})$, the eigenstate of $H_{\text{sc}}$ with energy $\epsilon_{\mu}$, and $|0\rangle$ is the vacuum. The product with a prime in eqn (B.5) extends over $N$ *occupied* single-particle states only. By convention, the creation operators in (B.5) are ordered in a sequence of increasing single-particle energies $\epsilon_{\mu}$. The index $M$ that distinguishes the various $N$-particle states can be understood as the collection of occupation numbers $n_1, n_2, \ldots$ (where $n_{\mu} = 0$ and $1$) of the various single-particle states; the set of occupation numbers is also known as a *configuration* in the occupation number representation.

The current density operator $\hat{j}_{\alpha}(\boldsymbol{r})$ is a *one-body operator*, i.e.,

$$\hat{j}_{\alpha}(\boldsymbol{r}) = \sum_{i} \hat{j}_{\alpha}^{(i)}(\boldsymbol{r}), \qquad (B.6)$$

$\hat{j}_{\alpha}^{(i)}(\boldsymbol{r})$ being the $i$th term in eqn (4.10); the index $i$ runs over all the conduction electrons in the system. Its second-quantized expression is thus

$$\hat{j}_{\alpha}(\boldsymbol{r}) = \sum_{\mu\nu} \left[\hat{j}_{\alpha}(\boldsymbol{r})\right]_{\mu\nu} c_{\mu}^{\dagger} c_{\nu}, \qquad (B.7)$$

where $\left[\hat{j}_{\alpha}(\boldsymbol{r})\right]_{\mu\nu}$ is the matrix element between the single-particle states $\psi_{\mu}(\boldsymbol{r})$ and $\psi_{\nu}(\boldsymbol{r})$ of any one of the terms appearing in eqn (B.6); using the term for particle 1, say, we have

$$\left[\hat{j}_{\alpha}(\boldsymbol{r})\right]_{\mu\nu} = \left(\psi_{\mu}(\boldsymbol{r}_1), \hat{j}_{\alpha}^{(1)}(\boldsymbol{r})\psi_{\nu}(\boldsymbol{r}_1)\right). \qquad (B.8)$$

As explained in the text and in connection with Exercise 4.3, the thermodynamic limit has to be taken (in the present case, this means making our system infinitely long in the $x$ direction) before the limit $\eta \to 0^+$ is evaluated, in order to find the conductivity tensor, as we shall do below. In that limit, $\mu$ becomes a continuous index (perhaps in combination with other discrete indices). However, for simplicity of notation, and with the above understanding, in what follows *we write summations over the index $\mu$ instead of the corresponding integrals*. We further remark that for a TRI system (as we are considering here) the states $\psi_{\mu}(\boldsymbol{r})$ can be taken to be real.

The current density operator being a one-body operator, it can only connect states (with the same number of particles) differing, at most, by two occupation numbers, i.e., by a particle–hole excitation. In other words, let one of the states $|N;M\rangle$ occurring in eqn (B.4) be of the form

$$|N;M\rangle = |N;n_1,n_2,\ldots,n_\sigma = 1,\ldots,n_\tau = 0,\ldots\rangle, \tag{B.9a}$$

where we have explicitly singled out the single-particle states $\sigma$ and $\tau$. We also recall that in eqn (B.4) we need only consider states with $M \neq M'$. Then a state $|N;M'\rangle$ for which the matrix element $\langle N;M'|\hat{j}_\beta(\boldsymbol{r}')|N;M\rangle$ is nonzero can only be of the form

$$|N;M'\rangle = |N;n_1,n_2,\ldots,n_\sigma = 0,\ldots,n_\tau = 1,\ldots\rangle, \tag{B.9b}$$

which is obtained by annihilating an electron in state $\sigma$, say, and creating one in state $\tau$. The matrix element of the current operator between the two states (B.9a) and (B.9b) is thus

$$\langle N;M'|\hat{j}_\beta(\boldsymbol{r}')|N;M\rangle$$
$$= \sum_{\mu\nu} [j_\beta(\boldsymbol{r}')]_{\mu\nu} \langle N;n_1,n_2,\ldots,n_\sigma = 0,\ldots,n_\tau = 1,\ldots|$$
$$c_\mu^\dagger c_\nu |N;n_1,n_2,\ldots,n_\sigma = 1,\ldots,n_\tau = 0,\ldots\rangle$$
$$= (-1)^\phi [j_\beta(\boldsymbol{r}')]_{\tau\sigma}, \tag{B.10}$$

where the real phase factor $(-1)^\phi$ depends upon the number of anti-commutations needed to take a creation operator from the place $\sigma$ to the place $\tau$. From the adjoint of the above equation, we find

$$\langle N;M|\hat{j}_\alpha(\boldsymbol{r})|N;M'\rangle = \langle N;M'|\hat{j}_\alpha(\boldsymbol{r})|N;M\rangle^*$$
$$= (-1)^\phi [j_\alpha(\boldsymbol{r})]_{\tau\sigma}^*$$
$$= (-1)^\phi [j_\alpha(\boldsymbol{r})]_{\sigma\tau}. \tag{B.11}$$

The energy difference $E_{NM'} - E_{NM}$ is now given by

$$E_{NM} - E_{NM'} = \epsilon_\sigma - \epsilon_\tau. \tag{B.12}$$

For a given $M$ occurring in eqn (B.4), we first perform the summation over $M'$, which amounts to a summation over all *occupied* states $\sigma$ and *unoccupied* states $\tau$ occurring in the given $|N;M\rangle$. The summation over $M$ is then carried out, and expressed as a summation over all of the occupation numbers $n_1, n_2, \ldots$ that define a configuration. We thus find, for the quantity $C_{\alpha\beta}(\boldsymbol{r},\boldsymbol{r}',\tau)$ of eqn (B.4),

$$C_{\alpha\beta}(\boldsymbol{r},\boldsymbol{r}',\tau) = \sum_N \sideset{}{'}\sum_{n_1,n_2,\ldots} \sum_{\sigma \neq \tau} n_\sigma(1-n_\tau) F(\sigma,\tau)$$
$$\times \Big[ P(N;n_1,n_2,\ldots,n_\sigma,\ldots,n_\tau,\ldots)$$
$$- P(N;n_1,n_2,\ldots,n_\sigma-1,\ldots,n_\tau+1,\ldots)\Big], \tag{B.13}$$

where we have defined

$$F(\sigma, \tau) = e^{(i/\hbar)(\epsilon_\sigma - \epsilon_\tau)\tau} [j_\alpha(\boldsymbol{r})]_{\sigma\tau} [j_\beta(\boldsymbol{r}')]_{\tau\sigma}. \tag{B.14}$$

The prime on the summation sign in eqn (B.13) indicates the restriction $n_1 + n_2 + \cdots = N$. The summation over the indices $\sigma$ and $\tau$ is unrestricted, due to the factor $n_\sigma(1 - n_\tau)$ that effectively restricts $\sigma$ to occupied levels and $\tau$ to unoccupied levels in the state $|N; M\rangle$. We can remove the sum over $N$ and, at the same time, make the sum over the $n_1, n_2, \ldots$ unrestricted, and write

$$C_{\alpha\beta}(\boldsymbol{r}, \boldsymbol{r}', \tau) = \sum_{n_1, n_2, \ldots} \sum_{\sigma \neq \tau} n_\sigma(1 - n_\tau) F(\sigma, \tau)$$
$$\times \Big[ P(n_1, n_2, \ldots, n_\sigma, \ldots, n_\tau, \ldots)$$
$$- P(n_1, n_2, \ldots, n_\sigma - 1, \ldots, n_\tau + 1, \ldots) \Big]. \tag{B.15}$$

The grand canonical weight of the configuration $n_1, n_2, \ldots$ is

$$P(n_1, n_2, \ldots) = \frac{e^{-\beta \sum_\rho n_\rho(\epsilon_\rho - \mu)}}{\mathcal{Z}(\beta, \mu)}, \tag{B.16}$$

where $\beta = 1/k_B T$ is the inverse temperature and $\mu$ (not to be confused with the single-particle state index $\mu$) is the chemical potential; $\mathcal{Z}(\beta, \mu)$ is the grand partition function

$$\mathcal{Z}(\beta, \mu) = \sum_{n_1, n_2, \ldots} e^{-\beta \sum_\rho n_\rho(\epsilon_\rho - \mu)}. \tag{B.17}$$

We then write the terms in the square brackets in eqn (B.15) as

$$P(n_1, n_2, \ldots, n_\sigma, \ldots, n_\tau, \ldots)$$
$$- P(n_1, n_2, \ldots, n_\sigma - 1, \ldots, n_\tau + 1, \ldots)$$
$$= \frac{1}{\mathcal{Z}(\beta, \mu)} \Big[ e^{-\beta n_1(\epsilon_1 - \mu)} \ldots e^{-\beta n_\sigma(\epsilon_\sigma - \mu)} \ldots e^{-\beta n_\tau(\epsilon_\tau - \mu)} \ldots$$
$$- e^{-\beta n_1(\epsilon_1 - \mu)} \ldots e^{-\beta(n_\sigma - 1)(\epsilon_\sigma - \mu)} \ldots e^{-\beta(n_\tau + 1)(\epsilon_\tau - \mu)} \ldots \Big]$$
$$= \frac{e^{-\beta \sum_\rho n_\rho(\epsilon_\rho - \mu)}}{\mathcal{Z}(\beta, \mu)} \Big[ 1 - e^{\beta(\epsilon_\sigma - \mu)} e^{-\beta(\epsilon_\tau - \mu)} \Big]. \tag{B.18}$$

We then write $C_{\alpha\beta}(\boldsymbol{r}, \boldsymbol{r}', \tau)$ of eqn (B.15) as

$$C_{\alpha\beta}(\boldsymbol{r}, \boldsymbol{r}', \tau) = \sum_{n_1, n_2, n_3, \ldots} \frac{e^{-\beta \sum_\rho n_\rho(\epsilon_\rho - \mu)}}{\mathcal{Z}(\beta, \mu)} \sum_{\sigma \neq \tau = 1}^{\infty} n_\sigma(1 - n_\tau) \Big[ 1 - e^{\beta(\epsilon_\sigma - \epsilon_\tau)} \Big] F(\sigma, \tau)$$

$$= \sum_{n_1,n_2,n_3,\ldots} \frac{e^{-\beta n_1(\epsilon_1-\mu)}}{1+e^{-\beta(\epsilon_1-\mu)}} \frac{e^{-\beta n_2(\epsilon_2-\mu)}}{1+e^{-\beta(\epsilon_2-\mu)}} \frac{e^{-\beta n_3(\epsilon_3-\mu)}}{1+e^{-\beta(\epsilon_3-\mu)}} \cdots$$

$$\times \Big\{ n_1(1-n_2)\left[1-e^{\beta(\epsilon_1-\epsilon_2)}\right]F(1,2) + [1 \leftrightarrow 2]$$

$$+ n_1(1-n_3)\left[1-e^{\beta(\epsilon_1-\epsilon_3)}\right]F(1,3) + [1 \leftrightarrow 3] + \cdots$$

$$+ n_2(1-n_3)\left[1-e^{\beta(\epsilon_2-\epsilon_3)}\right]F(2,3) + [2 \leftrightarrow 3] + \cdots$$

$$+ \cdots \Big\}. \tag{B.19}$$

A term indicated by $[\sigma \leftrightarrow \tau]$ means the term occurring immediately before it but with the indices $\sigma$ and $\tau$ interchanged. In the terms of this last equation containing the pair $1,2$, the sum over $n_3, n_4, \ldots$ gives 1; in the terms containing the pair $1,3$, the sum over $n_2, n_4, \ldots$ gives 1; etc. Thus

$$C_{\alpha\beta}(\mathbf{r},\mathbf{r}',\tau)$$

$$= \sum_{n_1,n_2} \frac{e^{-\beta n_1(\epsilon_1-\mu)}}{1+e^{-\beta(\epsilon_1-\mu)}} \frac{e^{-\beta n_2(\epsilon_2-\mu)}}{1+e^{-\beta(\epsilon_2-\mu)}} n_1(1-n_2)\left[1-e^{\beta(\epsilon_1-\epsilon_2)}\right]F(1,2)$$

$$+ [1 \leftrightarrow 2]$$

$$+ \sum_{n_1,n_3} \frac{e^{-\beta n_1(\epsilon_1-\mu)}}{1+e^{-\beta(\epsilon_1-\mu)}} \frac{e^{-\beta n_3(\epsilon_3-\mu)}}{1+e^{-\beta(\epsilon_3-\mu)}} n_1(1-n_3)\left[1-e^{\beta(\epsilon_1-\epsilon_3)}\right]F(1,3)$$

$$+ [1 \leftrightarrow 3] + \cdots$$

$$+ \sum_{n_2,n_3} \frac{e^{-\beta n_2(\epsilon_2-\mu)}}{1+e^{-\beta(\epsilon_2-\mu)}} \frac{e^{-\beta n_3(\epsilon_3-\mu)}}{1+e^{-\beta(\epsilon_3-\mu)}} n_2(1-n_3)\left[1-e^{\beta(\epsilon_1-\epsilon_3)}\right]F(2,3)$$

$$+ [2 \leftrightarrow 3] + \cdots$$

$$+ \cdots$$

$$= \sum_{\rho \neq \sigma} \sum_{n_\rho, n_\sigma} \frac{e^{-\beta n_\rho(\epsilon_\rho-\mu)}}{1+e^{-\beta(\epsilon_\rho-\mu)}} \frac{e^{-\beta n_\sigma(\epsilon_\sigma-\mu)}}{1+e^{-\beta(\epsilon_\sigma-\mu)}} n_\rho(1-n_\sigma)\left[1-e^{\beta(\epsilon_\rho-\epsilon_\sigma)}\right]F(\rho,\sigma)$$

$$= \sum_{\rho \neq \sigma = 1}^{\infty} \langle n_\rho \rangle \left[1 - \langle n_\sigma \rangle\right] \left[1 - e^{\beta(\epsilon_\rho-\epsilon_\sigma)}\right] F(\rho,\sigma). \tag{B.20}$$

The average $\langle n_\sigma \rangle$, also known as the *Fermi function* $f(\epsilon_\sigma)$, is given by

$$\langle n_\sigma \rangle = f(\epsilon_\sigma) = \frac{1}{1+e^{\beta(\epsilon_\sigma-\mu)}}. \tag{B.21}$$

We thus have

$$C_{\alpha\beta}(\boldsymbol{r},\boldsymbol{r}',\tau) = \sum_{\rho,\sigma=1}^{\infty} \frac{1}{1+e^{\beta(\epsilon_\rho-\mu)}} \frac{e^{\beta(\epsilon_\sigma-\mu)}}{1+e^{\beta(\epsilon_\sigma-\mu)}} \left[1 - e^{\beta(\epsilon_\rho-\epsilon_\sigma)}\right] F(\rho,\sigma)$$

$$= \sum_{\rho,\sigma=1}^{\infty} \frac{e^{\beta(\epsilon_\sigma-\mu)} - e^{\beta(\epsilon_\rho-\mu)}}{\left[1+e^{\beta(\epsilon_\rho-\mu)}\right]\left[1+e^{\beta(\epsilon_\sigma-\mu)}\right]} F(\rho,\sigma)$$

$$= \sum_{\rho,\sigma=1}^{\infty} \left[\frac{1}{1+e^{\beta(\epsilon_\rho-\mu)}} - \frac{1}{1+e^{\beta(\epsilon_\sigma-\mu)}}\right] F(\rho,\sigma). \quad (B.22)$$

Recalling the definition of $F(\rho,\sigma)$, eqn (B.14), we thus find

$$C_{\alpha\beta}(\boldsymbol{r},\boldsymbol{r}',\tau) = \sum_{\rho,\sigma=1}^{\infty} [f(\epsilon_\rho) - f(\epsilon_\sigma)] e^{(i/\hbar)(\epsilon_\rho-\epsilon_\sigma)\tau} \left\langle \rho \left| \hat{j}_\alpha^{(1)}(\boldsymbol{r}) \right| \sigma \right\rangle \left\langle \sigma \left| \hat{j}_\beta^{(1)}(\boldsymbol{r}') \right| \rho \right\rangle. \quad (B.23)$$

We now recall the definition (B.1) of $C_{\alpha\beta}(\boldsymbol{r},\boldsymbol{r}',\tau)$ and introduce it into eqn (4.124) for the conductivity tensor to obtain

$$\sigma_{\alpha\beta}^\omega(\boldsymbol{r},\boldsymbol{r}') = \frac{i}{\omega} \lim_{\eta\to 0^+} \sum_{\rho,\sigma=1}^{\infty} \frac{f(\epsilon_\rho) - f(\epsilon_\sigma)}{\hbar\omega + \epsilon_\rho - \epsilon_\sigma + i\hbar\eta} \left\langle \rho \left| \hat{j}_\alpha^{(1)}(\boldsymbol{r}) \right| \sigma \right\rangle \left\langle \sigma \left| \hat{j}_\beta^{(1)}(\boldsymbol{r}') \right| \rho \right\rangle$$
$$+ i\frac{e^2 n_{\text{sc}}(\boldsymbol{r})}{m\omega} \delta(\boldsymbol{r}-\boldsymbol{r}') \delta_{\alpha\beta}. \quad (B.24)$$

From eqn (B.24) it is clear that, for $\hbar\omega$ lying outside of the energy band defined by the range of $\epsilon_\rho - \epsilon_\sigma$, the denominator never vanishes and $\eta \to 0$ is ineffective; as reminded earlier, however, this is not usually the case of interest.

In terms of the single-particle states $\psi_\rho(\boldsymbol{r})$, we can write expression (B.24) as

$$\sigma_{\alpha\beta}^\omega(\boldsymbol{r},\boldsymbol{r}') = \frac{i}{\omega} \lim_{\eta\to 0^+} \sum_{\rho,\sigma=1}^{\infty} \iint d\epsilon\, d\epsilon' \iiiint d^3\boldsymbol{r}_1\, d^3\boldsymbol{r}_2\, d^3\boldsymbol{r}_3\, d^3\boldsymbol{r}_4$$
$$\times \delta(\epsilon-\epsilon_\rho)\delta(\epsilon'-\epsilon_\sigma) \frac{f(\epsilon) - f(\epsilon')}{\hbar\omega + \epsilon - \epsilon' + i\hbar\eta}$$
$$\times \psi_\rho^*(\boldsymbol{r}_1) \left\langle \boldsymbol{r}_1 \left| \hat{j}_\alpha^{(1)}(\boldsymbol{r}) \right| \boldsymbol{r}_2 \right\rangle \psi_\sigma(\boldsymbol{r}_2) \psi_\sigma^*(\boldsymbol{r}_3) \left\langle \boldsymbol{r}_3 \left| \hat{j}_\beta^{(1)}(\boldsymbol{r}') \right| \boldsymbol{r}_4 \right\rangle \psi_\rho(\boldsymbol{r}_4)$$
$$+ i\frac{e^2 n_{\text{sc}}(\boldsymbol{r})}{m\omega} \delta(\boldsymbol{r}-\boldsymbol{r}') \delta_{\alpha\beta}. \quad (B.25)$$

We now show that this last expression can be written in terms of the single-particle Green functions. The Green function for outgoing boundary conditions is given by

$$G^{(+)}(\epsilon;\boldsymbol{r},\boldsymbol{r}') = \lim_{\eta\to 0^+} \sum_\rho \frac{\psi_\rho(\boldsymbol{r})\psi_\rho^*(\boldsymbol{r}')}{\epsilon - \epsilon_\rho + i\eta}. \quad (B.26)$$

We recall again that for the system of interest, which extends from $x = -\infty$ to $x = +\infty$, the spectrum associated with the states $\rho$ forms a continuum.

The summation sign, instead of an integral, is used in eqn (B.26) and previous equations only for simplicity of notation. From eqn (B.26) we construct the combination

$$G^{(+)}(\epsilon;\bm{r},\bm{r}') - G^{(-)}(\epsilon;\bm{r},\bm{r}')$$
$$= \lim_{\eta \to 0^+} \sum_\rho \psi_\rho(\bm{r})\psi_\rho^*(\bm{r}') \left( \frac{1}{\epsilon - \epsilon_\rho + i\eta} - \frac{1}{\epsilon - \epsilon_\rho - i\eta} \right)$$
$$= -2\pi i \sum_\rho \delta(\epsilon - \epsilon_\rho) \psi_\rho(\bm{r})\psi_\rho^*(\bm{r}'), \tag{B.27}$$

from which we define the quantity

$$\mathcal{N}(\epsilon;\bm{r},\bm{r}') \equiv -\frac{1}{2\pi i} \left\{ G^{(+)}(\epsilon;\bm{r},\bm{r}') - G^{(-)}(\epsilon;\bm{r},\bm{r}') \right\}$$
$$= \sum_\rho \delta(\epsilon - \epsilon_\rho) \psi_\rho(\bm{r})\psi_\rho^*(\bm{r}')$$
$$= \mathcal{N}^*(\epsilon;\bm{r}',\bm{r}). \tag{B.28}$$

This result is consistent with the general relation (2.66), i.e.,

$$G^{(-)}(\epsilon;\bm{r},\bm{r}') = \left[ G^{(+)}(\epsilon;\bm{r}',\bm{r}) \right]^*. \tag{B.29}$$

The quantity $\mathcal{N}(\epsilon;\bm{r},\bm{r}')$ is the matrix element

$$\mathcal{N}(\epsilon;\bm{r},\bm{r}') = \langle \bm{r} | \hat{\mathcal{N}}(\epsilon) | \bm{r}' \rangle \tag{B.30}$$

of the operator

$$\hat{\mathcal{N}}(\epsilon) = \sum_\rho \delta(\epsilon - \epsilon_\rho) |\psi_\rho\rangle \langle \psi_\rho|. \tag{B.31}$$

For a system with *time-reversal invariance* (TRI) we can always find a basis of real wave functions, so that $\mathcal{N}(\epsilon;\bm{r},\bm{r}')$ of eqn (B.28) is real and symmetric, i.e.,

$$\mathcal{N}(\epsilon;\bm{r},\bm{r}') = \mathcal{N}(\epsilon;\bm{r}',\bm{r}). \tag{B.32}$$

This can also be seen by substituting eqn (B.29) into the first line of eqn (B.28) and using the relation (2.58), valid for TRI, i.e.,

$$G^{(+)}(\epsilon;\bm{r}',\bm{r}) = G^{(+)}(\epsilon;\bm{r},\bm{r}'). \tag{B.33}$$

We can now write (B.25) using the definition (B.28) as

$$\sigma_{\alpha\beta}^\omega(\bm{r},\bm{r}') = \frac{i}{\omega} \lim_{\eta \to 0^+} \iint d\varepsilon\, d\varepsilon' \iiiint d^3 r_1\, d^3 r_2\, d^3 r_3\, d^3 r_4$$
$$\times \mathcal{N}(\epsilon;\bm{r}_4,\bm{r}_1) \mathcal{N}(\epsilon';\bm{r}_2,\bm{r}_3)$$
$$\times \frac{f(\epsilon) - f(\epsilon')}{\hbar\omega + \epsilon - \epsilon' + i\hbar\eta} \left\langle \bm{r}_1 \left| \hat{j}_\alpha^{(1)}(\bm{r}) \right| \bm{r}_2 \right\rangle \left\langle \bm{r}_3 \left| \hat{j}_\beta^{(1)}(\bm{r}') \right| \bm{r}_4 \right\rangle$$
$$+ i\frac{e^2 n_{\mathrm{sc}}(\bm{r})}{m\omega} \delta(\bm{r} - \bm{r}') \delta_{\alpha\beta}. \tag{B.34}$$

From eqn (4.10) we find the matrix element of the current operator needed above to be (recall that $e > 0$)

$$\left\langle r' \left| j_\alpha^{(1)}(r) \right| r'' \right\rangle = i\frac{e\hbar}{2m}\left[\delta(r-r')\partial_\alpha\delta(r-r'') - \delta(r-r'')\partial_\alpha\delta(r-r')\right]. \quad (B.35)$$

Equation (B.34) thus takes the form

$$\sigma_{\alpha\beta}^\omega(r,r') = -\frac{i}{\omega}\lim_{\eta\to 0^+}\iint d\epsilon\, d\epsilon'\, \frac{f(\epsilon)-f(\epsilon')}{\hbar\omega+\epsilon-\epsilon'+i\hbar\eta}\frac{e^2\hbar^2}{4m^2}$$
$$\times \iiiint d^3r_1\, d^3r_2\, d^3r_3\, d^3r_4\, \mathcal{N}(\epsilon; r_4, r_1)\mathcal{N}(\epsilon'; r_2, r_3)$$
$$\times \left[\delta(r-r_1)\partial_\alpha\delta(r-r_2) - \delta(r-r_2)\partial_\alpha\delta(r-r_1)\right]$$
$$\times \left[\delta(r'-r_3)\partial'_\beta\delta(r'-r_4) - \delta(r'-r_4)\partial'_\beta\delta(r'-r_3)\right]$$
$$+ i\frac{e^2 n_{sc}(r)}{m\omega}\delta(r-r')\delta_{\alpha\beta}. \quad (B.36)$$

Performing the spatial integrals, we find

$$\sigma_{\alpha\beta}^\omega(r,r') = -i\frac{e^2\hbar^3}{4m^2\hbar\omega}\lim_{\eta\to 0^+}\iint d\epsilon\, d\epsilon'\, \frac{f(\epsilon)-f(\epsilon')}{\hbar\omega+\epsilon-\epsilon'+i\hbar\eta}$$
$$\times \Big\{\left[\partial'_\beta\mathcal{N}(\epsilon;r',r)\right]\left[\partial_\alpha\mathcal{N}(\epsilon';r,r')\right] - \left[\mathcal{N}(\epsilon;r',r)\right]\left[\partial_\alpha\partial'_\beta\mathcal{N}(\epsilon';r,r')\right]$$
$$- \left[\partial_\alpha\partial'_\beta\mathcal{N}(\epsilon;r',r)\right]\left[\mathcal{N}(\epsilon';r,r')\right] + \left[\partial_\alpha\mathcal{N}(\epsilon;r',r)\right]\left[\partial'_\beta\mathcal{N}(\epsilon';r,r')\right]\Big\}$$
$$+ i\frac{e^2 n_{sc}(r)}{m\omega}\delta(r-r')\delta_{\alpha\beta}. \quad (B.37)$$

We are actually interested in the *real part* of the conductivity tensor. We saw above that, under the condition of TRI, $\mathcal{N}(\epsilon; r, r')$ is real. Using the identity

$$\lim_{\eta\to 0^+}\frac{1}{\hbar\omega+\epsilon-\epsilon'+i\hbar\eta} = \frac{\mathcal{P}}{\hbar\omega+\epsilon-\epsilon'} - i\pi\delta(\hbar\omega+\epsilon-\epsilon'), \quad (B.38)$$

we have, in the limit as $\eta \to 0^+$,

$$\mathrm{Re}\,\sigma^{\omega}_{\alpha\beta}(\boldsymbol{r},\boldsymbol{r}') = -\pi\frac{e^2\hbar^3}{4m^2}\int d\epsilon\,\frac{f(\epsilon)-f(\epsilon+\hbar\omega)}{\hbar\omega}$$
$$\times\Big\{\big[\partial'_\beta\mathcal{N}(\epsilon;\boldsymbol{r}',\boldsymbol{r})\big]\big[\partial_\alpha\mathcal{N}(\epsilon+\hbar\omega;\boldsymbol{r},\boldsymbol{r}')\big]$$
$$-\big[\mathcal{N}(\epsilon;\boldsymbol{r}',\boldsymbol{r})\big]\big[\partial_\alpha\partial'_\beta\mathcal{N}(\epsilon+\hbar\omega;\boldsymbol{r},\boldsymbol{r}')\big]$$
$$-\big[\partial_\alpha\partial'_\beta\mathcal{N}(\epsilon;\boldsymbol{r}',\boldsymbol{r})\big]\big[\mathcal{N}(\epsilon+\hbar\omega;\boldsymbol{r},\boldsymbol{r}')\big]$$
$$+\big[\partial_\alpha\mathcal{N}(\epsilon;\boldsymbol{r}',\boldsymbol{r})\big]\big[\partial'_\beta\mathcal{N}(\epsilon+\hbar\omega;\boldsymbol{r},\boldsymbol{r}')\big]\Big\}.$$
(B.39)

In the static limit $\omega \to 0$, the first factor in the above integrand tends to the negative of the derivative of the Fermi function; we thus have

$$\lim_{\omega\to 0}\mathrm{Re}\,\sigma^{\omega}_{\alpha\beta}(\boldsymbol{r},\boldsymbol{r}') = \pi\frac{e^2\hbar^3}{2m^2}\int d\epsilon\left(-\frac{\partial f}{\partial\epsilon}\right)\Big\{\mathcal{N}(\epsilon;\boldsymbol{r},\boldsymbol{r}')\big[\partial_\alpha\partial'_\beta\mathcal{N}(\epsilon;\boldsymbol{r},\boldsymbol{r}')\big]$$
$$-\big[\partial_\alpha\mathcal{N}(\epsilon;\boldsymbol{r},\boldsymbol{r}')\big]\big[\partial'_\beta\mathcal{N}(\epsilon;\boldsymbol{r},\boldsymbol{r}')\big]\Big\}.$$
(B.40)

In the zero-temperature limit, we find

$$\lim_{T\to 0}\lim_{\omega\to 0}\mathrm{Re}\,\sigma^{\omega}_{\alpha\beta}(\boldsymbol{r},\boldsymbol{r}') = \pi\frac{e^2\hbar^3}{2m^2}\Big\{\mathcal{N}(\epsilon_F;\boldsymbol{r},\boldsymbol{r}')\big[\partial_\alpha\partial'_\beta\mathcal{N}(\epsilon_F;\boldsymbol{r},\boldsymbol{r}')\big]$$
$$-\big[\partial_\alpha\mathcal{N}(\epsilon_F;\boldsymbol{r},\boldsymbol{r}')\big]\big[\partial'_\beta\mathcal{N}(\epsilon_F;\boldsymbol{r},\boldsymbol{r}')\big]\Big\}.$$
(B.41)

# APPENDIX C

## THE CONDUCTANCE IN TERMS OF THE TRANSMISSION COEFFICIENT OF THE SAMPLE

Figure C.1 shows the system studied in Chapter 4, consisting of a sample (a cavity, say) and the two expanding horns which, for convenience in the discussion, are assumed to have a constant cross-section (each supporting $N'$ open channels) within some interval far away from the sample, and to be scatterer-free within that interval. We wish to calculate the scattering matrix $S^{(\mathrm{tot})}$ for the full system, combining the scattering matrices $S_1$, $S$ and $S_2$ for the left contact, the sample and the right contact, respectively. The sample is assumed to go into the expanding horns through very long, scatterer-free, quasi-one-dimensional, $N$-open-channel conductors, so that, according to the discussion presented in Section 3.1.7, we ignore closed channels and combine only the standard, open channel, $S$ matrices. The amplitudes of the various incoming and outgoing waves in the regions assumed to be scatterer-free are indicated in Fig. C.1. Every amplitude represents a column vector associated with open channels only, whose number appears in parentheses next to that amplitude.

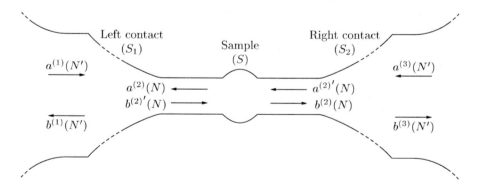

FIG. C.1. The system studied in Chapter 4, consisting of a sample and two expanding horns, which, for convenience in the discussion, are assumed to have constant cross-section within some interval far away from the sample. Shown in the figure are the amplitudes of the incoming and outgoing waves that constitute the most general wave function in the regions assumed to be scatterer-free. Every amplitude represents a column vector, with a dimension given by the number of *open* channels indicated in parentheses.

# CONDUCTANCE IN TERMS OF TRANSMISSION COEFFICIENT

We follow the same ideas as in Sections 2.1.6 and 3.1.4. The notation is also followed closely, with a few exceptions. Here, for instance, $S$ and $S^{(\text{tot})}$ indicate the scattering matrices of the sample and of the total system, respectively. The scattering matrices $S_1$, $S$ and $S_2$ associated with the left contact, the sample and the right contact, respectively, are defined to be

$$\begin{bmatrix} b^{(1)}(N') \\ b^{(2)'}(N) \end{bmatrix} = S_1 \begin{bmatrix} a^{(1)}(N') \\ a^{(2)}(N) \end{bmatrix}, \tag{C.1a}$$

$$\begin{bmatrix} a^{(2)}(N) \\ b^{(2)}(N) \end{bmatrix} = S \begin{bmatrix} b^{(2)'}(N) \\ a^{(2)'}(N) \end{bmatrix}, \tag{C.1b}$$

$$\begin{bmatrix} a^{(2)'}(N) \\ b^{(3)}(N') \end{bmatrix} = S_2 \begin{bmatrix} b^{(2)}(N) \\ a^{(3)}(N') \end{bmatrix}. \tag{C.1c}$$

These matrices have the structure (the dimensionality being indicated in parentheses)

$$S_1((N'+N)\times(N'+N)) = \begin{bmatrix} r_1 & t'_1 \\ t_1 & r'_1 \end{bmatrix}, \tag{C.2a}$$

$$S(2N\times 2N) = \begin{bmatrix} r & t' \\ t & r' \end{bmatrix}, \tag{C.2b}$$

$$S_2((N+N')\times(N+N')) = \begin{bmatrix} r_2 & t'_2 \\ t_2 & r'_2 \end{bmatrix}. \tag{C.2c}$$

It is useful to keep the various $S$ matrices in their general form, i.e., without imposing TRI. We combine eqns (C.1a) and (C.1c) as

$$\begin{bmatrix} b^{(1)}(N') \\ b^{(2)'}(N) \\ a^{(2)'}(N) \\ b^{(3)}(N') \end{bmatrix} = S_{12}^0 \begin{bmatrix} a^{(1)}(N') \\ a^{(2)}(N) \\ b^{(2)}(N) \\ a^{(3)}(N') \end{bmatrix}, \tag{C.3}$$

where

$$S_{12}^0 = \begin{bmatrix} S_1 & 0 \\ 0 & S_2 \end{bmatrix} = \begin{bmatrix} r_1 & t'_1 & & 0 \\ t_1 & r'_1 & & \\ & & r_2 & t'_2 \\ 0 & & t_2 & r'_2 \end{bmatrix}. \tag{C.4}$$

We reorder the rows and columns of this last matrix and define

$$S_{12} \equiv \begin{bmatrix} r_1 & 0 & t'_1 & 0 \\ 0 & r'_2 & 0 & t_2 \\ t_1 & 0 & r'_1 & 0 \\ 0 & t'_2 & 0 & r_2 \end{bmatrix} \equiv \begin{bmatrix} S_{12}^{PP} & S_{12}^{PQ} \\ S_{12}^{QP} & S_{12}^{QQ} \end{bmatrix}. \tag{C.5}$$

This matrix has the property

$$\begin{bmatrix} b^{(1)}(N') \\ b^{(3)}(N') \\ b^{(2)'}(N) \\ a^{(2)'}(N) \end{bmatrix} = \begin{bmatrix} r_1 & 0 & t'_1 & 0 \\ 0 & r'_2 & 0 & t_2 \\ t_1 & 0 & r'_1 & 0 \\ 0 & t'_2 & 0 & r_2 \end{bmatrix} \begin{bmatrix} a^{(1)}(N') \\ a^{(3)}(N') \\ a^{(2)}(N) \\ b^{(2)}(N) \end{bmatrix}. \quad\text{(C.6)}$$

Just as in Sections 2.1.6 and 3.1.4, $P$ projects unto the 'external' regions (far left and far right regions with constant cross-section in Fig. C.1) and $Q$ unto the 'internal' regions (the scatterer-free, quasi-one-dimensional regions in Fig. C.1 that connect the sample to the horns). Thus $S_{12}^{PP}$ is the submatrix that connects the external region to itself, $S_{12}^{PQ}$ connects the internal region to the external region, $S_{12}^{QP}$ connects the external region to the internal region, and $S_{12}^{QQ}$ connects the internal region to itself. Equation (C.6) can be rewritten as

$$\begin{bmatrix} b^P(2N') \\ S^{-1}c^Q(2N) \end{bmatrix} = \begin{bmatrix} S_{12}^{PP} & S_{12}^{PQ} \\ S_{12}^{QP} & S_{12}^{QQ} \end{bmatrix} \begin{bmatrix} a^P(2N') \\ c^Q(2N) \end{bmatrix}. \quad\text{(C.7)}$$

We have used eqn (C.1b) that defines the matrix $S$, and we have defined

$$a^P(2N') = \begin{bmatrix} a^{(1)}(N') \\ a^{(3)}(N') \end{bmatrix},$$

$$b^P(2N') = \begin{bmatrix} b^{(1)}(N') \\ b^{(3)}(N') \end{bmatrix}, \quad\text{(C.8)}$$

$$c^Q(2N) = \begin{bmatrix} a^{(2)}(N) \\ b^{(2)}(N) \end{bmatrix}.$$

From eqn (C.7) we obtain the pair of coupled equations

$$b^P = S_{12}^{PP} a^P + S_{12}^{PQ} c^Q, \quad\text{(C.9a)}$$
$$S^{-1} c^Q = S_{12}^{QP} a^P + S_{12}^{QQ} c^Q. \quad\text{(C.9b)}$$

Eliminating $c^Q$ from this pair of equations, we obtain

$$b^P = S_{12}^{PP} a^P + S_{12}^{PQ} \frac{1}{S^{-1} - S_{12}^{QQ}} S_{12}^{QP} a^P. \quad\text{(C.10)}$$

The resulting matrix $S^{(\text{tot})}$ satisfies, by definition, the relation

$$b^P = S^{(\text{tot})} a^P, \quad\text{(C.11)}$$

and is thus given by

$$S^{(\text{tot})} = S_{12}^{PP} + S_{12}^{PQ} \frac{1}{S^{-1} - S_{12}^{QQ}} S_{12}^{QP}. \quad\text{(C.12)}$$

Now suppose that the sample goes into the horns following a very smooth profile, so that waves going from the sample into the horns are almost not reflected, i.e.,

$$r'_1 = r_2 = 0. \tag{C.13}$$

The above 'adiabaticity' will hold if $N' \gg N$. (On the other hand, under the same conditions, $r_1$ and $r'_2$ are not required to vanish.) This implies that $S_{12}^{QQ} = 0$ and hence

$$S^{(\text{tot})} = S_{12}^{PP} + S_{12}^{PQ} S S_{12}^{QP}, \tag{C.14}$$

or

$$\begin{bmatrix} r^{(\text{tot})}(N' \times N') & t'^{(\text{tot})}(N' \times N) \\ t^{(\text{tot})}(N \times N') & r'^{(\text{tot})}(N \times N) \end{bmatrix} = \begin{bmatrix} r_1 & 0 \\ 0 & r'_2 \end{bmatrix} + \begin{bmatrix} t'_1 & 0 \\ 0 & t_2 \end{bmatrix} \begin{bmatrix} r & t' \\ t & r' \end{bmatrix} \begin{bmatrix} t_1 & 0 \\ 0 & t'_2 \end{bmatrix}$$

$$= \begin{bmatrix} r_1 + t'_1 r t_1 & t'_1 t' t'_2 \\ t_2 t t_1 & r'_2 + t_2 r' t'_2 \end{bmatrix}. \tag{C.15}$$

The trace needed in eqn (4.181) is thus

$$\text{Tr}\left[t^{(\text{tot})} t^{(\text{tot})\dagger}\right] = \text{Tr}\left[t_2 t t_1 t_1^\dagger t^\dagger t_2^\dagger\right]$$

$$= \text{Tr}\left[\left(t_2^\dagger t_2\right) t \left(t_1 t_1^\dagger\right) t^\dagger\right]. \tag{C.16}$$

Using the conditions (C.13), unitarity of $S_1$ and $S_2$ implies that

$$S_1 S_1^\dagger = \begin{bmatrix} r_1 & t'_1 \\ t_1 & 0 \end{bmatrix} \begin{bmatrix} r_1^\dagger & t_1^\dagger \\ t'_1{}^\dagger & 0 \end{bmatrix}$$

$$= \begin{bmatrix} r_1 r_1^\dagger + t'_1 t'_1{}^\dagger & r_1 t_1^\dagger \\ t_1 r_1^\dagger & t_1 t_1^\dagger \end{bmatrix} = \begin{bmatrix} I(N' \times N') & 0(N' \times N) \\ 0(N \times N') & I(N \times N) \end{bmatrix} \tag{C.17}$$

and

$$S_2^\dagger S_2 = \begin{bmatrix} 0 & t_2^\dagger \\ t'_2{}^\dagger & r'_2{}^\dagger \end{bmatrix} \begin{bmatrix} 0 & t'_2 \\ t_2 & r'_2 \end{bmatrix}$$

$$= \begin{bmatrix} t_2^\dagger t_2 & t_2^\dagger r'_2 \\ r'_2{}^\dagger t_2 & t'_2{}^\dagger t'_2 + r'_2{}^\dagger r'_2 \end{bmatrix} = \begin{bmatrix} I(N \times N) & 0(N \times N') \\ 0(N' \times N) & I(N' \times N') \end{bmatrix}. \tag{C.18}$$

We have thus found that

$$t_1 t_1^\dagger = t_2^\dagger t_2 = I(N \times N), \tag{C.19}$$

so that eqn (C.16) gives

$$\text{Tr}\left[t^{(\text{tot})} t^{(\text{tot})\dagger}\right] = \text{Tr}\left[t t^\dagger\right], \tag{C.20}$$

which is the relation needed to obtain eqn (4.182) in the text.

# APPENDIX D

## EVALUATION OF THE INVARIANT MEASURE

We evaluate $\mathrm{d}S$ in the polar representation (3.86) and then the arc element $\mathrm{d}s^2$ of eqn (6.5). We keep $V$ and $W$ independent, as is the case for $\beta = 2$. We then use this same algebraic development and set $W = V^\top$ in the proper place to analyze the $\beta = 1$ case. Differentiating $S$ of eqn (3.86), we obtain

$$\begin{aligned}\mathrm{d}S &= (\mathrm{d}V)RW + V(\mathrm{d}R)W + VR(\mathrm{d}W) \\ &= V\left[(\delta V)R + \mathrm{d}R + R(\delta W)\right]W,\end{aligned} \tag{D.1}$$

where we have defined the matrices

$$\delta V = V^\dagger \mathrm{d}V, \qquad \delta W = (\mathrm{d}W)W^\dagger, \tag{D.2}$$

which are *anti-Hermitian*, as can be seen by differentiating the unitarity relations $V^\dagger V = I$ and $WW^\dagger = I$. The arc element of eqn (6.5) is thus

$$\begin{aligned}\mathrm{d}s^2 &= \mathrm{Tr}\left\{[R^\top(\delta V)^\dagger + \mathrm{d}R^\top + (\delta W)^\dagger R^\top] \cdot [(\delta V)R + \mathrm{d}R + R\delta W]\right\} \\ &= \mathrm{Tr}\left[R^\top(\delta V)^\dagger(\delta V)R + R^\top(\delta V)^\dagger \mathrm{d}R + R^\top(\delta V)^\dagger R(\delta W)\right. \\ &\quad + (\mathrm{d}R^\top)(\delta V)R + (\mathrm{d}R^\top)\mathrm{d}R + (\mathrm{d}R^\top)R(\delta W) \\ &\quad \left. + (\delta W)^\dagger R^\top(\delta V)R + (\delta W)^\dagger R^\top \mathrm{d}R + (\delta W)^\dagger R^\top R(\delta W)\right]. \end{aligned} \tag{D.3}$$

We have used $RR^\top = I$, $R$ being real and orthogonal. The various differentials occurring in the previous equation can be expressed as

$$\begin{aligned}\mathrm{d}R &= \begin{bmatrix} -\mathrm{d}\sqrt{\rho} & \mathrm{d}\sqrt{\tau} \\ \mathrm{d}\sqrt{\tau} & \mathrm{d}\sqrt{\rho} \end{bmatrix} \\ &= \frac{1}{2}\begin{bmatrix} \mathrm{d}\tau/\sqrt{\rho} & \mathrm{d}\tau/\sqrt{\tau} \\ \mathrm{d}\tau/\sqrt{\tau} & -\mathrm{d}\tau/\sqrt{\rho} \end{bmatrix},\end{aligned} \tag{D.4}$$

$$\delta V = \begin{bmatrix} v^{(1)\dagger}\mathrm{d}v^{(1)} & 0 \\ 0 & v^{(2)\dagger}\mathrm{d}v^{(2)} \end{bmatrix} = \begin{bmatrix} \delta v^{(1)} & 0 \\ 0 & \delta v^{(2)} \end{bmatrix}, \tag{D.5}$$

$$\delta W = \begin{bmatrix} (\mathrm{d}v^{(3)})v^{(3)\dagger} & 0 \\ 0 & (\mathrm{d}v^{(4)})v^{(4)\dagger} \end{bmatrix} = \begin{bmatrix} \delta v^{(3)} & 0 \\ 0 & \delta v^{(4)} \end{bmatrix}, \tag{D.6}$$

where we have used the abbreviation

$$\rho = 1 - \tau. \tag{D.7}$$

We now calculate the various terms in (D.3). The first term plus the ninth term gives

$$\mathrm{Tr}\left[(\delta V)^\dagger(\delta V) + (\delta W)^\dagger(\delta W)\right] = \mathrm{tr}\sum_{i=1}^{4}\left(\delta v^{(i)}\right)^\dagger \delta v^{(i)}. \tag{D.8}$$

The notation Tr and tr indicates the trace of $2N$- and $N$-dimensional matrices, respectively. The second term plus the fourth term in (D.3) gives

$$\mathrm{Tr}\left[R^T(\delta V)^\dagger \mathrm{d}R + (\mathrm{d}R^T)(\delta V)R\right]$$
$$= \mathrm{tr}\left[\sqrt{\rho}\left(\delta v^{(1)}\right)^\dagger \mathrm{d}\sqrt{\rho} + \sqrt{\tau}\left(\delta v^{(2)}\right)^\dagger \mathrm{d}\sqrt{\tau}\right.$$
$$\left. + \sqrt{\tau}\left(\delta v^{(1)}\right)^\dagger \mathrm{d}\sqrt{\tau} + \sqrt{\rho}\left(\delta v^{(2)}\right)^\dagger \mathrm{d}\sqrt{\rho} + \mathrm{H.\,c.}\right]$$
$$= \frac{1}{2}\mathrm{tr}\left[-\left(\delta v^{(1)}\right)^\dagger \mathrm{d}\tau + \left(\delta v^{(2)}\right)^\dagger \mathrm{d}\tau\right.$$
$$\left. + \left(\delta v^{(1)}\right)^\dagger \mathrm{d}\tau - \left(\delta v^{(2)}\right)^\dagger \mathrm{d}\tau + \mathrm{H.\,c.}\right] = 0, \tag{D.9}$$

where 'H. c.' stands for Hermitian conjugate. For the third term plus the seventh term in (D.3), we find

$$\mathrm{Tr}\left[R^T(\delta V)^\dagger R(\delta W) + (\delta W)^\dagger R^T(\delta V)R\right]$$
$$= 2\,\mathrm{tr}\left[\sqrt{\rho}\left(\delta v^{(1)}\right)^\dagger \sqrt{\rho}\delta v^{(3)} + \sqrt{\tau}\left(\delta v^{(2)}\right)^\dagger \sqrt{\tau}\delta v^{(3)}\right.$$
$$\left. + \sqrt{\tau}\left(\delta v^{(1)}\right)^\dagger \sqrt{\tau}\delta v^{(4)} + \sqrt{\rho}\left(\delta v^{(2)}\right)^\dagger \sqrt{\rho}\delta v^{(4)}\right]. \tag{D.10}$$

The two terms on the left-hand side of this last equation are equal, due to the anti-Hermiticity of $\delta V$ and $\delta W$. The fifth term in (D.3) gives

$$\mathrm{Tr}\,(\mathrm{d}R^T)(\mathrm{d}R) = \frac{1}{2}\sum_a \frac{(\mathrm{d}\tau_a)^2}{\tau_a \rho_a}. \tag{D.11}$$

Finally, the sixth term plus the eighth term in (D.3) gives

$$\mathrm{Tr}\left[(\mathrm{d}R^T)R(\delta W) + \mathrm{H.\,c.}\right]$$
$$= \frac{1}{2}\mathrm{tr}\left[-(\mathrm{d}\tau)\delta v^{(3)} + (\mathrm{d}\tau)\delta v^{(3)} + (\mathrm{d}\tau)\delta v^{(4)} - (\mathrm{d}\tau)\delta v^{(4)} + \mathrm{H.\,c.}\right] = 0. \tag{D.12}$$

Substituting these expressions into (D.3), we find

$$\mathrm{d}s^2 = 2\,\mathrm{tr}\left\{\frac{1}{2}\sum_{i=1}^{4}\left(\delta v^{(i)}\right)^{\dagger}\left(\delta v^{(i)}\right) + \frac{(\mathrm{d}\tau)(\mathrm{d}\tau)}{4\tau\rho} + \sqrt{\rho}\left(\delta v^{(1)}\right)^{\dagger}\sqrt{\rho}\delta v^{(3)}\right.$$
$$+ \sqrt{\tau}\left(\delta v^{(2)}\right)^{\dagger}\sqrt{\tau}\delta v^{(3)} + \sqrt{\tau}\left(\delta v^{(1)}\right)^{\dagger}\sqrt{\tau}\delta v^{(4)}$$
$$\left.+ \sqrt{\rho}\left(\delta v^{(2)}\right)^{\dagger}\sqrt{\rho}\delta v^{(4)}\right\}. \tag{D.13}$$

The anti-Hermitian matrix $\delta v^{(i)}$ can be expressed as

$$\delta v^{(i)} = \delta a^{(i)} + i\delta s^{(i)}, \tag{D.14}$$

where $\delta a^{(i)}$ is real and anti-symmetric, and $\delta s^{(i)}$ is real and symmetric. Substituting (D.14) into the expression for $\mathrm{d}s^2$ and rearranging terms, we find

$$\mathrm{d}s^2 = \sum_{a}\left\{\sum_{i=1}^{4}\left(\delta s_{aa}^{(i)}\right)^2 + 2\rho_a\left[\delta s_{aa}^{(1)}\delta s_{aa}^{(3)} + \delta s_{aa}^{(2)}\delta s_{aa}^{(4)}\right]\right.$$
$$\left.+ 2\tau_a\left[\delta s_{aa}^{(2)}\delta s_{aa}^{(3)} + \delta s_{aa}^{(1)}\delta s_{aa}^{(4)}\right] + \frac{(\mathrm{d}\tau_a)^2}{2\tau_a\rho_a}\right\}$$
$$+ 2\sum_{a<b}\left\{\sum_{i=1}^{4}\left(\delta a_{ab}^{(i)}\right)^2 + \sum_{i=1}^{4}\left(\delta s_{ab}^{(i)}\right)^2\right.$$
$$+ 2\sqrt{\rho_a\rho_b}\left[\delta a_{ab}^{(1)}\delta a_{ab}^{(3)} + \delta a_{ab}^{(2)}\delta a_{ab}^{(4)} + \delta s_{ab}^{(1)}\delta s_{ab}^{(3)} + \delta s_{ab}^{(2)}\delta s_{ab}^{(4)}\right]$$
$$\left.+ 2\sqrt{\tau_a\tau_b}\left[\delta a_{ab}^{(2)}\delta a_{ab}^{(3)} + \delta a_{ab}^{(1)}\delta a_{ab}^{(4)} + \delta s_{ab}^{(2)}\delta s_{ab}^{(3)} + \delta s_{ab}^{(1)}\delta s_{ab}^{(4)}\right]\right\}. \tag{D.15}$$

## D.1  The orthogonal case, $\beta = 1$

In this case, $v^{(3)} = v^{(1)\top}$ and $v^{(4)} = v^{(2)\top}$, so that $\delta v^{(3)} = \left(\delta v^{(1)}\right)^{\top}$ and $\delta v^{(4)} = \left(\delta v^{(2)}\right)^{\top}$, and hence

$$\delta a^{(3)} = -\delta a^{(1)}, \quad \delta s^{(3)} = \delta s^{(1)}, \quad \delta a^{(4)} = -\delta a^{(2)}, \quad \delta s^{(4)} = \delta s^{(2)}. \tag{D.16}$$

Substituting into (D.15), we have

$$\mathrm{d}s^2 = 2\sum_a \left\{ \left(\delta s_{aa}^{(1)}\right)^2 + \left(\delta s_{aa}^{(2)}\right)^2 + 2\tau_a \delta s_{aa}^{(1)} \delta s_{aa}^{(2)} \right.$$
$$\left. + \rho_a \left[ \left(\delta s_{aa}^{(1)}\right)^2 + \left(\delta s_{aa}^{(2)}\right)^2 \right] + \frac{(\mathrm{d}\tau_a)^2}{4\tau_a(1-\tau_a)} \right\}$$
$$+ 4\sum_{a<b} \left\{ \left(\delta a_{ab}^{(1)}\right)^2 + \left(\delta a_{ab}^{(2)}\right)^2 + \left(\delta s_{ab}^{(1)}\right)^2 + \left(\delta s_{ab}^{(2)}\right)^2 \right.$$
$$+ 2\sqrt{\tau_a \tau_b} \left[ \delta s_{ab}^{(1)} \delta s_{ab}^{(2)} - \delta a_{ab}^{(1)} \delta a_{ab}^{(2)} \right]$$
$$\left. + \sqrt{\rho_a \rho_b} \left[ \left(\delta s_{ab}^{(1)}\right)^2 + \left(\delta s_{ab}^{(2)}\right)^2 - \left(\delta a_{ab}^{(1)}\right)^2 - \left(\delta a_{ab}^{(2)}\right)^2 \right] \right\}, \quad (\text{D}.17)$$

or

$$\frac{1}{2}\mathrm{d}s^2 = \sum_a \left\{ (1+\rho_a)\left[\left(\delta s_{aa}^{(1)}\right)^2 + \left(\delta s_{aa}^{(2)}\right)^2\right] + 2\tau_a \delta s_{aa}^{(1)} \delta s_{aa}^{(2)} + \frac{(\mathrm{d}\tau_a)^2}{4\tau_a \rho_a} \right\}$$
$$+ 2\sum_{a<b}\left\{ (1+\sqrt{\rho_a \rho_b})\left[\left(\delta s_{ab}^{(1)}\right)^2 + \left(\delta s_{ab}^{(2)}\right)^2\right] \right.$$
$$+ (1-\sqrt{\rho_a \rho_b})\left[\left(\delta a_{ab}^{(1)}\right)^2 + \left(\delta a_{ab}^{(2)}\right)^2\right]$$
$$\left. + 2\sqrt{\tau_a \tau_b}\left[\delta s_{ab}^{(1)} \delta s_{ab}^{(2)} - \delta a_{ab}^{(1)} \delta a_{ab}^{(2)}\right] \right\}. \quad (\text{D}.18)$$

In eqn (D.18), the $\delta s_{aa}^{(1)}$, $\delta s_{aa}^{(2)}$ and $\tau_a$, $a = 1,\ldots,N$, each contribute $N$ independent variations, while $\delta s_{ab}^{(1)}$, $\delta s_{ab}^{(2)}$, $\delta a_{ab}^{(1)}$ and $\delta a_{ab}^{(2)}$ ($a < b$) contribute $N(N-1)/2$ variations each, giving a total of

$$\nu = 3N + 4\frac{N(N-1)}{2} = 2N^2 + N, \quad (\text{D}.19)$$

which is the correct number of independent parameters for a $2N$-dimensional unitary symmetric matrix.

From eqn (D.18), the metric tensor defined in eqn (6.3) has a simple block structure, consisting of $1 \times 1$ and $2 \times 2$ blocks along the diagonal, as follows. There are $N$ two-dimensional blocks with rows and columns labeled $\delta s_{aa}^{(1)}$ and $\delta s_{aa}^{(2)}$ as follows:

$$\begin{array}{c} \phantom{\delta s_{aa}^{(1)}} \quad \delta s_{aa}^{(1)} \quad \delta s_{aa}^{(2)} \\ \begin{array}{c} \delta s_{aa}^{(1)} \\ \delta s_{aa}^{(2)} \end{array} \left[ \begin{array}{cc} 1+\rho_a & \tau_a \\ \tau_a & 1+\rho_a \end{array} \right], \end{array} \quad (\text{D}.20)$$

and $N$ one-dimensional blocks $1/4\rho_a \tau_a$ labeled by $\mathrm{d}\tau_a$, giving, altogether, the contribution

$$\prod_{a=1}^{N} \frac{(1+\rho_a)^2 - \tau_a^2}{4\rho_a \tau_a} = \prod_{a=1}^{N} \frac{1}{\tau_a} \tag{D.21}$$

to $\det g$. There are $N(N-1)/2$ two-dimensional blocks with rows and columns labeled $\delta s_{ab}^{(1)}$ and $\delta s_{ab}^{(2)}$, i.e.,

$$\begin{array}{c} \\ \delta s_{ab}^{(1)} \\ \delta s_{ab}^{(2)} \end{array} \begin{array}{cc} \delta s_{ab}^{(1)} & \delta s_{ab}^{(2)} \\ \left[ \begin{array}{cc} 1 + \sqrt{\rho_a \rho_b} & \sqrt{\tau_a \tau_b} \\ \sqrt{\tau_a \tau_b} & 1 + \sqrt{\rho_a \rho_b} \end{array} \right] \cdot 2, \end{array} \tag{D.22}$$

and also $N(N-1)/2$ two-dimensional blocks with rows and columns labeled $\delta a_{ab}^{(1)}$ and $\delta a_{ab}^{(2)}$, i.e.,

$$\begin{array}{c} \\ \delta a_{ab}^{(1)} \\ \delta a_{ab}^{(2)} \end{array} \begin{array}{cc} \delta a_{ab}^{(1)} & \delta a_{ab}^{(2)} \\ \left[ \begin{array}{cc} 1 - \sqrt{\rho_a \rho_b} & -\sqrt{\tau_a \tau_b} \\ -\sqrt{\tau_a \tau_b} & 1 - \sqrt{\rho_a \rho_b} \end{array} \right] \cdot 2; \end{array} \tag{D.23}$$

this gives, altogether, a contribution proportional to

$$\left\{ \left[ 1 + \sqrt{(1-\tau_a)(1-\tau_b)} \right]^2 - \tau_a \tau_b \right\} \left\{ \left[ 1 - \sqrt{(1-\tau_a)(1-\tau_b)} \right]^2 - \tau_a \tau_b \right\}$$
$$= \left[ 2 - \tau_a - \tau_b + 2\sqrt{(1-\tau_a)(1-\tau_b)} \right] \left[ 2 - \tau_a - \tau_b - 2\sqrt{(1-\tau_a)(1-\tau_b)} \right]$$
$$= (\tau_a - \tau_b)^2. \tag{D.24}$$

Multiplying eqns (D.21) and (D.24) we find, for $|\det g_{\mu\nu}|^{1/2}$,

$$|\det g_{\mu\nu}|^{1/2} = \frac{\prod_{a<b} |\tau_a - \tau_b|}{\prod_c \sqrt{\tau_c}}. \tag{D.25}$$

Now consider the unitary group $U(N)$ formed by $N$-dimensional unitary matrices $v$, for which we define $\delta v = v^\dagger dv$ as above and, as in eqn (D.14),

$$\delta v = \delta a + i\delta s. \tag{D.26}$$

Here $\delta a$ contributes $N(N-1)/2$ independent parameters and $\delta s$ another $N(N+1)/2$, giving, altogether, a total of $N^2$ independent parameters of $U(N)$. We recall [67, 87] that Haar's measure of $U(N)$ can then be expressed as

$$d\mu(v) = \prod_{a \leqslant b} \delta s_{ab} \prod_{a<b} \delta a_{ab}. \tag{D.27}$$

The product $\prod dx_\mu$ appearing in eqn (6.4) is thus, in the present context,

$$\prod_{a=1}^{N} d\tau_a \prod_{a \leqslant b} \delta s_{ab}^{(1)} \prod_{a<b} \delta a_{ab}^{(1)} \prod_{a \leqslant b} \delta s_{ab}^{(2)} \prod_{a<b} \delta a_{ab}^{(2)} = \prod_{a=1}^{N} d\tau_a d\mu(v^{(1)}) d\mu(v^{(2)}). \tag{D.28}$$

Equations (D.25) and (D.28) finally give the result of eqns (6.6) for $\beta = 1$.

## D.2 The unitary case, $\beta = 2$

From eqn (D.15) we can verify that the determinant of the block occurring in the metric tensor with rows and columns labeled $\delta s_{aa}^{(1)}$, $\delta s_{aa}^{(2)}$, $\delta s_{aa}^{(3)}$ and $\delta s_{aa}^{(4)}$ vanishes. This is a consequence of the redundancy in the polar parametrization for $\beta = 2$ discussed in relation to eqns (6.9) in the text. Indeed, we notice that we can write the single summation in eqn (D.15) (except for its last term) as

$$\sum_a \left\{ \left[ \delta s_{aa}^{(1)} + \delta s_{aa}^{(3)} \right]^2 + \left[ \delta s_{aa}^{(2)} + \delta s_{aa}^{(4)} \right]^2 - 2\tau_a \left[ \delta s_{aa}^{(1)} - \delta s_{aa}^{(2)} \right] \left[ \delta s_{aa}^{(3)} - \delta s_{aa}^{(4)} \right] \right\}$$
$$= \sum_a \left[ (\delta x_a)^2 + (\delta y_a)^2 - 2\tau_a (\delta z_a)(\delta x_a - \delta y_a - \delta z_a) \right], \quad \text{(D.29)}$$

where we have defined the combinations

$$\delta x_a = \delta s_{aa}^{(1)} + \delta s_{aa}^{(3)}, \quad \text{(D.30a)}$$
$$\delta y_a = \delta s_{aa}^{(2)} + \delta s_{aa}^{(4)}, \quad \text{(D.30b)}$$
$$\delta z_a = \delta s_{aa}^{(1)} - \delta s_{aa}^{(2)}. \quad \text{(D.30c)}$$

In other words, the $4N$ quantities $\delta s_{aa}^{(1)}$, $\delta s_{aa}^{(2)}$, $\delta s_{aa}^{(3)}$ and $\delta s_{aa}^{(4)}$ appear in $ds^2$ only through the $3N$ combinations $\delta x_a$, $\delta y_a$ and $\delta z_a$; these quantities, together with the $d\tau_a$, contribute $4N$ independent variations. The $\delta s_{ab}^{(i)}$ for $i = 1, \ldots, 4$ and $a < b$ contribute $4N(N-1)/2$ variations, and so do the $\delta a_{ab}^{(i)}$ for $i = 1, \ldots, 4$ and $a < b$; we thus have a total of $4N^2$ variations, which is the correct number of independent parameters for a $2N$-dimensional unitary matrix.

In terms of independent variations, we can thus write the $ds^2$ of eqn (D.15) as

$$ds^2 = \sum_a \left\{ \left[ (\delta x_a)^2 + (\delta y_a)^2 - 2\tau_a (\delta z_a)(\delta x_a - \delta y_a - \delta z_a) \right] + \frac{(d\tau_a)^2}{2\tau_a \rho_a} \right\}$$
$$+ 2 \sum_{a<b} \left\{ \sum_{i=1}^{4} \left( \delta a_{ab}^{(i)} \right)^2 + \sum_{i=1}^{4} \left( \delta s_{ab}^{(i)} \right)^2 \right.$$
$$+ 2\sqrt{\rho_a \rho_b} \left[ \delta a_{ab}^{(1)} \delta a_{ab}^{(3)} + \delta a_{ab}^{(2)} \delta a_{ab}^{(4)} + \delta s_{ab}^{(1)} \delta s_{ab}^{(3)} + \delta s_{ab}^{(2)} \delta s_{ab}^{(4)} \right]$$
$$\left. + 2\sqrt{\tau_a \tau_b} \left[ \delta a_{ab}^{(2)} \delta a_{ab}^{(3)} + \delta a_{ab}^{(1)} \delta a_{ab}^{(4)} + \delta s_{ab}^{(2)} \delta s_{ab}^{(3)} + \delta s_{ab}^{(1)} \delta s_{ab}^{(4)} \right] \right\}. \quad \text{(D.31)}$$

The metric tensor appearing in eqn (6.3) has a simple block structure, consisting of $1 \times 1$, $3 \times 3$ and $4 \times 4$ blocks along the diagonal, as follows. There are $N$ three-dimensional blocks with rows and columns labeled $\delta x_a$, $\delta y_a$ and $\delta z_a$, i.e.,

$$\begin{array}{c} \phantom{\delta x_a} \quad \delta x_a \quad \delta y_a \quad \delta z_a \\ \begin{array}{c} \delta x_a \\ \delta y_a \\ \delta z_a \end{array} \left[ \begin{array}{ccc} 1 & 0 & -\tau_a \\ 0 & 1 & \tau_a \\ -\tau_a & \tau_a & 2\tau_a \end{array} \right], \end{array} \qquad (\text{D}.32)$$

and $N$ one-dimensional blocks $1/2\rho_a\tau_a$ labeled by $\mathrm{d}\tau_a$, giving, altogether, the contribution

$$\prod_{a=1}^{N} \frac{2\tau_a(1-\tau_a)}{2\tau_a(1-\tau_a)} = 1 \qquad (\text{D}.33)$$

to $\det g$. There are $N(N-1)/2$ four-dimensional blocks with rows and columns labeled $\delta s_{ab}^{(1)}$, $\delta s_{ab}^{(2)}$, $\delta s_{ab}^{(3)}$ and $\delta s_{ab}^{(4)}$ (for $a < b$), i.e.,

$$\begin{array}{c} \phantom{\delta s_{ab}^{(1)}} \quad \delta s_{ab}^{(1)} \quad \delta s_{ab}^{(2)} \quad \delta s_{ab}^{(3)} \quad \delta s_{ab}^{(4)} \\ \begin{array}{c} \delta s_{ab}^{(1)} \\ \delta s_{ab}^{(2)} \\ \delta s_{ab}^{(3)} \\ \delta s_{ab}^{(4)} \end{array} \left[ \begin{array}{cccc} 1 & 0 & \sqrt{\rho_a\rho_b} & \sqrt{\tau_a\tau_b} \\ 0 & 1 & \sqrt{\tau_a\tau_b} & \sqrt{\rho_a\rho_b} \\ \sqrt{\rho_a\rho_b} & \sqrt{\tau_a\tau_b} & 1 & 0 \\ \sqrt{\tau_a\tau_b} & \sqrt{\rho_a\rho_b} & 0 & 1 \end{array} \right], \end{array} \qquad (\text{D}.34)$$

and also $N(N-1)/2$ four-dimensional blocks with rows and columns labeled $\delta a_{ab}^{(1)}$, $\delta a_{ab}^{(2)}$, $\delta a_{ab}^{(3)}$ and $\delta a_{ab}^{(4)}$ which are identical to the matrices of (D.34); this gives, altogether, a contribution

$$(\tau_a - \tau_b)^4. \qquad (\text{D}.35)$$

We multiply eqns (D.33) and (D.35) and take the square root as required by eqn (6.4); this, together with the product of the independent variations occurring in eqn (D.31), gives

$$\mathrm{d}\mu^{(2)}(S) = C_2' \prod_{a<b} (\tau_a - \tau_b)^2 \prod_{a=1}^{N} \mathrm{d}\tau_a \prod_{a=1}^{N} \delta x_a \delta y_a \delta z_a \prod_{i=1}^{4} \prod_{a<b} \delta s_{ab}^{(i)} \delta a_{ab}^{(i)}. \qquad (\text{D}.36)$$

Under the operation of eqn (6.9), $\delta v^{(1)}$ transforms as

$$\delta v^{(1)} = v^{(1)\dagger} \mathrm{d}v^{(1)} \to g^* v^{(1)\dagger} \left( \mathrm{d}v^{(1)} g + v^{(1)} \mathrm{d}g \right)$$
$$= g^* \left( \delta v^{(1)} \right) g + g^* \mathrm{d}g, \qquad (\text{D}.37)$$

so that

$$\delta s_{aa}^{(1)} \to \delta s_{aa}^{(1)} + \mathrm{d}\eta_a. \qquad (\text{D}.38a)$$

Similarly

# THE UNITARY CASE, $\beta = 2$

$$\delta s_{aa}^{(2)} \to \delta s_{aa}^{(2)} + \mathrm{d}\eta_a, \tag{D.38b}$$

$$\delta s_{aa}^{(3)} \to \delta s_{aa}^{(3)} - \mathrm{d}\eta_a, \tag{D.38c}$$

$$\delta s_{aa}^{(4)} \to \delta s_{aa}^{(4)} - \mathrm{d}\eta_a. \tag{D.38d}$$

In contrast, the combinations $\delta x_a$, $\delta y_a$ and $\delta z_a$ of eqns (D.30) are insensitive to this transformation. Writing, for instance, $\delta s_{aa}^{(1)} \to \delta s_{aa}^{(1)} + \mathrm{d}\eta_a'$ and choosing $\mathrm{d}\eta_a' = -\delta s_{aa}^{(1)}$, we could eliminate $\delta s_{aa}^{(1)}$. Writing, subsequently, $\delta s_{aa}^{(1)} \to \delta s_{aa}^{(1)} + \mathrm{d}\eta_a' + \mathrm{d}\eta_a$, we could identify $\delta s_{aa}^{(1)}$ with $\mathrm{d}\eta_a$. We multiply both sides of eqn (D.36) by $\prod_a \delta s_{aa}^{(1)}/2\pi$ and use this identification on the left-hand side to obtain

$$\mathrm{d}\mu^{(2)}(S) \prod_{a=1}^{N} \frac{\mathrm{d}\eta_a}{2\pi} = C_2 \prod_{a<b} (\tau_a - \tau_b)^2 \prod_{a=1}^{N} \mathrm{d}\tau_a \prod_{a=1}^{N} \delta x_a \delta y_a \delta z_a \delta s_{aa}^{(1)} \prod_{i=1}^{4} \prod_{a<b} \delta s_{ab}^{(i)} \delta a_{ab}^{(i)}. \tag{D.39}$$

Using the definitions of eqns (D.30), we find that the Jacobian of the transformation from $\delta x_a$, $\delta y_a$, $\delta z_a$, $\delta s_{aa}^{(1)}$ to $\delta s_{aa}^{(1)}$, $\delta s_{aa}^{(2)}$, $\delta s_{aa}^{(3)}$, $\delta s_{aa}^{(4)}$ is 1. On the right-hand side of eqn (D.39) we see that we can complete the Haar measures $\mathrm{d}\mu(v^{(i)})$; on the left-hand side we use eqn (6.8). We finally find the result of eqns (6.7) for $\beta = 2$.

# REFERENCES

[1] Abrahams, E., Anderson, P. W., Licciardello, D. C. and Ramakrishnan, T. V. (1979). *Phys. Rev. Lett.*, **42**, 673.

[2] Abramowitz, M. and Stegun, I. E. (1964). *Handbook of mathematical functions.* National Bureau of Standards, Washington.

[3] Agassi, D., Weidenmüller, H. A. and Mantzouranis, G. (1975). *Phys. Rep.*, **22**, 145.

[4] Albada, M. P., van der Mark, M. B. and Lagendijk, A. (1990). In *Scattering and localization of classical waves in random media* (ed. P. Sheng), p. 97. World Scientific, Singapore.

[5] Altland, A. and Zirnbauer, M. R. (1996). *Phys. Rev. Lett.*, **76**, 3420.

[6] Altshuler, B. L., Aronov, A. G., Khmelnitskii, D. E. and Larkin, A. I. (1982). In *Quantum theory of solids* (ed. I. M. Lifshitz), p. 130. MIR Publications, Moscow.

[7] Altshuler, B. L., Lee, P. A. and Webb, R. A. (ed.) (1991). *Mesoscopic phenomena in solids.* North-Holland, Amsterdam.

[8] Altshuler, B. L. and Simons, B. D. (1994). *Universalities: from Anderson localization to quantum chaos.* Elsevier, Amsterdam.

[9] Anderson, P. W. (1958). *Phys. Rev.*, **109**, 1492.

[10] Balian, R. (1968). *Nuovo Cimento B*, **57**, 183.

[11] Baranger, H. U., DiVincenzo, D. P., Jalabert, R. A. and Stone, A. D. (1991). *Phys. Rev. B*, **44**, 10637.

[12] Baranger, H. U., Jalabert, R. A. and Stone, A. D. (1993). *Phys. Rev. Lett.*, **70**, 3876.

[13] Baranger, H. U., Jalabert, R. A. and Stone, A. D. (1993). *Chaos*, **3**, 665.

[14] Baranger, H. U. and Mello, P. A. (1994). *Phys. Rev. Lett.*, **73**, 142.

[15] Baranger, H. U. and Mello, P. A. (1995). *Phys. Rev. B*, **51**, 4703.

[16] Baranger, H. U. and Mello, P. A. (1996). *Europhys. Lett.*, **33**, 465.

[17] Baranger, H. U. and Stone, A. D. (1989). *Phys. Rev. B*, **40**, 8169.

[18] Bargmann, V. (1947). *Annals Math.*, **48**, 568.

[19] Barrera, R. Private communication.

[20] Bauer, M., Mello, P. A. and McVoy, K. W. (1979). *Zeit. Physik A*, **293**, 151.

[21] Beenakker, C. W. J. (1997). *Rev. Modern Phys.*, **69**, 731.

[22] Beenakker, C. W. J. and Rejaei, B. (1993). *Phys. Rev. Lett.*, **71**, 3689.

[23] Beenakker, C. W. J. and Rejaei, B. (1994). *Phys. Rev. B*, **49**, 7499.

[24] Beenakker, C. W. J. and van Houten, H. (1991). In *Solid state physics* (ed. H. Ehrenreich and D. Turnbull), p. 1. Volume 44. Academic Press, New York.

[25] Bergmann, G. (1984). *Phys. Rep.*, **107**, 1.

[26] Berry, M. J., Baskey, J. H., Westervelt, R. M. and Gossard, A. C. (1994). *Phys. Rev. B*, **50**, 8857.

[27] Berry, M. J., Katine, J. A., Westervelt, R. M. and Gossard, A. C. (1994). *Phys. Rev. B*, **50**, 17721.

[28] Binder, K. (ed.) (1992). *The Monte Carlo method in condensed matter physics*. Topics in Applied Physics, Volume 71. Springer, New York.

[29] Blümel, R. and Smilansky, U. (1988). *Phys. Rev. Lett.*, **60**, 477.

[30] Blümel, R. and Smilansky, U. (1989). *Physica D*, **36**, 111.

[31] Blümel, R. and Smilansky, U. (1990). *Phys. Rev. Lett.*, **64**, 241.

[32] Bohigas, O., Giannoni, M. J. and Schmit, C. (1984). *Phys. Rev. Lett.*, **52**, 1.

[33] Bohr, N. (1936). *Nature*, **137**, 344.

[34] Bohr, N. (1937). *Science*, **86**, 161.

[35] Bonch-Bruevich, V. L., Mironov, A. G. and Zviagin, I. P. (1973). *Rivista Nuovo Cimento*, **3**, 321.

[36] Breit, G. (1940). *Phys. Rev.*, **58**, 506.

[37] Breit, G. (1946). *Phys. Rev.*, **69**, 472.

[38] Breit, G. and Wigner, E. P. (1936). *Phys. Rev.*, **49**, 519.

[39] Breit, G. and Wigner, E. P. (1936). *Phys. Rev.*, **49**, 642.

[40] Brody, T. A., Flores, J., French, J. R., Mello, P. A., Pandey, A. and Wing, S. S. M. (1981). *Rev. Modern Phys.*, **53**, 385.

[41] Brouwer, P. W. (1995). *Phys. Rev. B*, **51**, 16878.

[42] Brouwer, P. W. (1997). *On the random-matrix theory of quantum transport*. Ph.D. thesis, University of Leiden, The Netherlands.

[43] Brouwer, P. W. and Beenakker, C. W. J. (1995). *Phys. Rev. B*, **51**, 7739.

[44] Brouwer, P. W. and Beenakker, C. W. J. (1997). *Phys. Rev. B*, **55**, 4695.

[45] Brouwer, P. W., Furusaki, A., Gruzberg, I. A. and Mudry, C. (2000). *Phys. Rev. Lett.*, **85**, 1064.

[46] Brouwer, P. W., Mudry, C., Simons, B. D. and Altland, A. (1998). *Phys.*

*Rev. Lett.*, **81**, 862.

[47] Buck, B. and Macaulay, V. A. (ed.) (1991). *Maximum entropy in action.* Clarendon Press, Oxford.

[48] Büttiker, M. (1986). *Phys. Rev. B*, **33**, 3020.

[49] Büttiker, M. (1988). *IBM J. Res. Devel.*, **32**, 317.

[50] Büttiker, M. (1992). *Phys. Rev. B*, **46**, 12485.

[51] Büttiker, M. (1993). *J. Phys.: Condensed Matt.*, **5**, 9361.

[52] Büttiker, M., Imry, Y., Landauer, R. and Pinhas, S. (1985). *Phys. Rev. B*, **31**, 6207.

[53] Caselle, M. cond-mat/9610017.

[54] Caselle, M. and Magnea, U. (2004). *Phys. Rep.*, **394**, 41.

[55] Caselle, M. and Magnea, U. (2006). *J. Stat. Mech.*, P01013.

[56] Chan, I. H., Clarke, R. M., Marcus, C. M., Campman, K. and Gossard, A. C. (1995). *Phys. Rev. Lett.*, **74**, 3876.

[57] Chandrasekhar, S. (1943). *Rev. Modern Phys.*, **15**, 1. [Reprinted in *Selected papers on noise and stochastic processes* (ed. N. Wax), p. 3. Dover Publications, New York. (1954)].

[58] Chang, A. M., Baranger, H. U., Pfeiffer, L. N. and West, K. W. (1994). *Phys. Rev. Lett.*, **73**, 2111.

[59] D'Amato, J. L. and Pastawski, H. M. (1990). *Phys. Rev.*, **41**, 7441.

[60] Datta, S. (1989). *Phys. Rev. B*, **40**, 5830.

[61] Datta, S. (1995). *Electronic transport in mesoscopic systems.* Cambridge University Press.

[62] DiVincenzo, D. P. (1993). *Phys. Rev. B*, **48**, 1404.

[63] Doniach, S. and Sondheimer, E. H. (1974). *Green's functions of solid state physics.* Frontiers in Physics, Volume 44. Benjamin, New York.

[64] Dorokhov, O. N. (1982). *Pis'ma Zh. Eksp. Teor. Fiz.*, **36**, 259. [*JETP Lett.*, **36**, 318. (1982)].

[65] Doron, E., Smilansky, U. and Frenkel, A. (1990). *Phys. Rev. Lett.*, **65**, 3072.

[66] Doron, E., Smilansky, U. and Frenkel, A. (1991). *Physica D*, **50**, 367.

[67] Dyson, F. J. (1962). *J. Math. Phys.*, **3**, 140.

[68] Economou, E. N. and Soukoulis, C. M. (1981). *Phys. Rev. Lett.*, **46**, 618.

[69] Edwards, S. F. (1958). *Phil. Mag.*, **3**, 1020.

[70] Efetov, K. B. (1983). *Adv. Phys.*, **32**, 53.

[71] Ericson, T. and Mayer-Kuckuk, T. (1966). *Ann. Rev. Nuclear Sci.*, **16**,

183.

[72] Feshbach, H. (1973). In *Reaction dynamics* (ed. E. W. Montroll, G. H. Vineyard, M. Levy and P. T. Matthews), p. 169. Gordon and Breach, New York.

[73] Feshbach, H., Porter, C. E. and Weisskopf, V. F. (1954). *Phys. Rev.*, **96**, 448.

[74] Fisher, R. A. (1997). *Statistical inference and analysis: selected correspondence of R. A. Fisher.* Edited by J. H. Bennett. Clarendon Press, Oxford.

[75] French, J. B., Mello, P. A. and Pandey, A. (1978). *Phys. Lett. B*, **80**, 17.

[76] Friedman, W. and Mello, P. A. (1985). *Annals Phys. (NY)*, **161**, 276.

[77] Froufe-Pérez, L. S., García-Mochales, P., Serena, P. A., Mello, P. A. and Sáenz, J. J. (2002). *Phys. Rev. Lett.*, **89**, 246403.

[78] Froufe-Pérez, L. S., Yépez, M., Mello, P. A. and Sáenz, J. J. (2007). *Phys. Rev. E*, **75**, 031113.

[79] García-Martín, A. and Sáenz, J. J. (2001). *Phys. Rev. Lett.*, **87**, 116603.

[80] Genack, A. (1990). In *Scattering and localization of classical waves in random media* (ed. P. Sheng), p. 207. World Scientific, Singapore.

[81] Gopar, V. A. (1999). *Electronic transport in mesoscopic systems. An approach based on random S matrices at one and various energies.* Ph.D. thesis, Universidad Nacional Autónoma de México, México.

[82] Gutzwiller, M. C. (1983). *Physica D*, **7**, 341.

[83] Hamermesh, M. (1962). *Group theory and its applications to physical problems.* Addison-Wesley, Reading, MA.

[84] Hastings, M. B., Stone, A. D. and Baranger, H. U. (1994). *Phys. Rev. B*, **50**, 8230.

[85] Heinrichs, J. (1986). *Phys. Rev. B*, **33**, 5261.

[86] Hershfield, S. (1991). *Phys. Rev. B*, **43**, 11586.

[87] Hua, L. K. (1963). *Harmonic analysis of functions of several complex variables in the classical domains.* American Mathematical Society, Providence, RI. Translated by L. Ebner and A. Koranyi.

[88] Huang, K. (1963). *Statistical mechanics.* Wiley, New York.

[89] Huibers, A. G., Patel, S. R., Marcus, C. M., Brouwer, P. W., Duruöz, C. I. and Harris, Jr, J. S. (1998). *Phys. Rev. Lett.*, **81**, 1917.

[90] Iida, S., Weidenmüller, H. A. and Zuk, J. A. (1990). *Annals Phys. (NY)*, **200**, 219.

[91] Imamura, T. and Hikami, K. (2001). *J. Phys. Soc. Japan*, **70**, 3312.

[92] Imamura, T. and Wadati, M. (2002). *J. Phys. Soc. Japan*, **71**, 1511.

[93] Imry, Y. (1997). *Introduction to mesoscopic physics.* Oxford University Press.

[94] Jackson, J. D. (1962). *Classical electrodynamics.* Wiley, New York.

[95] Jalabert, R. A., Baranger, H. U. and Stone, A. D. (1990). *Phys. Rev. Lett.*, **65**, 2442.

[96] Jalabert, R. A., Pichard, J.-L. and Beenakker, C. W. J. (1994). *Europhys. Lett.*, **27**, 255.

[97] Jaynes, E. T. (1957). *Phys. Rev.*, **106**, 620.

[98] Jensen, R. (1991). *Chaos*, **1**, 101.

[99] Kalos, M. H. and Whitlock, P. A. (1986). *Monte Carlo methods.* Wiley, New York.

[100] Kane, C. L., Serota, R. A. and Lee, P. A. (1988). *Phys. Rev. B*, **37**, 6701.

[101] Kapur, P. L. and Peierls, R. E. (1938). *Proc. Roy. Soc. (London) A*, **166**, 277.

[102] Katz, A. (1967). *Principles of statistical mechanics. The information theory approach.* W. H. Freeman, San Francisco.

[103] Kendall, M. and Stuart, A. (1977). *The advanced theory of statistics* (4th edn). Griffin, London.

[104] Khinchin, A. I. (1957). *Mathematical foundations of information theory.* Dover Publications, New York. Translated by R. A. Silverman and M. D. Friedman.

[105] Kirczenow, G. (1990). *Solid State Comm.*, **74**, 1051.

[106] Kittel, C. (1963). *Quantum theory of solids.* Wiley, New York.

[107] Kubo, R. (1957). *J. Phys. Soc. Japan*, **12**, 570.

[108] Kubo, R. (1958). *Lectures Theor. Phys. (Boulder)*, **1**, 120.

[109] Kumar, N. (1985). *Phys. Rev. B*, **31**, 5513.

[110] Kumar, N. and Jayannavar, A. M. (1986). *J. Phys. C*, **19**, L85.

[111] Landauer, R. (1970). *Phil. Mag.*, **21**, 863.

[112] Lane, A. M. and Thomas, R. G. (1958). *Rev. Modern Phys.*, **30**, 257.

[113] Langer, J. S. and Neal, T. (1966). *Phys. Rev. Lett.*, **16**, 984.

[114] Langreth, D. C. and Abrahams, E. (1981). *Phys. Rev. B*, **24**, 2978.

[115] Lee, P. A. and Ramakrishnan, T. V. (1985). *Rev. Modern Phys.*, **57**, 287.

[116] Lee, P. A. and Stone, A. D. (1985). *Phys. Rev. Lett.*, **55**, 1622.

[117] Levine, R. D. and Bernstein, R. B. (1976). In *Modern theoretical chemistry* (ed. W. H. Miller), p. 323. Volume II. Plenum Press, New York.

[118] Levinson, I. B. (1989). *Sov. Phys. JETP*, **68**, 1257.

[119] Levinson, Y. B. and Shapiro, B. (1992). *Phys. Rev. B*, **46**, 15520.

[120] Lewenkopf, C. H. and Weidenmüller, H. A. (1991). *Annals Phys. (NY)*, **212**, 53.

[121] López, G., Mello, P. A. and Seligman, T. H. (1981). *Zeit. Physik A*, **302**, 351.

[122] Mahan, G. D. (1981). *Many-particle physics*. Plenum Press, New York.

[123] McVoy, K. W., Heller, L. and Bolsterli, M. (1967). *Rev. Modern Phys.*, **39**, 245.

[124] Mehta, M. L. (1991). *Random matrices*. Academic Press, New York.

[125] Mello, P. A. (1986). *J. Math. Phys.*, **27**, 2876.

[126] Mello, P. A. (1988). *Phys. Rev. Lett.*, **60**, 1089.

[127] Mello, P. A. (1990). *J. Phys. A*, **23**, 4061.

[128] Mello, P. A. (1995). In *Mesoscopic quantum physics* (ed. E. Akkermans, G. Montambaux and J.-L. Pichard), p. 435. Les Houches Summer School, Session LXI. Elsevier, Amsterdam.

[129] Mello, P. A., Akkermans, E. and Shapiro, B. (1988). *Phys. Rev. Lett.*, **61**, 459.

[130] Mello, P. A. and Baranger, H. U. (1995). *Physica A*, **220**, 15.

[131] Mello, P. A. and Baranger, H. U. (1999). *Waves Random Media*, **9**, 105.

[132] Mello, P. A., Pereyra, P. and Kumar, N. (1988). *Annals Phys. (NY)*, **181**, 290.

[133] Mello, P. A., Pereyra, P. and Seligman, T. H. (1985). *Annals Phys. (NY)*, **161**, 254.

[134] Mello, P. A. and Pichard, J.-L. (1991). *J. de Physique I*, **1**, 493.

[135] Mello, P. A. and Seligman, T. H. (1980). *Nuclear Phys. A*, **344**, 489.

[136] Mello, P. A. and Shapiro, B. (1988). *Phys. Rev. B*, **37**, 5860.

[137] Mello, P. A. and Stone, A. D. (1991). *Phys. Rev. B*, **44**, 3559.

[138] Mello, P. A. and Tomsovic, S. (1992). *Phys. Rev. B*, **46**, 15963.

[139] Mel'nikov, V. I. (1980). *Pis'ma Zh. Eksp. Teor. Fiz.*, **32**, 244. [*JETP Lett.*, **32**, 225. (1980)].

[140] Mel'nikov, V. I. (1981). *Fis. Tverd. Tela (Leningrad)*, **23**, 782. [*Sov. Phys. Solid State*, **23**, 444. (1981)].

[141] Merzbacher, E. (1970). *Quantum mechanics*. Wiley, New York.

[142] Messiah, A. (1986). *Quantum mechanics*, Volume II. North-Holland, Amsterdam.

[143] Mott, N. F. (1967). *Adv. Phys.*, **16**, 49.

[144] Mudry, C., Brouwer, P. W. and Furusaki, A. (1999). *Phys. Rev. B*, **59**, 13221.

[145] Mudry, C., Brouwer, P. W. and Furusaki, A. (2000). *Phys. Rev. B*, **62**, 8249.

[146] Muttalib, K. A. and Gopar, V. A. (2002). *Phys. Rev. B*, **66**, 115318.

[147] Newton, R. G. (1966). *Scattering theory of waves and particles* (1st edn). McGraw-Hill, New York.

[148] Nishioka, H. and Weidenmüller, H. A. (1985). *Phys. Lett. B*, **157**, 101.

[149] Peierls, R. E. (1947). *Proc. Cambridge Phil. Soc.*, **44**, 242.

[150] Pichard, J.-L. (1991). In *Quantum coherence in mesoscopic systems* (ed. B. Kramer), p. 369. NATO ASI Series B, **254**. Plenum, New York.

[151] Politzer, H. D. (1989). *Phys. Rev. B*, **40**, 11917.

[152] Porter, C. E. (1965). *Statistical theories of spectra: fluctuations*. Academic Press, New York.

[153] Preston, M. A. (1962). *Physics of the nucleus*. Addison-Wesley, Reading, MA.

[154] Rammal, R. and Doucot, B. (1987). *J. Phys. (Paris)*, **48**, 509.

[155] Roman, P. (1965). *Advanced quantum theory*. Addison-Wesley, Reading, MA.

[156] Sáenz, J. J. and García Mochales, P. Private communication.

[157] Santilli, R. M. (1978). *Foundations of theoretical mechanics*. Springer-Verlag, New York.

[158] Shannon, C. E. (1948). *Bell System Tech. J.*, **27**, 379.

[159] Shannon, C. E. (1948). *Bell System Tech. J.*, **27**, 623.

[160] Shapiro, B. Private communication.

[161] Sheng, P. (ed.) (1990). *Scattering and localization of classical waves in random media*. World Scientific, Singapore.

[162] Smilansky, U. (1991). In *Chaos and quantum physics* (ed. M.-J. Giannoni, A. Voros and J. Zinn-Justin), p. 371. North-Holland, New York.

[163] Smith, F. T. (1960). *Phys. Rev.*, **118**, 349.

[164] Stone, A. D., Mello, P. A., Muttalib, K. and Pichard, J.-L. (1991). In *Mesoscopic phenomena in solids* (ed. B. L. Altshuler, P. A. Lee and R. A. Webb), p. 369. Elsevier, Amsterdam.

[165] Stone, A. D. and Szafer, A. (1988). *IBM J. Res. Devel.*, **32**, 384.

[166] Stratonovich, R. L. (1963). *Topics in the theory of random noise*, Volume I. Gordon and Breach, New York.

[167] Takane, Y. (2004). *J. Phys. Soc. Japan*, **73**, 9.

[168] Thouless, D. J. (1977). *Phys. Rev. Lett.*, **39**, 1167.

[169] Thouless, D. J. (1981). *Phys. Rev. Lett.*, **47**, 972.

[170] Tsallis, C. (1988). *J. Statist. Phys.*, **52**, 479.

[171] Tsallis, C., Mendes, R. S. and Plastino, A. R. (1998). *Physica A*, **261**, 534.

[172] Wigner, E. P. (1946). *Phys. Rev.*, **70**, 15.

[173] Wigner, E. P. (1946). *Phys. Rev.*, **70**, 606.

[174] Wigner, E. P. (1951). *Annals Math.*, **53**, 36.

[175] Wigner, E. P. (1955). *Phys. Rev.*, **98**, 145.

[176] Wigner, E. P. (1959). *Group theory and its applications to the quantum mechanics of atomic spectra*. Academic Press, New York.

[177] Wigner, E. P. and Eisenbud, L. (1947). *Phys. Rev.*, **72**, 29.

[178] Wolf, K. B. (1980). In *Group theory and its applications in physics* (ed. T. H. Seligman), p. 1. Latin American School of Physics, Mexico City. [AIP Conference Proceedings No. 71. AIP, New York. (1981)].

[179] Wolf, P. E. and Maret, G. (1985). *Phys. Rev. Lett.*, **55**, 2696.

[180] Wurm, J., Rycerz, A., Adagideli, I., Wimmer, M., Richter, K. and Baranger, H. U. (2009). *Phys. Rev. Lett.*, **102**, 056806.

[181] Yaglom, A. M. (1962). *An introduction to the theory of stationary random functions*. Prentice-Hall, New York. Translated by R. A. Silverman.

[182] Zirnbauer, M. R. (1996). *J. Math. Phys.*, **37**, 4986.

# INDEX

analytic continuation
  wave function, 124
analyticity–ergodicity (AE), 250, 251
Anderson
  insulator, 332
  localization, 330
  transition, 4
anisotropic scattering, 357
anti-localization, 361
applied field, *see* external field
asymptotic wave function, 20, 129
atomic nuclei, 2
averages
  energy, 111, 113
  ensemble, 112, 113
  products of $S$-matrix elements, 253
avoided level crossing, 10

back-scattering enhancement, 257, 328
ballistic, 6
  cavities, 7
  open cavities, 244, 245
  regime, 6, 320
Bayes'
  hypothesis, 242
  theorem, 241
beta function, 342, 361
billiard
  Sinai, 265
  stadium, 265
Boltzmann kinetic transport theory, 357
Born approximation, 338
  self-consistent (to $\Sigma$), 349
Born series, 18
bound state, 71
  continuum (in the) (BSC), 152
  poles, 89
boundary conditions, 17
  incoming, 17, 23
  outgoing, 17, 23
Breit–Wigner form, 87
building block (BB), 280, 288, 289, 298

canonical ensemble, 236, 239
cavity
  ballistic, 7, 245
  chaotic, 244
  open, 244
cavity scattering theory, 168

central-limit theorem (CLT), 11, 12, 226, 234
channels, 122
  arbitrary number, 138
  closed, 123
  one open and one closed, 132, 136
  open, 122, 123, 125
  open and closed, 155
  two open, 131, 134
chaos
  classical, 1, 3, 7, 10, 244, 264, 265
  quantum, 244, 265
chaotic, 1, 3, 7, 10, 226
  cavity, 244, 265
  classical dynamics, 1
characteristic function
  $\lambda$ probability density, 293, 294
charge density, 190, 194
  equilibrium, 190
  induced, 201, 204, 207
charge reservoirs, 203
circular ensemble (CE)
  orthogonal, 247
  unitary, 247
classical
  chaos, 1, 244, 265
  conductance fluctuations, 7
  dynamics, 244, 264, 265
    chaotic, 1, 3, 244, 265
closed channels, 155, 158, 159, 161, 162, 376
coherent back-scattering (CBS), 257, 329, 330, 337, 338, 351
complex
  many-body system, 1
  system, 1
conductance, 4, 187, 221, 222, 225, 340–342
  average, 255, 257, 322
  d.c., 221
  dimensionless, 224
  distribution, 258, 323
    crossover regime, 323
    diffusive regime, 323
    localized regime, 323
  fluctuations, 253, 255, 258, 274, 322
    classical, 7
    universal, 8
  moments, 319

conductivity
 bubble, 355, 356, 358
 Drude, 330, 354, 356
  diagrams for, 355
 minimum metallic, 337
 weak localization (diagrams for), 358
conductivity tensor, 213, 216
 d.c. limit, 217, 220, 221, 365, 375
 in RPA, 213, 362, 367
 internal, 213
 zero-temperature limit, 375
configuration, 368
constraints in maximizing the entropy, 228, 231, 289, 298
 lifting one, 233, 236, 239
constriction, 189, 203
continuity equation, 213
convex function, 229
convolution, 287, 291, 301
Cooperons, 357
correlations, 187
critical-point phenomena, 11
current density, 208, 212, 213
 equilibrium, 191
 operator, 190, 208, 210
cut-off
 wave vector, 360

d.c. limit, 217, 220, 221, 365, 375
decoherence, 6
delayed response, 3
dense-weak-scattering limit, 289, 291, 298, 344
density matrix, 189
density of states
 glocal, 334
 local, 334
dephasing, 270, 360, 361
diagrammatic technique, 12, 331, 342
diagrams
 fan, 357
 ladder, 356, 357
 maximally-crossed, 357–359
diffusion equation, 295
 orthogonal case, 302, 310
 unified form, 318
 unitary case, 310, 317
  exact solution, 323
diffusive regime, 6, 320, 327
direct process, 251, 252, 262, 268
 direct reflection, 263
 direct transmission, 264
disorder
 bulk (quasi-one-dimensional), 324, 325
 quenched, 332

site-diagonal, 332
surface (quasi-one-dimensional), 324, 325
disordered conductors, 2
 conductance, 4
  fluctuations, 4
 one-dimensional, 288
 quasi-one-dimensional, 279, 298
  bulk disorder, 324, 325
  correlations in transmission and reflection, 325
  diffusion equation in general form, 318
  diffusive regime, 327
  expectation values, 318
  Fokker–Planck equation in general form, 318
  metallic regime, 325
  surface disorder, 324, 325
dissipative or resistive part of the response, 200
distribution of the electrical conductance
 cavity, 258
Drude conductivity, 330, 354, 356
duality, 70
Dyson's equation, 348

electric-dipole approximation, 352
electrical conductance, 187, 221, 222, 225
 average, 255, 257, 322
 d.c., 221
 dimensionless, 224
 distribution, 258, 323
  crossover regime, 324
  diffusive regime, 323
  localized regime, 323
 fluctuations, 253, 255, 258, 274, 322
 moments, 319
electrical conductivity, 187
 disordered metal
  classical, Drude, 350, 354
  quantum corrections; coherent back-scattering; weak localization, 350, 357
electron–electron interaction, 204
electronic charge density
 equilibrium, 190
 induced, 201, 207
electronic transport
 linear response theory (LRT), 187
energy averages, 112
ensemble
 dots (of), 246
 quasi-one-dimensional disordered conductors (of), 279

# INDEX

ensemble averages, 3, 112
ensemble of $M$ matrices, 280, 286
  orthogonal, 280
  unitary, 280
ensemble of $S$ matrices, 112, 245
  orthogonal, 247
  symplectic, 247
  unitary, 247
entropy, 11, 228
  conditional, 230
  continuous variables, 233
  convexity, 229
  Gibbs' inequality, 231
  maximum, 11
    equilibrium statistical mechanics, 235
    equilibrium statistical mechanics, canonical ensemble, 236, 239
    equilibrium statistical mechanics, microcanonical ensemble, 236
    Lagrange multipliers, 231, 237, 290, 299
    lifting a constraint, 233
  Poisson's kernel, 252
  prior, 227
    role of symmetries, 234
  properties, 229
  Shannon, or information-theoretic, 11, 228
  uniqueness theorem, 231
equal-a-priori probabilities, 12, 227, 247
equilibrated response, 3, 252
ergodic time, 6
ergodicity, 3, 113, 247, 249, 251, 279
Ericson fluctuations, 244
expectation values for quasi-one-dimensional disordered conductors, 318
extensive transport properties, 187
external field, 191, 202

fake lead, 270, 271
fan diagrams, 357
Fano resonance, 154
Fermi
  function, 221, 352, 371
  golden rule, 352
  wave vector, 5
field
  external, 191, 202
  self-consistent, 205
  total, 213
fluctuations, 2, 4
  level spacings, 11
  mesoscopic, 11
flux conservation, 33, 139

Fokker–Planck equation, 288, 298
  orthogonal case, 310
  unified form, 318
  unitary case, 317
following function, 110

gauge
  invariance, 193, 213
  scalar potential, 193
  transformation, 191
  vector potential, 193, 209
Gaussian white noise, 344
Gibbs' inequality, 231
glory effect, 340
grand canonical ensemble, 189, 364, 367, 370
Green function, 16, 343, 349
  advanced, 345
  full, 25, 30, 43, 224
    symmetry property, 26–28
  retarded, 344
  single-particle, 345, 372
  unperturbed, 17, 23, 125, 126
    symmetry property, 26
group
  compact, 325
  non-compact, 325

Haar's measure on the unitary group, 248, 286, 325, 384, 387
half-width at half-maximum, 85, 87, 88
harmonic oscillators (linear chain), 196, 199
heat bath, 199, 201
Heisenberg time, 6, 199
horns, 189

impurity diagram technique, 346
induced field, 202
integrable system, 10
intensive transport properties, 187
interaction picture, 195, 198
invariant arc element for $S$ matrices, 247, 380
invariant imbedding, 12, 68
invariant measure
  for $M$ matrices, 284
  for $S$ matrices, 245, 247, 380
    orthogonal, 247, 382
    symplectic, 247
    unitary, 247, 385
inverse participation ratio, 334
irreducible self-energy, 348, 349
irreversibility, 199

$K$ matrix, 110

following function, 110
Kane, Serota and Lee theorem (KSL), 217, 222, 362
Kubo linear response theory, *see* linear response theory (LRT)

ladder diagrams, 356, 357
Lagrange multipliers, 231, 237, 290, 299
Landauer's formula, 225, 292, 379
lead index notation, 62, 171
length-scale, 5
  elastic mean free path, 5
  inelastic mean free path, 5
  magnetic, 360
  Thouless length, 5
level
  repulsion, 10
  spacing ($\Delta$), 199
level-spacing distribution, 10
likelihood function, 242
linear response theory (LRT), 187, 207, 212, 342, 351
linearity
  left and right amplitudes (between), 44, 142
  outgoing and incoming amplitudes (between), 22, 135, 137
  Schrödinger's equation, 22, 44, 135, 136
Lippmann–Schwinger equation, 16, 18, 19, 120, 125, 129
localization, 4, 330
  Anderson, 330
  exponential, 332
  length, 333
    quasi-one-dimensional, 320
  strong, 330
  weak, 253, 257, 330, 342, 350, 360
localized regime, 320
long-wavelength approximation, 193

$M$ matrix, 44, 45, 141
  current conservation, 45
  ensemble, 280, 286
  extended, 66, 155, 159, 161
  polar representation, 47, 144, 280, 293, 303, 310
  probability distribution, 279, 301
  serial multiplicativity, 47, 145
    presence of open and closed channels, 162
  transformation under a translation, 51, 146
  TRI, 46, 47, 143
macroscopic
  sample, 6

system, 203
magnetic length-scale, 360
marginal distribution
  $\lambda$ variable, 296, 297
Markovian random processes, 288
maximally-crossed diagrams, 342, 357–359
maximum entropy, 11
  approach (MEA), 12, 227, 241, 244, 246, 289, 298
  method (MEM), 11, 241
Maxwell–Boltzmann distribution, 228, 231
mean free path, 5
  elastic, 5, 359
  inelastic, 5
measure, 234
  invariant, 234
  symmetries, 234
Mel'nikov's equation, 296, 297
mesoscopic
  conductors, 5
    ballistic, 7
    diffusive, 7
  fluctuations, 8
  system, 5, 203
  transport, 11
metal–insulator transition, 330, 335, 337
metallic regime, 320, 325
microcanonical ensemble, 236
microelectronic heterostructures, 245
microwave cavities, 2
minimum metallic conductivity, 337
mixed state, 239
mobility edge, 4, 335, 336, 340
modes, 122
  evanescent, 123
  running, 122, 123
moment generating function
  $\lambda$ probability density, 293
Mott's minimum metallic conductivity, 337

non-integrable, 244
nuclear
  Hamiltonian, 9
  optical model, 2
  physics, 1, 2, 8
numerical simulations
  Monte Carlo for quasi-one-dimensional disordered systems, 324
  tight-binding model for quasi-one-dimensional disordered systems, 325
two-dimensional cavities, 245, 265

Occam's razor, 12

INDEX 401

occupation number representation, 368
off-resonance, 78
one-body
  operator, 204, 368
  problem, 1
one-dimensional scattering theory, 15
one-parameter scaling, 342, 361
open systems, 11
optical $S$ matrix, 112, 113, 250, 253, 262
optical model, 2
  optical $S$ matrix, 112, 113

persistent property, 291, 301, 302
perturbation theory (time-dependent), 194
phase
  breaking, 6, 270, 271, 277
    length, 359
  coherence, 1
  shift, 72, 73
plane wave, 16
Poincaré or recurrence time, 6
Poisson's kernel, 112, 249, 251, 252
  entropy, 252
polar parameters, 282, 283, 303, 310
polar representation
  $M$ matrix, 280, 293, 303, 310
  $S$ matrix, 42, 140, 248, 258
potential, 16, 20
  barrier, 63
  delta potential, 53
    $M$ matrix, 54
    $S$ matrix, 53
    wave function, 54
  delta potential (two channels), 147
    $M$ matrix, 149
    $S$ matrix, 148
  left–right symmetric, 148, 151
  local, 16, 20
  non-local, 20
  profile, 194
    equilibrium, 190
  step, 57
  TRI, 40
  velocity dependent, 20, 39, 40
  wall and delta potential, 71
    phase shift, 73
    resonances, 75
    $S$ matrix, 72
    wave function, 78
prior, 227
  role of symmetries, 234
probability density for $M$ matrices, 286, 301
probability distribution

  of $M$, 279
  of $S$, 279
prompt response, 3, 251, 252, 262, 268
pseudounitary group
  $U(1,1)$, 46
    unimodular, 46
  $U(N,N)$, 143
pure state, 239

quantum chaos, 244
quantum dots, 7, 245
  experiments, 270, 276
quasi-one-dimensional scattering theory, 120
quasi-static approximation, 193
quasi-universal properties, 226

$R$ matrix, 93, 181, 185
  background levels, 107
  boundary, 95, 96, 101, 102, 106
    change of, 106
  isolated resonances, 105
  levels, 115
  meromorphic function, 100
  poles, 100, 103
  poles of the $S$ matrix (and), 105
  reduced
    amplitudes, 100, 185
    widths, 100
  residues, 100, 103
  single-level approximation, 105, 107
  single-level-plus-background
    approximation, 107, 108, 115, 116, 119
  theory, 93, 181
radio waves, 3
random-matrix theory (RMT), 8, 244
random-phase approximation (RPA), 204, 207, 209, 216, 220
Rayleigh distribution, 328
reactive part of the response, 200
reduced amplitudes, 100, 185
reduced widths, 100
reflection, 16
  amplitudes, 16, 120, 129, 130, 171
    analytic continuation, 66
    combination (imbedding), 68
    polar parameters (in terms of), 47
    polar parameters of $M$ (in terms of), 144
    potential (relation to), 21, 129
    $S$ matrix (relation to), 32, 139
    $T$ operator (relation to), 25
  coefficients, 255, 257, 279
    average for the orthogonal case, 328

average for the unitary case, 329
correlations, 325
regime
  ballistic, 320
  crossover, 323
  diffusive, 320, 323, 327
  localized, 320, 323
  metallic, 325
relevant parameters, 8, 226, 228, 229
relevant scaling variable, 341
reproducing property, 250
resonance, 70, 75, 78, 79, 84, 87
  barrier top, 64
  energy, 71, 249
  Fano, 154
  poles, 88
  two-dimensional cavity (in), 245
  width, 85, 88, 249
response function, 198, 212, 351

$S$ matrix, 10, 30, 43, 138, 171, 181
  analytic structure
    complex-energy plane, 90
    complex-momentum plane, 87
  average, 111, 250
  bound-state poles, 89
  branch point, 91
  cavities, 171, 181
  combination for system and horns, 376
  combination for two scatterers, 48, 145
    presence of open and closed channels, 162
  energy averages, 112
  ensemble, 112
  ensemble averages, 112
  ergodicity, 113
  extended, 66, 155, 157, 158, 161
  Green function (full, relation to), 43
  invariant measure, 245, 247, 380
    orthogonal, 247, 382
    symplectic, 247
    unitary, 247, 385
  measure, 112
  motion on the unitarity circle, 109
  on shell, 32, 138
  one-pole approximation, 92
  optical, 112, 113, 250, 253
  physical sheet, 91
  Poisson's kernel, 112, 249, 251, 252
  polar representation, 42, 140, 248, 258
  poles, 87
    in $R$ matrix language, 105
  probability distribution, 279
  resonance poles, 88, 89
  self-dual, 10

stationary random process, 112, 249, 251
statistical ensemble, 245
  orthogonal, 247
  symplectic, 247
  unitary, 247
subunitary, 111
symmetric, 10, 37, 140
symmetry, 140
  transformation under a translation, 51, 146
  unitary, 10, 32, 33, 140
  virtual-state poles, 89
scaling theory, 4, 340
  one-parameter scaling, 342, 361
scattering
  anisotropic, 357
  approach to electronic transport, 187
  elastic, 149
  inelastic, 149
scattering matrix, see $S$ matrix
scattering theory, 15, 120, 168
  cavities, 168
  one-dimensional systems, 15
  quasi-one-dimensional systems, 120
Schrödinger picture, 198
screening length, 208
second quantization, 368
self-averaging, 8, 335, 350
self-consistent field, 205
self-energy (irreducible), 348, 349
Shannon entropy, 11, 12
Sinai billiard, 265
Smoluchowski equation, 288, 291, 300
spin–orbit
  coupling, 360
  scattering, 9
stadium billiard, 265
stationarity, 249
stationary random process, 112, 249, 251
  ergodicity, 113, 249, 251
statistical ensemble of $S$ matrices, 245
  orthogonal, 247
  symplectic, 247
  unitary, 247
statistical inference, 241
statistical mechanics and maximum entropy, 235
  canonical ensemble, 236, 239
  microcanonical ensemble, 236
statistically independent $M$ matrices, 287
symmetries, 8, 11, 226
symplectic group (real)
  $Sp(2, \mathcal{R})$, 47
  $Sp(2N, \mathcal{R})$, 144

# INDEX

$T$ matrix (transition operator), 23, 25
thermodynamic limit, 187
Thomas–Fermi approximation, 208
Thouless length, 5
tight-binding Hamiltonian, 331
time
  average, 3
  delay, 74, 75, 77
  inversion, 9, 19
    anti-unitary operator for, 9, 19
time-reversal
  invariance (TRI), 9, 19, 37, 38, 140,
    143, 181, 188, 221, 245, 257,
    338, 350, 351, 360, 366, 368,
    373, 374
  symmetry, 10
time-scale, 5
  ergodic time, 6
  Heisenberg time, 6
  Poincaré time, 6
total field, 213
transfer matrix, *see* $M$ matrix
transition
  metal–insulator, 330
  probability, 288
transmission, 16
  amplitudes, 16, 120, 130, 171
    analytic continuation, 66
    combination (imbedding), 68
    polar parameters of $M$ (in terms of),
      47, 144
    potential (relation to), 21, 130
    $S$ matrix (relation to), 32, 139
    $T$ operator (relation to), 25
  coefficients, 224, 255, 257, 271, 279
    average and fluctuations, 326
    correlations, 325
transport
  electronic
    linear response theory (LRT), 187
    scattering approach, 187
  mesoscopic, 11
two-body operator, 204

uniqueness theorem for entropy, 231
unitary group
  Haar's measure, 248
universal
  conductance fluctuations (UCF), 4, 8,
    255, 323, 361
  properties, 1, 226
universality, 250
universality classes, 8, 10
  orthogonal, 9, 10, 140, 181, 247, 256,
    280, 302, 328

symplectic, 9, 10, 140, 247, 361
unitary, 9, 10, 140, 181, 247, 253, 280,
  310, 329

vector potential, 191
  longitudinal part, 191
  transverse part, 191
virtual state
  poles, 89

wave function
  asymptotic form, 20, 43, 129
  incoming, 17
  outgoing, 17
  perturbed, 18
  plane, 16
wave packet, 74, 250
weak localization, 4, 253, 322, 328, 330,
  342, 350, 360
weak-localization correction (WLC), 257,
  275, 339, 342
weight function, 111
  Lorentzian, 111, 112
  rectangular, 113
white-noise limit, 344
Wigner's time delay, 74

zero-temperature limit, 375